Natural Gums and Resins: Botany and Sustainable Uses in Medicine, Nutrition, Perfumery and Cosmetics

This book is dedicated to Professor Dr Sukh Dev, a renowned chemist who has become world-famous for his work in the field of natural resources, in particular resins and gums. His research interests cover a wide spectrum, ranging from natural product chemistry to drug development from Ayurvedic lead compounds and technology development. His research has contributed significantly to our understanding of the chemical composition and potential application of natural products in resins and gums. Thus, his group was able to isolate two steroidal ketones from guggul, the gum-resin of Commiphora wightii, well-known in Ayurveda for the treatment of lipid disorders, with pronounced hypocholesterolemic and hypolipaemic activity.

Natural Gums and Resins: Botany and Sustainable Uses in Medicine, Nutrition, Perfumery and Cosmetics

Edited by

Hartwig Schulz
Consulting and Project Management for Medicinal and Aromatic Plants
Waltraudstraße 4, Stahnsdorf, Germany

Victor Sam Jayakumar
St Thomas College of Engineering & Technology, Kozhuvalloor PO 689521
Chengannur, Kerala, India; Deepthi, Amaravila PO, Thiruvananthapuram 695122,
Kerala, India

Niranjan Prasad
ICAR-National Institute of Secondary Agriculture, Namkum, Ranchi, India

CABI

CABI is a trading name of CAB International

CABI
Nosworthy Way
Wallingford
Oxfordshire OX10 8DE
UK

Tel: +44 (0)1491 832111
E-mail: info@cabi.org
Website: www.cabi.org

CABI
200 Portland Street
Boston
MA 02114
USA

T: +1 (617)682-9015
E-mail: cabi-nao@cabi.org

A catalogue record for this book is available from the British Library, London, UK.

Library of Congress Control Number: 2025944441

ISBN-13: 9781800626676 (hardback)
9781800626683 (ePDF)
9781800626690 (ePub)

DOI: 10.1079/9781800626690.0000

Commissioning Editor: Rebecca Stubbs
Editorial Assistant: Emma McCann
Production Editor: Rosie Hayden

Typeset by Exeter Premedia Services Pvt Ltd, Chennai, India
Printed and bound in the UK by CPI Group (UK) Ltd, Croydon, CR0 4YY

Contents

Contributors

Maryam Akaberi, Department of Pharmacognosy, School of Pharmacy, Mashhad University of Medical Sciences, Mashhad, Iran.

Lisa Bradbury, International Species Protection in International Trade, Federal Food Safety and Veterinary Office (FSVO), Schwarzenburgstrasse 155, 3003 Bern, Switzerland.

Anthony B. Cunningham, School of Veterinary and Life Sciences, Murdoch University, 90 South St., Murdoch, WA 6150, Australia, and School of Life Sciences, University of KwaZulu-Natal, King Edward Avenue, Pietermaritzburg 3209, South Africa.

Seyed Ahmad Emami, Department of Pharmacognosy, School of Pharmacy, Mashhad University of Medical Sciences, Mashhad, Iran.

Anita Ghaffariannia, Department of Laboratory Sciences, School of Paramedical Sciences, Mashhad University of Medical Sciences, Mashhad, Iran.

Saroj Kumar Giri, Agriculture Structures & Process Engineering Division, National Institute of Secondary Agriculture (ICAR), Namkum, Ranchi, India.

Victor S. Jayakumar, St Thomas College of Engineering & Technology, Kozhuvalloor PO 689521 Chengannur, Kerala, India; Deepthi, Amaravila P O, Thiruvananthapuram 695122, Kerala, India.

Sakharam Kale, ICAR-National Institute of Secondary Agriculture, Namkum, Ranchi, India.

Mohammad Reza Kanani, Shahid Beheshti University, Medicinal Plants and Drugs Research Institute, Tehran, Iran.

Michael Keusgen, Philipps University Marburg, Institute for Pharmaceutical Chemistry, Marbacher Weg 8, 35032 Marburg, Germany.

Vaibhav Daulatrao Lohot, ICAR-National Institute of Secondary Agriculture, Namkum, Ranchi, India.

Mathias Lörtscher, International Species Protection in International Trade, Federal Food Safety and Veterinary Office (FSVO), Schwarzenburgstrasse 155, 3003 Bern, Switzerland.

Bruno Mainini, International Species Protection in International Trade, Federal Food Safety and Veterinary Office (FSVO), Schwarzenburgstrasse 155, 3003 Bern, Switzerland.

Arefeh Mollazadeh, Department of Traditional Pharmacy, School of Pharmacy, Mashhad University of Medical Sciences, Mashhad, Iran.

Ursula Moser, International Species Protection in International Trade, Federal Food Safety and Veterinary Office (FSVO), Schwarzenburgstrasse 155, 3003 Bern, Switzerland.

Farsad Nadjafi, Shahid Beheshti University, Medicinal Plants and Drugs Research Institute, Tehran, Iran, and MateriaMed GmbH, Wien, Josefstädterstrasse 43-45, Austria.

Sanjay Kumar Pandey, ICAR-National Institute of Secondary Agriculture, Namkum, Ranchi, India.

Satish Chandra Sharma, ICAR-National Institute of Secondary Agriculture, Namkum, Ranchi, India.

Nandkishore Thombare, ICAR-National Institute of Secondary Agriculture, Namkum, Ranchi, India.

About the Editors

Hartwig Schulz (PhD) studied chemistry in Bonn and Kiel (Germany). After completing his studies, he carried out several years' research in the field of human nutrition (vitamin E, phytosterins, plant lipids) at the university of Kiel, Germany. Then, as head of an analytical department, he worked 10 years for one of the world's largest perfume and flavour manufacturers (Symrise). From 1996 to 2019, he was the director and professor of a research institute under the responsibility of the Federal Ministry of Agriculture in Germany. Hartwig Schulz is author of 259 scientific articles and book chapters and, as the president of the German Society for Quality Research (DGQ), he was also engaged for 8 years in food research on a national and international level. Since his retirement, he has been working as a consultant, especially in the field of 'medicinal and aromatic plants', and is also supporting various development aid projects on a voluntary basis.

Victor Sam Jayakumar (PhD) completed his masters and doctoral studies in physics at the University of Kerala (India). After graduation, he worked at Mar Ivanios College, University of Kerala, as a lecturer in the Department of Physics, where he taught for more than 3 decades and carried out extensive research in the field of vibrational spectroscopy, including several years of studies on natural products with more than 65 research publications in reputed international journals. From 2005 to 2009, he was Professor and Head of the Postgraduate and Research Department of Physics at Mar Ivanios College, University of Kerala. From 2011, he worked as Director of Mar Baselios Institute of Technology Anchal (MBIT) for 10 years. Presently he is associated with St Thomas College of Engineering and Technology, Chengannur, India as Adjunct Professor and is also engaged in promoting research innovations and collaborative project initiatives, supporting various projects relevant to social development through science.

Niranjan Prasad (PhD) did his BE (agricultural engineering) at CTAE, Udaipur, India (1988), and his MTech (1991) and PhD at IIT, Kharagpur, India (1996). He obtained his postdoctoral training in smart/active packaging from Michigan State University, USA. He worked as a lecturer at NERIST, Nirjuli, Arunachal Pradesh, during 1996–1997. Later in 1997, he joined Indian Lac Research Institute, Ranchi, as an agricultural research service scientist. He currently works as a principal scientist and the head of Agri Bio-resources Augmentation Division at ICAR-National Institute of Secondary Agriculture, Ranchi, India. He also works as a project coordinator of the All India Network Project on Conservation of Lac Insect Genetic Resources. He worked as a consortium principal investigator of the National Agricultural Innovation Project (World Bank) on the Lac Value Chain. During his 28 years of experience in research and extension, he published 65 research papers in national and international journals. He has developed several tools and machines for gum and resin production and processing which have been commercialized. Dr. Prasad is a member of several professional societies and recipient of many awards.

Preface

In ancient times, the Middle and Near East played a significant role in the utilization of resins and gums, contributing to various aspects of daily life, trade and cultural practices. Mostly, resins and gums were derived from plants, particularly trees, and held multifaceted importance in these regions. Trade routes like the Silk Road played a crucial role in the exchange of resins and gums between the East and West. The demand for these precious substances led to the development of extensive trade networks, fostering cultural exchange and economic prosperity in the region. Cities like Petra in Jordan thrived as major trade hubs, showcasing the economic significance of resins and gums in the ancient world. The cities along the Incense Trade Route became flourishing centres of commerce, where merchants from different civilizations converged to engage in the trade of resins, spices and other valuable commodities. The profits generated from the incense trade contributed to the prosperity and development of these cities, fostering cultural exchange and the blending of diverse traditions.

Two of the most renowned natural resins and gums (NRGs) were frankincense, extracted from various *Boswellia* species, and myrrh, obtained from different *Commiphora* tree species. These aromatic resins were highly valued for their fragrant properties and were frequently used in religious ceremonies, particularly in the worship practices of ancient civilizations such as the Sumerians, Babylonians, Egyptians and, later, the Greeks and Romans. The aromatic smoke produced by burning these resins was believed to have spiritual significance, symbolizing purification and connection with the divine. Furthermore, resins and gums were used in ancient Egyptian burial practices to preserve mummies, which reflects their diverse uses in burial customs.

Even today, it is still not fully understood why exactly plants exude resins. The plant exudates consist mainly of secondary metabolic substances that apparently play no role in the primary physiology of the individual plants. While some scientists regard the resins merely as waste products, there is also evidence that numerous pathogens and predators are warded off by the resin. Today, NRGs are mainly used in food materials (as stabilizers, emulsifiers and flavouring agents), cosmetics and pharmaceuticals. Within the coming decades, enormous changes are expected in these sectors, which should have a positive impact on the demand for these raw materials. Furthermore, the development of new areas of consumption will also boost the demand, which will require a significant increase in NRG production. In this context, it will be of fundamental importance that production is environmentally friendly and that the raw materials used are renewable and sustainably produced.

However, agriculture today faces several challenges as risks associated with climate change increase, new forms of biotic and abiotic stresses emerge, production becomes more energy-intensive and biosafety concerns grow. In addition, increasing intellectual property rights and trade regulations, a decreasing preference for agricultural labour and changing dietary habits are creating enormous challenges for producers and industrial processors of NRGs worldwide. It is therefore important to utilize the potential of modern life sciences, innovations in technology development and provision as well as support from policy and commercial investment for the upcoming paradigm shift.

The most important NRG-producing countries in the world are currently the United States, China, Brazil, India, Russia, Indonesia and Thailand. While India, Nigeria and Sudan are the main producers of gums, Afghanistan, Iran, Spain, Ethiopia and Somalia are mainly responsible for the production of gum-resins. It is assumed that the NRG sector will face major challenges on several fronts in the future. Global climate change not only brings new threats that affect productivity and sustainability, but is also likely to have a positive effect on the production yields of plant exudates due to the increase in atmospheric temperature. In general, future production systems must be resource-efficient and the use of pesticides must be significantly minimized. In addition, more attention must be paid to improving the efficiency of increasingly scarce natural resources, including water and nutrients. In this context, it will also be necessary to develop a better understanding of the genetic diversity of gum and resin plants in terms of their chemical composition in order to be able to utilize the full genetic potential of the respective plant species in an economically viable manner.

In order to meet these requirements, individual conservation concepts (*ex situ* and *in situ* gene banks for the most important NRGs) would have to be established. In recent decades, not only the deforestation of the Brazilian rainforest, particularly in the Amazon region, but also over-exploitation in other countries has significantly reduced the availability of some plant species. The destruction of tropical forests thus threatens the indigenous population living in these regions, who still have extensive knowledge of the various uses of resins. By preserving genetic resources, the NRGs would become more important overall.

The development of comprehensive, globally networked databases for economically important gums and resins could also make the ecological significance of NRGs more transparent and link it to global demand, thus enabling the sustainable development of national strategies in the producer countries. It will also be crucial to continuously improve the industrial processing of NRGs and to apply new biotechnological technologies. The particular challenges in this context are to improve the shelf life of natural products and to use by-products as efficiently as possible. In addition, suitable quality management systems and sustainable packaging technology for the gums and resins produced will be of central importance. In order to further reduce the carbon footprint, it will be necessary in the future to use predominantly alternative energy sources such as solar and wind energy or 'green' hydrogen for industrial production processes. The aim must therefore be to develop international quality standards (in accordance with ISO) and analytical protocols for the most important NRGs, but also for less popular gums and resins, which will enable adulteration of these high-quality natural products to be recognized quickly and reliably and also the prohibited use of protected species to be recognized in this way.

It is reasonable to expect that in the future, numerous new natural products will be developed for the food, phytopharmaceutical and cosmetics industries, as well as for various biomedical applications. In particular, nanotechnology, but also the chemical/biochemical derivatization of gum and resin substances for various niche applications, will be of particular importance in this context. The technological revolution and the constant change in our way of life will also lead to the emergence of completely new areas of application in which NRGs could make a meaningful contribution to the sustainable development of industrial products.

In order to effectively meet the enormous challenges that climate change and the increasing scarcity of natural resources, competing economic systems and growing social and political tensions worldwide bring with them, it will be extremely necessary to scientifically accompany current developments at an early stage at national and international symposia/workshops and to develop and present effective measures to political decision-making bodies.

In the future, the profits from the sale of gum and resin should be used primarily to improve living conditions in the producing countries, in order to improve the socio-economic situation there.

In this context, it will also be increasingly necessary to establish sustainable concepts to counteract the continuous decline of many gum and resin trees and the considerable ecological destruction of the environment. Future research and development efforts will therefore need to focus on achieving higher and sustainable productivity with optimal inputs, ensuring high quality of NRGs in general and developing novel and specialized applications, in particular to ensure that the required paradigm shift can be implemented in the individual application areas of NRGs. In particular, conventional production systems must be gradually supplemented by highly organized plantations with advanced parameter monitoring and automated management systems for precision agriculture for sustainable production.

In this context, extensive literature studies were carried out for this book, which not only refer to various ethnobotanical reports and the individual traditional uses of gums and resins, but also present the occurrence of the individual ingredients, their analysis and technological processing. Finally, important aspects of sustainability and species protection are presented and discussed.

The editors would like to take this opportunity to thank all chapter authors for bringing their individual expertise to the various aspects. We would also like to thank CABI for the fruitful and pleasant collaboration.

Hartwig Schulz
Victor S. Jayakumar
Niranjan Prasad

1 Historical Significance of Resins and Gums in Cultural Practices, Cosmetics and Nutrition in the Middle and Far East

Arefeh Mollazadeh[1], Anita Ghaffariannia[2], Hartwig Schulz[3], Maryam Akaberi[4] Saroj Kumar Giri[5] and Seyed Ahmad Emami[1]*

[1]Department of Traditional Pharmacy, School of Pharmacy, Mashhad University of Medical Sciences, Mashhad, Iran; [2]Department of Laboratory Sciences, School of Paramedical Sciences, Mashhad University of Medical Sciences, Mashhad, Iran; [3]Consulting and Project Management for Medicinal and Aromatic Plants, Waltraudstraße 4, Stahnsdorf, Germany; [4]Department of Pharmacognosy, School of Pharmacy, Mashhad University of Medical Sciences, Mashhad, Iran; [5]ICAR-National Institute of Secondary Agriculture, Namkum, Ranchi, India

Abstract

Generally, the use of resins and gums is deeply rooted in the various cultural practices of the Middle and Far East and contributed significantly to the cultural heritage of the various communities there. The historical trade routes, such as the Silk Road, facilitated the exchange of resins and gums between the civilizations of the time. In general, these plant-based raw materials were highly valued in ancient times and sometimes contributed significantly to the economic prosperity of the regions along the trade routes. Resins and gums have played a central role in religious and spiritual practices throughout history. In particular, frankincense and myrrh were used in religious rituals in ancient Egypt, Greece and the Middle East. In addition, selected plant exudates such as gum arabic and gum tragacanth were used in traditional cooking and food preparation as they were already known as thickening, stabilizing and binding agents. Furthermore, gums and resins have been traditionally used for therapeutic, cosmetic and culinary purposes in different traditional systems of medicine. In Traditional Persian Medicine (TPM), gums, resins, oleoresins and oleo-gum resins are called *ṣamgh*, which is an Arabic general name for 'exudates of certain plants'. In this chapter, various *ṣamghs*, their individual plant sources and their applications in TPM are presented, although strictly speaking not all of them can be categorized as gums and resins according to the International Organization for Standardization (ISO) classification. Nevertheless, these traditional uses continue to influence numerous applications, as these plant-based raw materials are still important in various industries and cultural practices.

Introduction

Ṣamgh is a common Arabic name for the exudates obtained from certain plants used for medicinal, industrial and edible applications. *Ṣamghs* are liquid when secreted, but they solidify in air. In Traditional Persian Medicine (TPM) (Table 1.1), *ṣamgh* is referred to as either gum, resin, oleo-resin or oleo-gum resin. The ancient Egyptian word *kemayt* is the origin of the equivalents for

*Corresponding author: emamia@mums.ac.ir

DOI: 10.1079/9781800626690.0001

Table 1.1. Information regarding 10 major TPM books that described traditional effects of various gums. Author's own table.

Author	Book	Language	Year and place of publication
Mohammad Ibn Zakariyā Rāzi (Rhazes)	*Al-Hāwi fi aṭ-Ṭibb* (Comprehensive Book of Medicine)	Arabic	1967–1968, Hyderabad
Mowaffaq ad-din Abu Manṣur Heravi	*Al-Abniyah an Ḥaqāyeq al-Adwiyah* (Basics of Realities on Drugs)	Persian	1992, Tehran
Hossein Ibn Abdullah Ibn Sinā (Avicenna)	*Al-Qānun fi aṭ-Ṭibb* (Canon of Medicine)	Arabic	2015, Tehran
Ḍia ad-Din Ibn al-Beyṭār (Greatest botanist and pharmacist of the world of Islam)	*Al-Jām'e Le-Mofradāt al-Adwiah wa al-Aghḏiyah* (Comprehensive book on Simple Drugs and Foods)	Arabic	2015, Cairo
Dāwoud Anṭāki	*Taḏkereh Oulol-Albāb* (Memorandum Book)	Arabic	2000, Beirut
Mohammad Momen Tonekāboni	*Toḥfat al-Momenin* (Rarity of the Faithful)	Persian	2008, Tehran
Moḥammad Ḥussein Aqili Khorāsāni	*Makhzan al-Adwiah* (Drug Treasure)	Persian	2014, Tehran
Ibn al-Nafis al-Qarashi	*ash-Shāmil fi aṣ-Ṣina'a aṭ-Ṭibbiyyah* (Comprehensive Book on the Medical Industry)	Arabic	2000, Abu Dhabi
Ibn Al-Jazzār, J.A.	*Kitāb al-'Itimād fī al-Adwiyah al-Mufradah* (The Reliable Book on Simple Drugs)	Arabic	2019, London
Abu Reyhān al-Bīrūnī	*Kitab aṣ-Ṣaydana fī aṭ-Ṭibb* (Pharmacology in Medicine)	Arabic	1991, Tehran

the word "gum" in European languages (Forbes, 1965). In Middle Persian, the terms *zatak*, *zatuk* or *zadug* were used for gums or resins, but sometimes the term *angu* can also be found in ancient literature. In Modern Persian, it is referred to as *angum*, *angiyan*, *angožad* and *žad*. The term *žad* in Persian is derived from the Avestan word *jatara* (Tabrizi, 1982). It is possible that the Persian words, as well as the Sanskrit term *hingu*, are derived from the Sogdian word *inku* or at least share a common root. In the texts of the Islamic golden era, the general term for *ṣamgh* included several types of substances, such as *ṣamgh* itself (gum), *'ilk* (possibly mastic resin), *meghfār* (likely manna) and *lathā* (a soft and chewable gum (Ayati *et al.*, 2024)). Gums were considered fragrant, bright white or yellow pieces that were naturally, or through incisions made in the trunk, secreted from the trees, eventually solidifying and becoming brittle when crushed. Gums were classified into various types based on their size and colour and named differently (al-Bīrūnī, 1991; Dīnawarī, 1974) (Table 1.2). The Romans and Chinese used various types of gums for lighting torches and lamps, sealing gaps in ship hulls to prevent water ingress and for colouring clay pots using transparent and high-quality gums (Hamadānī, 1989). In ancient Egypt, gums were used for embalming bodies (mummification) (Carruthers, 1897; Plinius/Pliny [the Elder], 1951, 1956; Herodotus, 2009). In antiquity, Arabia Felix (one of the three regions of Arabia, including Yemen, Asir and Ḥadhramaut) was famous for its gum-bearing plants (Diodorus Siculus, 2005). The Nabataeans knew many types of plant gums and used them to treat diseases, purify contaminated water, repel pests and in the composition of *dukhna* (substances thrown into fire for medical or ritual purposes). They also exported some of these gums to

Table 1.2. Various gums in TPM. Author's own table.

Ṣamgh[a] (gum[b])	Plant source	Family
ʿAnbar (amber)	*Pinus* spp.	Pinaceae
Anzarut (sarcocolla gum)	1. *Astragalus fasciculifolius* Boiss. 2. *Astragalus glaucacanthos* Fisch. 3. *Astragalus sarcocolla* Dymock	Fabaceae
Damm al-akhawain (Dragon's blood)	*Dracaena cinnabari* Balf.f.	Asparagaceae
Heltit (asafoetida)	*Ferula* spp. mainly *Ferula assa-foetida* L.	Apiaceae
ʿIlk al-anbāṭ (pistachio gum)	*Pistacia vera* L.	Anacardiaceae
ʿIlk al-boṭm, Saghez (terpentine)	1. *Pistacia terebinthus* L. 2. *Pistacia atlantica* Desf. 3. *Pistacia eurycarpa* Yalt.	Anacardiaceae
ʿIlk al-boṭm (Persian turpentine)	*Pistacia khinjuk* Stocks	Anacardiaceae
Jāwshir (opoponax)	*Opopanax chironium* (L.) W.D.J.Koch Basionym: *Laserpitium chironium* L.	Apiaceae
Kathirā (gum tragacanth)	*Astragalus* spp. mainly: 1. *A. brachycalyx* Fisch. subsp. *brachycalyx* 2. *A. cerasocrenus* Bunge 3. *A. creticus* Lam. 4. *A. cylleneus* Fisch. 5. *A. echidna* Bunge 6. *A. floccosus* Boiss. 7. *A. gossypinus* Fisch. 8. *A. gummifer* Labill. 9. *A. hypsogeton* Bunge 10. *A. keratensis* Bunge 11. *A. kurdicus* Boiss. 12. *A. longistylus* Bunge 13. *A. microcephalus* Willd. 14. *A. verus* Olivier	Fabaceae
Lac (lac)	*Ficus altissima* Blume	Moraceae
Lobān (frankincense)	*Boswellia sacra* Flück. Syn. *Boswellia carteri* Birdw.	Burseraceae
Maṣṭakā (mastic gum)	*Pistacia lentiscus* L.	Anacardiaceae
Meyʿah sāʿelah (Oriental sweet gum)	*Liquidambar orientalis* Mill.	Altingiaceae
Meyʿah yābesah (storax)	*Styrax officinalis* L.	Styracaceae
Moql (mukul, gum guggul)	*Commiphora wightii* (Arn.) Bhandari Basionym: *Balsamodendrum wightii* Arn. Syn. *Commiphora mukul* (Hook. ex Stocks).Engl.	Burseraceae
Morr (myrrh)	*Commiphora myrrha* (T. Nees) Engl. Basionym: *Balsamodendrum myrrha* T. Nees Syn. *Commiphora molmol* (Engl.) Engl. ex Tschirch	Burseraceae
Oshaqq (gum ammoniacum)	*Ferula ammoniacum* (D. Don) Spalik, M. Panahi, Piwczyński & Puchałka (Basionym: *Dorema ammoniacum* D. Don)	Apiaceae

Continued

Table 1.2. Continued

Şamgh[a] (gum[b])	Plant source	Family
Qennah (galbanum)	*Ferula gummosa* Boiss.	Apiaceae
Sakbīnaj (sagapenum)	*Ferula persica* Willd.	Apiaceae
Şamgh ʿArabi (gum arabic)	*Vachellia nilotica* (L.) P.J.H. Hurter & Mabb. Basionym: *Mimosa nilotica* L. Syn. *Acacia arabica* var. *nilotica* (L.) Benth.	Fabaceae
Şamgh Farsi (Persian gum)	1. *Prunus avium* L. 2. *Prunus cerasus* L.	Rosaceae
	Prunus domestica L.	Rosaceae
	Prunus amygdalus Batsch	Rosaceae
Şamgh al-khaṭmī (mallow gum)	*Althaea officinalis* L.	Malvaceae
Şamgh al-qouqā (cedar resin)	*Cedrus libani* A.Rich.	Pinaceae
Şamgh as-sarw (Italian cypress resin)	*Cupressus sempervirens* L.	Cupressaceae
Şamgh as-soḏāb (rue resin)	*Ruta graveolens* L.	Rutaceae
Şamgh as-somāq (sumac resin)	*Rhus coriaria* L.	Anacardiaceae
Şamgh az-zaytoun (olive resin)	*Olea europaea* L.	Oleaceae
Şamgh lātī (stone pine resin)	*Pinus pinea* L.	Pinaceae
Şamgh satr'moulia (Norway spruce resin)	*Picea abies* var. *abies*	Pinaceae
Sāndārāus (sandarac)	*Tetraclinis articulata* (Vahl) Mast. Basionym: *Thuja articulata* Vahl	Cupressaceae
Sqamunyā (scammony gum)	*Convolvulus scammonia* L.	Convolvulaceae
Thafsia (tapsia)	*Thapsia garganica* L.	Apiaceae

[a]Arabic
[b]English

neighbouring lands (Ibn Waḥshiyya, 1993–1998). Some of the well-known gums in the Islamic world, including *lubān* (frankincense), *katirā* (tragacanth) and *lakjk* (gums from *Ficus laccifera* Roxb.), were exported to other countries for various purposes (Heyd, 1886; Stannard, 1999).

Natural gums are generally divided into two categories: plant-based and fossilized gums. The plant-based gums and resins are directly obtained from plants, but the fossilized gums and resins are composed of plant exudates after fossilization. Plant-based gums are gums from (i) *Ferula* spp. including ammoniac (Oshagh, obtained from *Ferula ammoniacum* (D. Don), galbanum (obtained from *Ferula gummosa* Boiss.), asafoetida (from *Ferula assa-foetida* L.) and sagapenum, (ii) coniferous gums including secretions from certain coniferous trees, (iii) gums from the Rosaceae family (known as Persian gums) including almond gum (from *Prunus amygdalus* Batsch), plum gum (from *Prunus domestica* L.) and some other *Prunus* species, (iv) gums from the *Pistacia* genus including terebinth (from *Pistacia terebinthus* L.), *ʿilk ol- anbāṭ* (from *Pistacia vera* L.), mastic (from *Pistacia lentiscus* L.), *Pistacia khinjuk* resin (from *Pistacia khinjuk* Stocks) and *ʿilk ol-boṭm* (from *Pistacia atlantica* Desf.), (v) gums from the *Astragalus* genus including *anzarut* (from *Astragalus sarcocolla* Dymock) and tragacanth (from *Astragalus tragacantha* L.) and (vi) other gums, namely mukul (gums secreted from *Commiphora wightii* (Arn.) Bhandari) and *meyʿah* (storax), secreted from *Commiphora myrrha* (T. Nees) Engl. One of the most important types is acacia gum (commonly known as gum arabic), which is secreted from trees of the *Acacia* genus belonging to the family Fabaceae.

Fossilized resins include copal and amber; it is believed that the resin of coniferous trees like *Pinus* (pine) first transforms into "copal" and then into "amber" (Mueller, 1891; Zargari, 1988; Smolinske, 1992; De Rovira, 1999; Jokār, 2006; Mohammadifar, 2022) (Table 1.2).

Historical Trade Routes

Numerous resins and gums, such as gum arabic, have been used in traditional medicine due to their various therapeutic properties. In addition, various resins, especially those with antimicrobial properties, were also used in ancient Egyptian mummification to preserve corpses. Myrrh and other resins were used in these elaborate rituals to prepare the deceased for the afterlife. For more than a millennium, trade caravans delivered frankincense gum-resins and other luxury goods from the southern coast of Arabia to the insatiable markets of the Mediterranean and Mesopotamia. In the heyday of imperial Rome, the frankincense trade reached its peak alongside the trade in ivory, pepper and silk (Spekke, 1957). The two Burseraceae genera *Boswellia* and *Commiphora* were mainly harvested in southern Arabia and the neighbouring regions. Hoots (1993) surmises that the original spark for the trade apparently came from Pharaonic Egypt in 3000 BC because myrrh oil was used to embalm the dead. As *Commiphora* species are not found in Egypt, he assumed that the product came via trade.

The Egyptians were not the only early users of frankincense. The Phoenicians valued frankincense so much that they named their main deity 'Baah Hamman', which means 'Lord of the Fragrant Atar'. The Babylonians, Sumerians and Assyrians also held frankincense in high esteem, but Groom (1981) doubted that these peoples used frankincense and myrrh, or rather other resins from native plants and aromatic woods, for incense. He conceded, however, that small quantities of frankincense and myrrh may have been brought into these areas.

In an expedition that is now known as the world's first plant-collecting expedition, 31 frankincense trees were brought back and planted in the temple of Karnak on the upper bank of the Nile in Egypt (Fig. 1.1). The carving on one of the walls of the temple shows how the plants were loaded onto a ship and transported in wicker baskets. Although it has not yet been possible to determine whether these were actually frankincense or myrrh trees, Groom (1981) suggested that they were probably myrrh, a species that is now common in Somalia.

The domestication of the camel enabled an organized trade in frankincense overland from southern Arabia to the Mediterranean (Groom, 1981). At the time, camel transport made it possible to move larger quantities of goods more quickly across the Arabian deserts, thus speeding up trade as a whole to meet the demand of the Greeks and Romans, who used frankincense in particular to win the favour of the gods. In addition, frankincense was also used as an offering to deceased people to drive away evil spirits, as a symbol of honour for a living person and to accompany festivals and processions (Atchley and Cuthbert, 1909). Among some peoples, such as the Egyptians, Persians and others, however, the right to offer frankincense was reserved for the priesthood (including kings). The Greeks and Romans, on the other hand, regarded the use of frankincense as a duty incumbent on all people.

Use of Resins and Gums in Ancient Cultures

The perfume industry dates back thousands of years, as perfumes were used in ancient civilizations for various purposes. After that, the perfume industry expanded significantly over time, as new methods for extracting and mixing essential oils were introduced, and new types of perfumes emerged. Ancient Egypt was a source of many essential oils, such as sandalwood, amber, musk and rose. It was also home to many experts in perfumery, who developed new methods of extracting and blending essential oils. Oriental perfumes were popular in the Middle Ages, as they were characterized by their strong and distinctive scents. Furthermore, oriental perfumes were often used in religious rituals and ceremonies. But scents had already held significant cultural and commercial value long before the Middle Ages. Generally, scents have probably played an important role throughout human history. The fact that earlier voyages

Fig. 1.1. Relief depicting men transporting large baskets containing frankincense and myrrh trees. (By Ismoon, 2020, CC BY-SA 4.0, via Wikimedia Commons. https://upload.wikimedia.org/wikipedia/commons/d/d6/Incense_trees_transport._Punt_relief._Mortuary_Temple_of_Hatshepsut%2C_Deir_el-Bahari.jpg.)

of discovery, wars and long-distance trade often depended on the search for various materials with strong characteristic odour properties, such as frankincense and various spice plants, points to this outstanding importance. However, new research into the properties of various aromatic compounds also shows that many of the mostly plant-based raw materials that people harvested, traded and used in the past have bioactive, psychological and medicinal properties. Furthermore, it is known that odours are fully embedded in social and historical contexts and are therefore both culturally and neurologically constituted (Cao *et al*., 2019).

In ancient times, frankincense was primarily burned as part of ritual practices to honour specific deities (Fig. 1.2). However, the use of incense extended beyond religious ceremonies. Resins used in incense also played a central role in the creation of perfumes, as reflected in the Latin term *per fumum*, meaning 'through smoke'. Ancient Egypt is often regarded as the cradle of perfumery, with its rich tradition in fragrance-making dating back as far as 4500 BC. Perfumes in Egypt held a multifaceted significance and were used for various purposes, including religious practices, various medicinal applications and funerary rites. The ancient Egyptians perfected the art of extracting scents from natural ingredients such as flowers, herbs, spices and resins like myrrh and frankincense. Their knowledge and techniques were highly valued and later influenced other cultures, laying the foundation for the global perfumery traditions we know today. Since the time of the Pharoahs, large timber trees such as *Acacia nilotica* subsp.

Fig. 1.2. Frankincense is mentioned many times in the Bible and is closely tied with myrrh – also a lovely incense ingredient. The gift of the magi to the Christ Child of gold, frankincense and myrrh is thought to symbolize: his kingship – gold; his spiritual nature – frankincense; and his death – myrrh (by Bosch (c.1475). *The Adoration of the Magi*. https://www.metmuseum.org/art/collection/search/435724).

adstringens (Syn. *Vachellia Nitolica* subsp. *adstringens*) have been exploited from the riverine forests of the Nile (Middeke-Conlin, 2014). Extracting gums was sometimes a difficult task, requiring collectors not only to maintain the hygiene and quality of the gums but also to safeguard their own health. According to Abū Rayḥān al-Bīrūnī, when collecting *thapsia* (the gum of *Thapsia garganica* L. from the Apiaceae family), they were advised to cover their faces or rub garlic on their bodies to avoid blisters and swelling caused by this gum (al-Bīrūnī, 1991).

Indian culture is one of the oldest cultures in the world that used perfume. In India, perfume compositions were used for various purposes: they were not only used to pay respect to the gods in religious rituals, but also to treat some ailments such as headaches and colds. They were also used to get rid of unpleasant odours from the body and to keep the skin soft and supple. India has always been a source of many essential oils, such as rose, jasmine and agarwood oil. Moreover, the country is home to many well-known perfumers who, among other things, have developed new methods for extracting and using essential oils in different cosmetic formulations.

Islamic civilization played a major role in the development of perfumery during the Middle Ages. Muslims were among the first to develop new methods of extracting essential oils, such as steam distillation, and also developed new methods of preserving them. Perfumes were widely used in Islamic civilization and for a variety of purposes, including religious rituals. Gums have also been used in the preparation of perfumes. In the book *Kitāb Kīmiyā' al-'Aṭr wa at-Taṣ'īdāt*, Kindī, in several instances, referred to the use of various gums, especially gum arabic, in the production of perfumes and oils and in sublimation processes (*duhn*) (Kindī, 1948).

Dioscorides described various species of *Acacia*, mentioning the characteristics and medicinal properties of the resins derived from them. He identified gum arabic, which is obtained from Egyptian acacia, as useful for certain skin diseases and especially beneficial in eye medications (Dioscorides, 2012). Avicenna considered the best gum to be gum arabic with minimal wood debris and recommended it for respiratory diseases, such as coughs and lung ulcers, as well as for strengthening the stomach (Bokhāri, 1992; Ibn al-Beyṭār, 2015; Ibn Sinā, 2015). Even today, various gums, especially extracts and resins from the *Acacia* genus, are widely used in the pharmaceutical industry (Zargari, 1988; Muller and Okoro, 2004).

Religious rituals

Perfumes were used in religious rituals to show respect to the gods, and they were also used in mummification to preserve the dead bodies. The historical importance of gums and resins is particularly evident in ancient civilizations such as ancient Egypt. There, it was believed that the scents of the substances used in embalming represented the food of the gods and could therefore enable communication beyond the threshold of consciousness. Over the past few years, Egyptian mummies have been the subject of numerous scientific studies (Maurer *et al.*, 2002). In particular, gas chromatography-mass spectrometry (GC-MS) has made it possible to identify many of the components used in ancient embalming materials. Phenols, guaiacols, naphthalenes, monoterpenes, sesquiterpenoids, oxidized diterpene resin acids and triterpenoids were identified; through these intergradients, the materials used in mummification could be determined. These compounds also have antibacterial and antifungal effects and prevent against deterioration caused by insects. Meanwhile analytical investigations have revealed a reasonably clear picture of the process of mummification and the materials used. Various Egyptian mummies were analysed to learn more about the main ingredients used in embalming (Lucas, 1914), and it was found that mainly NaCl, $NaCO_3$, $NaHCO_3$ and Na_2SO_4 were used for this purpose. Later, Sandison was able to prove through chemical analyses that artificial mummification was mainly achieved by dehydration of the body (Sandison, 1963).

Even in ancient times, it was known that unsaturated oils, which are able to polymerize spontaneously, protect the brittle tissue of the deceased from decomposition by forming a physico-chemical barrier that inhibits the activities of microorganisms (Buckley and Evershed, 2001). It is also reported that resins from conifers, especially cedars, junipers and pines, were mainly used for the mummification process (Kłys *et al.*, 1999).

Even in Roman times, pine oil and juniper oil were used for embalming processes due to their antibacterial and antifungal properties. In addition, the use of mastic oil was also practised due to certain antiseptic and antimicrobial ingredients of the essential oil, such as verbenone, α-terpineol, linalool and pentacyclic triterpenes. The use of myrrh and cassia was also propagated, and today we know that certain mono- and sesquiterpene hydrocarbons in particular are responsible for the defence against pests. In addition, various lichen species were also used for mummification, which, according to current research, is associated with ingredients such as usnic acid, sphaerophorin, pannarin and paraconic acid (Abdel-Maksoud and El-Amin, 2011).

Among herbal remedies, resins and gums were always of particular importance. Around the middle of the second millennium BC, trade routes were established that brought resins from Africa and Asia to the Mediterranean region. For example, amber from the Baltic Sea was traded with the Mycenaeans, and the Phoenicians brought the material by ship from northern Europe. Frankincense or myrrh was probably brought from Sudan, Ethiopia or Somalia to Egypt (Serpico, 2000; Langenheim, 2003). Not only were resins brought to Mediterranean countries, but the region also produced resins and gums from various resin-producing trees, shrubs and herbs (Howes, 2003). In his work *De Materia Medica*, Dioscorides (1st century AD) mentioned styrax, mastic, scammon, various pine resins, labdanum and the resin obtained from terebinth (*Pistacia terebinthus*) in the eastern Mediterranean. Archaeological excavations at Pyrgos-Mavrorachi, a Bronze Age metallurgy site on the southern coast of Crete dating to the 19th century BC, brought to light several relics of plant exudates, of which some could be identified as pine or terebinth resins (Lentini, 2004). A Mycenaean tablet from the same period mentions a substance called 'ki-ta-no', which has been interpreted as terebinth resin and was apparently imported to Crete from Cyprus or Syria (Karageorghis, 1996). Chemical analyses have clearly confirmed the presence of pine resin based on its characteristic diterpenoid components, including 7-oxodehydroabietic acid and 15-hydroxy-7-oxodehydroabietic acid (Buckley and Evershed, 2001). Although pine resins were used in the mummification process as early as 2200 BC, evidence suggests that their application became more widespread in later periods (Fig. 1.3). It has been shown that both the wrappings of mummies from the Roman period (30 BC to 395 AD), made of fabric, contain up to 37% diterpenoids from conifers. The increasing use of coniferous resin clearly indicates that the embalmers were aware of the protective effects of the individual plant substances.

Applications of Different *Ṣamghs* in TPM

Anbar (amber)

Amber is the fossilized resin of trees found in cold, snowy regions such as Eurasia, south-eastern Europe and the Maghreb. In traditional sources, the tree from which this gum is obtained is called *Hawz*. Amber forms from honey-like exudates of coniferous and other resin-producuing trees. Sometimes, small insects, plant material or pebbles become trapped inside amber. Table 1.3 provides the detailed information on amber in TPM. There are two types of amber: (i) Roman amber: high quality, hard, transparent, golden hue, emits lemony aroma when rubbed, electrostatic properties and (ii) Nabatean amber: lower quality, often containing impurities, less transparent and weaker than Roman amber. Amber is moderately neutral in heat and cold, dry in the second degree. Its colour is gold and it attracts lightweight objects when heated or rubbed. When burned, it has a mastic-like fragrance. It can be found in regions like central Italy and Western Andalusia. The finest amber can be found along the coasts of Western Andalusia, often buried under the soil (Dioscorides, 2012; Aqili Khorāsāni, 2014).

Anzarut (sarcocolla gum)

Anzarut, nearly spherical resin pieces with whitish-yellow or reddish colour, can be obtained from *Astragalus sarcocolla*, *Astragalus fasciculifolius* and *Astragalus glaucacanthos*. *Anzarut* is a gum similar to frankincense, has a brittle, transparent and crystalline texture with liquorice-like, slightly sweet, bitter and sharp taste. It has two types: one is white and exudes from the plant at night or in the shade, and the other one forms when the sun shines on it and is

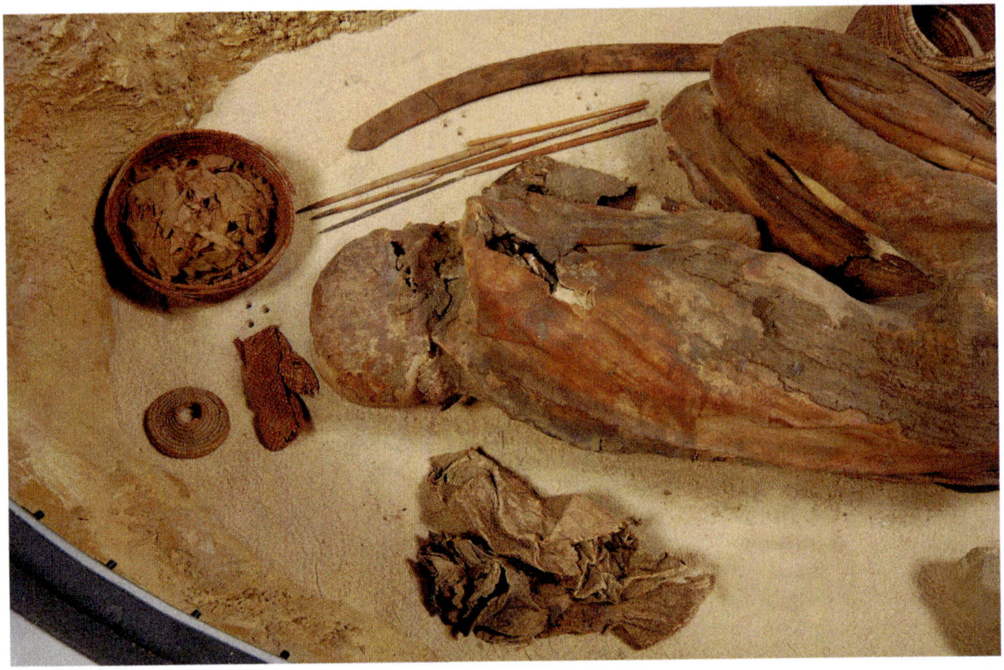

Fig. 1.3. Research has been conducted on funerary textiles of ancient Egypt which shows that gums and resins played a crucial role as part of the compounds used in these wrappings during early mummification processes. The results of these studies indicate that the use of these natural materials in burial practices was common at least 1500 years before mummification became an established tradition in the Old Kingdom (Jones *et al.*, 2014; Jones *et al.*, 2018). (From Museo Egizio (n.d.), Turin, via Wikimedia Commons. https://commons.wikimedia.org/wiki/File:Arrow,_associated_with_mummy_S293_-_Museo_Egizio,_Turin_S_298_p01_p03.jpg.)

slightly yellow. If the sun shines on it too much or it becomes old, it turns red (al-Bīrūnī, 1991). *Anzarut* is soluble in alcohol and forms a colloidal and sticky solution in water. Table 1.4 shows the traditional information on *anzarut*.

Damm al-akhawain (Dragon's blood)

'Dragon's blood' is a red-coloured resin obtained from the Dragon's blood tree (*Dracaena cinnabari* Balf.) (Ibn Sinā, 2015). The detailed traditional applications of Dragon's blood are presented in Table 1.5.

Heltit (asafoetida)

Asafoetida, with a strong, tenacious and sulfurous odour, is an oleo-gum resin obtained from the rhizome and root of *Ferula* spp., mainly *F. assa-foetida*. The plant is native to Iran, Afghanistan, and parts of India (Punjab and Kashmir). The plant and its oleo-gum resin have been known since ancient times. Asafoetida, with high economic value in the world, has medicinal and nutritional importance. Furthermore, it has been consumed as a spice and a folk medicine for centuries. It has also been widely used in India as a spice and in Indian medicinal systems such as Ayurveda (Mahendra and Bisht, 2012). Dioscorides mentioned the habitat of the plant in Syria, Armenia, Iran and Libya, emphasizing its medicinal use since antiquity (Dioscorides, 2012). The oleo-gum resin, known as *heltit* in TPM, is available on the market in two types: high-quality oleo-gum resin in the shape of tears and a low-quality one as bulk. It has been mentioned in traditional texts that the best quality oleo-gum resin was from Iran (*Heltit mediqus*) (Maimonides, 1940). Table 1.6 shows the traditional information on asafoetida.

Table 1.3. Anbar (Amber) (a fossilized gum). (From Dioscorides, 2012; Aqili Khorāsāni, 2014.) Author's own table.

Name	Amber	
Traditional names	Persian: *kahrobā* Greek: *dia-magnitis* Syriac: *hamrma* Roman: *meigherus* Hindi: *kapur* Arabic: *miṣbāḥ al-rūm, qarn al-baḥr*	
Traditional applications	Wound healing Plague prophylactic Ocular healing Anti-haemorrhoid Pharmaceutical preparations	
	Body odour reduction	hyperhidrosis management
	Gastrointestinal	antiemetic, anti-haemorrhagic, gastric tonic, dysentery treatment, hepatoprotective
	Respiratory	pulmonary moisture control, haemoptysis prevention, anti-arrhythmic
	Urinary	diuretic, jaundice
	Haemostasis	epistaxis prevention, menorraghia treatment
	Reproductive medicine	anti-abortive
	Psychocardiology	cardiotonic, antidepressant
Unique features	– Suspension (amulets): prevents nosebleeds, strengthens organs, protects against plague and jaundice – Astrological use: enhances vitality when carved into specific shapes under certain zodiac signs	
Side effects and correctives	Side effects: overuse may cause headaches Corrective: violet oil mitigates adverse effects	
Dosage and substitutes	Maximum dosage: ½ mithqal (~2.3 g). Substitutes: sandarac resin, Armenian clay, cinnamon	
Traditional formulation forms	Used in oils, powders and medicinal tablets	

Medicinally, asafoetida has been used for problems of infections. A large number of modern studies have discovered that asafoetida has various biological properties, particularly antimicrobial properties. Additionally, recent studies on asafoetida have shown several promising properties, particularly relaxant, neuroprotective, memory-enhancing, digestive enzyme, antioxidant, antispasmodic, hypotensive, hepatoprotective, antimicrobial, anticarcinogenic, anticancer, anti-obesity, anthelmintic and antagonistic effects (Iranshahy and Iranshahi, 2011; Amalraj and Gopi, 2017; Upadhyay *et al.*, 2017; Kazemi *et al.*, 2020; Niazmand *et al.*, 2020; Huang *et al.*, 2022). A recent review article investigating the beneficial effects of asafoetida for respiratory ailments found that it might be useful for respiratory viral infections like coronavirus and influenza, the effects of cigarette smoking, asthma, cough and cancer (Salavatifar *et al.*, 2023). In addition, asafoetida has been used in the perfume industry.

Table 1.4. *Anzarut* (sarcocolla gum). (From Maimonides, 1940; Rāzi (Rhazes), 1967–1968; al-Bīrūnī, 1991; Anṣārī Shīrāzī (1992); Heravi (1992); Anṭāki (2000); Ibn al-Nafis al-Qarashi, 2000; Solṭāni, 2004–2016, 2005–2016; Dioscorides, 2012; Ibn Sinā, 2015; Ibn Al-Jazzār, 2019.) Author's own table.

Common names	*Anzarut*, sarcocolla (Latin)	
Traditional names	Greek	*sārkokollā*
	Roman	*serququlā*
	Syriac	*sārāququlā*
	Hindi	*janjar*
	Sistani	*zanjro*
	Persian	*anzarut*
Medicinal applications	Nervous system	Neuroprotective
	Ocular system	Keratopathy, leukoma, eye itching, watery eyes
	Respiratory system	Expectorant, mucolytic, bronchodilator, dyspnea
	Digestive system	Digestive, cholagogue, hepatoprotective
	Immune system	Antiseptic, antidote, detoxifier, scrofula
	Reproductive system	Emmenagogue, menstrual regulator, abortifacient
	Urinary system	Bladder ulcers
	Endocrine and metabolic system	Increases weight
	Musculoskeletal system	Anti-rheumatic, anti-arthritic, joint mobilizer, orthopedic plaster application, anti-inflammatory (for gout and sciatica), joint dislocations
	Integumentary system	Wound healing, tissue regeneration stimulant, epithelialization stimulant, hyperpigmentation, depigmentation therapy, face-clearing lotion, pediculicide, deodorant and antiperspirant, partial leukopathy, vitiligo

Table 1.5. *Damm al-akhawain* (Dragon's blood). (From Nafisi, 1899; Maimonides, 1940; al-Bīrūnī, 1991; Anṣārī Shīrāzī, 1992; Heravi, 1992; Ibn al-Nafis al-Qarashi, 2000; Solṭāni, 2004–2016, 2005–2016; Momen Tonekāboni, 2011; Aqili Khorāsāni, 2014; Ibn Sinā, 2015; Ibn al-Beyṭār, 2015.) Author's own table.

Common names	Arabic	*damm al-akhawain*
	Persian	*khon-e-seyashān*
Geographical distribution	India, eastern and western Sumatra and western Borneo	
Temperament	Cold and dry (as per traditional humoral theory)	
Traditional uses	Tissue and skin	Wound healing, tissue regeneration
	Digestive	Anti-diarrhoeal, haemostatic (intestinal), antiemetic, gastric tonic
	Respiratory	Infections, anti-inflammatory
	Renal protection	Nephroprotective
	Eye	Eyesight enhancer, ocular tonic
	Oral	Gingival tonic, astringent
	Cardiac	Infections
Non-medical applications	Dyeing agent (industrial), pharmaceutical additive, cosmetic enhancer	

Table 1.6. Asafoetida oleo-gum resin (black anjedān). (From Nafisi, 1899; Maimonides, 1940; Rāzi (Rhazes), 1967–1968; Anṣārī Shīrāzī (1992); Heravi (1992); Anṭāki (2000); Solṭāni, 2004–2016, 2005–2016; Dioscorides, 2012; Aqili Khorāsāni, 2014; Ibn al-Beyṭār, 2015; Ibn Sinā, 2015; Ibn Al-Jazzār, 2019.) Author's own table.

Common names	Asafoetida, black anjedān, *heltit*	
Traditional Names	Persian	*angodān*
	Arabic	*hheltit*
	Sanskrit	*hingu*
	Syriac	*ankzāna, kāmā*
Traditional applications	Nervous system	Relaxant, neuroprotective, memory enhancing, antiepileptic, anti-convulsant, sedative
	Gastrointestinal system	Digestive enzyme enhancing Carminative, digestive, anti-colic, laxative Obesity, inflammation, obstruction, fistula
		Liver Hepatoprotective
	Cardiovascular system	Hypotensive, anti-coagulant
	Respiratory system	Expectorant, anti-tussive, bronchodilator
	Immune system	Antioxidant Antimicrobial, antibacterial, anti-inflammatory Antidote, antivenom Anti-infective Anticarcinogenic
	Urinary system	Diuretic, detoxifier
	Integumentary system	Alopecia, hair growth, wound healing
	Ocular system	Eyesight enhancer, cataract
	Reproductive system	Emmenagogue, dysmenorrhoea, abortifacient, aphrodisiac
	Pain and relief	Analgesic, toothache, myalgia relief

ʿIlks (gum-resins)

In TPM, gums were called *ʿilks* and one of the most well-known types of *ʿilks* was *ʿilk al-anbāṭ*, the secretion of trees of the genus *Pistacia*. In ancient times, the people of Iraq referred to all types of *ʿilk* as *ʿilk al-anbāṭ*, with the exception of mastic. Another important type of these gums is *ʿilk al-boṭm*, which is obtained from certain coniferous trees.

ʿIlk al-anbāṭ (pistachio gum)

The whitish resin *ʿilk al-anbāṭ* of the pistachio tree has a slightly bitter taste. The Nabataean resin freshens the breath and cleanses the mouth of impurities. Furthermore, it is useful in the treatment of chronic itching, absorbs internal moisture, increases diuresis and is effective in the treatment of coughs and chest pains caused by excess moisture. It also helps to eliminate the dampness that flows into the chest from above in children. As an alternative to *ʿilk al-anbāṭ*, the resin *ʿilk al-boṭm* or the resin of the cypress tree is also used. Table 1.7 provides extensive information on this oleoresin described in the TPM.

ʿIlk al-boṭm (Persian terpentine)

In order to obtain *ʿilk al-boṭm* during the warm season, the trunk of the tree is scratched, and the sap oozes from the incision. Initially, this tree sap, which is in a semi-liquid, dough-like

Table 1.7. *'Ilk al-anbāṭ, Saghez* (turpentine). (From Ibn Al-Jazzār, 2019; Mir Heydar, 2019.) Author's own table.

Local names in Iran	Tehran: *chatanqosh*; Hamadan: *shaqlanqosh*; Arasbaran: *chatangoo*; Karaj, Zanjan, Kurdistan: *saqez*; Fars and Baluchistan: *baneh*; Lorestan: *golkhongkaleh*; Kermanshah: *venoshk*; Marivan: *qasaqan*; Oshnavieh: *darben*; Sardasht: *benshet*	
French name	*Pistachier de l'Atlas*	
English name	Wild pistachio	
Different varieties in Iran	1. *Pistacia atlantica* Desf. subs. *cabulica* (Stocks) Rech	Syn. *Pistacia cabulica* Stocks Distribution: Kohgiluyeh, Shahkouh (Kerman), Jebal Barez, Taftan foothills (Baluchistan), Birjand (Khorasan)
	2. *Pistacia atlantica* Desf. subs. *kurdica* (Zohary) Rech	Distribution: Kermanshah, Gilan-e Gharb, Ilam, Sanandaj, Lorestan, Bakhtiari regions (Gachsaran, Sisakht), Fars, and Baluchistan
	3. *Pistacia atlantica* Desf. subs. *mutica* (Fisch. & C.A.Mey.) Rech. f	Syn. *Pistacia mutica* Fisch. & C.A.Mey. Distribution: Azerbaijan, Bakhtaran, Lorestan, Fars, Kerman, Baluchistan, Khorasan, Karaj
Resin applications	Traditional medium for pigments; dyeing: used in textile dye preparation	

Table 1.8. *'Ilk al-boṭm* (Persian turpentine). (From Mir Heydar, 2019, Ibn al-Nafis al-Qarashi, 2000.) Author's own table.

Common names	Traditional and historical names: *aqozon, banah, banak, bot'm akhdar, darban* and *kasoon* In Iran: *kolkhounag* in Fars, *kolkhanak-e narmeh* and *kolkhounak-e narmeh* in Sepid-Dasht, Kuh-Gazeh and Takht-e Dehreh in Lorestan, *khanjak* in Kerman and Afghanistan, and *kasour* in Sirjan and the Jabal Yarez region	
Traditional applications	Digestive system	Digestive stimulant, diuretic, gastrointestinal tonic, carminative
	Gastrointestinal	Gastric tonic, mucolytic
	Fractured limbs	Orthopedic plaster agent, bone healing stimulant
	Wounds	Tissue regeneration stimulant, wound healing agent

form, closely resembles Venetian turpentine. In Kurdistan (Iran), because the sap that comes out is mixed with dirt and debris and is dirty when collected, it is referred to as 'bitter Saghez'. To prepare it for the market and produce clean *'ilk al-boṭm*, it is boiled in water and purified before being sold under the name 'Ghandaroon'. *'Ilk al-boṭm*, which is a type of oily resin, is used to produce colophony after its essential oil is extracted (Table 1.8).

Table 1.9. *Jāwshir* (opopanax). (From Rāzi (Rhazes), 1967–1968; Anṣārī Shīrāzī (1992); Heravi (1992); Anṭāki (2000); Momen Tonekāboni, 2011; Dioscorides, 2012; Aqili Khorāsāni (2014); Ibn al-Beyṭār, 2015; Ibn Sinā, 2015; Solṭāni, 2004–2016, 2005–2016.) Author's own table.

Common name	Opopanax (gum-resin)	
Traditional name	*Jāwshir*	
Medicinal properties and uses	Laxative and soothing	Demulcent, mild laxative
	Fever and fatigue	Antipyretic, tonic, malaria, hepatoprotective
	Digestive and urinary	Diuretic, anti-inflammatory (urinary), urinary decongestant, anti-splenomegaly, colic
	Reproductive system	Menstrual regulator, abortifacient, uterine decongestant, anti-inflammatory, uterotonic, emmenagogue, abortifacient (suppository), uterine decongestant
	Pain and inflammation	Analgesic, neuralgia, topical anti-inflammatory, antispasmodic
	Wound healing and skin	Antiseptic, wound healing, anti-inflammatory (cutaneous), depigmentation therapy, anti-lymphadenitis
	Respiratory	Bronchodilator, expectorant
	Dental	Analgesic (dental)
	Eye and ear	Ocular tonic, cataract, otological tonic
	Rabid dog, scorpion and snake bites	Antivenom, antiseptic
	Tremors and stroke	Neuroprotective, fatigue
Dosage and consumption	Recommended dosage	2.5–5 g
	Special applications	50g of leaves soaked in grape juice for 20 days, strained and consumed for ascites and spleen disorders
	For fevers	5g for tertian fevers and jaundice
	For paralysis and tremors	Half a dirham (approx. 1.5 g) mixed with an ounce of marjoram decoction taken for 3 days

Jāwshir (opoponax)

High-quality opoponax should be bitter, oily, brittle and soluble in water. Its colour should be white inside and yellow outside. Black *jāwshir* that is soft has no therapeutic value. Sometimes it is adulterated with honey and wax. To test the purity of *jāwshir*, its solubility can be checked in water. If it is not adulterated, it will dissolve in water and turn a milky colour. It is warm in the third degree and dry in the second degree. To extract the gum, the root of the plant is split open at the beginning of the plant's growth. A white sap comes out of it and can be collected on the leaves spread out under the plant. While the fresh sap of the plant is white, it turns saffron yellow after the outer layer hardens. The resin of the plant can be obtained in the same way from the stem of the plant at the time of year when wheat is harvested (Dioscorides, 2012; Aqili Khorāsāni, 2014). The detailed traditional application of opopanax is shown in Table 1.9.

Kathirā (tragacanth)

Tragacanth is obtained from various *Astragalus* species (Table 1.2). Table 1.10 shows the traditional information about tragacanth. Katirā

Table 1.10. *Kathirā* (tragacanth). (From Rāzi (Rhazes), 1967–1968; Heravi (1992); Anṭāki (2000); Solṭāni, 2004–2016, 2005–2016; Momen Tonekāboni, 2011; Dioscorides, 2012; Aqili Khorāsāni, 2014; Ibn Sinā, 2015.) Author's own table.

Common names	*Katirā* (gum tragacanth), *draga canthae gummi*	
Historical names	Syriac	*halusia* or *kathyra*
	Greek	*trāgākāntā*
	Arabic	*tarāghāqanthā*
	Middle Persian	*katir* or *zol zada*
Traditional applications	Skin	Emollient, humectant, anti-hyperpigmentation, depigmentation therapy, antipruritic, antiparasitic, tissue repair stimulant, mucilage therapy for split ends
	Digestive and gastrointestinal	Gastroprotective, anti-ulcer, demulcent, laxative adjunct, mucolytic, anti-haemoptysis agent
	Eye health	Ocular anti-inflammatory, keratopathy treatment
	Weight gain	Nutritional supplement, anabolic agent
	Hair and scalp care	Anti-alopecia, mucilage-based hair conditioner
	Respiratory	Anti-tussive, pulmonary tissue repair, laryngeal demulcent, voice restorative
	Urinary and renal	Nephroprotective, anti-inflammatory (urinary tract), urinary anti-inflammatory, anti-dysuria agent
Pharmaceutical applications	Excipient, bioavailability enhancer, pharmaceutical stabilizer, binder	

is obtained by scraping the roots and stems of the plant material and collecting the resin that emerges in small pits at the base of the plant. There are two tragacanth types on the market: one is *katirā moftoli*, which is offered as thread-like, white pieces and is taken orally, particularly in traditional medicine. The other is *katirā khormani*, which comes in the form of irregularly shaped, yellowish pieces and is mainly used for external applications. *Katirā* possess a slightly sweet taste and is odourless.

Lobān (frankincense)

According to TPM researchers, frankincense (*Boswellia sacra*, *B. serrata*) improves vision and heals deep ulcers and fresh wounds. When taken orally, it is effective for bloody vomiting, nosebleeds and cerebral haemorrhages. Frankincense is used for stomach weakness, bloody diarrhoea, stomach ulcers and fever, and has also been used as a memory aid. However, it has also been found that excessive consumption can cause headaches. According to Ibn Sina (Avicenna), frankincense is the best remedy for fresh wounds. A mixture of the gum-resin and duck fat treats impetigo, and a mixture of frankincense and lemongrass removes dandruff and is effective for uterine bleeding (Ibn Sinā, 2015). It is also effective for eye injuries, leukoma, ulcerative blepharitis, cancer and blood clots in the eye. A mixture of frankincense, cinnabar and rose oil helps with breast swelling in women who have just given birth; a mixture of the gum-resin, olive oil and honey helps with joint pain and chronic bone pain, toothache and mouth ulcers. A mixture of frankincense and marjoram oil has proven effective against hair loss. In addition, the gum-resin has also been used successfully to treat impotence, shortness of breath and bronchitis. The individual applications of frankincense in traditional herbal medicine are shown in detail in Table 1.11.

Table 1.11. *Lobān* (frankincense) (From Rāzi (Rhazes), 1967–1968; Anṭāki (2000); Solṭāni, 2004–2016, 2005–2016; Momen Tonekāboni, 2011; Dioscorides, 2012; Aqili Khorāsāni, 2014; Ibn Sinā, 2015; Ibn Al-Jazzār, 2019.) Author's own table.

Common names	Persian: *bostak*, *kondor* Arabic: *kondor*, *lobān* English: frankincense, olibanum Greek: *khondros libānou* Sanskrit: *kundrus*	
Cultural and religious uses	Hebrews	Mentioned in the Torah as part of sacred incense
	Catholic Church	Used in religious ceremonies as aromatic incense
	Prophet Muhammad (PBUH)	Recommended soaking frankincense in water overnight to enhance memory
	Dioscorides	Described best frankincense as '*stagonis*', sticky, quick to ignite; white variant named '*amonitos*'
Traditional applications	Eye	Vision enhancer, ocular anti-inflammatory
	Wound healing	Wound healing, tissue repair, topical wound repair
	Bleeding conditions	Haemostatic, anti-haemorrhagic
	Digestive System	Gastroprotective, anti-diarrhoeal, digestive, carminative
	Respiratory	Expectorant, anti-tussive, respiratory decongestant, bronchitis
	Memory enhancement	Nootropic, memory enhancer
	Headache	Analgesic, anti-cephalalgia
	Reproductive	Haemostatic, emmenagogue
	Musculoskeletal	Anti-inflammatory, analgesic, anti-rheumatic, dental analgesic
	Hair	Anti-alopecia, scalp tonic, anti-dandruff
	Breast and Skin Conditions	Anti-inflammatory, topical soothing agent
	Oral	Antimicrobial gargle, dental analgesic
	Ibn Sinā's view	• Best for treating fresh wounds • Useful for uterine bleeding and preventing vomiting (with wine or vinegar) • Effective for eye infections, cancer and breast inflammation (with clay and rose oil) • Treats joint pain and dental ulcers when mixed with olive oil and honey • Combats hair loss when mixed with myrtle oil
Dosage	Recommended daily dose: 1–2.5 g	

Maṣṭakā (mastic gum)

Mastic is a resin obtained from *Pistacia lentiscus* L. (Anacardiaceae) and is described in TPM textbooks as follows: 'Mastic gum is a yellow-coloured resin extracted from a shrub or small tree that is native to regions with a Mediterranean climate and is distributed in North Africa and some of the Greek islands known as the Mycenaean Islands. The shrub is evergreen and a resinous scent is evident from all parts of the plant. The people of the

Mycenaean Islands obtain an edible oil from the fruit of this tree, which is astringent and is used instead of olive oil in their food. However, the main product of these shrubs is mastic resin that exudes from the branches, wood and stems of this tree spontaneously or through cracks made. Mastic is released in the form of small droplets during the warm months of the year and hardens after a short time in the presence of air. In order to extract a larger amount of resin more quickly from the tree, scratches are made on the trunk and branches. With this method, an average of 4–5 kg of resin is extracted from each tree. The mastic available on the market is offered in the form of small, hard, oval teardrop pieces which possess a yellowish colour and are more or less translucent. When chewed, the gum gets soft, releasing an aromatic taste. According to traditional medicine practitioners, mastic gum is relatively warm and dry in nature, and is a tonic, especially for the major organs of the body such as the stomach, liver, heart, kidneys, nervous system and lungs (Table 1.12) (Ibn Al-Jazzār, 2019; Mir Heydar, 2019).

Mey'ah

Mey'ah is an oleoresin extracted from the bark of some trees. *Mey'ah* is classified into two main types: (i) *mey'ah sā'elah* or *lobnā* (*'anbar sā'el* in Persian) or oriental sweet gum, a yellow resin that naturally exudes from *Liquidambar orientalis* (Altingiaceae family), and (ii) *mey'ah yābesah* or *iṣṭarak* (*mey'ah khoshk* in Persian) or officinal storax from *Styrax officinalis* (Styracaceae family), a black resin that often contains fewer impurities than the liquid variety. Liquid and dry *mey'ah* both possess astringent and drying properties. Some believe that *mey'ah* is hot and dry, expelling moisture from the brain and purifying it (possibly referring to the reduction of excess mucus or decongestant effects). However, this view opposes the prevailing belief, as *mey'ah*, due to its intense heat, affects the temperament of the head, leading to a warm imbalance (similar to 'localized hyperthermia' or 'inflammatory reactions caused by thermal stimulation' in modern medicine). This heat and pressure, through expansion and tension, can cause severe headaches (potentially referring to heat-induced tension headaches). Nevertheless, if the smoke of *mey'ah* is inhaled, it expels moisture and waste from the head, purifying it (referring to potential decongestant and anti-inflammatory effects on the respiratory tract), and is beneficial for cold brain conditions, nasal discharges (*nazlah*) and cold-induced headaches (possibly related to sinusitis or allergic rhinitis). Dry *mey'ah* is particularly beneficial for stomach moisture, as it dries excess moisture (potentially comparable to the use of antacids or agents regulating gastric secretions) and is especially effective for this purpose. It strengthens the natural function of the digestive system, has astringent and drying qualities, promotes menstrual flow (emmenagogue effects in modern medicine, though requiring caution) and is useful in treating phlegmatic cough (mucolytic or expectorant properties), catarrh (likely referring to upper respiratory inflammation, e.g. rhinitis) and scabies (referring to scabies, a skin condition caused by mites, treatable with modern topical medications like permethrin) as presented in Table 1.13 (Ibn Sinā, 2015).

Moql **(mukul, gum guggul)**

Mukul has been known since ancient times and its name is mentioned in the Torah (Numbers 11:7). Indians have used it for therapeutic purposes since the 7th century BC. Table 1.14 shows the traditional applications of guggul.

Morr **(myrrh)**

According to the traditional texts, myrrh is an oleoresin obtained from a thorny tree, typically ranging in height from 5 to 7 m. This tree is native to regions in Egypt, Sudan, Yemen, Ethiopia, Thailand and Somalia. It is believed that the best quality myrrh is derived from trees in Somalia, where locals refer it to as *Mulmul*. Based on the botanical studies, the plant is *Commiphora myrrha* (T. Nees) Engl. (Syn. *Commiphora molmol* (Engl.) Engl. ex Tschirch). The former scientific name of the plant (*C. molmol*) originates from the Somalian term for the oleoresin. In traditional texts,

Table 1.12. *Maṣṭakā* (mastic gum). (From Ibn Al-Jazzār, 2019; Mir Heydar, 2019.) Author's own table.

Common names	Persian: *muṣṭakī*, *ramas*, Roman gum, *velemshak*; Arabic: *'ilk rūmī*, *muṣṭakā*; French: mastic, *mastique*; English: mastic; Hindi: Roman mastic; Syriac: *kiya*	
Resin (mastic gum)	Tear-shaped, semi-transparent, yellowish, aromatic resin droplets harvested during summer by making incisions in the stems and branches	
Types of mastic	Nabati type	White, transparent, fragrant, slightly sweet; obtained without cutting the branches; highest quality
	Qibti type	Darker, bitter, less adhesive; obtained via stem and branch incisions
Traditional applications	Vital organ support	Cardioprotective, neuroprotective, hepatoprotective, gastric tonic, systemic energizer, adaptogen
	Gastrointestinal	Gastrointestinal tonic, hepatic tonic, anti-ulcer, anti-inflammatory, carminative, antacid, bloating, diarrhoeal, irritable bowel syndrome
	Respiratory	Expectorant, mucolytic, anti-inflammatory, anti-infective
	Oral and dental	Dental tonic, analgesic, gingival astringent, anti-inflammatory, halitosis, antimicrobial
	Dermatological	Skin brightener, hyperpigmentation, anti-inflammatory, wound healing
	Musculoskeletal	Analgesic, anti-inflammatory, antispasmodic, muscle relaxant, tetanus
	Neurological and mental	Anti-obsessive, depression, nootropic, cognitive enhancer
Methods of mastic consumption	Chewing	Strengthens gums and teeth, clears oral phlegm
	Drinking	Used as a decoction or infusion for digestion and liver health
	Inhalation	The smoke clears respiratory passages
	Topical application	Joint pain and skin inflammation (ointment)
Traditional medicinal combinations	With agarwood	Phlegm expulsion stimulant
	With *Aloe vera*	Choleretic, bile detoxifier
	With *Terminalia* sp.	Melancholy, black bile regulator
	With sugar	Gastric tonic, carminative
Dosage and side effects	Dosage	Maximum daily dosage: 4.6 g
	Side effects	May harm the bladder
	Correctives	Combine with vinegar or gum arabic to mitigate adverse effects

it was believed that the best-quality myrrh had a nearly white colour with a reddish tint, without any visible tree fibers or materials (Rāzi (Rhazes), 1967–1968).

Historically, myrrh has been recognized by many ancient civilizations and is mentioned in Egyptian papyri. In Egypt, it was employed for embalming bodies, treating oral and throat disorders and managing skin diseases (Table 1.15). Egyptian women burned it to repel fleas from their homes (Dioscorides, 2012). The Babylonians and Akkadians referred it to as *murru*, the Hebrews called it *mor* and in the Aramaic and Syriac languages it was called *murra*. Babylonians mixed myrrh with honey for treating wounds and inflammations, while they also used it as incense. In the Bible, myrrh was one of the components of the 'Holy Anointing

Table 1.13. *Mey'ah* (storax and oriental sweet gum). (Heravi, 1992; Solṭāni, 2004–2016, 2005–2016; Ibn Sinā, 2015; Ibn Al-Jazzār, 2019.) Author's own table.

Oleoresin traditional names	English	Asiatic styrax, balsam styracis, liquid amber, liquid storax, Oriental sweet gum, storax, Turkish sweet gum
	Sanskrit	*silhaka, turuskak*
	Greek	*staktê*
	Arabic	*mey'ah*
	Persian	*mey'ah*
Traditional properties	Respiratory system	Decongestant, anti-inflammatory, mucolytic and expectorant, anticatarrhal
	Nervous system	Gastric moisture regulation, astringent, digestive stimulant.
	Reproductive system	Emmenagogue
	Immune and integumentary systems	Antiparasitic (scabies), anti-inflammatory
	Circulatory system	Blood flow regulation
	Digestive system	Gastric moisture regulation, astringent, digestive stimulant

Oil' (Exodus 30:23). In ancient Persia (during the Achaemenid era), myrrh was used for quarantine and disinfection purposes. The Chinese called it *moyao* (600 BC). Indians named it *rasagandhi* (a Sanskrit word) and used it as both a wound-healing and an emmenagogue agent. According to a hadith by Abdullah ibn J'afar, Prophet Muhammad (PBUH) recommended burning thyme, myrrh and mugworts in homes. The Greeks called it *smorna* (Diodorus Siculus, 2005).

Oshaqq (gum ammoniacum)

Gum ammoniac is a gum obtained from *Ferula ammoniacum*. The plant has been known since ancient times, and Hippocrates used the gum-resin for medicinal purposes. In Greek it is called *ammoniakon*, which is derived from the name of the temple of Jupiter Ammon in the ancient land of Lydia. Istfan and Hunain transformed the name to *ammoniaqan* and called it *oshagh* (*oshaqq*) in Arabic. Traditional physicians have mentioned its properties in textbooks. Table 1.16 shows the traditional information on *oshaqq*.

Qennah (galbanum)

Galbanum is an oleo-gum resin obtained from roots and stems of *Ferula gummosa* Boiss., which is a well-known plant growing in the northern and north-eastern regions of Iran and Turkmenistan. In traditional texts, galbanum is referred to as *khalbani*, *qennah* and *bārazd*. According to Jorjānī, 'Qennah is the gum of a plant called *Bārzad*' (Jorjānī, 2005). Hakim Mo'men stated that *bārazd* is the Persian name of *bārijah*, the Arabic name is *qennah* and the Turkish name is *qasni*. Heravi believed that the gum was warm in the third degree and dry in the second degree. The highest-quality resin is tear-shaped, resembling frankincense, without wooden particles and with a whitish-yellow or yellowish-green colour. The lowest-quality gum is shapeless and in bulk form with a greenish-yellow or brown colour. If the gum is of the bulk type quality, it should be put in hot water and the oil and woods floating on the water removed; the rest should then be filtered through a cloth without pressing. What remains should be dissolved in bitter almond oil or juice of *Ruta gravolens* L. It has been believed that its dosage should be up to 3.5 g orally or 5 g as an antidote. Its substitute is half of its weight of sagapenum

Table 1.14. *Moql* (mukul, gum guggul) (From Anṣārī Shīrāzī, 1992; Heravi, 1992; Solṭāni, 2004–2016, 2005–2016; Momen Tonekāboni, 2011; Aqili Khorāsāni, 2014; Ibn Sinā, 2015; Ibn Al-Jazzār, 2019.) Author's own table.

Common name	Guggul, mukul resin	
Traditional names	Sanskrit	*devadhupa* (divine smoke), *guggula*
	Hindi	*guggul*
	Greek	*bdellium*
	Arabic	*moql*
Traditional applications	Wound and skin health	Wound healing, anti-verruca, alopecia
	Digestive and gastrointestinal system	Mucolytic, haemorrhoids, thrombolytic, intestinal epithelial repair, hernia, hepatoprotective, carminative, anti-eructation
	Respiratory health	Anti-edematous, inflammation (pharyngeal and laryngeal), anti-tussive, bronchodilator, pulmonary anti-inflammatory
	Urinary health	Diuretic, lithotriptic
	Neurological and musculoskeletal system	Analgesic, anti-convulsant, anti-inflammatory, nerve relaxant, gout, myalgia
	Gynaecological	Parturition, emmenagogue, anti-edematous, anti-inflammatory
	Anti-toxin and antiparasitic properties	Antidote, antivenom
	Oral health	Gingivitis, periodontitis, mucosal wound healing
	Fever and plague treatment	Anti-pyretic, anti-plague
	Cardiovascular system	Hypolipidemic, antithrombotic
	Endocrine	Thyroid stimulant, obesity
	Sexual	Aphrodisiac
	Pediatric applications	Hydrocele
Precautions	1. Pregnancy and breastfeeding: • Use with caution during pregnancy and lactation. 2. Children: • Avoid in paediatric populations unless under medical supervision. 3. Allergic Reactions: • Contraindicated in individuals with a known sensitivity to guggul. 4. Pre-surgical patients: • Discontinue use 1–2 days before surgery due to potential effects on blood clotting.	
Adverse effects	Mild effects	Dizziness, nausea, loss of appetite, diarrhoea (high doses)
	Severe effects	Kidney inflammation (high doses)

Table 1.15. *Morr* (myrrh). (From Rāzi (Rhazes), 1967–1968; al-Bīrūnī, 1991; Anṣārī Shīrāzī (1992); Heravi (1992); Anṭāki (2000); Solṭāni, 2004–2016, 2005–2016; Momen Tonekāboni, 2011; Dioscorides, 2012; Aqili Khorāsāni, 2014; Ibn al-Beyṭār, 2015; Ibn Sinā, 2015; Ibn Al-Jazzār, 2019.) Author's own table.

Traditional applications	Skin	Depigmentation therapy, anti-hyperpigmentation, antiparasitic, antipruritic
	Gynaecological	Haemostatic, uterine tonic, emmenagogue, abortifacient
	Eye	Leukoma, ocular decongestant, anti-inflammatory, conjunctivitis, blepharitis, antimicrobial, lachrymation, cataract, ocular protective agent, eyesight enhancer
	Musculoskeletal	Anti-inflammatory, tissue repair stimulant, anti-dislocation
	Hair	Hair growth, alopecia, hair tonic
	Wound healing	Wart, wound-healing stimulant, antibacterial, impetigo, gangrene, antiseptic
	Kidney and bladder	Nephroprotective, anti-inflammatory
	Circulatory	Vasodilatory agent
	Digestive	Carminative, gastric tonic, cholagogue
	Spleen	Anti-inflammatory, spleen protective
	Insect bites	Antivenom properties
	Oral and dental	Dental astringent, gingivitis
	Ear	Antimicrobial, anti-inflammatory
	Cosmetic	Deodorant
	Respiratory	Expectorant, mucolytic
	Liver	Hepatoprotective
	Antiparasitic	Anthelmintic, antiparasitic
	Reproductive	Abortifacient
	Neurological	Sedative, anxiolytic, antispasmodic
Therapeutic properties	Antiseptic, antimicrobial, astringent, anti-inflammatory	
Contraindications	Toxicity risk at high doses (7 g); not recommended in pregnancy due to abortifacient effects	
Dosage and preparation	Used as ointments, suppositories or orally (dissolved in honey or water)	

and a quarter of its weight of opoponax (Aqili Khorāsāni, 2014). Table 1.17 shows the traditional information on galbanum.

Sakbīnaj (sagapenum)

Sagapenum, with a pungent and garlic-like smell similar to asafoetida and galbanum, is obtained from *Ferula persica*, native to Iran and the Transcaucasus. According to TPM textbooks, sagapenum is hot and dry in the third degree. The gum-resin was exported to India, Babylon (modern-day Iraq) and Greece from ancient times. The outer surface of good-quality gum is smooth and red, and the inside is white with a bitter, sharp and unpleasant taste (Dioscorides, 2012). Table 1.18 shows the traditional information on sagapenum.

Table 1.16. *Oshaqq* (gum ammoniacum). (From Nafisi, 1899; Anṣārī Shīrāzī, 1992; Heravi, 1992; Solṭāni, 2004–2016, 2005–2016; Momen Tonekāboni, 2011; Dioscorides, 2012; Aqili Khorāsāni, 2014; Ibn Sinā, 2015.) Author's own table.

Common names	English: Ammoniacum plant, Bombay sumbul, dorema, gum ammoniacum plant	
Traditional applications	Nervous system	Neuroprotective, epilepsy, paralysis, pelvic and buttock pain, anti-mastitis, breast inflammation (hormonal and nervous regulation)
	Ocular system	Anti-leukoma, ocular inflammation, blepharitis
	Respiratory system	Asthma, orthopnea, expectorant and bronchodilator, pulmonary and nasal decongestant, chronic bronchitis, chronic nasal catarrh, goiter
	Digestive system	Laxative, digestive, carminative and mucolytic
		Liver and spleen: Hepatoprotective, symptoms of splenomegaly and hepatosplenomegaly, spleen and liver stiffness (traditionally mixed with vinegar)
	Skin	Vitiligo, keratopathy, wound healing
	Musculoskeletal system	Joint pain, sciatica (mixed with vinegar), joint inflammation (as a poultice with tar and honey), arthritic pain, joint mobilizer
	Reproductive system	Abortifacient, emmenagogue, leucorrhoea, orchitis
	Urinary system	Diuretic (excessive consumption may cause bloody urine), dropsy
	Immune system and antidote	Detoxifier and anthelmintic, scrofula, tuberculosis

Ṣamgh arabi (gum arabic)

Gum arabic is one of the most well-known gums used in both traditional and modern medicine and has a wide range of applications. African countries, especially Sudan, which accounts for around 70% of the global supply, and the entire Sahel region from Senegal to Somalia, are the main producers of gum arabic. In general terms, gum arabic is often referred to as the gum from any *Acacia* species. In this case, it is sometimes referred to as 'Acacia gum'. For instance, 'gum arabic' from Zimbabwe is derived from *Vachellia karroo* (Hayne) Banfi & Galasso. Table 1.19 shows the traditional information about gum arabic. According to traditional texts, gum arabic imported from Pontus (Turkey) and the southern Arabian Peninsula to Egypt was also known in Mesopotamia and used in the manufacture of papyrus and ink (Forbes, 1965). Dioscorides mentioned Egypt as the habitat of one of the tree species producing gum arabic and believed that the habitat of another species (possibly *Lebeckia sepiaria* (L.) Thunb.) producing this gum was Cappadocia and Pontus (Dioscorides, 2012). Al-Biruni described the method of gathering gum arabic and considered it the best chewable gum (al-Bīrūnī, 1991). There exist three types of gum arabic on the market, including white, red and sweet-tasting yellow-coloured gums. Historically, Egypt and Hejaz were the main exporters of gum arabic.

Ṣamgh Farsi (Persian gum)

In TPM, this term refers to gums obtained from various *Prunus* species, namely almond, plum, peach, cherry and sour cherry trees. Persian

Table 1.17. *Qennah* (galbanum). (From Anṣārī Shīrāzī, 1992, 2004–2016, 2005–2016; Momen Tonekāboni, 2011; Aqili Khorāsāni, 2014; Ibn al-Beyṭār, 2015; Ibn Sinā, 2015; Ibn Al-Jazzār, 2019.) Author's own table.

Traditional names	Persian	*bārazd*
	Arabic	*qennah*
	Turkish	*kasni*
	European	*galbanum*
	Greek	*metopion*
	Syriac	*hhelbānithā*
	Torah	*hhelbenah*
Traditional applications	Nervous system	Epilepsy, chorea, headaches, memory enhancement
	Digestive and gastrointestinal system	Gastrointestinal infections, gastric inflammations, anti-flatulent, haemorrhoids, gout
	Respiratory system	Asthma
	Immune system and detoxification	Scorpion venom, inflammation, scrofula
	Musculoskeletal system	Muscle spasms, muscle regeneration Sciatica, Muscle amputation
	Renal and urinary system	Kidney obstruction, kidney stones, urethral irritation
	Reproductive system	Abortifacient, uterotonic
	Integumentary system	Wounds, earache, ear swelling

Table 1.18. *Sakbīnaj* (sagapenum). (From Rāzi (Rhazes), 1967–1968; Anṣārī Shīrāzī (1992); Heravi (1992); Solṭāni, 2004–2016, 2005–2016; Momen Tonekāboni, 2011; Dioscorides, 2012; Aqili Khorāsāni, 2014; Ibn Sinā, 2015.) Author's own table.

Common names	Greek: *sagapenon*; Arabic: *sāghāphinon*; Persian: *sakbīnaj*; Hindi: *kundel*	
Therapeutic applications	Nervous system	Hysteria-induced fainting, headaches, epilepsy
	Ocular system	Cataracts, blurred vision, styes, leukoma
	Respiratory system	Chronic cough, expectorant, thick phlegm, asthma
	Digestive system	Anti-flatulent, demulcent, bile removal, colic, haemorrhoids, stomach pain, phlegmatic diarrhoea, gout
	Renal and urinary system	Kidney and bladder stones
	Reproductive system	Uterine pain, emmenagogue, aphrodisiac (with honey)
	Musculoskeletal system	Flank pain, joint pain, back pain, sacroiliac joint pain, stiffness, ankylosis, sciatica
	Integumentary system	Insect bites
	Immune system	Phlegmatic and Nubian fevers

Table 1.19. *Ṣamgh arabi* (gum arabic). (From Maimonides, 1940; Forbes, 1965; Rāzi (Rhazes), 1967–1968; al-Bīrūnī, 1991; Solṭāni, 2004–2016, 2005–2016; Jorjānī, 2005; Dioscorides, 2012; Aqili Khorāsāni, 2014; Ibn Al-Jazzār, 2019.) Author's own table.

Traditional names	Persian	*bozghond*
	Greek	*akanta*
	Syriac	*du'atha*
Traditional applications	Eye	Conjunctivitis, redness and ulcerations of eyelids, eye inflammation
	Gynaecological health	Uterine discharges, metrorrhagia, leukorrhoea, emmenagogue, sexual tonic (aphrodisiac)
	Haemostasis	Stops bleeding, including haemoptysis, pulmonary bleeding, and bloody diarrhoea; strong haemostatic agent for both internal and external bleeding
	Hair, nails and skin	Hair loss, hair tonic, hair darkener
		Excessive sweating (with rose petals or myrtle leaves), hyperhidrosis
		Nail bed infections (paronychia), nail tonic
	Proctology	Rectal prolapse
	Wound healing	Dries wounds, stops bleeding, prevents infection, haemostasis, wound healing
	Respiratory	Chronic cough, asthma, pleurisy
	Pain	Toothache, antiseptic, joint pain

gum is warmer than gum arabic and causes less constipation. It is beneficial for treating coughs and bladder stones and for wound healing. Rubbing it on the eyes improves eyesight, and when used as a poultice with vinegar, it is effective for rashes on children's bodies.

Ṣamgh al-qouqā, ṣamgh lātī, ṣamgh satr'moulia

The resins of various conifer species, including cedar resin (*Cedrus libani* A. Rich.), stone pine resin (*Pinus pinea* L.) and Norway spruce resin (*Picea abies* var. *abies*), share common characteristics. These resins are warm, laxative and dissolvent, making them suitable for treating coughs, lung wounds and chest bleeding. When consumed alone or with honey, they help clear the chest. In terms of potency and effectiveness, these resins are stronger than balsamic resins, with Norway spruce resin being the strongest and stone pine resin being milder (Ibn al-Beyṭār, 2015).

Ṣamgh as-sarw (Italian cypress resin)

According to Soleimān ibn Ḥasan, the resin of the cypress tree possesses a particular acridity and bitterness, and it is considered weaker in effect and benefit compared to other resins (Ibn al-Beyṭār, 2015). When inhaled through the nose, it clears sinus congestion. Its potency is similar to the resins of rue and spruce, though less intense. As a result, the resin-like tar extracted from the cypress tree is also weaker than the tar obtained from spruce, which is commonly known as *sharbin*. According to Ḥobaīsh ibn Hasan, if this resin is sprinkled together with cypress seeds on wounds located on the head or other parts of the body, it aids in healing them (Ibn al-Beyṭār, 2015).When mixed with pine resin, it is used for tumour reduction (Mir Heydar, 2019).

Ṣamgh as-soḏāb (rue resin)

The taste of rue resin is bitter, and its scent is sharp, resembling the smell of rue. According to Abu Jarij, it is classified as hot in the third degree and dry in the second degree (Ibn al-Beyṭār, 2015). Its benefits include dispelling flatulence and clearing cerebral phlegm, especially when used nasally as a snuff in combination with other medications. If applied to eye wounds, it helps treat them. It is also beneficial for scrofulous swellings in the neck and underarms. Nasal application in a dose of one daniq (approx. 0.65 g) yields therapeutic effects. If rue resin is unavailable, 1.5 dirhams (approx. 4.5 g) of sagapenum can be used as a substitute for each dirham of rue resin (Ibn Al-Jazzār, 2019).

Ṣamgh as-somāq (sumac resin)

In traditional medical texts, sumac is referred to by various names, including *samāq*, *samāqīl*, *tumtum* and *sumāq* al-*dabbāghīn*. In France, it is known as *corroyère* or *sumac des tanneurs*, while in England, it is called *tanner's sumac*. Sumac resin possesses astringent and strengthening properties. Additionally, due to its warmth and sharpness, it has dispersing properties and can help eliminate waste substances. When placed on decayed and worm-eaten teeth, it relieves pain. If dripped into the ear, its drying and astringent qualities help reduce pus and infections (Ibn Sinā, 2015). Sumac resin is also used in some suppositories intended to improve vision (Ibn al-Beyṭār, 2015). Another species from the same genus is *Rhus verniciflua* Stokes (this name is a synonym of *Toxicodendron vernicifluum* (Stokes) F.A. Barkley). It is a small tree or shrub native to countries between northern Pakistan and Korea and introduced into Japan, Tajikistan and Uzbekistan. Some researchers believe that lac is obtained from this plant and not from *Ficus altissima*.

Sqamunyā (scammony gum)

Scammony is obtained from *Convolvulus scammonia*. The gum-resin of the plant is extracted from the root (Table 1.20).

Astragalus fasciculifolius Boiss.

English name: flesh-glue plant, sarcocolla plant
Arabic name: *shajarat-al anzarut*
Persian name: *anzarut*
Native to: Gulf States, Iran (Govaerts, 2024)
Morphology: spiny shrub, 20–70 cm high, branching from base and above, old bark brownish to dark grey; stems and branches densely furnished with persistent alternate spines to 5 cm long. Leaves of two kinds; those of the young shoots large, usually 1–2 pairs of obovate leaflets on upper side of young spines, 4–9 mm long and 3–6 mm broad, hairy to glabrous; those on the old wood reduced to fascicles of 2–5 cuneate-obovate leaflets. Flowers solitary on short pedicel, 1–2 mm long. Bract oval, hairy, 3–4 mm long. Calyx 9–12 nerved, hairy; calyx tube soon inflated in fruit becoming membranous and glabrescent, 12–20 mm long. Corolla white, pink-tinged above; standard 15–22 mm long, oval to oblong, emarginate. Pod stiptate, ovoid, yellowish and hairy with shorter, straight and rigid beak included in the inflated calyx. Seeds ovoid, not smooth, 3–4 mm in diameter (UAE Flora, 2023).

Astragalus gossypinus Fisch.

English name: astragal, milkvetch
Arabic name: *qatad kathirā*
Persian name: *gavan-e- sefid*
Native to: Iran, Iraq, Lebanon-Syria, Turkey (Govaerts, 2024)
Morphology: dwarf cushion-forming shrub, 10–15 cm. Leaf rachis spiny, 1.5–3.5 cm, straight. Leaflets 5–6(–9) mm, oblong to widely ovate, densely white-tomentellous, spine-tipped, 4–8-paired. Stipules *c.*6 mm, triangular-ovate, tomentose. Flowers sessile, 3–4 per leaf axil. Inflorescence globose, 1–2 cm wide, 15–40-flowered. Bracts *c.*6 mm, lanceolate to oblong, pilose in the upper half. Bracteoles absent. Calyx 7–8 mm, densely white-villous; lobes divided to the base, incurved at the apex. Flowers pink; standard 10–12 mm. Fl. 7–8. Alt. 500–1000 m (Chamberlain and Matthews, 1970).

Table 1.20. Sqamunyā. (From Nafisi, 1899; Rāzi (Rhazes), 1967–1968; Anṣārī Shīrāzī, 1992; Heravi, 1992; Solṭāni, 2004–2016, 2005–2016; Aqili Khorāsāni, 2014; Ibn Sinā, 2015.) Author's own table.

Common names	Greek: *skammonia*; Arabic: *saqmunyā*; Persian: *saqmunyā*	
Extraction process	Funnel-shaped incision made at the root base; latex is collected in a container and processed	
Traditional applications	Skin	Anti-hyperpigmentation, depigmentation therapy, anti-inflammatory, eczema treatment
	Digestive and gastrointestinal	Laxative, prokinetic agent, anti-uremic, decongestant, anti-biliary, colic, ulcer
	Cardiovascular and circulatory	Anti-edematous, antihypertensive
	Neurological	Neuralgia, anti-inflammatory, decongestant, neurostimulator, analgesic, sedative
	Gynaecological	Abortifacient
	Anti-inflammatory and pain relief	Analgesic, anti-inflammatory
	Antivenom and antiparasitic properties	Ant venom, topical detoxifier, anthelmintic
	Fever and biliary disorders	Anti-biliary, antipyretic
Side effects and mitigation	Potential side effects	Suppresses appetite, induces melancholy, nausea and vomiting
	Remedy for side effects	Bake scammony inside an apple or quince to neutralize harmful effects; consume after cooling in shade
Precautions and dosage	Toxicity	Consuming ~1.75 g may cause severe symptoms like bowel obstruction, nausea, cold sweats and fatal diarrhoea
	Safe dosage	Adults: 0.4–1 g (0.3–0.6 g as resin pills); lethal dose: ~1.6 g
		Children: 0.05 g per year of age, administered as powdered scammony
	Overdose treatment	Consume whey with flour, apples, quince extract, sumac syrup (*rob-e-somagh*) or rhubarb syrup (*rob-e-rivas*)
Dosage forms	1. Powdered form	0.3–1 g for adults, mixed with milk
	2. Resin-based pills	0.3–0.6 g, prescribed in similar dosages for laxative effects
Substitutes	Replace scammony with 1.5 times the weight of *halila* or *laqia*	

Vachellia nilotica (L.) P.J.H. Hurter & Mabb.

Basionym: *Mimosa nilotica* L.
Syn. *Acacia arabica* var. *nilotica* (L.) Benth.
English name: babul, black babool, Indian gum arabic tree (Fig. 1.4)
Arabic name: *ammoghaylān*
Persian name: *chesh, keret*
Native to: Algeria, Angola, Assam, Bangladesh, Benin, Botswana, Burkina, Cameroon, Cape Verde, Caprivi Strip, Central African Republic, Chad, Djibouti, Eastern Himalayas, Egypt, Eritrea, Ethiopia, Gambia, Ghana, Guinea, Guinea-Bissau, Gulf States, India, Iran, Kenya, KwaZulu-Natal, Libya, Malawi, Mali, Mauritania, Mozambique, Myanmar, Namibia, Nepal, Niger, Nigeria, Northern Provinces, Pakistan, Senegal, Somalia, Sri Lanka, Sudan, Swaziland, Tanzania, Togo, Uganda, Western Himalayas, Yemen, Zambia, Zimbabwe.
Introduced into: Andaman Is., Argentina (north-eastern), Bahamas, Bolivia, California, Cuba, Ecuador, Galápagos, Gulf of Guinea Is., Iraq, Jamaica, Leeward Is., Madagascar, Mauritius, Mexico (central), Mexico (south-west), New Caledonia, New Guinea, Northern Territory (Australia), Peru, Puerto Rico, Queensland (Australia), Rodrigues (Mauritius), South Australia, Trinidad and Tobago, Western Australia, Windward Is. (Forbes, 1965).
Morphology: a small tree with dark brown or black longitudinally-fissured bark; branchlets slender, terete, pubescent when young. Leaves 2-pinnate, 5–10 cm long; main rhachis downy, often furnished with glands; petioles 2.5–5 cm long; stipular spines very variable, 0.6–5 cm long, smooth, usually whitish, straight, sharp, often absent; pinnae 4–9 pairs, 1.9–5 cm long, shortly stalked. Leaflets subsessile, 10–25 pairs, 0.3–0.6 cm by 0.13–0.21 cm oblong, subobtuse, glabrous or nearly so. Flowers yellow, in globose

Fig. 1.4. *Vachellia nilotica* L. (By Valke, 2007. *Vachellia nilotica* subsp. *tomentosa* (Benth.) [Online], via Wikimedia Commons. https://commons.wikimedia.org/wiki/File:Acacia_nilotica_ssp._tomentosa_(2135464178).jpg.)

heads; peduncles axillary, in fascicles of 2–6, terete, pubescent; bracteoles 2 above the middle of the peduncle, broadly ovate, acute, pubescent. Calyx campanulate, 1/20 in. long; teeth very short. Corolla 1/8 in. long; lobes short, triangular. Pods stalked, 3–6 by 1/2–5/8 in., moniliform, compressed, constricted at the sutures between the seeds, densely and persistently grey, downy. Seeds 8–12 (Kyalangalilwa *et al.*, 2013).

English name: Atlas pistachio, Atlantic pistachio, Atlantic terebinth, Cyprus turpentine, Persian turpentine tree (Fig. 1.5)
Arabic name: *shajaratal-boṭm aṭlasi*
Persian name: *derakht-e banah*
Native to: Afghanistan, Algeria, Canary Is., Crimea (Krym), Cyprus, East Aegean Is., Greece, Iran, Iraq, Lebanon-Syria, Libya, Morocco, North Caucasus, Pakistan, Palestine, Saudi Arabia, Sinai, Transcaucasus, Tunisia, Turkey, Turkey-in-Europe, Western Himalayas
Introduced into: California, Tajikistan, Utah (Govaerts, 2024)
Morphology: deciduous tree up to 7 m tall, with a dense, subglobose crown. Leaves imparipinnate; leaflets 2–5-paired, ovate-oblong or lanceolate, 2.5–8 cm by 0.8–2.2 cm, obtuse, mucronate, dark green above, paler beneath, glabrous except for ciliolate margin; rachis narrowly winged. Fruits paniculate, as long as or longer than broad, 5–8 mm by 5–6 mm (Yaltrik, 1972).

Pistacia khinjuk stocks

English name: East Indian mastiche
Arabic name: *shajarat al-boṭm akhdar*
Persian name: *kasur, khonjak*
Native to: Afghanistan, Egypt, India, Iran, Iraq, Lebanon-Syria, Nepal, Pakistan, Palestine, Saudi Arabia, Sinai, Sudan, Tajikistan, Turkey, Western Himalayas (Govaerts, 2024)
Morphology: tree or shrub up to 6 m tall. Young shoots puberulous. Leaves imparipinnate. Leaflets 3–7 in number, coriaceous, opposite to subopposite, subsessile, 45–76 by 26–55 mm, ovate to ovate-oblong, glabrous (young leaves puberulous), apiculate or subacuminate, base often oblique; terminal leaflet larger than the laterals; rachis ± terete, sparsely puberulous. Panicles extra-axillary, suberect, 14–16(–22) cm long. Drupe 4–6 mm long, suborbicular in outline, compressed, apiculate, epicarp nervose, but not prominently so (Flora of Pakistan, 2025).

Pistacia eurycarpa Yalt.

Syn. *Pistacia atlantica* subsp. *kurdica* (Zohary) Rech.f.
English name: Persian turpentine tree
Arabic name: *shajarat al-boṭm*
Persian name: *saqez*
Native to: Afghanistan, Iran, Iraq, Lebanon-Syria, Transcaucasus, Turkey (Govaerts, 2024)
Morphology: shrub or small tree up to 5 m. Leaves deciduous, imparipinnate; leaflets 2–3-paired, ovate or ovate-oblong, 5–8.5 by 2–5 cm, obtuse, not mucronate, glabrous except for ciliolate margins, light green on both sides; rachis not winged or very narrowly and inconstantly so. Fruits paniculate, always depressed-globose, 6–7 by 8–9 mm (Yaltrik, 1972).

Pistacia vera L.

English name: pistachio tree (Fig. 1.6)
Arabic name: *shajarat al-fustuq*
Persian name: *derakht-e pesteh*
Native to: Afghanistan, Kazakhstan, Kirgizstan, Tajikistan, Turkmenistan, Uzbekistan
Introduced into: Bulgaria, East Aegean Is., France, Greece, Iran, Italy, Libya, Morocco, Spain, Turkey (Govaerts, 2024)
Morphology: tree up to 10 m tall. Leaves deciduous, trifoliolate or imparipinnate; leaflets large, 1–2(–3)-paired, ovate or broadly lanceolate, acute or obtuse, mucronulate, 5–10(–12) cm by 3–6 cm, conspicuously reticulate-veined, terminal leaflet as large or larger than lateral ones, sparsely puberulent; rachis not winged, hairy. Fruits paniculate, ovate-oblong, 16–29 mm by 9–12 mm, seed containing large green cotyledons, edible (Yaltrik, 1972).

Pistacia lentiscus L.

English name: lentisk, mastic tree (Fig. 1.7)
Arabic name: *fustuq sharqî, shajar al-maṣṭikā*

Fig. 1.5. *Pistacia atlantica* Desf. (By Kenraiz, 2016. *Pistacia atlantica* Desf. [Online]. CC BY-SA 4.0, via Wikimedia Commons. https://commons.wikimedia.org/wiki/File:Pistacia_atlantica_kz7.jpg.)

Fig. 1.6. *Pistacia vera* L. (By Missvain, 2023. *Pistacia vera* L. [Online]. CC BY-SA 4.0, via Wikimedia Commons. https://commons.wikimedia.org/wiki/File:Pistachios_in_Coalinga_-_September_2023_-_Sarah_Stierch_03.jpg.)

Fig. 1.7. *Pistacia lentiscus* L. (By Hillewaert, 2009. *Pistacia lentiscus* L. (male flowers) [Online], via Wikimedia Commons. https://commons.wikimedia.org/wiki/File:*Pistacia_lentiscus*_(male_flowers).jpg.)

Fig. 1.8. *Rhus coriaria* L. (By Langenheim, 2003. *Rhus coriaria* L. [Online], via Wikimedia Commons. https://commons.wikimedia.org/wiki/File:თუთუბო_Rhus_coriaria_Gerbersumach,_Gewürzsumach.JPG.)

Persian name: *maṣṭaki*
Native to: Albania, Algeria, Balearic Is., Canary Is., Corsica, Crete, Cyprus, East Aegean Is., Egypt, France, Greece, Italy, Lebanon-Syria, Libya, Morocco, Palestine, Portugal, Sardinia, Sicily, Spain, Tunisia, Turkey, Balkan region (Govaerts, 2024)
Morphology: mall evergreen tree or shrub, 1–3(–5) m. Leaves paripinnate, coriaceous; leaflets 2–4(5–7)-paired, oblong or oblong-lanceolate, 1.5–3.5 by 0·6–1.5 cm, obtuse, shortly mucronate, glabrous; rachis distinctly and constantly winged. Fruits borne in short racemes (Yaltrik, 1972).

Rhus coriaria L.

English name: tanner's sumach (Fig. 1.8)
Arabic name: *shajarat as-somāq*
Persian name: *derakht-e somāq*
Native to: Afghanistan, Albania, Algeria, Azores, Bulgaria, Canary Is., Crete, Crimea (Krym), Cyprus, East Aegean Is., France, Greece, Iran, Iraq, Italy, Lebanon-Syria, Madeira, Morocco, North Caucasus, Palestine, Portugal, Sicily, Sinai, Spain, Tajikistan, Transcaucasus, Turkey, Turkey-in-Europe, Turkmenistan, Uzbekistan, Balkan region
Introduced into: Germany, Great Britain, South European Russia, Tunisia (Govaerts, 2024)
Morphology: shrub, up to 3.5 m., young branches densely brownish-tomentose. Leaves with 9–15, broadly lanceolate to elliptic, serrate to deeply serrate leaflets, sparsely pilose above and beneath, with small setae along the veins beneath. Petals greenish white, 3–4.5 mm. Drupe globose, villose, reddish (Rechinger, 1969; Yaltrik, 1972; Khatamsaz, 2007).

Styrax officinalis L.

English name: officinal storax (Fig. 1.9)
Arabic name: *mey ʿah yābesah*
Persian name: *mey ʿah khosk*
Native to: Albania, Cyprus, East Aegean Is., Crete, Greece, Italy, Lebanon-Syria, Palestine, Turkey, Turkey-in-Europe, Balkan region

Fig. 1.9. *Styrax officinalis* L. (By Flogaus-Faust, 2016. CC BY-SA 4.0, via Wikimedia Commons. https://commons.wikimedia.org/wiki/File:Styrax_officinalis_RF.jpg.)

Introduced into: France (Govaerts, 2024)
Morphology: stellate-pubescent shrub or small tree 2–7 m. Leaves 3–7 cm, broadly to ovate-oblong, entire, obtuse, rounded or shortly cunate at base. Inflorescence usually 3–6-flowered, pedicels 1–2 cm. Calyx nearly entire, corolla *c*.2 cm, white, campanulate (Tutin, 1972).

Conclusion

The historical significance of resins, oleoresins and gums, which has been comprehensively presented in this chapter, also provides valuable insights for modern medicine, cosmetics and nutrition in the present. The numerous natural products that have been used for centuries in different cultures for a variety of purposes not only form an important basis for current product development in industry, but also contribute significantly to the scientific knowledge base in the sense of a 'One Health' concept. Traditional knowledge also emphasizes the importance of the sustainable availability of natural products, thus pointing to the importance of ecological balance. The active principles found in natural plant products are very often of great importance even today and offer natural, sustainable and effective solutions for various industries. In addition, resins, gums and oleoresins generally offer safer, biocompatible options for the medical, cosmetic and food industries, and modern research serves as a bridge between this ancient knowledge and evidence-based, new insights.

References

Abdel-Maksoud, G. and El-Amin, A.-R. (2011) A review on the materials used during the mummification processes in Ancient Egypt. *Mediterranean Archaeology and Archaeometry* 11(2), 129–150.

al-Bīrūnī (1991) *Kitab aṣ-Ṣaydanah fi aṭ-Ṭibb* (Pharmacology in Medicine). Zaryāb, A. (ed.). Tehran University Press, Tehran, pp. 67–68, 76, 79–80, 115–116, 142–143, 198, 218–219, 272, 322, 338–340, 394–395, 399–400, 403, 426–427, 435, 449, 457–458, 512–513, 547–549, 551–553, 559–560, 573–574, 581–585, 594–595. [in Arabic].

Amalraj, A. and Gopi, S. (2017) Biological activities and medicinal properties of asafoetida: A review. *Journal of Traditional and Complementary Medicine* 7, 347–359.

Anṣārī Shīrāzī, I.H. (1992) *Ikhtiyārāt Badī'ī: Section on Simple Drugs*. Mīr, M.T. (ed.). Razi Pharmaceutical Distribution Company, Tehran, pp. 33, 44–45, 50–51, 91–92, 168–169, 225, 227–228, 410–411,4 22–423. [in Persian].

Anṭāki, D. (2000) *Taḏkereh Oulol-Albāb* (Memorandum Book). Shamsuddin, A. (ed.). Dār al-Kotob al-'Ilmiyyah, Beirut, Vol. 1, pp. 60–61.

Aqili Khorāsāni, M.H. (2014) *Makhzan al-Adwiah* (Drug Treasure). Ardakānī, M.R.S., Rahimī, R. and Farjādmad, F. (eds.). Sabzārang Publications in collaboration with the Traditional Medicine Faculty of Tehran University of Medical Sciences, Tehran, pp. 170–171, 187–188, 201–204, 291–292, 299–300, 306, 355, 409–410, 479–482, 678, 703–704, 706–707, 751–752, 765–766. [in Persian].

Atchley, E.G. and Cuthbert, F. (1990) *A History of the Use of Incense in Divine Worship*. Longmans Green, London.

Ayati, Z., Azizi, N., Amiri, M.S., Ramezani, M., Nikakhtar, Z. *et al.* (2024) Ethnobotany, phytochemistry and medicinal properties of plants contain manna in Iran. In: Kumar, L., Bharadvaja, N., Singh, R. and Anand, R. (eds) *Medicinal and Aromatic Plants*. Springer, Cham, pp. 101–146.

Bokhāri, A.A. (1992) *Hidayat al-Muta'allemin fi aṭ-Ṭibb* (A Guide to Medical Learners). Matīnī, J. (ed.), Mashhad University Press, Mashhad, Iran, pp. 312, 322, 338. [in Persian].

Bosch, H. (c.1475) *The Adoration of the Magi*. Available at: https://www.metmuseum.org/art/collection/search/435724 (accessed 27 April 2025).

Buckley, S.A. and Evershed, R.P. (2001) Organic chemistry of embalming agents in Pharaonic and Graeco-Roman mummies. *Nature* 413(6858), 837–841.

Cao, B., Wei, X.-C., Xu, X.-R., Zhang, H.-Z., Luo, C.-H. *et al.* (2019) Seeing the unseen of the combination of two natural resins, frankincense and myrrh: Changes in chemical constituents and pharmacological activities. *Molecules* 24, 3076. DOI: 10.3390/molecules24173076.

Carruthers, W. (1897) Plants. In: Wright, W. (ed.) *The Illustrated Bible Treasury*. Thomas Nelson and Sons, London, pp. 311–312.

Chamberlain, D.F. and Matthews, V.A. (1970) *Asteragalus*. In: Davis, P.H. (ed.) *Flora of Turkey and the East Aegean Islands*, Vol. 3. Edinburgh University Press, Edinburgh, 142 pp.

De Rovira, A.D. (1999) *The Dictionary of Flavors and General Guide for Those Training in the Art and Science of Flavor Chemistry*. Wiley, Oxford.

Dīnawarī, A.D. (1974) *Kitāb an-Nabāt: Plants*, Vol. 3. Lewin, B. (ed.). Franz Steiner Verlag, Wiesbaden, Germany, pp. 86–87, 91–95, 104. [in Arabic].

Diodorus Siculus (2005) In Persian. *Iran and the Ancient East in the Historical Library*. Jami Publishing, Tehran, pp. 169. [in Persian].

Dioscorides, P. (2012) *Hyolā aṭ-Ṭibb fī al-Ḥashā'ish wa as-Sumūm* (Materia Medica on Herbs and Poisons). Tabṭabai, S.M. (ed.). Tehran University of Medical Sciences Press, Tehran, Vol. 1, pp. 62, 100, 105–107, 110, 136, 161 and Vol. 3, pp. 27–28, 66–68, 103–108, 111–113 [in Arabic].

Flogaus-Faust, R. (2016) *Styrax officinalis* L. Wikimedia Commons. Available at: https://commons.wikimedia.org/wiki/File:Styrax_officinalis_RF.jpg (accessed 27 April 2025).

Flora of Pakistan (2025) *Pistacia khinjuk Stocks*. Available at: http://www.efloras.org/florataxon.aspx?flora_id=5&taxon_id=242424411 (accessed 27 April 2025).

Forbes, R.J. (1965) Chemical, culinary, and cosmetic arts. In: Singer, C., Holmyard, E.J. and Hall, A.R. (eds) *A History of Technology*, Vol. 1. Oxford University Press, Oxford, UK, pp. 243–244.

Govaerts, R. (2024) *WCVP: World Checklist of Vascular Plants*. Facilitated by the Royal Botanic Gardens, Kew, UK. Available at: http://sftp.kew.org/pub/data-repositories/WCVP/ (accessed 21 May 2024).

Groom, N. (1981) *Frankincense and Myrrh: A Study of the Arabian Incense Trade*. Longman, New York.

Hamadānī, R.F. (1989) *Āthār wa Eḥyā': A Persian Text on Agriculture*. Sotūdeh, M. and Afshār, I.(eds). Institute of Islamic Studies, McGill University, in collaboration with the University of Tehran, Tehran [in Persian], 92 pp.

Heravi, A.M. (1992) *Al-Abniyah an Ḥaqāyeq al-Adwiyah* (Basics of Realities on Drugs). Bahmanyār, A. (ed.). University of Tehran Press, Tehran, pp. 12–13, 35–38, 95–96, 157–158, 189–192, 195, 270–272, 315, 319–320. [in Persian].

Herodotus (2009) *The Histories*. Strassler, R.B. (ed.). Anchor Books, New York, pp. 152–153.

Heyd, W. (1886) *Histoire du commerce du Levant au moyen âge*, Vol. 2. O. Harrassowitz, Leipzig, Germany, pp. 580–581, 623–626.

Hillewaert, H. (2009) *Pistacia lentiscus* L. (male flowers). Wikimedia Commons. Available at: https://commons.wikimedia.org/wiki/File:*Pistacia_lentiscus*_(male_flowers).jpg (accessed 27 April 2025).

Hoots, C. (1993) Retracing the incense route. *Mosaic* (February), 43–45.

Howes, D. (2003) *Sensual Relations: Engaging the Senses in Culture and Social Theory*. University of Michigan Press, Ann Arbor, Michigan.

Huang, Q., Zhang, C., Dong, S., Han, J., Qu, S. *et al*. (2022) Asafoetida exerts neuroprotective effect on oxidative stress induced apoptosis through PI3K/Akt/GSK3β/Nrf2/HO-1 pathway. *Chinese Medicine* 17, e83.

Ibn al-Beyṭār, A.A. (2015) *Al-Jam'e Le-Mofradāt al-Adwiah wa al-Aghḏiyah* (Comprehensive Book on Simple Drugs and Foods). Badawī, M.M. (ed.). Center for Research on Islamic and Arab Heritage, Misr University for Science and Technology, Cairo, Vol. 1, pp. 89, 549, 553–556; Vol. 2, pp. 251, 603–609; Vol. 3, pp. 85–87, 93, 124–129; Vol. 4, pp. 31–33, 321–327, 375–377. [in Arabic].

Ibn Al-Jazzār, J.A. (2019) *Kitāb al-'Itimād fī al-Adwiyah al-Mufradah* (The Reliable Book on Simple Drugs). Ibrahim Ben Mrad. (ed.). Al-Furqān Islamic Heritage Foundation, London, pp. 212–215, 245–247, 270–273, 335–338, 377–380, 429–436, 457–459, 575–572, 640–643, 692–695, 725–726, 882–886. [in Arabic].

Ibn al-Nafis al-Qarashi (2000) *ash-Shāmil fi aṣ-Ṣina'a aṭ-Ṭibbiyyah* (Comprehensive Book on the Medical Industry). Ziedan, Y. (ed.). Cultural Foundation Publications, Abu Dhabi, Part 1, pp. 189–203; Part 2, pp. 429–436; Part 3, pp. 597–605, 635–644, 740; Part 5, pp. 363–368; Part 6, pp. 777; Part 8, pp. 7–19; Part 10, pp. 107–115, 207–219, 256; Part 12, pp. 142; Part 14, pp. 167–180; Part 15, pp. 33–47, 51–61, 105–110, 129–139, 159–169; Part 16, pp. 87–92; Part 18, pp. 1–33, 83–97, 101–105. [in Arabic].

Ibn Sinā, H.A. (2015) *Al-Qānun fi aṭ-Ṭibb* (Canon of Medicine). Masoudi, A. (ed.). Alma'ee, Tehran, Vol. 2, pp. 158–160, 172–175, 185, 275–279, 360–362, 377–379, 401–406, 443–445, 452, 486, 495–497, 521–523, 542–546, 583, 604–606, 609, 816–817. [in Arabic].

Ibn Waḥshiyya, A.A. (1993–1998) *Al-Filāḥa al-Nabaṭiyyah (L'agriculture nabatéenne)*. Fahd, T. (ed.). Institut Français d'Études Arabes de Damas, Damascus, Vol. 1, pp. 100, 354, 747–748, 769, 1145, 1149, 1257, 1261; Vol. 2, pp. 769, 1145, 1149, 1257, 1261. [in Arabic].

Iranshahy, M. and Iranshahi, M. (2011) Traditional uses, phytochemistry and pharmacology of asafoetida (*Ferula assa-foetida* oleo-gum-resin) - a review. *Journal of Ethnopharmacology* 134, 1–10.

Ismoon (2020) Wikimedia Commons. Available at: https://upload.wikimedia.org/wikipedia/commons/d/d6/Incense_trees_transport._Punt_relief._Mortuary_Temple_of_Hatshepsut%2C_Deir_el-Bahari.jpg (accessed 27 April 2025).

Jokār, J. (2006) *Techniques of Color-Making in the Realm of Art and Painting*. Institute for Compilation, Translation, and Publication of Artistic Works (Matn), Academy of Arts, Tehran, pp. 99–101. [in Persian].

Jones, J., Higham, T.F.G., Oldfield, R., O'Connor, T.P. and Buckley, S.A. (2014) Evidence for prehistoric origins of Egyptian mummification in Late Neolithic burials. *PLoS One* 9, e103608. DOI: 10.1371/journal.pone.0103608.

Jones, J., Higham, T.F.G., Chivall, D., Bianucci, R., Kay, G.L, *et al*. (2018) A prehistoric Egyptian mummy: Evidence for an 'embalming recipe' and the evolution of early formative funerary treatments. *Journal of Archaeological Science* 100, 191–200. DOI: 10.1016/j.jas.2018.07.011.

Jorjānī, I. (2005) *Al-Aghrāḍ aṭ-Ṭibbīa val Mabāheth al-'laiiah*. Tājbakhsh, H. (ed.). University of Tehran, Academy of Sciences, Tehran, Vol. 1, pp. 286, 304, 319. [in Persian].

Karageorghis, V. (1996) Some aspects of the maritime trade of Cyprus during the late Bronze Age. In: Karageorghis, V. and Michaelides, D. (eds) *The Development of the Cypriot Economy, from the Prehistoric Period to the Present Day*. University of Cyprus and Bank of Cyprus, Nicosia, pp. 61–67.

Kazemi, F., Mohebbati, R., Niazmand, S. and Shafei, M.N. (2020) Antihypertensive effects of standardized asafoetida: effect on hypertension induced by angiotensin II. *Advanced Biomedical Research* 9, e77.

Kenraiz (2016) *Pistacia atlantica* Desf. [Online]. Wikimedia Commons. Available at: https://commons.wikimedia.org/wiki/File:Pistacia_atlantica_kz7.jpg (accessed 27 April 2025).

Khatamsaz, M. (2007) Anacardiaceae. In: Assadi, M. (ed.) *Flora of Iran*. Research Institute of Forests and Rangelands, Tehran, No 3 [in Persian].

Kindī, Y.I. (1948) *Kitāb Kīmiyā' al-'Aṭtr wa at-Taṣ'īdāt* (The Book of Perfume Chemistry and Distillation). Garbers, K. (ed.). Deutsche Orientalische Gesellschaft, komissionseverlage F.A. Brockhaus, Leipzig, Germany, pp. 1–10, 26, 41–47, 50, 53–55.

Kłys, M., Lech, T., Zieba-Palus, J. and Białka, J. (1999) A chemical and physicochemical study of an Egyptian mummy 'Iset Iri Hetes' from the Ptolemaic period III–I B.C. *Forensic Science International* 99, 217–228.

Kyalangalilwa, B., Boatwright, J.S., Daru, B.H., Maurin, O. and Van der Bank, M. (2013) Phylogenetic position and revised classification of *Acacia* s.l. (Fabaceae: Mimosoideae) in Africa, including new combinations in *Vachellia* and *Senegalia*. *Botanical Journal of the Linnean Society* 172(4), 500–523.

Langenheim, J.H. (2003) *Plant Resins: Chemistry, Evolution, Ecology, and Ethnobotany*. Timber Press, Portland.

Lentini, A. (2004) Fragrant substances and therapeutic compounds. In: Belgiorno, M.R. (ed.) *Pyrgos-Mavroraki, Advanced Technology in Bronze Age Cyprus*. Archaeological Museum, Nicosia, pp. 45–47.

Lucas, A. (1914) The use of natron by the ancient Egyptians in mummification. *The Journal of Egyptian Archaeology* 1, 119–137.

Mahendra, P. and Bisht, S. (2012) *Ferula asafoetida*: Traditional uses and pharmacological activity. *Pharmacognosy Reviews* 6, 141–146.

Maimonides, M. (1940) *Sharḥ Asmā' al-'Uqqār* (Explanation of the Names of Drugs). Meyerhof, M. (ed.). Imprimerie de l'Institut Français d'Archéologie Orientale, Le Caire, Egypt, pp. 4, 5, 6, 13. [in Arabic].

Maurer, J., Möhring, T. and Rullkötter, J. (2002) Plant lipids and fossil hydrocarbons in embalming material of Roman period mummies from the Dakhleh Oasis, Western Desert, Egypt. *Journal of Archaeological Science* 29, 751–762.

Middeke-Conlin, R. (2014) The scents of Larsa: A study of the aromatics industry in an Old Babylonian kingdom. *Cuneiform Digital Library Journal* 2014(1).

Mir Heydar, H. (2019) *Ma'aref-e Giyahi* (Encyclopedia of Medicinal Plants). Daftar Nashr Farhang Eslami, Tehran, Vol. 1, pp. 424; Vol. 2, pp.11–14, 20–25, 111–112, 151–156, 373; Vol. 3, pp. 40–42, 58–60; Vol. 6, pp. 146–150, 154–157. [in Persian].

Missvain (2023) *Pistacia vera* L. Wikimedia Commons. Available at: https://commons.wikimedia.org/wiki/File:Pistachios_in_Coalinga_-September_2023_-_Sarah_Stierch_03.jpg (accessed April 2025).

Mohammadifar, S. (2022) Ṣamgh. In: Haddād'Ādel, G.A. (ed.) *Encyclopedia of the World of Islam*, Vol. 30. Encyclopedia Islamica Foundation, Tehran, pp. 88–91. [in Persian].

Momen Tonekāboni, H.M. (2011) *Toḥfat al-Momenin* (Rarity of the Faithful). Nūr Vaḥy Publications, Qom, Iran, Vol. 1, pp. 116–117, 148–149, 151,171, 284–285, 401–402, 498, 697–698, 724–725, 779–780, 797–798. [in Persian].

Mueller, F.V. (1891) *Select Extra-Tropical Plants, Readily Eligible for Industrial Culture or Naturalisation, with Indications of Their Native Countries and Some of Their Uses*. C, Troedel and Co, Melbourne, pp. 2–3, 5–7, 11, 13–14.

Muller, D. and Okoro, C. (2004) *Production and Marketing of Gum Arabic*. Network for Natural Gums and Resins in Africa (NGARA), Nairobi, pp. 2–3, 5–7, 11, 13–14.

Nafisi, A.A. (1899) *Pezeshki-Nāmeh* (The Book of Medicine). Matba'e Tehran, Tehran, pp. 121–122, 125, 128, 269, 285. [in Persian].

Niazmand, R., Razavizadeh, B.M. and Sabbagh, F. (2020) Simulating release model and antimicrobial efficiency of LDPE film carrying *Ferula asafetida* leaf and gum extracts. *Polymer Bulletin* 79, 1151–1174.

Plinius/Pliny [the Elder] (1951, 1956) *Natural History*. Harvard University Press, Cambridge, Massachusetts, Vol. 6, pp. 31, 121, 1299 (1951); Vol. 7, pp. 37, 79, 399 (1956); Vol. 8, pp. 315 (1963).

Rāzi (Rhazes) (1967–1968) *Al-Hāwi fi aṭ-Ṭibb* (Comprehensive Book of Medicine). 'Abd al-Mo'eid Khān (ed.). Osmania Oriental Publications Bureau, Osmania University, Hyderabad, India, Vol. 20, pp. 25–27, 41–43, 86–92, 238–240; Vol. 21, pp. 78–81, 264, 308–309; Vol. 22, pp. 43, 500–504. [in Arabic].

Rechinger, K.H. (1969) Anacardiaceae. In: Rechinger, K.H. (ed.) *Flora Iranica*. Akademische Druck-u, Graz, Austria, p. 2.

Salavatifar, M., Mahboubi, M. and Nia, S.S. (2023) Asafetida (*Ferula asafoetida*) and its potential risks on respiratory disorders. *Current Traditional Medicine* 9, e281222212243.

Sandison, A.T. (1963) The use of natron in mummification in ancient Egypt. *Journal of Near Eastern Studies* 22, 259–267.

Serpico, M. (2000) Resins, amber and bitumen. In: Nicholson, P.T. (ed.) *Ancient Egyptian Materials and Technology*. Cambridge University Press, Cambridge, UK, pp. 432–474.

Smolinske, S.C. (1992) *Handbook of Food, Drug, and Cosmetic Excipients*. CRC Press, Boca Raton, p. 7.

Solṭāni, A. (2004–2016) *Medicinal Plants*. University of Medical Sciences, Tehran. [in Persian].

Solṭāni, A. (2005–2016) *Dictionary of Medicinal Plants*. Iran University of Medical Sciences, Tehran. [in Persian].

Spekke, A. (1957) *The Ancient Amber Routes and Geographic Discovery of the Eastern Baltic*. M. Goppers, Stockholm.

Stannard, J. (1999) *Pristina Medicamenta: Ancient and Medieval Medical Botany*. Stannard, K.E. and Kay, R. (eds). Variorum, Aldershot, UK, pp. 41–45.

Tabrizi, M.H. (1982) *Borhan-e Qaat'e*. Moein, M.(ed.). Amir Kabir Publishing House, Tehran, Vol. 1, p. 175. [in Persian].

Tutin, G. (1972) Styracaceae. In: Tutin, T.G., Heywood, V.H., Burges, N.A., Moore, D.M. and Valentine, D.H. *et al.* (eds) *Flora Europeae*, Vol. 3. Cambridge University Press, Cambridge, UK, p. 52.

UAE Flora (2023) *Astragalus fasciculifolius* Boiss. Available at: https://uaeflora.ae/plant/96 (accessed 27 April 2025).

Upadhyay, P.K., Singh, S., Agrawal, G. and Vishwakarma, V.K. (2017) Pharmacological activities and therapeutic uses of resins obtained from *Ferula asafoetida* Linn.: A review. *International Journal of Green Pharmacy* 11, S240–S247.

Valke, D. (2007) *Vachellia nilotica subsp. tomentosa (Benth.)*. Wikimedia Commons. Available at: https://commons.wikimedia.org/wiki/File:Acacia_nilotica_ssp._tomentosa_(2135464178).jpg (accessed 27 April 2025).

Yaltrik, E. (1972) Anacardiaceae. In: Davis, P.H. (ed.) *Flora of Turkey and the East Aegean Islands*, Vol. 2. Edinburgh University Press, Edinburgh, UK, pp. 543–548.

Zargari, A. (1988) *Medicinal Plants*. University of Tehran Press, Tehran, Vol. 2, pp. 144, 146, 150–151. [in Persian].

2 Taxonomy and Distribution of Botanical Resources Providing Most Important Gums and Resins

Nandkishore Thombare[1], Vaibhav Daulatrao Lohot[1*], Mohammad Reza Kanani[2], Farsad Nadjafi[2,3], Niranjan Prasad[1] and Hartwig Schulz[4]

[1]ICAR-National Institute of Secondary Agriculture, Namkum, Ranchi, India; [2]Shahid Beheshti University, Medicinal Plants and Drugs Research Institute, Tehran, Iran; [3]MateriaMed GmbH, Wien, Josefstädterstrasse 43-45, Austria; [4]Consulting and Project Management for Medicinal and Aromatic Plants, Waltraudstraße 4, Stahnsdorf, Germany

Abstract

Natural resins and gums (NRGs) are crucial to the livelihoods of millions of people, particularly in developing regions of Asia and Africa. However, the economic potential of many NRG-producing plants is not always fully exploited because of difficulties in correct botanical identification. For example, individual tree species in different regions sometimes have several different local names, which often leads to confusion and misunderstandings when trading these NRGs. Furthermore, a lack of knowledge sometimes leads to protected species being confused with other common species of the same genus, which can sometimes lead to significant ecological problems. But the adulteration of high-quality gums with low-quality, similar-looking NRGs also causes significant economic damage, as the gums produced in this way are of inferior quality. Bringing together information about NRG plants and raising awareness among harvesters, traders and users of resins and gums can therefore potentially help to solve these problems. This chapter aims to contribute to these efforts by providing detailed insights into the morphology, distribution and use of NRG plants, and by offering clear instructions for identifying important plant parts such as leaves, flowers and fruits. The chapter is designed to highlight commercially important NRG plants, provide insights into their common and botanical names and show the potential applications of their gum and resin products. This compilation is intended to improve trade and ensure the sustainability of these valuable natural products by providing accurate details for their identification and use.

Introduction

Natural resins and gums (NRGs), whose special significance is derived in particular from their various local cultural backgrounds, are increasingly securing the livelihoods of many millions of people worldwide. In developing and newly industrializing countries in particular, many families depend on the economic use of natural resins and gums to earn a living from their sale and processing. To further optimize the economic potential of NRGs and at the same time ensure sustainable management of plant resources, clear botanical identification of the

*Corresponding author: vaibhav.icar@gmail.com

DOI: 10.1079/9781800626690.0002

individual species used is essential. The same tree is sometimes known to the local population in different countries/regions under different common names, and in some genera the respective species and subspecies are referred to by the same name. In India, for example, the name 'kikar' is used for various *Acacia* species such as *Acacia nilotica, A. senegal, A. tortolis, A. mollisima* and *A. fernaciana*. In addition, different plant species belonging to the same genus look morphologically similar, which may lead to confusion when identifying the gums made from them. Furthermore, high-quality gums are sometimes knowingly or unknowingly adulterated with inferior, similar-looking gums, which makes it difficult to maintain uniform quality standards. In any case, the resulting differences in quality cause an overall negative impact on the trade and export of these natural products. Although it is generally recognized that herbal products have a higher variability of plant substances than synthetic products due to different environmental conditions of the plants, it should in principle be possible to maintain specified parameters within certain limits.

Although resins and gums are produced by a large number of different plant species, their commercial use is limited to comparatively few members of the botanical families Leguminosae, Sterculiaceae and Combretaceae. Resins are mainly obtained from Pinaceae (rosin, amber), Leguminosae (copal), Dipterocarpaceae (damar), Burseraceae (elemi, incense, gugul) and Umbelliferae (asafoetida, galbanum). Gum resins and resins are considered to be by products of certain metabolic pathways, with soil erosion, drought and other unfavourable conditions promoting the production of gum-resins and resins. Indonesia, India, Iran, China and Sudan are the major producers of gums and resins. Millions of people around the world, particularly in developing countries, depend on the production of gums, resins and related products.

Natural gums and resins are found either in the intercellular space (channels or cavities) of the plant parts or as exudate resulting from injuries. Gums and resins mainly consist of secondary metabolites that do not appear to play a role in the basic physiology of the plant. When plants suffer external injuries, resins can be induced at the site of injury without having to be formed in a specialized secretory structure beforehand. It was once thought that resins could serve as a source of bioenergy for the plant, but it is now assumed that their main function is to protect sensitive plant surfaces.

The first scientific studies on the biosynthesis and storage of resins in various plant species were carried out by Tschirch (1906) towards the end of the 19th century. With the growing knowledge of the chemical composition of plant secretions, an increasing economic interest developed in the 1930s, which led to further extensive scientific publications (Barry, 1932; Tschirch and Stock, 1933–1936) and, in particular, enabled enormous progress in the identification of the chemical constituents of resins and the quantification of their composition. Thus, numerous exudates that previously could not be clearly classified could now be better categorized based on their individual chemical fingerprints. Furthermore, many of the biosynthetic pathways leading to resin production have been elucidated. In addition, electron microscopy and various ecological studies have provided important information about resin-producing structures. Most gums are exudated as metabolic by-products of plant tissues either in the ordinary course of life or often due to disease or injury to the bark or wood of certain plants. The most popular justification for the production of these exudates by plants would be that the gums are produced as natural antibiotics to protect the plants from infection due to injury or to gain relief from stress. At the same time, the resins are secreted by plants to conserve moisture, protect themselves from freezing.

Today, it is well known that poor soil composition, dryness and other unfavourable environmental conditions support their formation. Gums and resins are not part of the metabolism of the plant in which they are formed, which is why they are regarded as by-products of certain metabolic changes in the plants. Mucilage formation is thought to be a pathological reaction of the plant that protects an injured part by sealing it to prevent water loss and infection. The development of the product is schizogonic (cleavage of the product initiated by dissolution of the middle lamellae), schizolysogenic (cleavage of the product initiated by lysis of epithelial cells) and lysogenic (cleavage of the product initiated by lysis of epithelial cells).

Unfortunately, there are still sometimes differences in the terminology of resins and gums in the individual producing countries, which do not always correspond to the above-mentioned botanical definition. The Technical Committee (TC 54) of the ISO (International Organization for Standardization) has therefore endeavoured to define internationally valid standards for the resins and gums available on the market. The most important natural resins and gums that comply with this standard are listed below:

- **gums:** exudates consisting mainly of polysaccharides;
- **gum-resins:** exudates consisting mainly of resinous compounds and gums;
- **oleo-gum resins:** exudates consisting mainly of resinous compounds, gums and certain quantities of volatile compounds (essential oil substances);
- **oleoresins:** exudates consisting mainly of resinous and volatile compounds (essential oil substances);
- **resins:** products obtained from oleoresin by as complete as possible evaporation of the volatile substances.

In addition, there are several natural raw materials on the market that do not meet the ISO definition or the botanical classification for the category 'resins and gums'. In the cosmetics, flavouring and phytopharmaceutical industries these products are called 'absolutes', 'concretes', 'resinoids' or 'tinctures', depending on the individual applied method of preparation. In accordance with the definition of the ISO in cooperation with the European Committee for Standardization (CEN), the following manufacturing conditions have been defined for these natural products:

- **absolute:** product obtained by extraction with ethanol from a concrete, a resinoid or a supercritical fluid extract (SFE);
- **concrete:** extract obtained from a fresh natural raw material by extraction with one or several solvents;
- **resinoid:** extract obtained from a dry natural plant raw material by extraction with one or several solvents;
- **SFE:** extract obtained by treating a natural plant raw material in a supercritical fluid followed by a separation by expansion.

Furthermore, in the case of essential oils obtained from resinous trees, a distinction must be made between those that are obtained by hydro-distillation from the individual gum-resin (e.g. from various *Boswellia* species) and those that are produced by distillation directly from the wood or other parts of the plant.

This chapter deals in particular with the resins and gums obtained by exudation from the respective species. In addition, a few particularly important latex plants are listed, but these are not discussed further in the text. Here, too, reference is made to standard articles that deal specifically with the extraction of latex from plants (Murthy, 2022; Murthy *et al.*, 2024). The so-called 'lac-resins' produced by various lac insects (especially species of the genus *Kerria*) are not discussed in detail here. The extraction of these very specific lac-resins is briefly mentioned in Chapters 4 and 5.

This book also does not go into detail about the extraction of essential oils from the crushed wood of various tree species such as *Santalum album* L., *Bulnesia sarmienti* Lor. or *Juniperus virginiana* L. by steam distillation. For readers who want to learn more about these essential oils, we recommend relevant monographs such as that edited by Surburg and Panten (2006). The botany and chemistry of resin-forming plant species has already been extensively covered in various books (Langenheim, 2003; Murthy, 2022; Murthy *et al.*, 2024), which is why only the most important species and their global distribution and economic use will be briefly presented here.

Botany, Taxonomy, Distribution and Use

For the taxonomic classification of the individual species, which are listed below in alphabetical order according to their botanical names, both selected review articles and botanical textbooks were used (Roxburgh, 1832; Kirtikar, 1918; Rechinger and Cullen, 1966; Browicz *et al.*, 1969; Rechinger, 1969; Davis, 1972; Rechinger, 1987; Brandis, 1990; Coppen, 1995; Chevallier, 1996; Gachathi, 1997; Zheng-Yi and Raven, 1999; Nair, 2000; Pandey, 2001; Rustiami *et al.*, 2004; Farjon, 2010; Mozaffarian, 2013;

Bhat, 2014; Lohot *et al.*, 2016; Thombare *et al.*, 2018; Vázquez *et al.*, 2023; Thombare *et al.*, 2023; Schulz *et al.*, 2023; Lohot *et al.*, 2023; WorldFlora Online, 2024; POWO, 2024; Flora of China, n.d.).

Acacia catechu (Khair)

Botanical name: *Acacia catechu* (L.f.) Willd.
Synonyms: *Senegalia catechu; Acacia catechu* var. *catechuoides* (Roxb.) Prain; *Acacia catechuoides* (Roxb.) Benth.; *Acacia sundra* (Roxb.) Bedd.; *Acacia wallichiana* DC.; *Mimosa catechu* L.f.; *Mimosa catechuoides* Roxb.
Classification: Plantae > Tracheophyta > Magnoliopsida (Angiosperms) > Fabales > Fabaceae > Mimosoideae > Acacia > catechu.
Common names: Cutch tree, Catechu (English); Khair (Hindi); Khadira, Rakta Sara (Sanskrit).
Distribution: native to South and South-east Asia, including the Indian subcontinent, Myanmar, Thailand and China.
Morphology: medium-sized deciduous tree with a light, irregular crown and twisted branches. Its bark is typically dark brown or grey with a reddish-brown interior. The branchlets feature paired, hooked spines located beneath the petioles. The tree's leaves are bipinnate, and leaf drop typically occurs between January and April. The flowers are pale yellow, arranged in sessile, axillary spikes on short peduncles, and the tree continues to bloom until July or August. The pods are narrow, flat, strap-like and taper at both ends, initially green, turning reddish-green and eventually brown as they mature. The seeds are shiny, broad and ovoid with a brown hue. The wood of this tree is used to extract 'Kath' or 'Cutch', which is highly sought after in the production of masticatory products (Fig. 2.1).
Use: Khair gum is highly regarded for its superior quality and as an excellent alternative to true gum arabic. The gum droplets, or 'tears', can reach sizes up to 3 cm in diameter and vary in colour from pale yellow to dark amber. It is commonly used in products such as masticatories and hair dyes. However, it is not typically

Fig. 2.1. Khair (*Acacia catechu*), flowering plant. (By Nandkishore Thombare.)

harvested on its own but is often found mixed with other acacia gums.

Acacia nilotica (Babool)

Botanical name: *Acacia nilotica* (Linn.) Willd. ex Del.
Synonyms: *Acacia arabica* (Lam.) Willd.; *Acacia scorpioides* W. Wight; *Mimosa arabica* Lam.; *Mimosa nilotica* L.; *Mimosa scorpioides* L.; *Vachellia nilotica* subsp. *indica*.
Classification: Plantae > Tracheophyta > Magnoliopsida > Fabales > Fabaceae > Mimosaceae > Acacia > nilotica.
Common names: Babul, Babool, Gum arabic tree (English); Babul, Kikar (Hindi); Babbula, Babula, Barburah (Sanskrit).
Distribution: *Africa*: Algeria, Angola, Benin, Botswana, Burkina Faso, Cameroon, Cape Verde, Chad, Djibouti, Egypt, Eritrea, Ethiopia, Gambia, Ghana, Guinea, Guinea-Bissau, Kenya, Libya, Malawi, Mali, Madagascar, Mauritania, Mauritius, Mozambique, Namibia, Niger, Nigeria, Senegal, Somalia, Sudan, Tanzania, Togo, Uganda, Zambia, Zimbabwe. *Asia*: Bangladesh, India, Iran, Iraq, Myanmar, Nepal, Oman, Pakistan, Saudi Arabia, Sri Lanka, Yemen. *Americas*: Argentina, Bahamas, Bolivia, Cuba, Ecuador, Galápagos, Jamaica, Mexico, Peru, Puerto Rico, Trinidad and Tobago. *Oceania and Australasia*: Australia, New Guinea.
Morphology: perennial shrub or tree 2.5–10 m high with a dense spherical crown, stems and branches usually dark to black-coloured, branchlets purple-brown, bark thin, rough, fissured, shortly or densely pubescent, with lenticels, exuding a reddish low-quality gum. Thorns thin, straight, light grey, usually in 3–12 pairs, 5–7.5 cm long in young trees, mature trees commonly without thorns. Leaves bipinnate, 3–4 cm long, often with 1–2 petiole glands and other glands between all or only the uppermost pinnae, pinnae 2–11 pairs, with 7–25 pairs of leaflets (1.5–7 mm long) per pinnae, tomentose, rachis with a gland at the bottom of the last pair of pinnulae. Peduncles clustered at nodes of leafy and leafless branchlets. Flowers in globulous heads 1.2–1.5 cm in diameter, golden-yellow, set up either axillary or whorly on peduncles 2–3 cm long located at the end of the branches. Pods straight or lightly curved, 5–15 cm long, 0.5–1.2 cm wide, strongly constricted, white-grey, thick and softly tomentose, fleshy when young, indehiscent, becoming black and hard at maturity. Seeds deep blackish-brown, smooth, subcircular, compressed, areole 6–7 mm long, 4.5–5 mm wide (Fig. 2.2).
Use: The gum derived from *Acacia nilotica* has been traditionally used in indigenous medicine to treat various ailments, including diarrhoea, dysentery, diabetes, sore throat, constipation and slow digestion. Edible in nature, it is a popular ingredient in India, particularly in making sweets such as laddus, halwa and other confections. This gum acts as a natural thickening agent, improving the texture of food products and helping to stabilize items like syrups, ice creams and beverages. Beyond its culinary applications, it is also valuable in several industries, including calico printing and dyeing and as a sizing agent for silk and cotton fabrics. Additionally, it is used to create mucilages and adhesives, functioning as a simple domestic glue for affixing paper, cardboard and other materials.

Acacia senegal (Gum arabic)

Botanical name: *Acacia senegal* (L.) Willd.
Synonyms: *Acacia circummarginata* Chiov; *Acacia cufodontii* Chiov; *Acacia glaucophylla* Sensu Brenan; *Acacia kinionge* Sensu Brenan; *Acacia oxyosprion* Chiov; *Acacia rupestris* Boiss; *Acacia senegal* subsp. *modesta* (Wall.) Roberty; *Acacia senegal* subsp. *senegalensis* Roberty; *Acacia somalensis* Sensu Brenan; *Acacia spinosa* Marloth & Engl; *Acacia thomasii* Sensu Brenan; *Acacia volkii* Suess; *Mimosa senegal* L.; *Senegalia senegal*.
Classification: Plantae > Tracheophyta > Magnoliopsida > Fabales > Fabaceae > Mimosaceae > Acacia > senegal.
Common names: Gum arabic tree, Gum acacia (English); Kumttha, Kumatiyo, Kumat (Hindi); Babul, Shveta khadira (Sanskrit).
Distribution: native to Angola, Botswana, Burkina Faso, Eritrea, Ethiopia, Gambia, Kenya, Mali, Mozambique, Namibia, Niger, Nigeria, Senegal, Sudan, Tanzania, Uganda, Zambia, Zimbabwe.

Fig. 2.2. Gum arabic (*Acacia nilotica*), flowering plant. (By Nandkishore Thombare.)

Fig. 2.3. *Acacia senegal*, flowering plant. (By Nandkishore Thombare.)

Morphology: small, deciduous, spiny shrub that typically branches from the base. Its branches repeatedly bifurcate, forming a rounded, flat-topped crown when fully mature. The trunk is covered in greyish-white bark, which can become dark, scaly and thin in older trees. The plant has three strong, hooked thorns just below the nodes, some pointing forwards and others backwards. The leaves are pinnately compound, featuring two glands located between the uppermost and lowermost leaflets. The leaflets are small, stiff, leathery and smooth, and they range from linear to elliptic-oblong in shape. Their edges are ciliate, and the colour is a pale, glaucous green with a blunt or slightly pointed tip. The flowers grow on cylindrical spikes, are yellowish-white and emit a fragrant scent. The pods are thin, flat and oblong, turning green when mature and then shiny bronze with dark spots marked by prominent veins. The seeds are smooth, flat, shiny and dark brown (Fig. 2.3).

Use: Gum arabic is widely used across various industries, including food, pharmaceuticals and cosmetics. It is commonly added to sweet dishes such as laddu and halwa in India. Known for its digestive benefits, it has gained popularity in diabetic and metabolic foods. In the food industry, gum arabic acts as a binder in sweets, a stabilizer in dairy products, beverages and confectionery, and an emulsifier in juices and flavourings. In pharmaceuticals, it is a binder in tablets, a stabilizer in syrups and emulsions and an emollient and demulcent in cough syrups and lozenges. In cosmetics, it is used to enhance the texture of creams and lotions and as an adhesive in facial masks and similar products. Additionally, the gum functions as a mucilage, simple adhesive and glue for various materials like paper, glass and metal. It is also a sensitizer for lithographic plates and a protective colloid and suspending agent. Due to its superior suspension properties, gum arabic is used in paints as a flocculant and emulsifier, in sizing and finishing agents for textiles and in printing formulations.

Acacia sieberiana (Paperbark acacia)

Botanical name: *Acacia sieberiana* DC.

Synonyms: *Acacia vermoesenii* De Wild.; *Acacia woodii* Burtt Davy; *Vachellia sieberiana* (DC.) Kyal. & Boatwr. var. *woodii* (Burtt Davy) Kyal. & Boatwr.; *Acacia abyssinica* sensu auct.; *Acacia amboensis* Schinz; *Acacia davyi* sensu auct.; *Acacia purpurascens* Vatke; *Acacia sieberana* DC.; *Acacia sieberiana* DC. subsp. *vermoesenii* (De Wild.) Troupin.

Classification: Plantae > Tracheophyta > Magnoliopsida > Fabales > Fabaceae > Mimosaceae > Acacia > sieberiana.

Common names: Umbrella thorn, Paperbark thorn, Paperbark acacia, Flat top thorn (English).

Distribution: located in various African habitats, including woodlands, wooded grasslands and riverbanks. Its distribution spans from tropical West Africa to Sudan and Ethiopia, extending southward through East Africa to Namibia, Zimbabwe and South Africa.

Morphology: umbrella-shaped tree that can grow up to 20–25 m tall. Its bark is corky and peels away to reveal a yellowish layer beneath. The branches are thick and bear pairs of long thorns at the nodes. The compound leaves are typically sparsely hairy and grouped in small clusters. These bipinnate leaves consist of up to 30 pairs of pinnae, each with as many as 45 pairs of narrow, oblong leaflets. The leaflets have rounded tips, a central midrib and stalks that end in a tiny spine. The petiole is short, up to 1 cm long, hairy, glandular and may be swollen. The stipules are spiny. The flowers are bisexual, pale yellowish-white and arranged in heads about 13 mm in diameter, borne on slender stalks up to 5 cm long with tiny bracts near the top. These heads can appear either singly or in small clusters near the branchlets. The fruit is a woody, thick pod that is straight or slightly curved. The seeds are olive to grey-brown, elliptical to nearly round, compressed somewhat and smooth (Fig. 2.4).

Use: The tree's stem yields a clear, high-quality, pale yellow to yellowish-brown soluble gum. When combined with soot, it can be used to create ink. The tree's leaf, bark and resin serve as an astringent and are traditionally used to treat colds, chest issues, diarrhoea, haemorrhages

Fig. 2.4. *Acacia sieberiana*, flowering plant. (By Vaibhav D. Lohot.)

and eye inflammation. Additionally, the edible gum acts as an effective adhesive.

Acacia tortilis (Israeli babool)

Botanical name: *Acacia tortilis* (Forssk.) Hayne
Synonym: *Mimosa tortilis* Forssk.
Classification: Plantae > Tracheophyta > Magnoliopsida > Fabales > Fabaceae > Mimosaceae > Acacia > tortolis.
Common names: Umbrella thorn, Israeli babool (English); Israeli babool, Kikar (Hindi).
Distribution: native to Angola, Botswana, Egypt, Eritrea, Ethiopia, Iran, Israel, Kenya, Mozambique, Namibia, Qatar, Saudi Arabia, Somalia, South Africa, Sudan, Swaziland, Tanzania, Uganda, United Arab Emirates, Zambia, Zimbabwe.
Morphology: small to medium-sized evergreen tree, occasionally growing as a shrub or bush. It typically has a flat, spreading crown. The trunk features rough, fissured bark that ranges from grey to dark brown. The tree has numerous branchlets, each with two types of spines – straight and hooked. The leaves are pinnately compound with a linear to linear-oblong shape. The flowers are white or yellowish-white, fragrant and form round, fluffy, globular heads. The pods are twisted or spiraled, either indehiscent or dehiscent over time. The seeds are positioned either slanting or parallel to the long axis of the pod (Fig. 2.5).

Fig. 2.5. *Acacia tortilis*, flowering plant. (By Nandkishore Thombare.)

Use: The gum produced by the tree is edible and serves as a substitute for gum arabic, sometimes referred to as 'poor man's gum arabic' It is commonly used as a food additive and also has medicinal applications.

Albizia lebbeck (Siris)

Botanical name: *Albizia lebbeck* (L.) Benth.
Synonyms: *Acacia lebbeck* (L.) Willd.; *Acacia lebbek* (L.) Willd.; *Acacia macrophylla* Bunge; *Acacia speciosa* (Jacq.) Willd.; *Albizia latifolia* B. Boivin; *Albizia lebbeck* var. *leucoxylon* Hassk.; *Albizia lebbeck* var. *pubescens* Haines; *Albizia lebbeck* var. *rostrata* Haines; *Albizia lebbek* sensu auct.; *Feuilleea lebbeck* (L.) Kuntze; *Inga borbonica* Hassk.; *Inga leucoxylon* Hassk.; *Mimosa lebbeck* L.; *Mimosa lebbek* L.; *Mimosa sirissa* Roxb.; *Mimosa speciosa* Jacq.; *Pithecellobium splitgerberianum* Miq.
Classification: Plantae > Tracheophyta > Magnoliopsida > Fabales > Fabaceae > Mimosaceae > Albizia > lebbeck.
Common names: East India walnut, Kalshish, Woman's tongue, Lebbek tree, Siris tree, Black siris (English); Siris (Hindi); Shirish (Sanskrit).
Distribution: native to tropical Africa, Asia and Northern Australia.
Morphology: deciduous tree with grey, fissured, corky bark that can grow up to 25 m tall. Its leaves are bipinnate and alternate, with stipules present. Each leaf has 2–6 pairs of pinnae, and each pinna contains 4–12 pairs of elliptic-oblong leaflets measuring 2–4 cm in length. The inflorescences are subglobose heads, 3.5–3.7 cm in diameter, appearing singly or in small clusters of 2–4 within axillary corymbose racemes. Each flower has many long stamens. The fruits are reddish-brown, flat pods measuring about 15–25 cm long and 4–5 cm wide, containing 5–15 flat, rounded, free-moving seeds. These seeds are orbicular or elliptic, approximately 8–10 mm × 6–7 mm in size (Fig. 2.6).
Use: The stem produces a reddish gum, often used as an adulterant in gum arabic. Additionally, a red dye can be extracted from the bark of the Siris tree.

Fig. 2.6. Albizia lebbeck, flowering plant. (By Nandkishore Thombare.)

Albizia lucidior (Potka siris)

Botanical name: *Albizia lucidior* (Steud.) I.C. Nielsen
Synonyms: *Albizia gamblei* Prain.; *Albizia lucida* Benth.; *Albizia meyeri* Ricker.; *Inga lucida* Wall.; *Inga lucidior* Steud.; *Mimosa lucida* Roxb.; *Acacia lucida* Baill.; *Albizia bracteata* Dunn.; *Pithecellobium bigeminum* Hassk.; *Feuilleea lucida* Kuntze.
Classification: Plantae > Tracheophyta > Magnoliopsida > Fabales > Fabaceae > Mimosaceae > Albizia > lucidior.
Common names: Potka siris tree (English); Potka siris (Bengali); Silkoroi, Moj, Micha-goch, Katabhi (Assamese); Sirisa (Sanskrit).
Distribution: native to the moist forests of North-east India, but also found in Thailand, South China and Myanmar.
Morphology: deciduous tree with an open, airy crown of dark green foliage. The large, compound leaves are distinctive, featuring numerous linear, sessile leaflets. The rachis is pubescent with a few recurved spines, and the petiole has an oblong gland near its base. The leaves have 9–11 pairs of pinnae, and the flower heads are arranged in axillary panicles that can reach up to 25 cm long. The flowers are small and yellow and resemble powder-puff clusters, similar to those of the Siris tree. The calyx is tubular with faint teeth, and the corolla measures 3–4 mm. The stamens are numerous with a deciduous gland at the apex, and the ovary is sparsely pubescent at the top. The fruit pods are thin, flexible and dark brown, growing up to 25 cm long (Fig. 2.7).
Use: The tree hosts the lac insect (*Kerria lacca*), whose female secretes the resin known as 'lac', which was historically used to produce shellac. The gum from the tree is edible and can be used as an alternative to other types of gum.

Alhagi pseudalhagi subsp. *persarum*

Botanical name: *Alhagi pseudalhagi* subsp. *persarum* (Boiss. & Buhse) Takht.
Synonym: *Alhagi persarum* Boiss. & Buhse.
Classification: Plantae > Streptophyta > Equisetopsida > Magnoliidae > Fabales > Fabaceae > Alhagi > pseudalhagi > persarum.
Common name: Toranjebin.
Distribution: *Africa*: Algeria, Chad, Libya, Niger, Sudan. *Asia*: Afghanistan, China, Iran, Iraq, Kazakhstan, Kirgizstan, Kuwait, Lebanon-Syria, Mongolia, Nepal, Oman, Palestine, Saudi Arabia, Tadzhikistan, Transcaucasia, Turkey, Turkmenistan, Uzbekistan, Yemen. *Europe*: Cyprus, Egypt, Greece.
Morphology: glabrum, glaucous; arcuate-ascending branches, strong enough; spines 2–6 cm long, rigid, subrectangularly open, usually producing only 2–3(–4) flowers. Leaves 10–15 × 3–5 mm, oblong or lanceolate, obtuse, lower spines hardly shorter, upper spines three or four times shorter, persistent after anthesis, glabrous. Calyx glabrous; teeth almost obsolete. Legume 4–7-seeded, glabrous, ring-shaped.
Use: According to Ayurvedic studies, extracts of the plant species are used in folk medicine as a remedy for rheumatic pain, gastrointestinal complaints and urinary tract and liver diseases (Srivastava *et al.*, 2014).

Anacardium occidentale (Cashew nut)

Botanical name: *Anacardium occidentale* L.
Synonyms: *Acajuba occidentalis* (L.) Gaert.; *Anacardium amilcarianum* Machado.; *Anacardium kuhlmannianum* Machado.; Anacardium *microcarpum* Ducke.; *Anacardium othonianum* Rizzini; *Anacardium rondonianum* Machado.; *Cassuvium pomiferum* Lam.; *Cassuvium reniforme* Blanco.; *Cassuvium solitarium* Stokes.
Classification: Plantae > Tracheophyta > Magnoliopsida > Sapindales > Anacardiaceae > Anacardium > occidentale.
Common names: Cashew nut, Cashew apple (English); Kaju (Hindi); Agnikrita (Sanskrit).
Distribution: originally occurring in Central and South America (Costa Rica to Brazil and Ecuador) and cultivated commercially in semi-arid tropical areas of Africa and India.
Morphology: medium-sized, evergreen tree that typically grows in association with other trees. Its bark ranges from pale grey to brown. The leaves are simple, alternate, leathery and smooth with an obovate shape and rounded tips, typically clustered at the ends of the twigs. The tree produces small, pale green flowers

Fig. 2.7. *Albizia lucidior*, flowering plant. (By Vaibhav D. Lohot.)

Fig. 2.8. *Anacardium occidentale*, flowering plant. (By Nandkishore Thombare.)

arranged in panicles or corymbs, eventually turning reddish. The cashew tree bears a false oval or pear-shaped fruit, while the true fruit, a drupe, develops at the tip of the false fruit. The true fruit contains a single seed (Fig. 2.8).

Use: The bark of the cashew tree produces a bitter sap that forms a thick brown resin, which turns black when exposed to air. This resin is secreted from the plant's stem and is commonly used as a substitute for gum arabic or in similar applications, such as varnish. It is used as an indelible ink for marking and printing on fabrics like linens and cotton. The varnish made from the resin is also used for preserving fishing nets and as a flux in metal soldering. Additionally, the stem produces an amber-coloured gum with insecticidal qualities that is valued for its adhesive properties in woodwork, plywood and bookbinding.

Anogeissus latifolia (Ghatti)

Botanical name: *Anogeissus latifolia* (Roxb.ex DC.) Wall.ex Guill & Perr.

Synonym: *Conocarpus latifolia* Roxb. ex-DC.

Classification: Plantae > Tracheophyta > Magnoliopsida > Myrtales > Combretaceae > Anogeissus > latifolia.

Common names: Axle wood, Ghatti tree, Indian gum tree, Indian sumac, Dindiga tree, Dhawa, Button tree (English); Bakli, Dhaura (Hindi); Dhurandhara, Indravruksha, Shakataaya, Dhavaha (Sanskrit).

Distribution: the tree is found in India, Myanmar, Nepal, Sri Lanka and Pakistan.

Morphology: tall deciduous tree with smooth, greyish bark marked by whitish patches. The bark sheds in irregular, thin flakes. The leaves

Fig. 2.9. *Anogeissus latifolia*, flowering plant. (By Nandkishore Thombare.)

are simple and can be opposite or alternate in arrangement, varying in shape from elliptic and elliptic-oblong to suborbicular or oblong-obovate. Young leaves have a reddish-brown hue. The flowers are bisexual, pale green or yellow and are borne on dense, axillary globose heads. The fruit is a small drupe, forming clusters in a globular head. It is glabrous and orbicular with lateral wings, and the fruit's beak is long, equal to or longer than the main body. The fruit is yellowish-brown and encloses a single obovate seed. The tree's leaves produce significant amounts of gallotannins, used in tanning in India. Additionally, the tree exudes a gum known as gum ghatti (Fig. 2.9).

Use: The gum is used as an emulsifier, stabilizer and thickener in various industries, including ceramics, food and pharmaceuticals. It is an effective alternative to gum arabic and finds applications in calico printing, sweetmeats, confectionery and dyeing processes. The gum stabilizes ice creams, table syrup emulsions and more. It also functions as a binder in tablet manufacturing and is incorporated into hair-setting gels. In the petroleum and mining sectors, gum ghatti is used to emulsify drilling fluids and waxes and in acidizing oil wells. Additionally, it plays a role in explosives, acting as a water blocker, absorber and desiccant.

Anthoshorea javanica

Botanical name: *Anthoshorea javanica* (Koord. & Valeton) P.S. Ashton & J. Heck.

Synonyms: *Shorea javanica* Koord. & Valeton; *Shorea vandekoppeli* Parijs; *Shorea vandekoppeli* var. *grandifolia* Parijs.

Classification: Plantae > Streptophyta > Equisetopsida > Magnoliidae > Malvales > Dipterocarpaceae > Anthoshorea > javanica.

Common names: Dammar, Damar gum, White meranti, Damar mata kuching.

Distribution: *Asia:* Java, Sumatra.

Morphology: large tree, growing up to 40 or 50 m tall, with a diameter of up to 1.5 m; trunk is straight and cylindrical, bark is smooth and

Fig. 2.10. *Anthoshorea javanica* trees in Caringin District, West Java, Indonesia. (By Afrogindahood, CC BY-SA 4.0, https://creativecommons.org/licenses/by-sa/4.0, via Wikimedia Commons.)

greyish-brown, often with shallow fissures as the tree matures. Leaves are simple, alternate and pinnately veined. They are typically dark green and glossy on the upper surface, while the underside is paler. Flowers are small, white and fragrant. They usually bloom in January and are insect-pollinated. The fruit is a round, winged capsule that splits open to release winged seeds, which are dispersed by the wind. It generally flowers in January and is pollinated by insects (Fig. 2.10).

Use: Damar gum is used in the food and cosmetic industries and in the manufactures of paper, wood varnishes, lacquers, polishes and additives for beverages. Furthermore, it is used for water-resistant coating and in the pharmaceutical and dental industries for its strong binding properties. Some new research has shown its potential as a novel microencapsulating material.

Astragalus brachycalyx subsp. *brachycalyx*

Botanical name: *Astragalus brachycalyx* subsp. *brachycalyx*.

Synonyms: *Astracantha adscendens* (Boiss. & Hausskn.) Podlech; *Astracantha leioclados* (Boiss.) Podlech; *Astracantha oechtoeranensis* (Freyn) Podlech; *Astragalus adscendens* Boiss. & Hausskn.; *Astragalus belangerianus* Fisch.; *Astragalus brachycalyx* var. *umbraciformis* Boiss.; *Astragalus leioclados* Boiss.; *Astragalus luristanicus* Freyn; *Astragalus oechtoeranensis* Freyn; *Astragalus turrillii* Eig; *Tragacantha leioclados* (Boiss.) Kuntze.

Classification: Plantae > Streptophyta > Equisetopsida > Magnoliidae > Fabales > Fabaceae > Astragalus > brachycalyx > brachycalyx.

Common names: Gazangebin, Gaz khansar.

Distribution: *Asia*: Iran, Iraq, Turkey.

Morphology: shrub, sometimes umbraciform. Leaf rachis spiny, 25–45 mm, recurved. Leaflets elliptic, muticous, 4–6-paired, 3–10 mm; stipules lanceolate, *c.*6 mm, velutinous below, united for two-thirds of their length. Flowers sessile, 2–3 per leaf axil. Inflorescence globose, 10–20-flowered. Bracts broadly ovate, white-hairy on the dorsal surface, 2–4 mm. Calyx *c.*4 mm, tomentose; tube *c.*3 mm; teeth narrowly triangular. Corolla white; 7–9 mm, platonychioid.
Use: Resins collected from various species of the genus *Astragalus* are used to produce tragacanth, a natural product used as a thickening and stabilizing agent in the food industry, especially in dairy products and confectionery.

Astragalus fasciculifolius

Botanical name: *Astragalus fasciculifolius* Boiss.
Synonyms: *Tragacantha fasciculifolia* (Boiss.) Kuntze; *Astragalus afghanopersicus* Kitam.; *Astragalus arbusculinus* Bornm. & Gauba; *Astragalus cornutus* Bunge; *Astragalus fasciculifolius* subsp. *arbusculinus* subsp. (Bornm. & Gauba) Tietz; *Tragacantha tebesiensis* Kuntze.
Classification: Plantae > Streptophyta > Equisetopsida > Magnoliidae > Fabales > Fabaceae > Astragalus > fasciculifolius.
Common names: Anzaroot, Milkvetch.
Distribution: *Asia*: Bahrain, Iran, Iraq, Kuwait, Oman, Qatar, Saudi Arabia, United Arab Emirates.
Morphology: shrub 30–120 cm wide, puffed up with elongated branches, most appositely grey-tomentous, armed with numerous long white thorns. Rameal leaves remotely 2–3-paired, axillary to bunch of 5–7-leaflets reduced, leaflets sub lens sparingly and appressedly puberulous or glabrate, fleshy glaucous, obovate or linear-cuneate, obtuse or notched, mucronate. Flowers of solitary bundles short; pedicellate, bracteolis very small, squamiform; calyx flowery tubular, appressed and sparingly hirsute, with teeth four times shorter than the lanceolate-subulate tube, ovate-oblong or globose, glistening in fruit, of an oblong standard, calyces of fruit smaller than those of the preceding ones; petals 1.5–3 lines long.
Use: The natural plant exudate gum tragacanth is obtained from the stem of the bush-like plant 'goat's-horn', *Astragalus* species. Gum tragacanth has been used commercially for well over 2000 years and is currently widely used as an emulsifier and thickener in emulsion systems in different food, pharmaceutical and cosmetic applications (Whistler, 1993).

Azadirachta indica (Neem)

Botanical name: *Azadirachta indica* A. Juss.
Synonyms: *Antelaea azadirachta* (L.) Adelbert; *Azedarach deleteria* Medik.; *Azedarach fraxinifolia* Moench; *Melia azadirachta* L.; *Melia fraxinifolia* Salisb.; *Melia indica* (A. Juss.) Brandis; *Melia pinnata* Stokes.
Classification: Plantae > Tracheophyta > Magnoliopsida > Sapindales > Meliaceae > Azadirachta > indica.
Common names: Indian lilac, Margosa tree (English); Neem (Hindi).
Distribution: throughout tropical Asia: *Indian subcontinent*: India, Bangladesh. *Indochina*: Cambodia, Laos, Myanmar, Thailand, Vietnam.
Morphology: evergreen tree of small to medium size. It typically has a straight trunk with spreading branches that form a rounded crown. The bark is blackish, moderately thick, deeply fissured and flakes off in older trees. The leaves are light green, pinnate and arranged alternately, clustering at the tips of the branches. They have two pairs of glands at the base, and the leaflets feature short petioles, serrated edges and an acuminate tip with an unequal base. The tree produces numerous white or pale yellow, strongly honey-scented flowers in axillary clusters. These flowers are bisexual, although male flowers can also be on the same tree. The fruit is a drupe containing a single seed. It is greenish-yellow, ellipsoidal, with a pulpy mesocarp. The seeds are ovoid or spherical with a pointed apex and consist of a hard shell surrounding a kernel (Fig. 2.11).
Use: The gum is widely used in South-east Asia, where it is known as 'Neem glue'. It serves as a useful excipient in tablet binding. The edible gum, which is considered a substitute for gum arabic, also acts as a laxative and helps with constipation.

Fig. 2.11. Fruiting neem tree (*Azadirachta indica*) (By Nandkishore Thombare.)

Boswellia serrata (*Salai guggul*)

Botanical name: *Boswellia serrata* (Roxb.) ex Colebr.

Synonyms: *Boswellia glabra* Roxb.; *Boswellia serrata* var. *glabra* (Roxb.) A.W. Bennett; *Boswellia thurifera* Roxb. ex Flem.; *Chloroxylon dupada* Buch. -Ham.; *Libanotus asiaticus* Stackh.; *Libanus thuriferus* Colebr.

Classification: Plantae > Tracheophyta > Magnoliopsida > Sapindales > Burseraceae > Boswellia > serrata.

Common names: Indian olibanum, Indian frankincense, incense tree, olibanum, Indian olibanum tree, Indian frankincense tree (English); Shallaki, Salai (Hindi); Bhishan, Kurunda, Hradini, Guggula, Sallaki (Sanskrit).

Distribution: widely distributed throughout the dry regions of tropical India (Indian subcontinent).

Morphology: medium to large deciduous tree with a spreading crown and drooping branches. Its bark is papery and may appear greyish-green, ashy or reddish, peeling off in thin flakes. Beneath the outer bark layer is a chlorophyll-rich layer. The tree's leaves are alternate with leaflets arranged oppositely. The leaflets are sessile, and the basal pairs are typically smaller than the others. The leaf margins are serrated, resembling a saw, which is the origin of the species name 'serrata'. The leaves tend to cluster near the tips of the branches. The white flowers are arranged near the branch tips but are not terminal. The fruit is a three-angled drupe. When injured, the tree exudes a yellowish-green oleo-gum resin from its bark, known as *Salai guggul* (Fig. 2.12).

Use: *Salai guggul* is often used as a substitute for Canada balsam and has a pleasant aroma when burned. Referred to as Indian frankincense, it is commonly used in medicines, perfumes and as a scented incense to freshen the air. In traditional medicine, it is used to treat rheumatism. The gum also serves as a thickening agent in

Fig. 2.12. *Boswellia serrata* leaves. (By Nandkishore Thombare.)

paints and varnishes. For centuries, Ayurvedic medicine has utilized the gum-resin to treat arthritis. Additionally, its extract has been clinically studied for its effectiveness in managing osteoarthritis, particularly in the knee.

Buchanania lanzan (Chironji)

Botanical name: *Buchanania lanzan* Spreng.
Synonyms: *Buchanania cochinchinensis*; *Buchanania latifolia* (Roxb.).
Classification: Plantae > Tracheophyta > Magnoliopsida > Sapindales > Anacardiaceae > Buchanania > lanzan.
Common names: Almondette tree, Cuddapah almond, Cheronjee (English); Piyal, Chironji, Chiraunji, Char, Pra-savak, Priyal (Hindi); Priyala, Akhatth, Prasavakh, Piyala, Rajanadanha, Muni, Upavatth (Sanskrit).
Distribution: found in the tropical part of Asia (*Indian subcontinent*: Bangladesh, India, Nepal, western Himalayas. *Indochina*: Cambodia, Laos, Myanmar, Thailand, Vietnam) as well as in the temperate part of Asia (*China*: China South-Central, Hainan).
Morphology: medium-sized tree with dark grey or nearly black bark, which is rough and marked with small quadrangular plates and deep, narrow fissures, resembling the texture of crocodile skin. The young branches are covered in silky hairs. The leaves are simple, broadly oblong with a notch at the apex (emarginate) and have a leathery (coriaceous) texture. While the upper side of the leaves is smooth and hairless (glabrous), the underside is densely covered with hairs. The greenish-white flowers grow in terminal and axillary panicles without a peduncle (sessile). Initially green, the fruits mature to black, each containing a single seed encased in a hard, stone-like seed coat (Fig. 2.13).
Use: The gum extracted from the bark is edible and known for its medicinal benefits. It is commonly used to treat diarrhoea, and when combined with goat's milk, it helps relieve intercostal pain due to its analgesic properties. In traditional medicine, this gum is also utilized to treat leprosy. Additionally, high-quality gum

Fig. 2.13. *Buchanania lanzan*, flowering tree. (By Vaibhav D. Lohot.)

is used as a colouring agent for soft drinks and other food items.

Bursera bipinnata

Botanical name: *Bursera bipinnata* (Moc. & Sessé ex DC.) Engl.
Synonyms: *Elaphrium bipinnatum* (Moc. & Sessé ex DC.) Schltdl.; *Terebinthus bipinnata* (Moc. & Sessé ex DC.) W. Wight ex Rose; *Amyris bipinnata* Moc. & Sessé ex DC.; *Elemifera bipinnata* (Moc. & Sessé ex DC.) Kuntze; *Bursera bipinnata* var. *ovatifolia* Donn.Sm.; *Bursera elemifera* (Royle) Baill., *Bursera gracilis* Engl.; *Bursera tenuifolia* Engl. ex Kuntze; *Bursera verapacensis* Pittier; *Elaphrium elemiferum* Royle; *Elaphrium gracile* (Engl.) Rose; *Terebinthus gracilis* (Engl.) Rose; *Terebinthus tenuifolia* (Engl. ex Kuntze) Rose.
Classification: Plantae > Streptophyta > Equisetopsida > Magnoliidae > Sapindales > Burseraceae > Bursera > bipinnata.
Common name: Black copal.
Distribution: *Americas*: El Salvador, Guatemala, Honduras, Mexico.
Morphology: tree or shrub 4–12 m in height, dioecious, deciduous, profusely branched and resinous; resin colourless, releasing a pleasant scent when squeezed; stem 30–50 cm in diameter; external bark greyish, smooth and not exfoliating. Leaves bipinnate; leaflets shiny in the upper surface, with prominent venation in the bottom surface. Inflorescences racemose or paniculate; male and female flowers, tetramerous, similar in form and size, white, yellow or greenish. Fruit drupe, bivalvated, obovoid and covered by a red pseudo-aril when mature. Seeds generally one per fruit (Fig. 2.14).
Use: *Bursera bipinnata* and other species of the Burseraceae family provide 'Black copal resins' used in traditional Central and South American rituals as well as in perfumery and incense.

Bursera graveolens

Botanical name: *Bursera graveolens* (Kunth) Triana & Planch.
Synonyms: *Elaphrium graveolens* Kunth; *Terebinthus graveolens* (Kunth) Rose.
Classification: Plantae > Streptophyta > Equisetopsida > Magnoliidae > Sapindales > Burseraceae > Bursera > graveolens.
Common names: Palo santo, Carano, Bija, Tatamaco, Chachigue, Sasafras, Crispin.
Distribution: *Americas*: Colombia, Costa Rica, Cuba, Ecuador, El Salvador, Galápagos, Guatemala, Honduras, Mexico, Nicaragua, Peru, Venezuela.
Morphology: tree 4–10 m high, bark smooth, grey, non-exfoliating. Leaves once pinnate, up to 27.5 cm long and 13 cm wide, glabrous to pubescent on the rachis, petioles and main nerves of the leaflets, winged rachis; leaflets 5–9, narrowly ovate to obovate or ovate, apex acuminate to broadly attenuated (acute), crenate, more or less of the same colour on both surfaces. Panicles up to 6 cm long (6.5–14 cm long in fruit); sepals 4, 1/4–1/3 of the length of the petals; petals 4, 2–3 mm long. Fruits obovoid to ellipsoid, 7–11 mm long, 2-valved, tinged with red when ripe (Fig. 2.15).
Use: *Burcera graveolens* is often used in ritual cleansing ceremonies and in folk medicine for stomach aches. The use of 'palo santo' (holy stick) from *B. graveolens* is traditional in South America, particularly in Peru and Ecuador.

Butea monosperma (Palas)

Botanical name: *Butea monsperma* (Lam.) Taub.
Synonyms: *Butea frondosa* Roxb.; *Erythrina monosperma* Lamk.
Classification: Plantae > Tracheophyta > Magnoliopsida > Fabales > Fabaceae > Leguminosae > Butea > monosperma.
Common names: Flame of the forest, Bastard teak (English); Palas, Dhak (Hindi); Brahmvriksha, Karaka, Krimighna, Lakshataru, Parasu (Sanskrit).
Distribution: found in tropical Asia (*Indian subcontinent*: India, Bangladesh, Nepal, eastern and western Himalayas, Pakistan. *Indochina*: Cambodia, Laos, Myanmar, Vietnam) and temperate Asia (China South-Central).
Morphology: medium-sized deciduous tree that loses its leaves in the spring. It typically has a short, twisted trunk that is elongated and cylindrical. The branches grow low and

Fig. 2.14. *Bursera bipinnata* tree. (CC BY 4.0, https://commons.wikimedia.org/wiki/File:Bursera_bipinnata.jpg#/media/File:Bursera_bipinnata.jpg.)

Fig. 2.15. *Burcera graveolens* tree. (By Haplochromis-Own work, CC BY-SA 3.0, https://commons.wikimedia.org/w/index.php?curid=4550456.)

irregularly, forming an open crown. They are crooked, green, angular and cylindrical. The bark is rough but not profoundly fissured with colours ranging from slate grey to pale brown or blackish. The leaves are pinnately trifoliate with the terminal leaflet being rhomboid, broadly obovate or nearly circular. The leaf apex is usually rounded or slightly notched (emarginate), and the base is rounded or wedge-shaped (cuneate). The leaf margins are smooth and entire without any indentations. The leaf surface is covered with soft, straight hairs, giving it a pubescent texture, and it is glutinous, appearing green and shiny. The leaflets are asymmetrical, stiff, leathery (coriaceous), ovate in shape and rough on both sides (Fig. 2.16).

Use: A red sap is released from natural cracks and cuts in the bark, which hardens into a vitreous, ruby-red substance known as *Bengal Kino*. In India, this gum is called *Kamarkas* and is a key ingredient in *Kamar kas ke laddoo*, a traditional Unani sweet believed to have aphrodisiac properties. In Indochina, folk medicine uses gum to treat sores, boils, ulcers and other skin conditions. It has astringent qualities and is used to treat diarrhoea, heartburn, dysentery, corneal opacity, syphilis and excessive sweating and as a stimulant. As a contraceptive, a solution made by dissolving 50 g of the gum in water is taken orally starting from the fifth day of menstruation for 1 week. The gum is also used to make a durable ink by adding alkali to its solution. The gum serves as a lotion or poultice in combination with other substances. It is incorporated into some food recipes and is applied to cracks on the soles of the feet before bedtime. For dysentery, two spoons of diluted gum are recommended until recovery.

Calamus draco

Botanical name: *Calamus draco* Willd.
Synonyms: *Daemonorops draco* (Willd.) Blume; *Palmijuncus draco* (Willd.) Kuntze; *Calamus draconis* Oken.
Classification: Plantae > Streptophyta > Equisetopsida > Magnoliidae > Arecales > Arecaceae > Calamus > draco.
Common name: Dragon's blood.
Distribution: Sumatra.
Morphology: clustering rattan, climbing to 8 m; stem without sheaths 8–14 mm in diameter, with sheaths up to 30 mm, internodes to 20 cm long; leaf sheaths brownish, armed by densely horizontal-triangular spines, flattened, in irregular length, 5–45 mm long, abundantly covered by blackish-brown indumentum, very conspicuous, numerous, easily fall off; leaf sheath mouth unarmed; knee present, conspicuously, armed as the rest of leaf sheath. Leaf to 3 m long, cirrus to 100 cm long; petiole to 10 cm long, adaxially yellow, armed with group of spines, scattered; leaf rachis armed as the rest of leaf sheath, especially on the cirrus armed with regularly arranged groups of grapnel-like spines; leaflets irregularly arranged at upper half, regularly arranged below, up to 20 on each side, leaflets linear lanceolate, leathery, terminate in an acute point and at the margin blackish, bristly, equidistant, slightly conspicuous, midribs prominent abaxially, longitudinally plicate. Leaflets to 37 cm long and 4 cm wide. Inflorescence pendulous, to 51 cm long, peduncle 9 cm long, armed along the margin with short spines; female prophyll deciduous, leathery, armed with very rare group of spines, 16 cm long ×1.5 cm wide, ellipsoid oblong, bearing up to 6 partial inflorescence (rachilla), each rachilla up to 10 cm long, consisting of up to 9 rachilla each; involucre short, 5 mm long × 6 mm wide, flat, papery, just above the involucrophore; involucrophore short, 6 mm long × 9 mm wide, papery, angular; male prophyll deciduous, leathery, armed with group of spines, scattered at irregular distance, 16 cm long × 2.5 cm wide, ellipsoid oblong, bearing 3 partial inflorescence (rachilla), each rachilla up to 10 cm long, consisting of up to 10 rachilla each; involucre short, 5 mm long × 6 mm wide, flat, papery, just above the involucrophore; involucrophore short, 6 mm long × 9 mm wide, papery, angular. Female and male flowers unknown. Infructescence covered by ferruginous detachable indumentum. Fruit covered by 10 scales arranged in vertical rows, broad, 1.3 × 1.3 cm, dark reddish line in the middle of scales, scattered with Dragon's blood. Seed ovate oblong, *c.*1 × 0.7 cm, reticulate on half surface and deep rugose on the other half surface (Fig. 2.17).
Use: *Calamus draco* and other *Calamus* species produce the so-called 'Dragon's blood resin'. The deep red-coloured resin is used in traditional medicine and as incense.

Fig. 2.16. *Butea monsperma*, flowering tree. (By Vaibhav D. Lohot.)

Fig. 2.17. Fruits of a *Calamus draco* tree. (CC0 1.0 public domain, https://commons.wikimedia.org/wiki/File:9093Yantok_07.jpg.)

Canarium strictum (Black dammar)

Botanical name: *Canarium strictum* Roxb.

Synonyms: *Canarium resiniferum* Bruce ex King.; *Canarium sikkimense* King; *Pimela stricta* (Roxb.) Bl.

Classification: Plantae > Tracheophyta > Magnoliopsida > Sapindales > Burseraceae > Canarium > strictum.

Common names: Doopa, Black dammar, Black dammar tree, Black dhup, Indian white mahogany (English); Dhuna-goch (Assamese); Kunthirikkam, Viraka, Thelli, Kungiliyium, Pantappayan, Karutha-kungiliyam, Pantham, Thellippayin (Malayalam); Ral-dhup, Karuppu dhamar, Dhoopam, Karunkungiliyam, Karangkunthrikam (Tamil).

Distribution: found in tropical Asia (*Indian subcontinent*: India, eastern Himalayas. *Indochina*: Cambodia, and Thailand) and temperate Asia (China South-Central)

Morphology: tall tree with a straight trunk and buttressed roots. Its bark is smooth, ranging from grey-brown to yellow with small, irregular exfoliations. The leaves are either ovate-lanceolate or elliptical with serrated edges and an acuminate apex, typically arranged singly at the tips of branches. The upper surface of the leaves is smooth, while the underside is covered with tiny, woolly hairs (tomentose). The young leaves and branches also have a rust-coloured, woolly texture. The leaflets are arranged oppositely along the branches. The tree produces axillary inflorescences with flowers clustered at the branch tips. Male and female flowers grow on separate plants with the male flowers being more abundant. The flowers are densely covered in rusty or yellow woolly hairs. The tree exudes a resinous, brownish-black secretion from its trunk. The fruit is a dark blue ellipsoid drupe containing 1 to 3 seeds (Fig. 2.18).

Fig. 2.18. Infrutescence of a black dammar tree (*Canarium strictum*). (By Nishad Sankar.)

Use: The resin exuded by the tree is known as Black dammar and is commonly used as incense, particularly in religious ceremonies, where it is burned to produce thick white fumes. It is also employed in producing wood varnishes, lacquers and low-quality paints and for caulking boats. In traditional medicine, this resin is recognized for its astringent properties and is used to treat diarrhoea and dysentery. Additionally, it serves as a binding agent for softer waxes, which are used in manufacturing shoe polish, carbon paper and typewriter ribbons, and as a cementing material for plywood and asbestos sheets.

Cedrus libani

Botanical name: *Cedrus libani* A. Rich.

Synonyms: *Abies cedrus* (L.) J. St.-Hil.; *Cedrus cedrus* (L.) Huth; *Cedrus effusa* (Salisb.) Voss; *Cedrus libanitica* Trew ex Pilg.; *Cedrus libanitica* subsp. *libani* (A. Rich.) O. Schwarz; *Cedrus libanotica* Link; *Cedrus libanotica* subsp. *libani* (A. Rich.) Holmboe; *Cedrus patula* K. Koch; *Larix cedrus* (L.) Mill.; *Larix patula* Salisb.; *Peuce cedrus* (L.) Rich.; *Picea cedrus* (L.) Peterm.; *Pinus cedrus* L.; *Pinus cedrus* var. *effusa* Voss; *Pinus effusa* Salisb.

Classification: Plantae > Streptophyta > Equisetopsida > Pinidae > Pinales > Pinaceae > Cedrus > libani.

Common name: Cedar of Lebanon.

Distribution: *Asia*: Lebanon-Syria, Turkey. *Europe*: Bulgaria, Cyprus, Great Britain.

Morphology: tree up to 40 m tall, with a single massive columnar trunk, or often forked low on the bowl; row first conical, later spreading horizontally. Bark cracking and becoming scaly with age, developing longitudinal fissures, blackish-brown. Primary branches massive, often erect, later spreading horizontally; second-order branches usually spreading in

horizontal planes. Twigs bear a long, slender leading shoot; otherwise, short, firm, pale brown turning grey, shallowly grooved, at first with dark brown pubescence; short shoots thick, scaly, of variable length with age (0.5–4 cm). Vegetative buds ovoid, not or slightly resinous; bud scales broadly ovate, free at tips, pale brown with dark apex, deciduous. Leaves on long shoots spirally arranged, radially spreading, more crowded at base of the shoot; leaves in short shoots densely crowded, 20–35, most numerous on older short shoots, spreading radially, 20–25 × 1–1.5 mm, narrowly linear, straight or curved, quadrangular, acute or acuminate, with at least a few stomata on all surfaces; light to dark green, sometimes glaucous. Pollen cones terminal on short shoots, erect, subtended by leaves, numerous, 4–5 cm long, cylindrical, curved when ripe, pale green ripening to pale brown. Seed cones terminal on short shoots, erect, sessile, becoming woody in second year, ovoid to cylindrical, with obtuse to depressed apex, (5–)8–12 × 3–6 cm, light green ripening to grey-brown; cone rachis persistent, narrowly conical. Seed scales broad rectangular flabellate, thin, coriaceous, 3–3.5 × 3.5–4 cm (wider than long); glabrous with orange-brown pubescent at base. Seeds 10–14 × 4–6 mm, brown, with a cuneate seed wings 20–30 mm long (Fig. 2.19).
Use: *Cedrus libani* is used as a herbal supplement for a variety of ailments, including digestive, respiratory and skin disorders. It is also said to have anti-inflammatory and antioxidant properties. The essential oil distilled from the wood is used in the perfume industry.

Ceratonia siliqua

Botanical name: *Ceratonia siliqua* L.
Synonyms: *Ceratonia coriacea* Salisb.; *Ceratonia inermis* Stokes; *Ceratonia siliqua* var. *latissima* Risso; *Ceratonia siliqua* var. *siccata* Risso; *Ceratonia siliqua* var. *sterilis* Risso; *Ceratonia siliqua* var. *vulgaris* Risso.
Classification: Plantae > Streptophyta > Equisetopsida > Magnoliidae > Fabales > Fabaceae (Leguminosae) > Mimosoideae > Ceratonia> siliqua.
Common names: Kharnoub, Carob tree, Bean gum.
Distribution: Albania, Algeria, Cyprus, East Aegean Islands, France, Greece, Iran, Italy, Crete, Lebanon-Syria, Libya, Morocco, Palestine, Sardinia, Sicilia, Sinai, Spain, Tunisia, Turkey, California, Canary Islands, Cape Provinces, Cape Verde, China (south-east), Corsica, Ethiopia, Germany, India, Iraq, Kenya, Madeira, Mexico (south-west), Mozambique, Namibia, New South Wales, Northern Provinces, Pakistan, Peru, Portugal, South Australia, Sudan, Tanzania, Transcaucasus, Trinidad and Tobago, Tubuai Islands, Yemen, Zambia, Zimbabwe.
Morphology: tree 15–16 m tall; leaves 8–17 cm; leaflets 2–4 pairs, shiny, obovate or suborbicular, 3.5–5.5 × 3–3.5 cm, leathery, glabrous, lateral veins conspicuously raised, base cuneate or broadly cuneate, margin entire, apex rounded, emarginate or conspicuously cordate depressed. Raceme with rachis densely yellowish-brown pubescent. Flowers reddish, legume curved, 10–25 × *c.*2.5 cm (Fig. 2.20).
Use: The gum, extracted from the seeds of the carob tree, is used in the food industry as a thickening and gelling agent, particularly in dairy products and ice cream.

Cistus ladanifer

Botanical name: *Cistus ladanifer* L.
Classification: Plantae > Streptophyta > Equisetopsida > Magnoliidae > Malvales > Cistaceae > Cistus > ladanifer.
Common names: Gum rockrose, Labdanum, Common gum cistus, Brown-eyed rockrose.
Distribution: *Africa*: Algeria, Morocco, Madeira. *Europe*: France, Cyprus, Portugal, Spain. *America*: California. *Oceania*: New Zealand.
Morphology: shrub up to 2.5 m. Leaves sessile; blade 3-veined from base, elliptic, lanceolate-elliptic, lanceolate-linear or oblong,

Fig. 2.19. Lebanon cedar (*Cedrus libani*). (CC BY-SA 2.0, https://commons.wikimedia.org/w/index.php?curid=50183767.)

Fig. 2.20. Ripe carob fruit pods on the tree. (CC BY-SA 3.0, https://commons.wikimedia.org/wiki/File:Garroves.JPG#/media/File:Garroves.JPG.)

40–80(–120) × 6–25 mm. Inflorescence solitary flowers. Flowers: sepals 3, petals white, sometimes with red or yellow spot near bases, 30–50 mm; styles 0–0.5 mm. Capsules 6–12-locular (Fig. 2.21).

Use: *Cistus* species such as *Cistus ladanifer* produce a resin known as 'labdanum', which is used in the perfume industry as a fixative and base note, giving depth and complexity to fragrances.

Cochlospermum religiosum

Botanical name: *Cochlospermum religiosum* (L.) Alston.

Synonyms: *Cochlospermum gossypium* DC.; *Cochlospermum balicum* Boerl.; *Maximilianea gossypium* Kuntze; *Wittelsbachia gossypium* Mart. & Zucc.; *Bombax gossypium* L.; *Bombax religiosum* L.; *Bombax congo* Burm.f.; *Bombax grandiflorum* Sonn.

Classification: Plantae > Streptophyta > Equisetopsida > Magnoliidae > Malvales > Bixaceae > Cochlospermum > religiosum.

Common names: Silk-cotton tree, Buttercup tree.

Distribution: *Asia*: Bangladesh, India, Myanmar, Himalayas, Cambodia, Java, Lesser Sunda Islands (Indonesia), Malaysia, Sri Lanka. *Americas*: Trinidad and Tobago.

Morphology: small deciduous tree, stem is smooth and ash coloured, bark contains gum. The leaves are 7–20 cm across, orbicular in outline, palmately 3–5 lobed, cordate at base, tomentose beneath, 5–7 nerved at base. Petiole is up to 25 cm long. Flowers appear after leaf fall. Flowers are large, cup shaped, up to 8 cm

Fig. 2.21. Flowering brown-eyed rockrose (*Cistus ladanifer*). (By Alvesgaspar, CC BY-SA 4.0, https://commons.wikimedia.org/w/index.php?curid=61974526.)

or more across, in terminal racemes or loosely branched panicles. Sepals are 5 in number, deciduous and imbricate. Petals are 5 in number, bright golden yellow in colour and contorted in bud. Stamens are free and numerous. Ovary is 1-celled or incompletely 3–5-celled. Fruit is a capsule, large, 9–13 cm long, obovoid. Seeds are brown in colour, reniform and covered with wooly hairs (Fig. 2.22).

Use: The gum obtained from the bark of the tree is used as a stabilizer and thickening agent, particularly in the cosmetics, pharmaceutical and food industries. It has a better emulsifying effect than acacia gum.

Commiphora guidottii

Botanical name: *Commiphora guidottii* Chiov. ex Guid.

Synonyms: *Commiphora sessiliflora* Vollesen.

Classification: Plantae > Streptophyta > Equisetopsida > Magnoliidae > Sapindales > Burseraceae > Commiphora > guidottii.

Common name: Scented myrrh.

Distribution: *Africa*: Ethiopia, Somalia.

Morphology: unarmed shrub or small tree up to 5 m tall; trunk and main branches unusually thick, with a smooth, pale, silvery greyish-brown bark, outer layer peeling off in irregular papery flakes; exudate colourless and odourless; branches tending to bend down, 4–8 mm in diameter when young. Flowers cream or pinkish, appearing with the young leaves. Leaves up to 20 cm long including a petiole of 1.5–10 cm, (1)–3–5(–7)-foliolate, glabrous, puberulous or pubescent, the upper surface often concave with small blisters when mature; leaflets suborbicular, ovate or broadly elliptic, 5–11 cm long, 3–8 cm wide, entire; the terminal tapering to the base on a petiolule 1–4 mm long, the laterals sessile, rounded or cordate at the base. Female

A

B

Fig. 2.22. *Cochlospermum religiosum*: A. trees; B. detailed view of a blossom. (https://commons.wikimedia.org/wiki/File:Cochlospermum_religiosum_(2).jpg#/media/Datei:Cochlospermum_religiosum_(2).jpg, and by Prenn, CC BY-SA 3.0, https://commons.wikimedia.org/w/index.php?curid=18946348.)

inflorescences similar but usually rather shorter. Male flowers in interrupted subspicate panicles 3–5.5 cm long on pedicels ± 1 mm long; calyx pubescent, 2–3 mm long on a saucer-shaped receptacle, the lobes ovate-triangular, as long as the tube; petals sparsely puberulous along the median line, ± 3 mm long; filaments 1.0 and 0.5 mm, anthers 1.0 and 0.8 mm long. Fruit ellipsoid to obovoid, puberulous, ± 13 × 11 × (4 + 5) mm; pericarp 2-valved, fleshy, up to 3 mm thick; pseudo-aril short, fleshy, cupular, very shallowly lobed; stone smooth with gently convex faces, at first white but perhaps finally black, 7.5 × 4.5 × (1.5 + 3) mm.

Use: The traditional use of the oleo-gum resin of *Commiphora guidottii* Chiov. ex. Guid., which is commonly called 'scented myrrh', for topical treatment of wounds is well documented (Gebrehiwot *et al.*, 2015).

Commiphora wightii (Guggul)

Botanical name: *Commiphora wightii* (Arn.) Bhandari.

Synonyms: *Commiphora caudata* (Wight & Arn.) Engl.; *Protium caudatum* (Wight & Arn.).

Classification: Plantae > Tracheophyta > Magnoliopsida > Sapindales > Burseraceae > Commiphora > wightii.

Common names: Guggul, Indian myrrh, Indian bdellium (English); Gugal, Guggul (Hindi); Guggulu (Sanskrit).

Distribution: China, India, Nepal, Sri Lanka.

Morphology: slow-growing, perennial woody shrub with multiple thorny branches. It has a long dormant period and sheds its leaves annually. The bark is papery, and the wood is soft with a greyish hue. The plant produces compound leaves with 1–3 leaflets, which are sessile and arranged opposite each other. The largest leaflet is located at the terminal end, while the lateral leaflets are smaller, about half the size of the terminal one. The leaflets have irregularly toothed edges. This species exhibits dimorphism, with some plants bearing bisexual and male flowers, while others have female flowers with staminodes. The flowers are small, pink and arranged in fascicles. The fruit is a drupe, typically pea-sized, and can be either ovate or acuminate in shape. When ripe, the fruit turns red and splits open to reveal a brightly colored pseudo-aril (Fig. 2.23).

Use: Guggul oleo-gum resin is recognized for its wide range of therapeutic properties, including its ability to lower lipid and cholesterol levels, combat obesity and offer anti-inflammatory, anti-arthritic and antimicrobial benefits. It is also known for its astringent, antiseptic and anticancer effects. Additionally, it is used in treating conditions such as thrombosis, chronic bronchitis, nodulocystic acne, spongy gums, chronic tonsillitis and dental caries. The resin has applications in perfumery, calico printing, fumigation and dyeing fabrics like silk and cotton, as well as being used in incense. Gugulipid, derived from guggul, is a popular natural supplement for reducing high cholesterol levels and has been a traditional treatment in India for years. Products derived from guggul include Abana (for heart care), Diabecon (for blood sugar control), Diakof (for cough relief), Koflet (cough care), Lukol, Pilex (for vein health), Reosto, Rumalaya Forte and Septilin (for immune support). Ayurvedic formulations such as Triphala Guggulu, Yogaraj Guggulu, Kaishor Guggulu and Punavadi Guggulu are commonly used for detoxification, weight management, arthritis treatment, muscle pain, rheumatism, gout, fluid retention, heart health and inflammation.

Copaifera guyanensis

Botanical name: *Copaifera guyanensis* Desf.

Synonyms: *Copaiba guyanensis* (Desf.) Kuntze; *Copaiba bijuga* (Willd. ex Hayne) Kuntze; *Copaiva guyanensis* (Desf.) E.H.L. Krause; *Copaifera beyrichii* Hayne; *Copaifera bijuga* Willd. ex Hayne.

Classification: Plantae > Streptophyta > Equisetopsida > Magnoliidae > Fabales > Fabaceae > Copaifera > guyanensis.

Common name: Copaiba balm.

Distribution: *Americas:* Brazil, Colombia, French Guiana, Guyana, Panama, Suriname, Venezuela.

Morphology: tree up to 50 m high; bark coumarin-scented; twigs smooth, the bark reddish or silver-grey, tight or cracking, essentially pubescent toward tip; petioles slender,

Fig. 2.23. *Commiphora wightii* tree in its natural habitat. (By Geethaka99, CC BY-SA 4.0, https://commons.wikimedia.org/w/index.php?curid=45728648.)

minutely golden-pubescent to subglabrous, 5–17 cm long, the leaves up to 25 cm long, glabrous above and below, the petiolules up to 7 mm long, the leaflets 3–4 pairs, usually subopposite, punctate, inequilateral or not distinctly so, 2–4.5 cm apart, oblong, oblong-lanceolate, obovate-oblong (rarely ovate-rotund), 4–12 cm long, 3–5.5 cm wide, usually tapering obtusely or acuminately into an attenuate-acuminate tip, the latter falcate, or simply cuneate or obtuse at base, the costa plane or subplane above, prominent beneath, pubescent (usually?) above and below, apparently subglabrous below, the principal parallel secondary veins 15–20+, simple for 3/4 of their length, usually bifurcating into unequal branches, each branch usually forming a part of a distinct submarginal and wide undulate or zigzag vein, irregular-parallel to margin, the reticulate areas patent, conspicuous, the ultimate veins prominulous, each reticulate area enclosing at least one subpapillary glandular punctation; inflorescence axillary, paniculate; flowers white (*in vivo*); sepals carnose, glandular-punctate under illumination, glabrous or subpubescent on outside, hirsute within, usually narrow-lanceolate, rarely ovate-rotund, 4.2–5.2 mm long, 1.2–2.5 mm wide, acute or vaguely obtuse at apex, tapering slightly at base, the veins 1–5+, evanescent or distinct, the margin usually distinct; stamens 10, the filaments unequal, the anthers smooth, oblong or ovate-oblong, 2.3–2.5 mm long, 1–1.2 mm wide, obtuse and usually apiculate at apex, cordate at base; ovaries sparse-pubescent on margins of valves, ovate-oblong, 1.8–2 mm long, 1–1.5 mm wide, the style (here immature) about 2.5 mm long; fruit drying red-black or dull-red, vaguely marcescent, the valves often thick, usually curled, 2.5–3.5 cm long, 2–3 cm wide, falcate-mucronate or obtuse-apiculate at apex, the endocarp drying yellow-brown or dull red-brown, the seeds drying a lustrous black, sometimes with an oblique ridge across surface, 1.7–2.4 cm long, 1–1.5 cm wide, the arillus red (*in vivo*), drying red, minutely peaked at apex, 1.8 cm long, 0.8 cm wide.

Use: Copaiba oleoresins are marketed in Brazil and other South American countries primarily as an anti-inflammatory and antiseptic. They are also used for various medicinal applications, particularly in traditional Amazonian medicine.

Cotoneaster nummularioides

Botanical name: *Cotoneaster nummularioides* Pojark.

Synonyms: *Cotoneaster racemiflorus* var. *nummularioides* (Pojark.) Kitam.; *Pyrus nummularioides* (Pojark.) M.F. Fay & Christenh.

Classification: Plantae > Streptophyta > Equisetopsida > Magnoliidae > Rosales > Rosaceae > Cotoneaster > nummularioides.

Common name: Shirkhesht.

Distribution: *Asia*: Iran, Kirgizstan, Tadzhikistan, Turkmenistan, Uzbekistan.

Morphology: erect shrub, 1.5–2.75 m high. Juvenile branches of tomentelli, finally glabrous, dark-purple. Leaves 5–10(–20) mm long, 4–8 mm wide, orbicular or rotundato-obovate or rotundato-elliptic, apex broadly rounded or narrowed, apiculate, sometimes emarginate, opaque above, appreciably hairy, then often subglabrous, below pale yellow or grey-tomentose. Inflorescence erect, 2–4 flowered, tomentose. Corolla explanted 6–8(–9) mm in diameter. Stamens up to 20. Styles 2. Fruit obovoid-globose, (5–)6–7.5 mm long, first dark purple, then frosty black, pyrene 2.

Use: In Iran, manna (locally called *Shri-e-Khesht*), i.e. the gum fraction obtained from the twigs of *Cotoneaster nummularioides*, orally administered to the mother or newborn baby, is considered an effective remedy for jaundice (Fakhri *et al.*, 2018). It is a sweet, yellowish-white substance known as Purgative manna. It is a mild laxative and it is used in Persian Traditional Medicine to remove toxins from the liver, bile ducts, gall bladder and stomach and is the best medicine for patients with fever (Nematzadeh *et al.*, 2023).

Delonix regia (Gulmohar)

Botanical name: *Delonix regia* (Hook.) Raf.

Synonyms: *Delonix regia* var. *flavida* Stehlé; *Delonix regia* var. *genuina* Stehlé; *Poinciana regia* Hook.; *Poinciana regia* Bojer.

Classification: Plantae > Tracheophyta > Magnoliopsida > Fabales > Fabaceae > Delonix > regia.

Common names: Gulmohar, Flambouyant, Flame tree, Peacock flower, Poinciana, Royal poinciana, red tree (English); Gulmohar (Hindi).

Distribution: native to Madagascar but widely planted worldwide in tropical and semi-tropical areas.

Morphology: fast-growing tree with a wide, spreading canopy. Typically, it grows up to 10 m tall but can occasionally reach 15 m. The leaves are arranged alternately and are bipinnate, consisting of a central stalk with 7–20 pairs of pinnae. The leaflets are 5–13 mm long and 2–5 mm wide, with bright green upper surfaces and paler, pubescent undersides that quickly become smooth. They are oblong or elliptical with smooth, entire edges. The flowers are bisexual and crimson, arranged in axillary racemes. Each flower has five lanceolate-ovate sepals that are acuminate and finely pilose on the outside. The five petals are red and elliptic-orbicular in shape, with the upper petal being distinct and streaked with white, red and yellow. The flowers also feature 10 free, long and exserted stamens, with villous filaments below. The ovary is topped with a style and stigma. The fruit is a flat, elongated, woody pod containing many oblong, transverse seeds (Fig. 2.24).

Use: The tree exudes gum in yellowish or reddish-brown warty tears, which dissolve in water to create a thick, opalescent mucilage. This gum is rich in calcium oxalate. The seeds also contain gum, utilized in the textile and food industries.

Dipterocarpus turbinatus

Botanical name: *Dipterocarpus turbinatus* C.F. Gaertn.

Synonyms: *Dipterocarpus jourdainii* Pierre; *Dipterocarpus laevis* Buch. -Ham.; *Dipterocarpus mayapis* (Blanco) Blanco; *Dipterocarpus*

Fig. 2.24. *Delonix regia*, flowering tree. (By Vaibhav D. Lohot.)

turbinatus var. *ramipiliferus* Y.K. Yang & J.K. Wu; *Mocanera mayapis* Blanco; *Shorea trinervia* Banks ex C.F. Gaertn.

Classification: Plantae > Streptophyta > Equisetopsida > Magnoliidae > Malvales > Dipterocarpaceae > Dipterocarpus > turbinatus.

Common name: Gurjun tree.

Distribution: Andaman Islands, Assam, Bangladesh, Cambodia, India, Laos, Myanmar, Sri Lanka, Thailand, Vietnam, China South-Central.

Morphology: tree lofty, *c.*35 m tall. Bark grey or dark brown, shallowly longitudinally fissured and flaky. Branchlets glabrescent; leaf buds falcate, buds and young twigs densely grey puberulous. Stipules 2–6 cm, densely, shortly dark greyish or dark yellow puberulous; petiole 2–3 cm, densely grey puberulous or glabrescent; leaf blade ovate-oblong, 20–30 × 8–13 cm, leathery, glabrous or sparsely stellate pubescent, lateral veins 15–20 pairs conspicuously raised abaxially, base rounded or somewhat cordate, margin entire or sometimes sinuate, apex acuminate or acute. Racemes axillary, 3–6-flowered. Calyx segments: 2 linear, 3 shorter, all glabrous, outside glaucous. Stamens *c.*30; anthers linear-lanceolate; connective appendages filiform. Ovary densely pubescent; style terete, silvery grey tomentose on lower half. Nut ovoid or narrowly ovoid, densely appressed tomentose; calyx tube to 2.8 cm in diameter, glabrous, glaucous; wing-like calyx segments linear-lanceolate,

12–15 × *c.*3 cm, glabrous, minutely papillate near much-ramified solitary midvein.

Dipterocarpus tuberculatus

Synonyms: *Dipterocarpus tuberculatus* Roxb.; *Dipterocarpus cordatus* Wall. ex A.DC.; *Dipterocarpus grandifolius* Teijsm. ex Miq.; *Dipterocarpus tuberculatus* var. *grandifolius* (Teijsm. ex Miq.) Craib; *Dipterocarpus tuberculatus* var. *tomentosus* Kerr.

Classification: Plantae > Streptophyta > Equisetopsida > Magnoliidae > Malvales > Dipterocarpaceae > Dipterocarpus > tuberculatus.

Common names: Gurjun tree, Khmer khlông, Dau Duong, Kung, Yaang Phluang.

Distribution: Bangladesh, Cambodia, Laos, Myanmar, Thailand, Vietnam.

Morphology: tree lofty, emergent, with greyish brown to orange flaky, prominently lenticellate bark and aromatic oily white resin, with stout buttresses. Stipules large, enclosing terminal bud, finally caducous and leaving an annular scar; leaf blade leathery, plicate in bud and ± corrugate when opened; lateral veins pinnate, straight; tertiary veins subscalariform, conspicuous, margin entire or sinuate-crenate. Raceme 3–9-flowered, hardly branched. Flowers large, sweetly scented. The tree is found in clear forests on plains, at altitudes up to 800–1,000 m. It grows to a height of 5–25 m. Calyx 1-leaved, permanent, body of the calyx spherical, with five knobs under five fissures on the outside, two of the five divisions of its border large, and growing with the pericarp into two very long, scariose wings. Corolla 5-petalled. Germ. superior, 3-celled; cells 2-seeded; attachment interior. Nut ovate, 1-celled, 1-seeded. Embryo inverse, no perisperm.

Use: Gurjun balsam, obtained from various *Dipterocarpus* species of the Dipterocarpaceae family (mainly *D. turbinatus* and *D. tuberculatus*), is used primarily to distil the essential oil from the exudate. The resulting volatile oil has a balsamic, plum-like odour with a slight hint of patchouli. In addition, the balm is also used in traditional herbal medicine.

Ferula assa-foetida

Botanical name: *Ferula assa-foetida* L.

Synonyms: *Peucedanum assa-foetida; Scorodosma assa-foetida* (L.) H. Karst.; *Ferula foetida* St.-Lag.; *Ferula hooshee* Lindl. ex Descourt.; *Narthex polakii* Stapf; *Narthex silphium* Oerst.; *Peucedanum hooshee* (Lindl. ex Descourt.) Baill.

Classification: Plantae > Streptophyta > Equisetopsida > Magnoliidae > Apiales > Apiaceae > Ferula > assa-foetida.

Common names: Asafoetida, 'Devil's dung', Anghuzah, Anjedān, Ḥiltit.

Distribution: *Africa*: Libya. *Asia*: Bangladesh, Iran, Laos, Turkmenistan, Uzbekistan.

Morphology: perennial plant, hairy stem 100–200 cm high, in middle *c.*15 mm thick, yellowish, grooved, hairless. Basal leaves *c.*2-ternato-pinnatisect, 30–35 mm × 15–25 mm; final segments up to 100 mm × 20 mm, entire or slightly lobed. Lower stem leaves 3–4-ternate-pinnate, pubescent, up to 30 cm × 20 cm; final segments smooth, up to 40 mm × 15 mm, coarsely pinnate-lobed; upper sheath *c.*9 cm × 4 cm, sub-membranaceous, sparingly pubescent. Panicle compact, a partial inflorescence from the central fruit umbel, briefly pedunculate and from the male umbel ± very pedunculate *c.*4 composite, leaf sheath with reduced ovate, supported by a spurious sheath for the anthers. Umbrella-bearing rays 20–25, *c.*5 cm long, glabrous; umbrella *c.*17–20-flowered. Small bracts, bracts or absent. Petals blonde, 1–5 mm long, glabrous, ovary glabrous. Mericarpia obovate or elliptic, up to 12 mm × 7 mm, glabrous (Fig. 2.25).

Use: The oleo-gum resin is obtained from the rhizome and root of the plant. It is used as a digestive aid, in food as a condiment and in pickles. It is used in modern herbalism in the treatment of hysteria, some nervous conditions, bronchitis, asthma and whooping cough. Ferula assa-foetida is mainly found in Afghanistan and Iran. The resin is obtained by extraction from the dried resin sap. The onion to garlic-like odour is mainly used in perfumery as a nuanceur.

Fig. 2.25. *Ferula assa-foetida* in Uzbekistan. (Public domain, https://commons.wikimedia.org/wiki/File:Ferula_assafoetida_(Uzbekistan)_1.jpg#/media/Datei:Ferula_assa-foetida_(Uzbekistan)_1.jpg.)

Ferula gabrielii

Botanical name: *Ferula gabrielii* Rech.f.
Classification: Plantae > Streptophyta > Equisetopsida > Magnoliidae > Apiales > Apiaceae > Ferula > gabrielii.
Common name: Anghuz-e Talkh & Shirin.
Distribution: Iran.
Morphology: perennial, hairy, raised plant. Leaves herbaceous, to 2-ternate-pinnate, pubescent; last segments 20–30 × 15–20 mm, decurrent, coarsely lobed. Panicles strictly branched, branches bent-hairy, striated. The central fruiting umbels are subsessile; rays *c.*15, 2–2.5 cm long, deflexed-hairy. Lateral umbels for the most part fruit-bearing, peduncles 8–9 cm long. Umbrellas *c.*15-flowers; small bracteoles. Petals 1.5–2 mm, slightly papillate. Ovary papillose. Mericarpia juvenile papillosa, *c.*6 × 4 mm; wings *c.*1 mm wide; dorsal cords 3–4 per valley; two commissural laces.
Use: Galbanum resin is used in perfumery as a green, fresh note and is known for its unique fragrance. The resin is obtained from the milky sap that flows from the cut roots and dries in the air.

Ferula gummosa

Botanical name: *Ferula gummosa* Boiss.
Synonyms: *Ferula erubescens* Boiss.; *Ferula galbaniflua* Boiss. & Buhse; *Peucedanum galbanifluum* (Boiss. & Buhse) Baill.

Classification: Plantae > Streptophyta > Equisetopsida > Magnoliidae > Apiales > Apiaceae > Ferula > gummosa.
Common names: Galbanum, Bārijah, Qāsni, Qinah.
Distribution: Iran, Turkmenistan.
Morphology: perennial plant, monocarpic, white. Stems 70–100 cm high, strawberry, 15 mm thick or thicker in the middle, solid, grooved. Basal leaves 30 cm × 20 cm, 4-pinna, densely pubescent; last pieces *c.*1.5 mm × 1 mm, pinnate; lobes 0.5–1 mm × 0.2 mm; upper sheath *c.*5 cm × 4 cm, leathery, glabrous. Panicle loose. Shadowy fruit rays 10–20; radius 2–4 cm long. Umbel 15–20-flowered; petals 1.5 mm long, yellowish, pubescent. Mericarpia 10 15 mm × *c.*8 mm.
Use: The resin is used in perfumery as a green, fresh note with fixative properties and is known for its unique aroma. The resin is obtained from the milky sap that flows from the cut roots and dries in the air.

Ferula persica

Botanical name: *Ferula persica* Willd.
Synonyms: *Peucedanum persicum* (Willd.) Baill.; *Ferula puberula* Boiss. & Buhse; *Ferula szowitziana* var. *kandavanensis* Bornm. & Gouba; *Prangos euryangiodes* Stapf & Wettst.
Classification: Plantae > Streptophyta > Equisetopsida > Magnoliidae > Apiales > Apiaceae > Ferula > persica.
Common names: Sakbinej, Komaye Biabani, Sakbinaj.
Distribution: Iran, Transcaucasus.
Morphology: perennial plant, polycarpic, hairy stems 30–100 cm high, thick, streaked. Basal leaves 3–4-pinnate, 30–35 cm × *c.*25 cm, above glabrous, below sparingly, with a broad-pubescent spine densely; final segments 4–20 mm ×3–10 mm, slightly to deeply lobed, margin entire or coarsely toothed; upper sheath *c.*8 cm × 2 cm, herbaceous, pubes, not inflated. Panicle loose. Umbel-bearing rays, 17–25; radius *c.*4 cm. Mericarpia 9–11 mm × 5–6 mm; wing *c.*0.5 mm wide; dorsal tubes 5–7 through the valley; joiners 16–18 (Fig. 2.26).
Use: Sagapenum, known as 'Sakbinaj' in Iran, is the oleo-gum resin of *Ferula persica*, which is obtained by cutting the plant stems. It is red or yellow with white inside, traditionally famous for its purgative, anti-hysteric and antidiabetic properties as well as its beneficial applications in the treatment of stiffness and spine pain (Ghadami *et al.*, 2024).

Ferula ammoniacum

Botanical name: *Ferula ammoniacum* (D. Don) Spalik, M. Panahi, Piwczyński & Puchałka.
Synonyms: *Dorema ammoniacum* D. Don; *Ferula ammonifera* Oken; *Peucedanum ammoniacum* (D. Don) Baill.; *Diserneston gummiferum* Jaub. & Spach; *Diserneston hirsutum* Lofius ex I.G. Borshch.; *Dorema gummiferum* (Jaub. & Spach) K.M. Korol.; *Dorema hirsutum* Lofius ex I.G. Borshch.
Classification: Plantae > Streptophyta > Equisetopsida > Magnoliidae > Apiales > Apiaceae > Ferula > ammoniacum.
Common names: vashagh, koma, kandal, vasha.
Distribution: Afghanistan, Iran, Pakistan, Turkmenistan.
Morphology: fibrous and hairy stem 180–250 cm high, base 3–6 cm in diameter, distally paniculate-branched, first strawberry-green, finally brownish, with nodes not thickened. All basal leaves ternate, with pinnate primary segments, more rarely cut bipinnate; outermost segments large, all largely adnate to the rachis, ± connate at the base, running down the base, sometimes unevenly bipartite, 4–10 cm × 1.5–4.5 cm, ± sharp; petioles and rachis thick, finely grooved in dry groove; stem leaves generally reduced to large amplexicaules with sheaths pointed at length. Umbels short (about 10 mm) thick, pedunculate or subsessile, mostly white-floccose-lanate or less long, glabrescent or globose. The ovary is like a tinum. Fruit up to 8 mm × 5–6 mm, widely elliptic, brownish-violet, first whitish-woolly, finally glabrous; ridges filiform, prominent; valle ± distinctly united; patch 2–4-bandaged (Fig. 2.27).
Use: *Dorema ammoniacum* is commonly known as 'Ushaq' or 'Vasha', being rich in ammoniacum, a medicinal gum-resin. It has been mentioned in Traditional Persian Medicine (TPM) as a treatment for various

Fig. 2.26. *Ferula persica*. (By Mohammad Reza Kanani.)

disorders, such as gastrointestinal, upper respiratory tract and central nervous system problems. Furthermore, numerous chemical compounds, including terpenes, coumarins and phenolic compounds, have been isolated from *Ferula* species, and a wide range of pharmacological activities, including antimicrobial, anti-inflammatory, antioxidant, cytotoxicity, anticonvulsant, antidiabetic and hypolipidemic activities, have been reported from this genus in modern medicine (Zibaee *et al.*, 2020).

Fig. 2.27. *Ferula ammoniacum*. (By Mohammad Reza Kanani.)

Ferula downieorum

Botanical name: *Ferula downieorum* Spalik, M. Panahi, Piwczyński & Puchałka.
Synonyms: *Dorema aureum* Stocks.
Classification: Plantae > Streptophyta > Equisetopsida > Magnoliidae > Apiales > Apiaceae > Ferula > downieorum.
Common name: Vasha.
Distribution: Afghanistan, Iran, Pakistan.

Morphology: stem 2 m high, thick; the panicles of the branches are very branched. Leaves shortly pubescent, finally glabrous, tripinnate; segments small, lanceolate, ± lobed, running down the base. Panicles pubescent branches. Umbrellas small, numerous, all thinly stalked. Peduncles 1–3 cm long. Pedicels 1–2 mm long, floccose-pubescent, equaling the flowers. Petals yellow, glabrous. Fruit glabrous, 5–6 × 3–4 mm, glaucous, elliptic, margined, with a margin six times narrower than the seed; ribbons thin, indistinct.
Use: *Ferula downieorum* is used similarly to *F. ammoniacum*, mainly in traditional Persian medicine.

Firmiana simplex

Botanical name: *Firmiana simplex* (L.) W. Wight.
Synonyms: *Clompanus simplex* (L.) Kuntze; *Culhamia simplex* (L.) Nakai; *Hibiscus simplex* L.; *Sterculia firmiana* J.F. Gmel.; *S. simplex* (L.) Druce; *Caucanthus platanifolia* (L.f.) Raf.; *Clompanus pyriformis* Kuntze; *Culhamia hadiensis* J.F. Gmel.; *C. platanifolia* Steud.; *C. triloba* Raf.; *Firmiana chinensis* Medik. ex Steud.; *F. platanifolia* (L.f.) Marsili; *F. platanifolia F. tomentosa* (Thunb.) H. Hara; *F. platanifolia* var. *tomentosa* (Thunb.) Sa. Kurata; *F. simplex* var. *glabra* Hatus.; *Sterculia fiorniana* J.F. Gmel.; *S. platanifolia* L.f.; *S. pyriformis* Bunge; *S. tomentosa* Thunb.
Classification: Plantae > Streptophyta > Equisetopsida > Magnoliidae > Malvales > Malvaceae > Firmiana > simplex.
Common names: Chinese parasol tree, Wutong.
Distribution: *Asia*: Bangladesh, China, Hainan, Japan, Nansei-shoto, Korea, Taiwan, Vietnam. *Americas*: Alabama, Arkansas, District of Columbia, Florida, Georgia, Louisiana, Mauritius, Mississippi, Carolina, Tennessee, Texas, Virginia.
Morphology: deciduous tree, up to 16 m tall; bark greenish, smooth. Petiole 15–30 cm; leaf blade cordate, palmately 3–5-lobed, 15–30 cm in diameter, lobes triangular, both surfaces glabrous or minutely puberulent, basal veins 7, base cordate, apex acuminate. Inflorescence paniculate, terminal, 20–50 cm. Calyx yellowish green, 7–9 mm, divided nearly to base, lobes linear, twisted outward, abaxially puberulent and yellowish, adaxially villous only basally; pedicel nearly as long as calyx. Male flower: androgynophore as long as calyx, thicker in lower part, glabrous. Anthers 15, irregularly fascicled, pistillode pear-shaped and very small. Female flower: ovary globose, hairy. Follicle membranous, stalked, foliaceous, 6–11 × 1.5–2.5 cm, 2–4-seeded, abaxially puberulent or nearly glabrous. Seeds globose, *c.* 7 mm in diameter, wrinkled (Fig. 2.28).
Use: The resin collected from tapped trees is used in the food industry as a thickening and gelling agent.

Gardenia resinifera (Dikamali)

Botanical name: *Gardenia resinifera* Roth.
Synonyms: *Gardenia lucida* Roxb.; *Genipa resinifera* (Roth) Baill.
Classification: Plantae > Tracheophyta > Magnoliopsida > Gentianales > Rubiaceae > Gardenia > resinifera.
Common names: White emetic nut (English); Dekamali (Hindi); Nadihingu (Sanskrit).
Distribution: tropical Asia (Indian subcontinent: India, Bangladesh; Indochina: Myanmar).
Morphology: small, unarmed tree that grows up to 3 m tall and secretes gum from its buds. The tree's leaves have very short stalks. The flowers are bisexual, initially white but turning yellow, and grow in axillary clusters. The flowers are oblong, oval or obovate, with a blunt or rounded tip, smooth, hard and shiny, featuring simple parallel veins. The flower tube is long, glabrous, striated and flares into five obovate-oblong petals slightly shorter than the tube. The fruit is an oblong drupe, with a hard, thick and bony nut inside (Fig. 2.29).
Use: The yellowish resin is widely used in Ayurvedic medicine. To make it pliable, the resin is chewed after being removed from the tip and then applied to children's foreheads as a remedy for headaches. Once the headache is relieved, the resin falls off. When the resin is heated in oil, its vapour is inhaled to treat a runny nose. The resin has antiseptic properties and is used in paste

Fig. 2.28. Chinese parasol tree (*Firmiana simplex*). (By KENPEI, CC BY-SA 3.0, https://commons.wikimedia.org/w/index.php?curid=2529218.)

Fig. 2.29. *Gardenia resinifera*, flowering twig. (By Nandkishore Thombare.)

and powder form to treat a variety of ailments, including loss of taste, wounds, indigestion, constipation, flatulence, colic, roundworms, coughs, respiratory diseases, fever and skin diseases. It is particularly popular in rural areas of India for treating tapeworms in children.

Grevillea robusta (Silver oak)

Botanical name: *Grevillea robusta* A. Cunn. ex R. Br.

Synonyms: *Grevillea umbratica* A. Cunn. ex Meissn.; *Grevillea venusta* A. Cunn. ex Meissn.; *Stylurus robusta*.

Classification: Plantae > Tracheophyta > Magnoliopsida > Proteales > Proteaceae > Grevillea > robusta.

Common names: silky oak, silver oak, Southern silky oak, Grevillea, River oak (English); Silver oak (Hindi).

Distribution: native to Australasia (Australia: New South Wales and Queensland), introduced in many parts of the world.

Morphology: large evergreen tree that reaches 10–25 m tall and has a straight trunk with a moderately spreading canopy. The bark is grey-brown and fissured. The leaves are fern-like, dark green on top and pinnately compound (almost bipinnately so), with 11–21 pairs of pinnae, each 4–9 cm long. The leaflets are divided into narrow, elongated, pointed lobes with recurved margins and silvery undersides arranged densely to create a deep shade. The flower clusters, measuring 7.5–15 cm in length, are unbranched and primarily arise from the trunk in simple or branched racemes, either terminal or axillary, and are protandrous.

Fig. 2.30. *Grevillea robusta*, flowering tree. (By Nandkishore Thombare.)

The sepals are hooded and 1.5–2 cm long. The petals are united into a recurved tube with a broadened apex. There are four stamens, sessile within the concave limb, and the disk is annular or semi-annular, sometimes bilobed. The ovary is glabrous, one-locular with two ovules, and the style is lemon yellow, ranging from 1 to 2.5 cm long, eventually straight, and persistent, with its apex free from the limb. The fruit is a small, flat, oval seed pod that darkens to brown or nearly black when mature. It contains one or two small, winged seeds. The seeds are thin, ovate and lack endosperm (Fig. 2.30).

Use: Silver oak produces small amounts of resin, which may have potential uses in various industrial applications due to its solubility, viscosity and notable resistance to hydrolysis.

Hevea brasiliensis (Rubber)

Botanical name: *Hevea brasiliensis* (Willd. ex A. Juss.) Müll.Arg.

Synonyms: *Hevea brasiliensis* f. acreana (Ule) Ducke; *Hevea brasiliensis* f. angustifolia (Ule ex Huber) Ule; *Hevea brasiliensis* f. latifolia (Ule ex Huber) Ule; *Hevea brasiliensis* f. randiana (Huber) Ducke; *Hevea brasiliensis* var. janeirensis (Müll.Arg.) Pax; *Hevea brasiliensis* var. stylosa Huber; *Hevea camargoana* Pires; *Hevea granthamii* Bartlett; *Hevea janeirensis* Müll.Arg.; *Hevea randiana* Huber; *Hevea sieberi* Warb.; *Siphonia brasiliensis* Willd. ex A. Juss.; *Siphonia janeirensis* (Müll. Arg.) O.F. Cook; *Siphonia ridleyana* O.F. Cook

Classification: Plantae > Tracheophyta > Magnoliopsida > Malpighiales > Euphorbiaceae > Hevea > brasiliensis.

Common names: para rubber tree, caoutchouc tree, rubber, hevea, rubber wood (English); rubber (Hindi).

Distribution: native to the tropical evergreen rainforest of the Amazon region of South America, including Brazil, Venezuela, Ecuador, Colombia, Peru and Bolivia. Rubber is successfully cultivated under humid lowland tropical conditions. Today, rubber is grown in Asia in Malaysia, Indonesia, Thailand, Vietnam, Sri Lanka, China and India and in Oceania in Papua New Guinea.

Morphology: deciduous tree that can grow up to 25 m in plantations and reach 43m in its natural habitats. The tree has a smooth, straight, grey trunk. The bark ranges from pale to dark brown, with a smooth texture, while the inner bark is pale brown. When the bark is cut or damaged, it exudes a sticky white latex. The leaves are alternate and palmate, typically consisting of three leaflets. The flowers are small, dioecious and pale yellowish, arranged in terminal panicled racemes. Female flowers tend to be larger than male flowers. In the female flower, the gynoecium consists of three fused carpels, forming a 3-lobed, 3-celled ovary with one ovule in each cell. Between 5 and 10 stamens are arranged in two whorls, with filaments fused into a column extending beyond the anthers. The fruit is a woody capsule containing three large seeds, about 5 cm in diameter, which explode open when ripe. The seeds are yellowish-brown and have a distinctive marbled pattern (Fig. 2.31).

Use: *Hevea brasiliensis* is the main source of natural rubber. Latex is obtained by tapping the tree trunks. The latex coagulates with acetic acid, formic acid and alum. After hardening, the rubber is used in many products, including car tyres, shoes, balls, rubber bands, erasers, toys, gloves, latex thread, catheters and swim caps. Locally, it is also used to make household items such as water bottles and balls.

Lannea coromandelica (Jhingan)

Botanical name: *Lannea coromandelica* (Houtt.) Merr.

Synonyms: *Calesiam grande* (Dennst.) Kuntze.; *Calsiama malabarica* Rafin.; *Dialium coromandelicum* Houtt.; *Haberlia grandis* Dennst.; *Lannea grandis* (Dennst.) Engl.; *Lannea wodier* (Roxb.) Adel.; *Odina gummifera* Bl., nom. illeg.; *Odina pinnata* Rottl.; *Odina wodier* Roxb.; *Rhus odina* Buch.-Ham. ex Wall.; *Spondias wirtgenii* Hassk.; *Tapiria wodier* (Roxb.) March.; *Wirtgenia octandra* Jungh.

Classification: Plantae > Tracheophyta > Magnoliopsida > Sapindales > Anacardiaceae > Lannea > coromandelica.

Common names: Indian ash tree, Jhingam, Wodier, Moi tree (English); Mohin (Hindi); Jhingini, Ajashringi (Sanskrit).

Distribution: found in tropical Asia (Indian subcontinent: India, eastern Himalayas, Nepal, Pakistan, Bangladesh and Sri Lanka) and temperate Asia (China: China South-Central, China Southeast).

Morphology: medium to large-sized deciduous tree with greyish, rough bark that features longitudinal furrows and sheds in thin, irregular flakes. The leaves are compound with a single terminal leaflet (imparipinnate) and are concentrated at the tips of the branches. The leaflets are opposite, arranged in pairs and either elliptic-oblong or ovate-elliptic shape with acuminate tips. The young leaves are covered in tiny white or greyish-green hairs. The tree blooms while it is leafless, producing small yellowish or purplish flowers grouped in cymose clusters at the ends of the branchlets. The fruit is a drupe, ovoid in shape and red in colour, and contains a hard stone. The seed is solitary and compressed (Fig. 2.32).

Use: The resin obtained from the trunk is of lower quality than other known plant resins. It is edible and widely used in confectionery. This resin also has medicinal properties and is used for conditions such as sprains and asthma, and as a tonic for lactating women. It is also used for sizing paper and for calico printing.

Fig. 2.31. Rubber tree (*Hevea brasiliensis*) in India. (By Vaibhav D. Lohot.)

Leucaena leucocephala (Subabul)

Botanical name: *Leucaena leucocephala* (Lam.) de Wit.

Synonyms: *Acacia frondosa* Willd.; *Acacia leucocephala* (Lam.) Link; *Leucaena glauca* (Willd.) Benth.; *Leucaena glauca* (sensu L.1763) Benth.; *Leucaena glauca* Benth.; *Mimosa glauca* Sensu L.; *Mimosa leucocephala* Lam.

Classification: Plantae > Tracheophyta > Magnoliopsida > Fabales > Fabaceae > Mimosaceae > Leucaena > leucocephala.

Common names: Black wood, Leucaena, Jumpy-bean, Wild tamarind, Lead tree, White popinac, White lead tree, Horse tamarind, White babool (English); Safed babool, Kadam, Toira, Subabul, Tagari, Koo babul, Lamtoro (Hindi).

Fig. 2.32. Indian ash tree (*Lannea coromandelica*). (By Nandkishore Thombare.)

Distribution: native to Colombia, Costa Rica, El Salvador, Guatemala, Honduras, Mexico, Nicaragua, Panama, Spain, United States of America.

Morphology: small to medium-sized tree, typically ranging from 2 to 10 m in height, although it can occasionally grow up to 15 m or more. Older stems have smooth, greyish or greyish-brown bark with noticeable lenticels. The leaves are bipinnate, arranged alternately along the stems, and are linear-oblong to weakly elliptic in shape. The leaf tips are acute, while the bases are rounded to obtuse, with a glabrous surface except for the margins. The leaves resemble those of the tamarind tree. The flowers are borne in dense, globular heads. They are uniform and tetramerous, with a campanulate calyx that is slightly 4-toothed, a funnel-shaped corolla, numerous white stamens, and a glabrous ovary. The fruit is a strap-shaped legume, flat with thickened margins, attenuated at the base, rounded at the apex, and dehiscent from the tip to the base. The elongated, flat fruits can reach up to 20 cm in length and 2 cm in width, changing colour from green to brown as they ripen. Each fruit contains 15–25 dark brown seeds with a hard, shiny seed coat (Fig. 2.33).

Use: The gum, secreted from the trunk of the tree, is used under unspecified conditions for injuries, illnesses or sterile hybrids, particularly of *L. leucocephala* and *L. esculenta*. It has some potential for commercial use. In addition, red,

Fig. 2.33. Wild tamarind (*Leucaena leucocephala*) tree with fruits. (By Nandkishore Thombare.)

brown and black dyes are extracted from the pods, leaves and bark of the tree.

Liquidambar orientalis

Botanical name: *Liquidambar orientalis* Mill.
Synonyms: *Liquidambar imberbis* Aiton; *L. orientalis* var. *integriloba* Fiori.
Classification: Plantae > Streptophyta > Equisetopsida > Magnoliidae > Saxifragales > Altingiaceae > Liquidambar > orientalis.
Common names: Oriental sweetgum, Turkish sweetgum.
Distribution: East Aegean Islands, Turkey.
Morphology: tree up to 20 m, deciduous. Leaves palmatifid, 5–10(–12) × 6–13 cm, divided into 5 lobes, each lobe undivided or sparsely lobulate, margin finely sinuate-dentate or -serrate, glabrous or rarely with a sparse tuft of short hairs beneath at the base of the main veins. Its leaves have a thin and long petiole and the blades have 3–5 and sometimes 7 lobes. Lobes have sharp-toothed margin. The flower: many flowers without petals are all clustered at the top of a long pedicel. Flowers unisexual. The male flowers are in terminal and cluster shaped. The female flowers are lateral in leaf axils. Ovary has two septa. The fruit is a septicide capsule. Fruiting head 2.5–3 cm in diameter, pendulous (Fig. 2.34).
Use: Storax (sweet gum) and its derivatives (resinoid, essential oil, absolute) are used as flavours, fragrances and in pharmaceuticals (Friar's Balsam). American storax resin (*Liquidambar styraciflua*) is chewed like gum to freshen breath and clean teeth (Aljesri *et al*., 2014).

Fig. 2.34. Fruits of an Oriental sweetgum tree (*Liquidambar orientalis*). (CC BY-SA 4.0, https://commons.wikimedia.org/wiki/File:Liquidambar_orientalis_Turkish_sweetgum_03.jpg#/media/Datei:Liquidambar_orientalis_-_Turkish_sweetgum_03.jpg.)

Moringa oleifera (Drum stick)

Botanical name: *Moringa oleifera* Lam.

Synonyms: *Anoma moringa* (L.) Lour.; *Guilandia moringa* L.; *Hyperanthera arborea* F.Gmel.; *Hyperanthera decandra* Willd.; *Hyperanthera moringa* (L.) Vahl; *Hyperanthera pterygosperma* (Gaertn.) Oken; Moringa *domestica* Buch. -Ham.; *Moringa edulis* Medic.; *Moringa erecta* Salisb.; *Moringa moringa* (L.) Millsp.; *Moringa nux-eben* Desf.; *Moringa octogona* Stokes; *Moringa parvifolia* Noronha; *Moringa polygona* DC.; *Moringa pterygosperma* Gaertn.; *Moringa robusta* Boj.; *Moringa sylvestris* Buch. -Ham.; *Moringa zeylanica* Pers.

Classification: Plantae > Tracheophyta > Magnoliopsida > Capparales > Moringaceae > Moringa > oleifera.

Common names: Horse-radish tree, Drum stick tree (English); Senjana, Hargua, Mungana, Mungna, Rasunna, Sahijan, Sainjna, Sajna, Sanan suhanjua, Sanan suhanjuna, Sanjano, Saonjna, Sargua, Segu, Segve, Shajna, Soanjna, Shevga, Sondna, Sunara (Hindi).

Distribution: native to India, Malaysia, Oman, Qatar, Saudi Arabia, United Arab Emirates, Yemen.

Morphology: small, charming deciduous tree with delicate foliage. The trunk is often crooked and typically forks near the base. The bark is smooth and corky, with a grey hue. The crown of the tree is wide and open, resembling an umbrella. Due to the softness of the wood, the branches are prone to breakage. The leaves are alternate and tripinnate, with leaflets arranged in opposite pairs. The terminal leaflet is slightly larger than the others and is rounded-elliptic. The small, whitish, honey-scented flowers bloom profusely in panicle clusters. The fruit is an elongated capsule containing about nine seeds. When dry, the capsule splits open and the wind carries the seeds away, aided by their three-cornered wings. When exposed to sunlight, the tree trunk exudes gum that starts as white to reddish, changing to brown or brownish-black (Fig. 2.35).

Use: The moringa tree produces a reddish-brown gum that is only slightly soluble in water but swells upon contact, forming a thick, viscous solution. This gum is commonly used in treating dental caries, as an astringent and for managing blood pressure. It also finds applications in calico printing, as a condiment and treating stomach and bladder conditions.

Myroxylon balsamum

Botanical name: *Myroxylon balsamum* (L.) Harms.

Fig. 2.35. Flowering twig of a horse-radish tree (*Moringa oleifera*). (By Vaibhav D. Lohot.)

Synonyms: *Myrospermum toluiferum* DC.; *Myrospermum balsamiferum* Ruiz & Pav.; *Myrospermum pereirae* Royle; *Myrospermum punctatum* (Klotzsch) Walp.; *Myrospermum sonsonatense* Pereira ex Oerst.; *Myrospermum balsamiferum* Ruiz & Pav.; *Myroxylon balsamum* var. *genuinum* Harms; *Myroxylon toluiferum* A. Rich.; *Myroxylon hanburyanum* Klotzsch; *Myroxylon balsamum* var. *pereirae* (Royle) Harms; *Myroxylon pereirae* (Royle) Klotzsch; *Myroxylon punctatum* Klotzsch; *Myroxylon sonsouatense* (Pereira ex Oerst.) Flück. & D. Hanb.; *Myroxylon balsamum* var. *punctatum* (Klotzsch) Harms; *Toluifera balsamum* L., *Toluifera balsamum* var. *pereirae* (Royle) Baill.; *Toluifera balsamum* var. *punctata* (Klotzsch) Baill.; *Toluifera pereirae* (Royle) Baill.; *Toluifera punctata* (Klotzsch) Baill.

Classification: Plantae > Streptophyta > Equisetopsida > Magnoliidae > Fabales > Fabaceae > Myroxylon > balsamum.

Common names: Peru balsam, Tolu balsam.

Distribution: *Africa*: Uganda, Zaïre, Gabon, Ghana, Sierra Leone, Tanzania. *Asia*: India, Sri Lanka. *Americas*: Belize, Bolivia, Brazil, Colombia, Costa Rica, Ecuador, El Salvador, Guatemala, Gulf of Mexico, Guyana, Honduras, Mexico, Nicaragua, Panama, Peru, Puerto Rico, Venezuela, Caroline Islands, Cuba, Dominican Republic, Florida, Trinidad and Tobago.

Morphology: tree up to 40 m high, inermes, grey bark, abundantly lenticellated, resinous and fragrant. Leaves commonly imparipinnate; leaflets 5–11, alternate, lanceolate, elliptic to ovate, 3–11 cm long and 1–6 cm wide, apex acute to acuminate, base obtuse, essentially glabrous at maturity, beam sharp, with hairy lines and dots, petioles 3–6 mm long, stipels and stipules absent. Inflorescences axillary clusters, up to 20 cm long, erect and above the foliage, puberulent shaft, pedicels 10–15 mm long, deltoid bracts up to 3 mm long, flowers 12–15 mm in length, obliques in pedicels; calyx turbinate-bell-shaped, 4–6 mm long, finely ribbed, minutely pubescent, grey-green, 5 lobes up to 1.5 mm long; corolla white, the inside of the nail of the standard yellow, petals 5, free and subequal except the standard, which is 2–3 times as wide as the other petals, orbicular, unguiculate, wings elliptic to narrowly spatulate, 10 mm long, keel sub-elliptical, 8.5 mm long, unguiculated; stamens 10, free, filaments essentially equal to each other and slightly longer than calyx, anthers 4–4.5 mm long; ovary glabrous, stippled. Fruits narrowly obovate, up to 11 cm long and 2–3 cm wide, compressed, coriaceous, glabrous, light brown, indehiscent, seminiferous portion 15–20 mm long and 5–8 mm wide, stipe 5–15 mm of width; seed 1 (2), apical, reniform, light brown, resinous.

Use: Peru balsam is a sticky resin used in the production of perfumes and as a flavouring in food and drinks. The essential oil, obtained by steam distillation from the resinoid, has a balsamic-sweet, vanilla-like smell with a smoky undertone. Nowadays, Peru balsam oil is sometimes used in perfumes, although it is known that skin irritation cannot be completely ruled out.

Pinus gerardiana

Botanical name: *Pinus gerardiana* Wall. ex D. Don.

Synonyms: *Pinus aucklandii* Lodd. ex Gordon & Glend.; *Pinus chilghoza* Knight; *Pinus gerardii* J. Forbes; *Pinus neosa* Gouan ex W.H. Baxter.

Classification: Plantae > Streptophyta > Equisetopsida > Pinidae > Pinales > Pinaceae > Pinus > gerardiana.

Common names: Chilgoza pine, Neja.

Distribution: Afghanistan, Pakistan, Tibet, western Himalayas.

Morphology: tree, bark white, fissured into irregular, thin plates; first-year branchlets pale green-yellow, glabrous, with projected leaf scars. Needles 3 per bundle, triangular-flabellate in cross section, 6–10 cm, stiff, vascular bundle 1, resin canals 5–7, marginal, base with sheath shed. Seed cones shortly pedunculate, almost brown at maturity, oblong or ovoid, 12–20 × 9–11 cm. Seed scales 4–5 cm; apophyses broad, swollen, ± recurved, obviously ridged; umbo dorsal, apex obtuse. Seeds cylindric, *c.*2.5 cm; wing rudimentary, usually adhering to adjacent scale (Fig. 2.36).

Use: Turpentine which is obtained from the oleoresin has a wide range of uses including as a natural solvent for waxes and medicinal products. Rosin is also obtained by distillation of pine resin. This is used in paper glue and soap manufacturing, as a constituent of varnishes and paints and for coating the bows of stringed

Fig. 2.36. Female cone of a *Pinus gerardiana* tree. (Public domain, https://commons.wikimedia.org/wiki/File:Pinus_gerardiana_female_cone.jpg#/media/File:Pinus_gerardiana_female_cone.jpg.)

musical instruments such as violins. The turpentine obtained from the resin of all pine trees is antiseptic, diuretic, rubefacient and vermifuge. It is a valuable remedy used internally in the treatment of kidney and bladder complaints and is used both internally and as a rub and with steam baths in the treatment of rheumatic affections (Singh *et al.*, 2021).

Pinus roxburghii (Chir pine)

Botanical name: *Pinus roxburghii* Sarg.
Synonym: *Pinus longifolia* Roxb. ex Lamb.
Classification: Plantae > Tracheophyta > Pinopsida > Pinales > Pinaceae > Pinus > roxburghii.
Common names: Chir pine, long-leaved Indian pine, Himalayan longleaf pine (English); Chir (Hindi).
Distribution: native to Europe and Asia (Bangladesh, China, eastern Himalayas, India, Nepal, Pakistan, Tibet and western Himalayas).
Morphology: upright, evergreen tree with a rounded canopy, typically reaching a height of around 9 m and a trunk diameter of up to 100 cm. The bark is thick, scaly, dark red-brown, deeply fissured and longitudinally cracked. The leaves are needle-like, arranged in bundles of three, slender and yellowish-green, measuring 20–30 cm long and 1.5 mm wide. The tree is monoecious, meaning it has both male and female flowers on the same plant, which are wind-pollinated. The cones are ovoid-conic, measuring 12–24 cm long and 5–8 cm in diameter at the base when closed. Initially green, the cones ripen to a glossy chestnut-brown after about 24 months. They open gradually over the following year. The seeds are winged, 8–12 mm long, with wings approximately 2.5 cm long. The seeds mature between October and November (Fig. 2.37).
Use: The tree is commercially tapped for its resin, which, when distilled, produces turpentine (an essential oil) and rosin, a non-volatile solid. Turpentine has medicinal properties and is primarily used as a solvent in pharmaceutical products. It is also widely employed in the perfume industry, disinfectants, insecticides and denaturants. Additionally, turpentine is a key raw material for producing terpene chemicals

Fig. 2.37. *Pinus roxburghii*. (By Nandkishore Thombare.)

used across various industries, such as adhesives, paper and rubber. Chir pine rosin is mainly used to make paper, soap, cosmetics, paint, varnish, rubber and polish. Other applications include its use in the manufacture of linoleum, explosives, insecticides and disinfectants. It is also used as a flux for soldering, in brewing and as a frothing agent in mineral beneficiation.

Pistacia atlantica

Botanical name: *Pistacia atlantica* Desf.
Synonyms: *Pistacia choulettei* Gand.; *Pistacia atlantica* subsp. *cabulica* (Stocks) Rech.f.; *Pistacia. atlantica* subsp. *Cypricola* H. Lindb.; *Pistacia atlantica* var. *latifolia* DC.; *Pistacia atlantica* subsp. *mutica* (Fisch. & C.A. Mey.) Rech.f.; *Pistacia*

atlantica f. *oxycarpa* Zohary; *Pistacia cabulica* Stocks; *Pistacia chia* Desf.; *Pistacia mutica* Fisch. & C.A. Mey.; *Pistacia mutica* subsp. *cabulica* (Stocks) Engl.; *Pistacia mutica* var. *cypricola* (H. Lindb.) H. Lindb.; *Pistacia mutica* f. *multijuga* Engl.; *Terebinthus atlanticus* (Desf.) Dum. Cours.; *Lentiscus atlantica* Kuntze; *Lentiscus mutica* (Fisch. & C.A. Mey.) Kuntze.

Classification: Plantae > Streptophyta > Equisetopsida > Magnoliidae > Sapindales > Anacardiaceae > Pistacia > atlantica.

Common name: Saqqez.

Distribution: *Africa*: Algeria, Libya, Morocco, Sinai, Tunisia. *Asia*: Afghanistan, Caucasus, Iran, Iraq, Lebanon, Pakistan, Saudi Arabia, Syria, Tajikistan, Transcaucasus, Turkey. *Europe*: Canary Islands, Cyprus, Greece. *Americas*: USA.

Morphology: deciduous tree up to 7 m tall, with a dense, subglobose crown. Leaves imparipinnate; leaflets 2–5-paired, ovate-oblong or lanceolate, 2.5–8 cm × 0.8–2.2 cm, obtuse, mucronate, dark green above, paler beneath, glabrous except for ciliolate margin; rachis narrowly winged. Fruits paniculate, as long as or longer than broad, 5–8 mm × 5–6 mm.

Use: The resin of the tree is used as a varnish or adhesive, but today it has hardly any economic significance.

Pistacia lentiscus

Botanical name: *Pistacia lentiscus* L.

Synonyms: *Lentiscus vulgaris* Fourr.; *Pistacia gummifera* Salisb.; *Terebinthus lentiscus* (L.) Moench; *Lentiscus massiliensis* Four; *Pistacia brevifolia* Gand.; *Pistacia lentiscus* var. *angustifolia* DC.; *Pistacia lentiscus* subvar. *brevifolia* (Gand.) Nyman; *Pistacia lentiscus var. chia Poir.; Pistacia lentiscus f. chia* (Poir.) Cout.; *Pistacia lentiscus var. genuina* Rouy; *Pistacia lentiscus* f. *genuina* Rouy ex Cout.; *Pistacia lentiscus* f. *lanceolata* Cout.; *Pistacia lentiscus* f. *latifolia* (Coss.) Bolzon; *Pistacia lentiscus* var. *latifolia* Coss.; *Pistacia lentiscus* f. *leptophylla* Albo ex Fiori; *Pistacia lentiscus* var. *massiliensis* (Mill.) Poir.; *Pistacia lentiscus* f. *massiliensis* (Mill.) Fiori; *Pistacia lentiscus* var. *multiflora* (Gand.) Nyman; *Pistacia lentiscus* subvar. *subfalcata* (Gand.) Nyman; *Pistacia massiliensis* Mill.; *P. multiflora* Gand.; *Pistacia subfalcata* Gand.

Classification: Plantae > Streptophyta > Equisetopsida > Magnoliidae > Sapindales > Anacardiaceae > Pistacia > lentiscus.

Common names: Mastic, Chios mastic.

Distribution: *Africa*: Algeria, Libya, Morocco, Tunisia. *Europe*: Albania, Balearics, Cyprus, Egypt, France, Greece, Italy, Portugal, Spain. *Asia*: Lebanon-Syria, Palestine, Turkey.

Morphology: small evergreen tree or shrub 1–8 m. Leaves paripinnate; leaflets 1–5 × 0.5–1.5 cm, (4–)8–12, lanceolate to obovate-lanceolate, mucronate, coriaceous; rhachis broadly winged; petioles glabrous. Inflorescence compact, spike-like. Flowers yellowish or purplish. Drupe *c.*4 mm, globose, apiculate, red becoming black (Fig. 2.38).

Uses: The resin obtained from the mastic tree is used in traditional medicine, in food and in cosmetics. It is also used to produce chewing gum.

Prunus persica (Peach)

Botanical name: *Prunus persica* (L.) Stokes.

Synonyms: *Amygdalus persica* L.; *Persica laevis* DC.; *Persica nucipersica* Borckh.; *Persica platycarpa* Decne.; *Persica vulgaris* Mill.; *Persica vulgaris* subsp. *laevis* (DC.) Bonnier & Layens; *Prunus daemonifuga* H. Lév. & Vaniot; *Prunus persica* (L.) Batsch.

Classification: Plantae > Tracheophyta > Magnoliopsida > Rosales > Rosaceae > Prunus > persica.

Common names: Nectarine, Akroot, Peach (English); Adoo (Hindi).

Distribution: native to China, introduced as a fruit tree in other parts of the world.

Morphology: small deciduous tree, typically growing up to 15 m in height. It has a broad crown with branches that spread or ascend. The bark is dark grey and features horizontal lenticels. The leaves are alternate, simple and lanceolate, measuring 3–6 inches in length, often curving along the midrib. They are shiny dark green on top and paler underneath, with serrated edges. The flowers appear solitary in the axils of fallen leaves, about 3.5 cm wide, and are subsessile, blooming before the foliage. Each flower consists of five spreading pink petals, a burgundy calyx with five rounded lobes, 15–30 stamens and a pistil with a single style. The

Fig. 2.38. Flowering mastic shrub (*Pistacia lentiscus*). (By Júlio Reis Tintazul, CC BY-SA 2.5, https://creativecommons.org/licenses/by-sa/2.5, via Wikimedia Commons.)

petals are oval to orbicular, with clawed bases. The filaments of the stamens are slender, light pink and glabrous, while the style is long and slender, usually light green, becoming hairy towards the base. The fruit is a drupe, ovoid in shape, measuring 6 × 4.5 cm. It has a hairy exocarp, a fleshy mesocarp and a bony endocarp with deep pits (Fig. 2.39).

Use: The gum obtained from the stem has astringent and soothing properties. It is commonly used for chewing.

Prunus scoparia

Botanical name: *Prunus scoparia* (Spach) C.K.Schneid.

Synonyms: *Amygdalus scoparia* Spach.

Classification: Plantae > Streptophyta > Equisetopsida > Magnoliidae > Rosales > Rosaceae > Prunus > scoparia.

Common name: Zodoo.

Distribution: Iran, Turkmenistan.

Morphology: shrub, rarely a low tree, very spiny, soboliferous. Juvenile branches usually pubescent, then often glabrous. Leaves up to 5 × 2.5 cm, often smaller, elliptic, obovate or obovate-oblong, apex acute or obtuse, base cuneate, crenate-serrate, when young on both sides briefly hairy, then glabrous or glabrous or pubescent beneath. Petioles up to 1 cm long, glabrous or pubescent. Early white flowers, singly or very rarely in pairs, *c.*15 mm in diameter. Sepals ciliated. Pedicels up to 1 cm long, glabrous or pubescent. Drupes globose, 10–15 mm in diameter, dark blue or almost black, frosty, green flesh, bitter. Cuticles ovate or elliptical, smooth or subscaly, adhering to the pericarp (Fig. 2.40).

Use: Recently, possible applications of the plant exudate (so-called 'Persian gum') in the food industry have been described (Dabestani *et al.*, 2018).

Fig. 2.39. Fruit-bearing peach (*Prunus persica*). (By Nandkishore Thombare.)

Samanea saman (Rain tree)

Botanical name: *Samanea saman* (Jacq.) Merr.
Synonyms: *Albizia saman* (Jacq.) Merr.; *Mimosa saman* Jacq.; *Albizia saman* (Jacq.) F. Muell.; *Inga saman* (Jacq.) Willd.; *Pithecellobium saman* (Jacq.) Benth; *Enterolobium saman* (Jacquin) Prain ex King; Inga *salutaris* Kunth.; *Calliandra saman* (Jacq.) Griseb.
Classification: Plantae > Tracheophyta > Magnoliopsida > Fabales > Fabaceae > Samanea > saman.
Common names: Rain tree, Coco tamarind, French tamarind, Saman, Monkey pod (English); Vilaiti siris (Hindi).
Distribution: native of North Tropical South America but introduced all over the tropics.
Morphology: medium-sized tree, either deciduous or evergreen, growing up to 30 m with an umbrella-shaped spreading canopy. The bark of mature trees is grey, rough and fissured into long plates or corky ridges, while younger trees have smoother, paler grey to brownish bark. The inner bark is light-coloured and has a bitter taste. The leaves are bipinnately compound, arranged alternately along the twigs, with a noticeable swelling (pulvinus) at the base of the petiole. Each leaf has 3–8 pairs of asymmetrical, oblong leaflets, measuring 2–4 cm in length and 1–1.8 cm in width. The leaflets are pubescent on the underside, shiny on the upper side, with half-rounded and half-truncated bases, and rounded or obtuse apices, often with a small notch and a mucronulate tip. The flowers are grouped in heads which are 5–6 cm wide and about 4 cm tall. The central flower in each head is larger, stalkless, has more petals and does not produce fruit. Both the calyx and corolla are covered in

Fig. 2.40. *Prunus scoparia* tree in the valley of Gilazard in the mountains of Piramagrun, Kurdistan. (By Ari Khorshid, CC BY-SA 4.0, https://creativecommons.org/licenses/by-sa/4.0, via Wikimedia Commons.)

yellowish hairs. The calyx is funnel-shaped and approximately 7 mm long, while the corolla is about 14 mm long. The staminal filaments are pink in the upper and white in the lower half, measuring 35–40 mm long. The mature pods are oblong, compressed and septate between seeds. The seeds are brown and ellipsoidal (Fig. 2.41).

Use: The bark of the rain tree is a valuable source of gums and resins. It produces a gum of lesser quality, which can serve as a more affordable alternative to gum arabic. This gum is also used as an adhesive.

Shorea robusta (Sal)

Botanical name: *Shorea robusta* Gaertn. F.

Synonyms: *Dryobalanops robusta*; *Shorea Roxb.* ex C.F.Gaertn.

Classification: Plantae > Tracheophyta > Magnoliopsida > Malvales > Dipterocarpaceae > Shorea > robusta.

Common names: Sal (English); Sal, Salwa, Sakhua, Sakher, Sakhuwa (Hindi); Salah, Avakarnah, Agnivallabha, Ashvakarna, Ashvakarnika (Sanskrit).

Distribution: native to the Indian subcontinent, ranging south of the Himalayas, from Myanmar in the east to Nepal, India and Bangladesh.

Morphology: large, slow-growing deciduous tree with smooth or longitudinally fissured reddish bark that is thick and varies in colour from brown to grey. Its crown spreads and is spherical in open spaces but becomes vertical in denser vegetation. The leaves are simple, shiny and glabrous, with an ovate-oblong shape, typically ranging from 10 to 25 cm in length and from 5 to 15 cm in width. In wetter regions, the tree remains evergreen, while in drier areas, it sheds most of its leaves from February to April and re-leafs in early summer. The flowers are yellowish-white, arranged in large terminal or axillary panicles, which may contain up to 50 stamens. The fruit is indehiscent, reaching a size of 1.3–1.5 cm long and 1 cm in diameter, surrounded by segments of the calyx that form 5

Fig. 2.41. Flowering rain tree (*Samanea saman*). (By Vaibhav D. Lohot.)

uneven wings, each measuring 5 7.5 cm. When tapped, the tree produces a large quantity of a whitish, aromatic, transparent resin called *lul dhuna* (Fig. 2.42).

Use: The wood of the tree is resinous, and the presence of resin enhances its durability and water resistance. The sal resin is commonly used to seal boats and ships. It possesses a range of medicinal properties, being astringent, sweet, acrid, cooling, anodyne, vulnerary, antibacterial, deodorizing, constipating, detergent, carminative, stomachic, aphrodisiac, expectorant, ophthalmic and tonic. The resin is beneficial in treating conditions such as excessive sweating, imbalances of *pitta* (digestive fire and metabolic energy in the body), wounds, ulcers, neuralgia, burns, itching, fractures, fever, diarrhoea, dysentery, haemorrhoids, gonorrhoea, menorrhagia, spleen enlargement, obesity, headaches, toothaches, eye irritation and eye pain.

Shorea roxburghii (Jalari)

Botanical name: *Shorea roxburghii* G. Don.

Synonyms: *Shorea talura; Shorea laccifera; Vatica laccifera.*

Classification: Plantae > Tracheophyta > Magnoliopsida > Malvales > Dipterocarpaceae > Shorea > roxburghii.

Common names: Jalari, Taloora lac tree, White meranti (English); Jalari, Jalari sal, Madrasi sakhua (Hindi).

Distribution: widespread species native to South and South-east Asia.

Morphology: deciduous tree with simple, alternate leaves. Its bark is brownish to greyish-brown, smooth or narrow-longitudinally fissured and the blaze is creamy-yellow. The leaves are simple and alternate, featuring large coriaceous stipules. The petioles are stout,

Fig. 2.42. Flowering Sal (*Shorea robusta*). (By Vaibhav D. Lohot.)

glabrous and 10–25 mm long. The leaf blades are 6–18 cm × 3.5–9 cm, with shapes ranging from elliptic, oblong, elliptic-ovate and elliptic-oblong to elliptic-lanceolate. The base is rounded or slightly heart-shaped, and the apex is obtuse or emarginate, with entire, glabrous and coriaceous margins. There are 12–16 pairs of lateral nerves, which are parallel and prominent, with faint intercostal scalariform veins. The flowers are bisexual and grow in terminal panicles, each holding up to three drooping, fragrant flowers. The flowers are lanceolate and white, measuring 8 mm × 4 mm, with a reddish tint at the base inside and 15 stamens. The fruit is a capsule, enclosed by accrescent sepals, with oblong, unequal wings, 4–7 cm long and prominently nerved. The seed is solitary, obovoid, measuring 7 × 4 mm. The tree produces resin naturally or when tapped and is also known for supporting lac cultivation by rearing lac insects (Fig. 2.43).
Use: The resin finds a similar application to *Shorea robusta* resin.

Sterculia urens (Karaya)

Botanical name: *Sterculia urens* Roxb.
Synonyms: *Cavallium urens; Clompanus urens.*
Classification: Plantae > Tracheophyta > Magnoliopsida > Malvales > Sterculiaceae > Sterculia> urens.

Fig. 2.43. Inflorescence of a Jalari tree (*Shorea roxburghii*). (By Vaibhav D. Lohot.)

Common names: Karaya, Indian tragacanth, Kateera, Kulu, Katira, ghost tree. (English); Kulu, Gulu, Karaya, Katira, Katila (Hindi).

Distribution: native to India, but also introduced into Myanmar.

Morphology: deciduous tree with a short, crooked trunk and distinctive smooth bark that appears white or greenish-grey. When illuminated by moonlight, the bark gives the tree a ghostly appearance. The bark is thick and powdery and peels off in large, thin, papery flakes with a reddish blaze beneath. Its branches spread horizontally, and the younger branchlets are initially pubescent. The tree secretes a yellowish-white, sticky, semi-solid gum from natural cracks or injuries. The leaves are clustered at the twig tips, alternate, simple, with entire margins. They are glabrous on the upper surface and velvety pubescent on the underside, with long petioles and 3–5 palmate lobes. The flowers are polygamous, greenish-yellow and grow in axillary panicles covered in stinging hairs. These bisexual flowers lack petals and are hairy on both surfaces. The fruit is an aggregate of five sessile, radiating, ovate-lanceolate carpels, each about 7.5 cm long. When ripe, the fruit turns red and is covered in stiff bristles on the outside (Fig. 2.44).

Use: Gum karaya is recognized in medicine for its laxative properties. It is also widely used as a thickening agent in cosmetics and medicines, and as an adhesive for dental prostheses. The gum is also used as a binder, emulsifier and stabilizer in food and beverage preparations.

Fig. 2.44. Infrutescence of a Karaya tree (*Sterculia urens*). (By Vaibhav D. Lohot.)

Styrax benzoin

Botanical name: *Styrax benzoin* Dryand.
Synonyms: *Lithocarpus benzoin* (Dryand.) Royle; *Plagiospermum benzoin* (Dryand.) Pierre; *Styrax benjuifer* Stokes.
Classification: Plantae > Streptophyta > Equisetopsida > Magnoliidae > Ericales > Styracaceae > Styrax > benzoin.
Common names: Gum benjamin tree, Loban, Kemenyan, Onycha, Sumatra benzoin tree.
Distribution: *Africa*: Guinea, Mauritius. *Americas*: Trinidad and Tobago. *Asia*: Bangladesh, Borneo, Cambodia, Java, Laos, Lesser Sunda Islands, Malaysia, Myanmar, Philippines, Sumatra, Thailand, Vietnam.
Morphology: tree up to 34 m tall, with small buttresses. The plant excretes an aromatic and resinous substance (benzoin) when bruised. Foliage spirally arranged, stalked leaves have leaf blades that are egg-shaped to oblong or lance-shaped and 6–20 × 2–9 cm. They also have white, woolly, star-like hair covering the undersides. Flower shoot is 13–20 cm long and develops at the end of leafy twigs or the leaf axils. Each flowering shoot bears stalked, fragrant, white flowers.
Use: Benzoe Sumatra resin, from *Styrax benzoin*, is highly valued in perfumery due to its fixative properties and its balsamic-sweet, powdery, vanilla-like fragrance. The wild-growing trees are mainly found on the Indonesian island of Sumatra.

Styrax tonkinensis

Botanical name: *Styrax tonkinensis* (Pierre) Craib ex Hartwich.
Synonyms: *Anthostyrax tonkinensis* Pierre; *Styrax hypoglaucus* Perkins; *Styrax macrothyrsus* Perkins; *Styrax subniveus* Merr. & Chun.

Classification: Plantae > Streptophyta > Equisetopsida > Magnoliidae > Ericales > Styracaceae > Styrax > tonkinensis.
Common name: Benzoe Siam resinoid.
Distribution: Cambodia, China, Laos, Thailand, Vietnam.
Morphology: tree 6–30 m tall. Trunk 8–60 cm diameter at breast height, bark dark grey, irregularly divided. Branchlets stellate tomentose, glabrescent. Leaves alternate; petiole 0.5–1.5 cm; leaf blade elliptic to ovate, 5–18 × 4–10 cm, papery to leathery, abaxially densely grey to glaucous stellate pubescent, adaxially glabrous except for veins, base rounded to cuneate, margin entire or apically serrate, apex shortly acuminate, secondary veins 5 or 6 pairs, tertiary veins subparallel. Racemes or panicles 3–10 cm. Pedicel 5–10 mm. Flowers 1.2–2.5 cm. Calyx cup-shaped, 3–5 mm, truncate to 5-toothed. Corolla tube 3–4 mm; lobes ovate-lanceolate, 10–16 × 3–4 mm, membranous. Filaments sparsely white stellate pubescent; anthers narrowly oblong. Style *c.*1.5 cm, glabrous. Fruit subglobose, 1–1.2 cm in diameter, densely grey stellate tomentose, apex apiculate to obtuse. Seeds chestnut brown, ovoid, densely tuberculate, stellate tomentose and pubescent (Fig. 2.45).
Use: The resin of this tree species is obtained by solvent extraction. The chocolatey, sweet and balsamic scent of the resin is used as a fixative in balsamic perfumes. It is also used as incense and in various traditional medical systems.

Tara spinosa

Botanical name: *Tara spinosa* (Molina) Britton & Rose.
Synonyms: *Caesalpinia spinosa* (Molina) Kuntze; *Caesalpinia horrida* (Kunth) Spreng.; *Caesalpinia mucronata* Willd.; *Caesalpinia pectinata* Cav.; *Caesalpinia tara* Ruiz & Pav.; *Caesalpinia tinctoria*

Fig. 2.45. Resin obtained from a Benzoe Siam tree (*Styrax tonkinensis*). (By Maša Sinreih in Valentina Vivod, CC BY-SA 3.0, https://creativecommons.org/licenses/by-sa/3.0, via Wikimedia Commons.)

Benth. ex Taub.; *Caesalpinia tinctoria* (Molina) Dombey ex DC.; *Caesalpinia tinctoria* var. *horrida* Pittier; *Coulteria chilensis* DC.; *Coulteria horrida* Kunth; *Coulteria tinctoria* (Molina) Kunth; *Poinciana spinosa* Molina; *Poinciana mucronata* (Willd.) Poir.; *Poinciana tara* Ruiz & Pav. ex DC.; *Tara tinctoria* Molina; *Tara tinctoria* var. *horrida* (Kunth) Pittier.

Classification: Plantae > Streptophyta > Equisetopsida > Magnoliidae > Fabales > Fabaceae (Leguminosae) > Mimosoideae > Tara> spinosa.

Common name: Tara gum.

Distribution: *Africa*: Ethiopia, Kenya, Namibia, Uganda, Zimbabwe. *Americas*: Bolivia, California, Chile, Colombia, Cuba, Ecuador, Peru, Venezuela. *Asia*: Java. *Europe*: Spain.

Morphology: tree up to 5 m high, branches armed with short prickles up to 5 mm long. Leaves: petiole and rhachis together (1)4–10 cm long, typically armed with short prickles at the insertions of the pinnae; pinnae 1–4 pairs; rhachillae 4–10 cm long, unarmed or with short prickles at the insertions of the leaflets; leaflets 4–7 pairs per pinna, 1.5–4.5 cm long, 0.6–2 cm wide (in our area), oblong-elliptic, obtuse or emarginate apically, glabrous or subglabrous. Racemes 10–20 cm long, many-flowered, usually sparsely prickly. Flowers pedicellate, pedicels 5–12 mm long, articulated near the apex. Sepals 4–6 mm long, margins ± erose or lacerate, the lower sepal much larger and fimbriate-pectinate. Petals up to 1 cm long. Stamens not or scarcely exserted, yellow; filaments up to 8 mm long, pubescent below, glabrous above. Pods oblong, thick, 5–9.5 cm long, 1.5–2.5 cm wide, pinkish-brown or crimson, indehiscent. Seeds brown, 8–10 × 6–7 mm, somewhat compressed (Fig. 2.46).

Use: The gum, extracted from the seeds of the tree, is used in the food industry as a thickening and stabilizing agent. It is commonly used in dairy and meat products.

Terminalia bellirica (Baheda)

Botanical name: *Terminalia bellirica* (Gaertn.) Roxb.

Synonyms: *Buceras bellirica* (Gaertn.) Lyons; *Myrobalanus bellirica* Gaertn.; *Myrobalanus laurinoides* Kuntze; *Terminalia attenuata* Edgew.; *Terminalia belirica* (Gaertn.) Wall.; *Terminalia biticaria* Roxb.; *Terminalia chebula* Willd. ex Flem.; *Terminalia eglandulosa* Roxb. ex C.B. Clarke; *Terminalia gella* Dalz.; *Terminalia laurinoides* Teijsm. & Binn.; *Terminalia moluccana* Roxb.; *Terminalia punctata* Roth.

Classification: Plantae > Tracheophyta > Magnoliopsida > Myrtales > Combretaceae > Terminalia > bellirica.

Common names: Bedda nut tree (English); Bahera, Baheda (Hindi).

Distribution: native to Bangladesh, Bhutan, Cambodia, China, Indonesia, Laos, Malaysia, Nepal, Pakistan, Sri Lanka, Thailand and Vietnam. In India the tree is found in the provinces Karnataka, Maharashtra, Kerala, Odisha and the north-east.

Morphology: tall tree that reaches heights between 12 and 50 m, distinguished by its unique bark. The leaves are arranged alternately or in clusters at the branch tips, with an elliptic or elliptic-obovate shape. They have a leathery texture, are dotted and feature entire margins. The leaf tip is either narrow and pointed or rounded. The leaves are typically 8–20 cm long and 7.5–15 cm wide, with petioles approximately 2.15 cm long. Flowers appear in spikes originating from the leaf axils, ranging from 5 to 15 cm long. These flowers are greenish-yellow, 5–6 mm wide and lack stalks. The upper flowers are male, while the lower ones are bisexual, with 3–4 mm stamens. The fruit is obovoid, measuring 1.5–2.5 cm in diameter, covered in tiny pale hairs and has a thick stone with an indistinct five-angled shape (Fig. 2.47).

Use: The gum is utilized as an ingredient in adhesive formulations based on water and natural rubber.

Terminalia tomentosa (Asan)

Botanical name: *Terminalia tomentosa* Willd.

Synonyms: *Terminalia alat; Terminalia elliptica; Terminalia macrocarpa; Terminalia ovate; Terminalia crenulata.*

Classification: Plantae > Tracheophyta > Magnoliopsida > Myrtales > Combretaceae > Terminalia > tomentosa.

Fig. 2.46. Inflorescences of a *Tara spinosa* tree. (By Consultaplantas, CC BY-SA 4.0, https://commons.wikimedia.org/w/index.php?curid=44961620.)

Common names: Asan, Saja, Crocodile bark tree, Silver grey wood, Indian laurel, Black murdah (English); Asana, Saja (Hindi); Raktarjun (Sanskrit).

Distribution: widely distributed in India, Myanmar, Bangladesh, Bhutan, Cambodia, China, Indonesia, Laos, Malaysia, Nepal, Pakistan, Sri Lanka, Thailand and Vietnam.

Morphology: deciduous tree with a grey-black, rough trunk that is deeply fissured vertically and cracked horizontally, resembling the bark of a crocodile. The blaze of

Fig. 2.47. Inflorescence of a bedda nut tree (*Terminalia bellirica*). (By Nandkishore Thombare.)

the tree is red. Its leaves are simple, arranged opposite or subopposite, without stipules, smooth and have entire or slightly crenulated edges. A distinct feature of this tree is the presence of two stalked glands near the base of the leaf's midrib on the underside. The flowers are bisexual, dull yellow and grow in terminal or axillary panicle-like clusters. The fruit is ovoid, about 3 cm long, with five wings that do not extend past the fruit's apex. The tree's bark is fire-resistant, and in winter it appears bare, making it identifiable mainly by its scaly and cracked bark (Fig. 2.48).

Use: The resin that drips from branches in large, spherical drops, especially on young branches, is edible and transparent. It resembles dhawra resin and is often sold under the name 'dhawra' or mixed with it in trade. The resin paste is used externally to treat burns and reduce swelling caused by inflammation.

Fig. 2.48. Flowering asan tree (*Terminalia tomentosa*). (By Nandkishore Thombare.)

Vateria indica (White damar)

Botanical name: *Vateria indica* L.
Synonym: *Vateria malabarica* Bl.
Classification: Plantae > Tracheophyta > Magnoliopsida > Malvales > Dipterocarpaceae > Vateria > indica.
Common names: White damar, Indian copal, Malabar tallow, Piney varnish (English); Badasal, Kahruba, Dammar, Safed dammar (Hindi); Svetasarja, Ajakarnah, Chagakarna, Salah, Mandadhupa, Sarja, Kundura (Sanskrit).
Distribution: endemic to the Western Ghats in Maharashtra, Karnataka, Kerala and Tamil Nadu (India).
Morphology: large, evergreen tree known for its resinous nature. The leaves are simple, alternate and vary from elliptic to oblong, heart-shaped or rounded. They have acuminate tips, entire margins and a leathery texture, arranged alternately in a spiral pattern. The stipules are prominent and caducous, and the petioles are swollen at the apex. The young branchlets are round, covered in hairs and have a coppery colour. The bark is smooth, pale grey with green blotches and has a bitter, acrid taste. It peels off in round flakes and exudes a whitish resin, either naturally or when injured. The blaze of the bark is dull brown, and the wood is white and hard. The flowers appear in panicles in the leaf axils, with dense stellate hairs. These flowers are bisexual, white and slightly fragrant. The panicles are robust, multi-branched and drooping. The fruit is a capsule measuring 4–6 cm × 2–4 cm, pale brown and fleshy, which becomes hard when dry. The fruit splits open into three valves when it ripens (Fig. 2.49).
Use: The resin known as white dammar is obtained by scraping the bark of the tree. It is used in India mainly as incense in traditional religious services and produces a fragrant white smoke when burned. In medicine, the resin is used to treat chronic bronchitis, throat problems, coughs, asthma, leprosy, skin rashes, infections, wounds and ulcers.

Fig. 2.49. Infructescence on a white dammar tree (*Vateria indica*). (By Nishad Sankar.)

References

Aljesri, Z., Hakkı Alma, M., Göçeri, A., Karaoğul, E. and Salan, T. (2014) Application area of sweet gum (*liquidambar orientalis* l.). In: *3rd International Non-wood Forest Products Symposium*, Kahramanmaraş, Turkey, May 8–10.

Barry, H.T. (1932) *Natural Varnish Resins*. Ernest Bonn, London.

Bhat, G.K. (2014) *Flora of South Kanara*. Akriti Prints, Mangalore, India.

Brandis, D. (1990) *Indian Trees: An Account of Trees, Shrubs, Woody Climbers, Bamboos and Palms Indigenous or Commonly Cultivated in the British Indian Empire*. International Book Distributors, Dehradun, India.

Browicz, K., Fröhner, S., Gilli, A., Nordborg, G., Riedl, H. *et al.* (1969) *Flora Iranica: Rosaceae I*, Vol. 66. Naturhistorisches Museum, Vienna.

Chevallier, A. (1996) *Encyclopedia of Medicinal Plants*. DK Publishers, Boston, Massachusetts.

Coppen, J.J. (1995) *Gums, Resins and Latexes of Plant Origin. Non-wood Forest Products 6*. Food and Agriculture Organization of the United Nations, Rome.

Dabestani, M., Kadkhodaee, R., Phillips, G.O. and Abbasi, B. (2018) Persian gum: a comprehensive review on its physicochemical and functional properties. *Food Hydrocolloids* 78, 92–99. DOI: 10.1016/j.foodhyd.2017.06.006.

Davis, P.H. (1972) *Flora of Turkey and the East Aegean Islands*, Vol. 4. Edinburgh University Press, Edinburgh, UK, pp. 263–264.

Fakhri, M., Davoodi, A., Hamzegardeshi, Z., Farhadi, R., Mousavinasab, N. *et al.* (2018) Is cotoneaster manna improving the treatment of neonatal jaundice? *Bangladesh Journal of Pharmacology* 13, 168–178.

Farjon, A. (2010) *A Handbook of the World's Conifers*. Brill Academic Publishers, Leiden, Netherlands.

Flora of China (n.d.) efloras.org. Available at: http://www.efloras.org/flora_page.aspx?flora_id=2 (accessed 1 December 2024).

Gachathi, F.N. (1997) Recent advances on classification and status of the main gum-resin producing species in the family burseraceae. In: Mugah, J.O., Chikamai, B.N., Mbiru, S.S. and Casadei, E. (eds) *Proceedings of a Regional Conference for Africa Held in Nairobi*, Kenya, October 6–10, 1997, pp. 18–22. Available at: https://openknowledge.fao.org/server/api/core/bitstreams/a0d8add4-fbd9-4dd0-8aa7-c00eeba1f600/content (accessed 24 April 2025).

Gebrehiwot, M., Asres, K., Bisrat, D., Mazumder, A., Lindermann, P. *et al.* (2015) Evaluation of the wound healing property of *Commiphora guidottii* Chiov.ex.Guid. *BMC Complementary and Alternative Medicine* 15(282), 2–11. DOI: 10.1186/s12906-015-0813-2.

Ghadami, S., Saeedi, M., Delnavazi, M.R., Eftekhari, M., Edraki, N. *et al.* (2024) Oleo-gum-resin of *Ferula persica*: Phytochemical analysis and enzyme inhibitory activity related to Alzheimer's disease. *Research Journal of Pharmacognosy* 11(4), 39–48. DOI: 10.22127/RJP.2024.454504.2438.

Kirtikar, K.R. (1918) *Indian Medicinal Plants*, Vol. 1–5. Bishen Singh Mahendra Pal Singh, Dehradun, India.

Langenheim, J.H. (2003) *Plant Resins: Chemistry, Evolution, Ecology, and Ethnobotany*. Timber Press, Portland, Oregon.

Lohot, V.D., Thamilarasi, K., Ghosh, J., Mohanasundaram, A. and Sharma, K.K. (2016) *Monograph on Palas* (Butea monosperma *Lam. Taubert*). ICAR-IINRG, Ranchi, India.

Lohot, V.D., Mohanasundaram, A., Thamilarasi, K., Ghosh, J., Thombare, N. *et al.* (2023) *Lac Host Plants of India: A Field Guide*. ICAR-NISA, Ranchi, India.

Mozaffarian, V. (2013) *Identification of Medicinal and Aromatic Plants of Iran*. Farhang Moaser, Tehran.

Murthy, H.N. (ed.) (2022) *Gums, Resins and Latexes of Plant Origin: Chemistry, Biological Activities and Uses*. Springer, Cham, Switzerland.

Murthy, H.N., Paek, K.Y. and Park, S.-Y. (eds) (2024) *Bioactive Compounds in the Storage Organs of Plants*. Springer, Cham, Switzerland.

Nair, K.K.N. (2000) *Manual of Non-wood Forest Produce Plants of Kerala* (No. 185). Kerala Forest Research Institute, Kerala, India.

Nematzadeh, H., Ekraminasab, S., Bahrami, R. and Asadian, F. (2023) The mechanism of action of Cotoneaster manna, an Iranian herbal medicine, in the treatment of neonatal jaundice. *World Journal of Peri & Neonatology* 6(1), 49–53. DOI: 10.18502/wjpn.v6i1.14252.

Pandey, B.P. (2001) *Taxonomy of Angiosperms. S*. Chand Publishing, New Delhi.

POWO (2024) Plants of the World Online. Facilitated by the Royal Botanic Gardens, Kew, UK. Available at: https://powo.science.kew.org (accessed 29 July 2024).

Rechinger, K.H. (1969) *Flora Iranica 63*. Akademische Druck- u.Verlagsanstalt, Graz, Austria.

Rechinger, K.H. (1987) *Flora Iranica 62*. Akademische Druck- u. Verlagsanstalt, Graz, Austria.

Rechinger, K.H. and Cullen, J. (1966) *Flora Iranica 34*. Akademische Druck- u. Verlagsanstalt, Graz, Austria.

Roxburgh, W. (1832) *Description of Indian Plants*, Vol. 2. Thacker and Co., Calcutta, India; Parbury, Allen and Co., London, pp. 611–615.

Rustiami, H.S.F.M., Setyowati, F.M. and Kartawinata, K. (2004) Taxonomy and uses of *Daemonorops draco* (Willd.) Blume. *Journal of Tropical Ethnobiology* 1(2), 65–75.

Schulz, H., Emami, S.A. and Nadjadi, F. (2023) *Medicinal Plants Used in Traditional Persian Medicine*. CAB International, Wallingford, UK.

Singh, G., Kumar, D. and Dash, A.K. (2021) *Pinus gerardiana* Wallichex. D. Don. - a review. *Phytomedicine Plus* 1, 100024. DOI: 10.1016/j.phyplu.2021.100024.

Srivastava, B., Sharma, H., Dey, Y.N., Wanjari, M.M. and Jadhav, A.D. (2014) *Alhagi pseudalhagi*: A review of its phyto-chemistry, pharmacology, folklore claims and Ayurvedic studies. *International Journal of Herbal Medicine* 2, 47–51.

Surburg, H. and Panten, J. (2006) *Common Fragrance and Flavor Materials*. Wiley-VCH, Weinheim, Germany.

Thombare, N., Lohot, V.D., Prasad, N. and Sharma, K.K. (2018) *Major Gum and Resin Plants of India: A Field Guide*. ICAR-IINRG, Ranchi, India.

Thombare, N., Lohot, V.D., Prasad, N., Gopal, K.S., Shiran, K. *et al.* (2023) *Gum and Resin Plants of India: A Field Guide*. ICAR- NISA, Ranchi, India.

Tschirch, A. (1906) *Die Harze und die Harzebehälter*. Bornträger, Leipzig, Germany.

Tschirch, A. and Stock, E. (1933–1936) *Die Harze: Die botanischen und chemischen Grund-lagen unserer Kenntnisse über die Bildung, die Entwicklung und die Zusammensetzung der pflanzlichen Exkrete*, 3rd edn, 2 vols. Gebrüder Bornträger, Berlin, Germany.

Vázquez, J.J.B., Abad-Fitz, I., Beltrán-Rodríguez, L., Casas, A., Maldonado-Almanza, B. *et al.* (2023) *Burseracopallifera* (Sessé & Moc. Ex DC.) Bullock, *Bursera fagaroides* (Kunth) *Burseraglabrifolia* (Kunth) *Bursera linanoe* (La Llave) Rzed., Calderón, and Medina *Bursera morelensis* Ramírez *Bursera simaruba* (L.) Sarg. BURSERACEAE. In: Casas, A. and Blancas Vázquez, J.J. (eds) *Ethnobotany of the Mountain Regions of Mexico*. Springer, Cham,Switzerland.

WFO (2024) World Flora Online. Available at: http://www.worldfloraonline.org (accessed 2 December 2024).

Whistler, R.L. (1993) Exudate gums. In: Whistler, R.L. and Berniller, J. (eds) *Industrial Gums*. Academic Press, New York, pp. 309–339.

Zheng-Yi, W. and Raven, P.H. (1999) *Flora of China*, Vol. 4. Science Press, Beijing; Missouri Botanical Garden, St. Louis, Missouri.

Zibaee, E., Amiri, M.S., Boghrati, Z., Farhadi, F., Ramezani, M. *et al.* (2020) Ethnomedicinal uses, phytochemistry and pharmacology of *Dorema* species (Apiaceae): A review. *Journal of Pharmacopuncture* 23(3), 91–123. DOI: 10.3831/KPI.2020.23.3.91.

3 Valuable Chemical Constituents and Properties

Niranjan Prasad[1]* and Hartwig Schulz[2]
[1]ICAR-National Institute of Secondary Agriculture, Ranchi, India;
[2]Consulting and Project Management for Medicinal and Aromatic Plants, Waltraudstraße 4, Stahnsdorf, Germany

Abstract

Natural resins and gums (NRG) are versatile biomaterials with significant industrial applications due to their unique chemical compositions and functional properties. This chapter delves into the detailed chemistry and properties of key NRG materials, including resins like rosin and shellac, exudate gums such as karaya, gum arabic, gum ghatti, frankincense, myrrh and seed-derived gums like guar, tamarind and *Cassia* (*Senna*) *tora*. These resins and gums are described for their structural components, their physico-chemical characteristics and their solubility behaviour. The text particularly highlights the diverse applications of the listed substances in the pharmaceutical, cosmetic and food industries. The study of their rheological, thermal and optical properties reveals their excellent adaptability to varied environmental and industrial conditions. Moreover, the influence of geographical and botanical factors on their quality and functionality is discussed. This chapter provides a comprehensive basis for a better understanding of the diverse roles of NRGs in modern technology and industrial innovation.

Introduction

Much has been reported in recent decades about the chemical composition and individual properties of resins and gums (Langenheim, 2003; Murthy, 2022). Therefore, this chapter only provides a current overview of today's economically most important NRGs, with references to the latest literature from recent years.

The main natural polymers found in NRGs are complex mixtures of organic compounds, each with unique properties that make gums and resins highly adaptable for various applications. Some important reasons are given below that particularly highlight the importance of their chemistry. The main chemical substances of NRG, such as polysaccharides, terpenoids and esters, largely define their physical and chemical properties. For example, the amphiphilic nature of shellac and the high molecular weight of gums such as guar and karaya enable their use as emulsifiers and stabilizers. The diverse properties that result from their chemical composition make NRG indispensable in the food, pharmaceutical and cosmetic industries. A deeper understanding of their chemical composition also enables customized modifications, such as improving solubility or other customer-oriented requirements.

*Corresponding author: niranjan_ilri_1999@yahoo.com

 Natural Gums and Resins: Botany and Sustainable Uses in Medicine, Nutrition, Perfumery and Cosmetics (eds Hartwig Schulz *et al.*)
DOI: 10.1079/9781800626690.0003

These properties, derived from their chemical composition and structural characteristics, essentially determine their functionality and adaptability in specific applications. Below a detailed overview is presented showing the individual areas of application.

Gums like guar and karaya exhibit exceptional viscosity, even at low concentrations, making them ideal for use as thickeners in food products, stabilizers in pharmaceutical formulations and binders in cosmetics. The shear-thinning properties of these gums facilitate their application in products requiring controlled flow, such as adhesives. Exudate gums, such as gum arabic and gum ghatti, are excellent emulsifiers due to their amphiphilic nature, which stabilizes oil–water emulsions. These properties are crucial for encapsulating flavours in food, stabilizing beverages and manufacturing emulsified pharmaceutical products. Resins like shellac are valued for their heat resistance and stability under UV light, making them suitable for coatings, varnishes and protective finishes. These properties also contribute to their use in electronics and preservation of artwork. Rosin and shellac are used in particular as coatings for tablet formulations and as protective coatings in the food industry due to their film-forming properties. Due to their natural origin and biodegradability, natural gums are considered safe for use in food, pharmaceuticals and personal care products. For example, guar gum is a common dietary fibre, while gum karaya is used as a bulk-forming laxative. The water-solubility and hydrophilic nature of gums like tamarind seed gum enhance their use as thickeners, humectants and stabilizers in foods and cosmetics.

Gum karaya's ability to swell and absorb water is particularly useful in applications requiring high water retention, such as dental adhesives and wound dressings. Properties like molecular weight, solubility and cross-linking potential allow for chemical modifications, enhancing their performance in specific roles. For instance, modified gums are used in drug delivery systems, while resins with tailored thermal properties find applications in specialized coatings.

Resins

Rosin

Chemistry

Rosin (colophony) is composed of about 50–70% resin acids. The oleoresins that exude from the trees are viscous solutions of mono- and diterpenoids. Evaporation of the volatile monoterpene fractions of the resins results in rosin (van den Berg *et al.*, 2000; van den Werf *et al.*, 2000). Pine rosin contains mostly diterpene acids and various monoterpenes. Rosin is used in a wide range of products extending from the chemical to the medical and cosmetic industries (van den Werf *et al.*, 2000). Gören *et al.* (2010) analysed the main monoterpene substances of rosin obtained using the acid paste method from the Ayvacık (Çanakkale) region and found α-pinene (20.5%), limonene (10.5%), linalyl acetate (7.6%), β-pinene (5.3%) and delta-3-carene (4.7%) in it. The main diterpenoids were palustric acid (12.9%), abietic acid (8.7%), kaur-9(11),16-dien-18-oic acid (4.7%) and neoabietic acid (1.3%).

Properties

Rosin is a brittle solid; its colour varies from light yellow to red, brown, blue or black. Its specific weight is 1.08 and its melting point is between 100 and 140°C.

Shellac

Chemistry

Shellac is a long-chain polyester type of resin consisting of inter- and intra-esters of polyhydroxycarboxylic acids, some of which are aliphatic long-chain hydroxy acids and some of which are sesquiterpene acids. The study to determine the typical structure of this complex biopolymer was carried out in the 1960s at the National Chemical Laboratory, Pune, India, and published in a series of papers in seven different parts (Wadia *et al.*, 1969; Singh *et al.*, 1974). Further confirmation of this structure

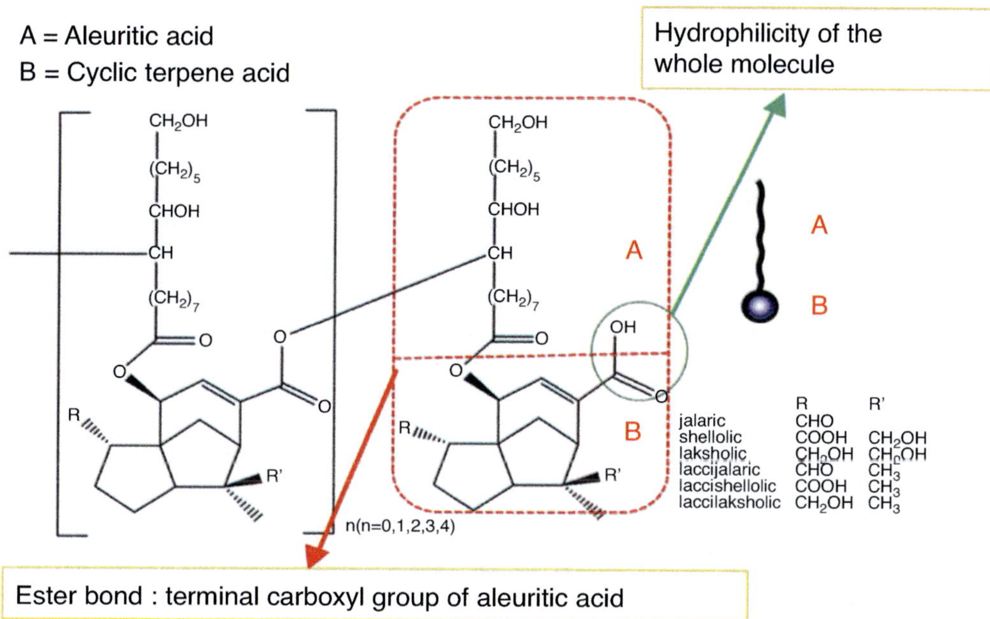

Fig. 3.1. Chemical structure of the typical building block of shellac. Author's own image.

and the molecular characterization of the individual acids was obtained using pyrolysis-GC (gas chromatography) (Sharma, 2017) and, more recently, using flow injection-LC (liquid chromatography), electrospray ionization and mass spectrometry (MS) (Tamburini *et al.*, 2017).

Shellac is a composite macromolecule with a combination of hard and soft resin as the essential components, along with wax and other odoriferous principles. Although shellac consists of several polyester types that are naturally linked together by complex bonds (Gardner and Whitmore, 1929; Sharma *et al.*, 1983), a typical isolated fraction with a homogeneous nature, known as pure lac resin, has been studied and is also referred to as the backbone of the shellac molecule (Khurana *et al.*, 1970). Of the aliphatic acid components of shellac, aleuritic acid (approx. 35%) is the most abundant, while jalaric acid, with a share of approximately 25%, is the most important component of the terpenic acid fraction.

Other isolated acids are butolic (approx. 8%), shellolic-epishellolic and laccijalaric (approx. 8%) acids (Sharma *et al.*, 1983). A typical shellac unit is shown in Fig. 3.1; the functional groups are chemically bound to each other by means of ester, acylal, acetal and ether bonds (Sharma *et al.*, 1983). Shellac is inherently amphiphilic, with the hydrophilic character largely imparted by the sesquiterpene acid moiety, while the long-chain aliphatic hydroxy acids account for the hydrophobic character (Fig. 3.1). This property makes shellac suitable for a wide range of applications in which the orientation of the molecules is of particular importance.

Properties

Shellac is a hard, brittle resinous solid. Its colour can range from light yellow to dark red. It is odourless in cold conditions, but it smells when heated.

Solubility. Shellac consists of dyes (0.5–1.0%), various fragrances, wax (4.5–5.5%), resin (90–94%), moisture (1–2%) and insect remains and other foreign matter (0.5–4%). The partial solubility of shellac in diethylether was

Table 3.1. Solubility of shellac in various organic solvents. Author's own table.

Solubility of shellac	Solvents
highly soluble	acetaldehyde, acetic acid, amyl alcohol, aniline, isobutyl alcohol, n-butyl alcohol, s-butyl alcohol, ter-butyl alcohol, isobutyl carbinol, sec-butyl carbinol, diethylcarbinol, diacetone alcohol, diethylene glycol, diethylene glycol butyl ether, diethylene glycol ethyl ether, diethylene glycol methyl ether, ethyl alcohol (anhydrous), ethyl alcohol (95%), ethyl lactate, ethylene glycol butyl ether, ethylene glycol ethyl ether, ethylene glycol methyl ether, formic acid (85%), fusel oil (refined), methyl alcohol, methyl propyl carbinol, propionic acid, isopropyl alcohol, propylene glycol, propylene glycol +ethyl ether (mixture), propylene glycol +methyl ether (mixture), mesityl oxide, monomethyl aniline
fairly soluble	acetone, acetophenone, butyric acid, cyclohexanes, dimethyl ethyl carbinol, ethyl acetate (85%), ethyl methyl ketone, ethyl oxybutyrate, ethylene glycol ethyl ether, lactic acid, propylene glycol
slightly soluble	n-butyl acetate (85%), citronellol diethyl ether, ethyl acetoacetone, geraniol
insoluble	amyl acetate, amyl chloride, benzene, bromobenzene, butyraldehyde, n-butyl acetate, isobutyl propionate, carbon disulfide, carbon tetrachloride, chloroform, dibutyl phthalate, diethyl aniline, diethyl carbonate, diethyl phthalate, dimethyl aniline, ethyl acetate, ethyl oxalate, ethyl sulfate, ethylene dichloride, ethylene glycol, ethylene glycol ethyl ether acetate, glycerol, palmitic acid, petroleum ether, toluene, xylene, nitrobenzene
swells	chloroform, ethylene dichloride, nitrobenzene, toluene, xylene

examined in more detail (Rasmussen, 2017) in order to separate the ether-insoluble from the ether-soluble part of the resin. Shellac is also soluble in acetic acid, ethyl and methyl alcohol, sodium hydroxide solution, sodium carbonate, borax solution and partially in ethyl acetate, chloroform, benzene and toluene. However, alcohols, organic acids and ketones have been found to be the best solvents for shellac. Table 3.1 lists various solubility classes of shellac in different solvents.

Solubility of shellac in alkaline solutions. Shellac is insoluble in water, but it dissolves in alkaline solutions when cooled due to its acidic nature. On heating, it forms solutions of alkaline salts such as sodium carbonate, borax and organic bases (triethanolamine, morpholine). Proper salt formation occurs with sodium carbonate and bicarbonate because free CO_2 has been observed to correspond to the acid value (Kamath and Mainkar, 1950). In diluted aqueous solutions, shellac behaves like a colloid. The hydrosol of shellac can be easily prepared by continuously adding alcoholic solution to water and then evaporating the alcohol in a vacuum. Aqueous alkaline shellac solutions generally have a higher viscosity than alcoholic solutions of the same concentration (Bose, 1963).

Softening and melting points. Shellac does not have a distinct softening or melting point like other resins. However, according to a study by the Indian Lac Research Institute, a temperature range of 65–70°C was determined for the softening point of shellac, and a range of 75–80°C was determined for the melting point (Sharma *et al.*, 1983).

Specific gravity and refractive index. Specific gravity values of shellac at a temperature of 15°C were found in the range of 1.035–1.114 (Banerjee *et al.*, 1982). Bhattacharya measured the refractive index with an Abbe refractometer at different temperatures (Bhattacharya, 1940). According to his studies, the refractive index of shellac is 1.5272 at 100°C. He also stated that the exact temperature for the refractive index determination of shellac is 70°C because this temperature is close to the melting point of shellac.

Optical properties. If the colour of the shellac ranges from a translucent pale orange to an opaque dark red, a higher price can be obtained when selling it. Shellac fluoresces orange-red in ultraviolet light and is resistant to yellowing under the influence of ultraviolet light (Kunkel and Seo, 1994).

Acid value and saponification value. Shellac undergoes continuous esterification-induced polymerization with ageing, resulting in the consumption of its acidic functional groups. These changes can be studied by estimating their acid value, which is an excellent indicator of the quality or condition of lac samples. Lac with a higher range of acid value often fetches better prices in the trade as with time, and during storage, the acid value keeps decreasing continuously (Weinberger and Gardner, 1933; Farag and Leopold, 2009). The acid value of shellac ranges from 60 to 65, as reported by most investigators (Bose, 1963). Still, some reporters have observed acid values in the range of 65–75 (Sharma *et al.*, 1983).

Basicity and iodine value. Pure resin shellac has a basicity of 2, and soft resin shellac has a basicity of 1. According to Wij's method, the iodine value of shellac ranges from 14 to 18 (Sarkar and Shrivastava, 1997).

Molecular weight. Several values are reported for the molecular weight of the two different components of shellac, such as 2210 (Singh *et al.*, 1974), 2000 (Azouka *et al.*, 1993) for the hard resin (ether insoluble portion) and 550 (Azouka *et al.*, 1993) and 500 (Bose, 1963) for the soft resin (ether soluble portion); whereas, for practical purposes, the average molecular weight of shellac is considered to be 1000–1006 (Bose, 1963; Sharma *et al.*, 1983). Recently, Bar and Bianco-Peled used flow injection analysis-electrospray ionization-quadrupole-time of flight (FIA-ESI-Q-ToF) technique by injecting the complete shellac sample to achieve greater accuracy in the characterization of shellac (Bar and Bianco-Peled, 2021). They identified a large number of combinations of compounds found in shellac and determined an average molecular weight of at least 5800.

Another study, published in Springer Nature's *Scientific Reports*, elaborately highlighted the molecular fragmentations of shellac by high-performance liquid chromatography-ESI-Q-ToF (HPLC-ESI-Q-ToF) (Tamburini *et al.*, 2017). They reported the shellac as a combination of free acids, esters, and di-, tri-, tetra-, penta-, hexa- and heptaesters made up of the building blocks of hydroxyl-aliphatic and sesquiterpenic acids. The combined molecular weight of these building blocks is >62,000. Moreover, self-esterification or polymerization occurs as the shellac ages as a result of cross-linking of its components (Coelho *et al.*, 2012). The molecular weight of shellac increases many times over during such cross-linking.

Influence of host plants on the quality of lac resin. As two different strains of lac insect, *Rangeeni* and *Kusmi*, feed on the selective host plants (*Rangeeni* specific to *Butea monosperma* and *Kusmi* on *Schleichera oleosa*), the properties of resin produced by them differ significantly. Overall, the resin produced by the *Kerria* sp. *kusmi* strain, especially on the host *S. oleosa*, is superior to its *Rangeeni* strain. Properties like flow, heat polymerization time, colour index etc., were found to be better for *Kusmi* lac compared to Rangeeni. According to Sharma *et al.* (2016), the resin-producing potential of the lac insect depends on the type of lac host it is feeding on. The average resin production of the *Kusmi* strain is higher than that of the *Rangeeni* strain. Additionally, *S. oleosa* is considered the most suitable host in terms of quality and quantity of resin. In the *Rangeeni* strain, lac production in *B. monosperma* and *Ziziphus mauritiana* is highest during the summer and rainy seasons, respectively.

Gums

Gum karaya

Chemistry

Gum karaya polysaccharide is a long-chain and branched structure with a high molecular mass of about 9–16 × 10^6 Da (Le Cerf *et al.*, 1990; Chauhan *et al.*, 2019). It is a partially acetylated polysaccharide of the ramnogalacturonan type, which mainly occurs in the form of calcium and magnesium salts.

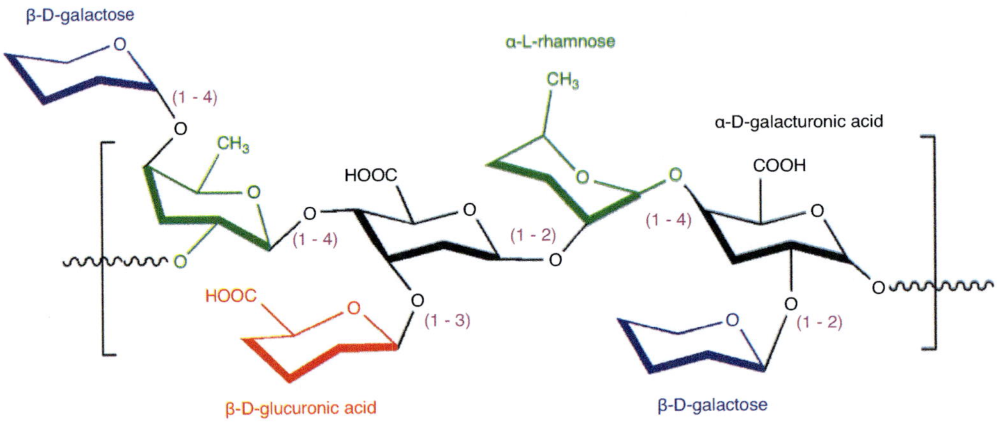

Fig. 3.2. Chemical structure of the typical gum karaya polysaccharides. Author's own image.

The backbone of the typical gum karaya molecule is made up of α-D-galacturonic acid and α-L-rhamnose units which are bonded to each other alternately through 1–2 and 1–4 linkages. This linear backbone has two types of side attachments. β-D glucuronic acid side chains are attached through 1,3-linkage to the galacturonic acid part and β-D-galactose side chains are linked to the rhamnose through 1,4-linkage, as well as to the galacturonic acid part through 1,2-linkage (Fig. 3.2). Almost half of the rhamnose units of the main chain are occupied by 1,4-linking with β-D-galactose side chains (Weiping and Branwell, 2000). This gum consists of 55–60% of neutral sugars, namely rhamnose and galactose, 37–40% acid sugars comprising glucuronic acid and galacturonic acids and the remaining portion of about 8% is occupied by the acetyl groups (Stephen, 1995; Panda, 2004; Raj *et al.*, 2021). Gum karaya obtained from different *Sterculia* species and produced in different places were found to possess similar chemical composition (Anderson *et al.*, 1983; Whistler, 1993).

Gum karaya contains 13–26% galactose and 15–30% rhamnose. The protein content of gum karaya has been determined to be about 1% and is thus lower than in other exudate gums. In addition, gum karaya contains uronic acid (40%) and acetyl groups (8%), which release free acetic acid when the resin ages. The acetyl groups make gum karaya insoluble in water but it has a very high swelling capacity (Imeson, 1992). Only 10% of native gum karaya is soluble in cold water, but this value increases to 30% in hot water.

The solubility of gum karaya becomes 90% when deacetylated with dilute ammonia. This shows that only molecules with a lower molecular weight are soluble in cold water, whereas molecules with a higher molecular weight can be made soluble by the deacetylation process (Le Cerf *et al.*, 1990).

Exposure to different growth conditions in different geographical regions may lead to variations in chemical composition of gum karaya (Kumbhare and Bhargava, 1999). Anderson and Bridgeman (1985) confirmed the presence of different amino acids like aspartic acid and glutamic acid as a major proportion and valine, leucine, proline and serine in minor proportion in gum karaya through GC–MS analysis.

Vinod *et al.* (2010) also carried out amino and fatty acid profile analysis of gum karaya by derivatization of amino acids and direct transesterification of methyl esters in the case of fatty acids. Gum karaya contained $64.2 \pm 2.44\,\mu g/g$ aspartic acid, and the ratio of saturated and unsaturated fatty acid was reported to be 6.6:1. The amino acids alanine, valine, methionine, tyrosine, tryptophan and arachidic acid could not be detected in the gum.

In the studies carried out by Marvelys *et al.* (2006) to identify the polysaccharides occurring in *Sterculia apetala* (chemical methods and nuclear magnetic resonance (NMR) spectroscopy), various uronic acids, galactose, arabinose and xylose were detected.

Properties

As already mentioned, gum karaya swells more than 60-fold due to the acetyl groups it contains (Le Cerf *et al.*, 1990), whereby the degree of acetylation is a decisive criterion for the solubility of the gum in water. In its aqueous solution, natural gum karaya forms a compact and branched conformation. Fully deacetylated gum, on the other hand, forms an extended conformation that looks like randomly distributed spirals.

Deacetylation of the gum using an alkali treatment makes it water-soluble. Coarse gum particles result in a granular dispersion, but fine gum powder hydrates quickly and results in a homogeneous dispersion.

Weiping and Branwell (2000) reported that viscosity of gum karaya dispersion depends upon its grade, and it generally ranged between 0.12 and 0.40 Pa s for 0.5% dispersion. Gum karaya dispersion exhibits non-Newtonian and pseudoplastic properties as the viscosity of the gum dispersion is inversely related to the shear rate (Walde *et al.*, 1997). During storage, a loss of viscosity was observed in the gum, which was more pronounced in the powder form with increasing aging and at higher temperatures (Imeson, 2010).

Gum dispersed in cold water had more viscosity compared to that dispersed in hot water (Dziezak, 1991). Heating increases the solubility of gum and allows the dispersion of up to 18% gum concentration, compared to 4–5% in the case of cold water. Strong electrolytes and acid/alkali addition in gum dispersion reduced the viscosity; however, such decrease was less for gum hydration before pH adjustment (Whistler, 1993). Compared to African gum karaya (pH 4.7–5.2), the corresponding gum karaya from India (pH 4.4–4.7) show a lower pH value in aqueous dispersion (Imeson, 1992), which is due to a higher uronic acid content. The Indian gum karaya is therefore more resistant to hydrolysis and can be used in an acidic environment. In general, the dispersion properties of gum karaya decrease with increasing degree of deacetylation.

Sahu *et al.* (2019) studied the physicochemical and rheological properties of gum karaya obtained from forests of Chhattisgarh (India) adopting a method followed for food materials (Sahu *et al.*, 2019). In the study, the gum was used in either grain or powder form. The results of the study are presented in Table 3.2. The viscosity of a 1% solution of gum karaya was 619 and 1586 cP, respectively, at a spindle speed of 100 and 20 rpm and an ambient temperature of 24–26°C.

Gum produced from *Sterculia* genus has unique emulsifying, stabilizing and thickening properties. *Sterculia opetala* gum absorbs water at its surface and produces dispersion of high viscosity and viscoelasticity. Pérez-Mosqueda *et al.* (2013) determined the rheological properties of this gum with respect to the individual gum concentration (0.5–3% w/v) at 25°C temperature. The flow properties of the aqueous dispersion of *S. opetala* gum change from Newtonian to shear thinning at a constant shear rate for the investigated range of gum concentration. The viscoelastic and dynamic surface tension properties studied in the range of gum concentration (0.1–3% w/v) demonstrate its special suitability as an emulsifier in oil–water emulsions

Silva *et al.* (2020) modified natural gum karaya with maleic anhydride. The developed derivative exhibited antibacterial properties and inhibited the growth of bacteria by almost 100%, without causing any significant cytotoxicity in mammalian cells. The derivatives can be used in various biomedical applications in controlling infectious diseases due to *Staphylococcus aureus*.

Gum arabic

Chemistry

The structure of gum arabic (GA) is a bit complicated compared to other natural gums. It comprises highly branched, high molecular weight and complex polysaccharides in the form of salts of magnesium, calcium and potassium (Yadav and Karthikeyan, 2019). The composition of GA can differ depending on species, age, habitat, climate, etc. from where it is collected (Tahir *et al.*, 2007; Dauqan and Abdullah,

Table 3.2. Physical and chemical properties of gum karaya. Author's own table.

Properties	Value
Moisture content	17.47 ± 0.44 (% wb)
Bulk density	0.79 ± 0.02 (g/cm^3)
True density	1.5 ± 0.017 (g/cm^3)
Tap density	0.90 ± 0.04 (g/cm^3)
Bulkiness	1.26 ± 0.03 (cm^3/gm)
Porosity	46.76 ± 6.12 (%)
Hausener's ratio	1.14 ± 0.06
Carr's compressibility index	12.31 ± 4.19
Angle of repose	47.27 ± 2.91°
Ash content	4.62 (%)
pH	4.26
Refractive index	1.336
Water activity	0.651
Water holding capacity (per 100 ml)	84.76
Nitrogen	0.16 (%)
Protein	1.06 (%)

2013). Although there are few studies on the genetic variability between different varieties of *Acacia senegal*, it is assumed that the chemical composition of the main chemical ingredients found in the gums varies only slightly (Ali *et al.*, 2009).

The structure of GA is revealed in the series of studies carried out from the 1960s to the recent past, and still there is scope for refinement. The wattle blossom model explains the structure of GA as a solid spherical assembly formed by attaching many arabinogalactan units to the protein chain (Connolly *et al.*, 1988; Goodrum *et al.*, 2000). These units were often classified into arabinogalactan, glycoprotein and arabinogalactan protein complex (Akiyama *et al.*, 1984; Idris *et al.*, 1998; Elmanan *et al.*, 2008). The possible structure for GA was given in different segments comprised of galactopyranose backbone with several side chains proposed based on the random interpretation of substituent sugars (Anderson and Stoddart, 1966; Anderson *et al.*, 1967; Chikamai *et al.*, 1996; Dror *et al.*, 2006). Later, the entire structure was interpreted based on modelling and recalculating the available literature (Street and Anderson, 1983). Nie and co-workers (2013) presented the more refined structure of GA (Fig. 3.3) using advanced techniques like analysis of methylated derivatives and 2D NMR spectroscopy.

This multifunctional polysaccharide comprises five central monomer units, including galactopyranose, rhamnopyranose, arabinofuranose, arabinopyranose and uronic acid (Islam *et al.*, 1997; Wang *et al.*, 2008). In the case of uronic acid, most studies describe the presence of glucuronic acid (Street and Anderson, 1983; Wang *et al.*, 2008). However, Nie and co-workers indicated that these uronic acids could be either glucuronic or galacturonic acids (Nie *et al.*, 2013). Along with these polysaccharides, the presence of proteins in small quantities is also reported in GA, which mainly include hydroxyproline, aspartic acid, proline and serine (Ali *et al.*, 2009). As shown in Fig. 3.3, the structure of GA consists of a backbone or main chain formed with 1,3-linked β-D-galactopyranosyl units with side chains containing different combinations of arabinofuranose, galactopyranose, rhamnopyranose and uronic acid (Nie *et al.*, 2013). Galacturonic acid is present in its β-D form, whereas arabinose and rhamnose exist in the α-L form. The molecular weight of GA determined by several studies lies in the range of 2.5×10^5 to 1.0×10^6 Da (Mahendran *et al.*, 2008).

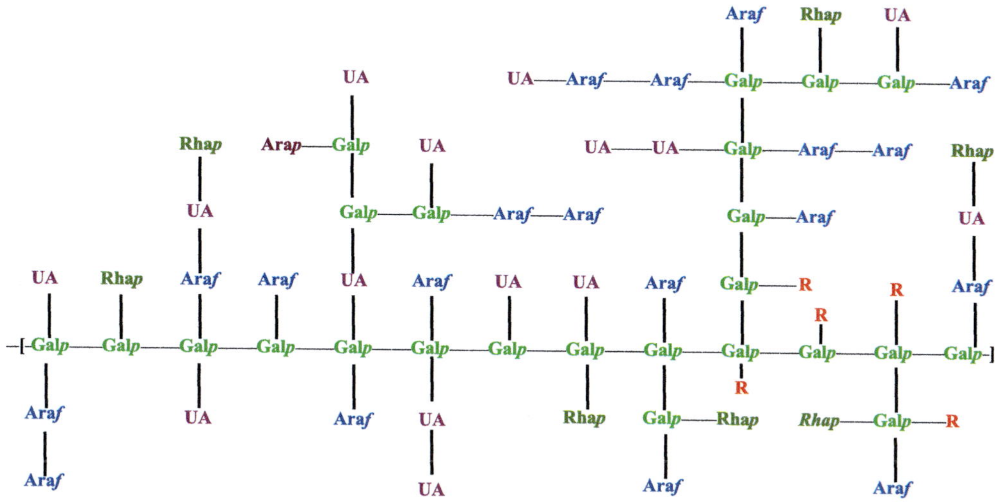

Galp: β-D-galactopyranose; **Rhap:** α-L-rhamnopyranose; **Araf:** α-L-arabinofuranose; **Arap:** α-L-arabinopyranose; **UA:** β-D-uronic acid (glucuronic/galacturonic); **R:** Rhap or Araf or UA or Araf—Araf or UA—UA.

Fig. 3.3. Typical structure of gum arabic polysaccharides. Author's own image.

Properties

It has been found that the properties of GA may vary significantly depending on its source (Chikamai and Banks, 1993). Ali and Daffalla (2018) compared physico-chemical and functional properties of GA collected from different regions of Sudan. The ash, protein and nitrogen content of the gum were constant to a great extent regardless of collection region and climatic condition. Contrary to that, the moisture content varied depending on climatic conditions at the time of exudation or storage. Significant variation was observed in viscosity and pH, possibly due to different preparation procedures, like the temperature and concentration of gum solutions and different laboratory conditions.

Anderson *et al.* (1968) found considerable variation in the viscosity and molecular weight of gums produced from trees growing in similar soil and climatic environments. Furthermore, variations in molecular masses were also found in gums produced by a single tree from different parts, showing both intra- and inter-tree variation.

Duvallet *et al.* (1993) observed wide variations in intrinsic viscosities ranging from 0.14 to 0.60 dl/g, specific rotations from −25 to −62° and nitrogen from 0.12 to 0.57% in 75 GA samples collected from different trees at the M'Biddi research farm in Senegal. The study, which was carried out on the basis of electrophoresis, ion-exchange chromatography (Jermyn, 1962) and fractional precipitation with sodium sulfate (Anderson and Stoddart, 1966), concluded that GA is heterogeneously composed and is therefore classified as 'hetero-polymolecular'. This could be due to variations in the monomer composition and/or the mode of linking and branching of the monomer units, as well as the molecular mass distribution.

Karamalla *et al.* (1998) analysed over 1500 gum samples of *Acacia senegal* to study properties like pH, viscosity, molecular weight and uronic acid content. The mean values for moisture, ash, nitrogen, specific rotation, pH, equivalent weight and uronic acid were 10.75%, 3.77%, 0.328%, −31.31°, 4.66, 1436 and 13.71%, respectively. The results further indicated large variations in intrinsic viscosity and viscosity

Table 3.3. Colour values of GA at different concentrations. Author's own table.

Colour parameters	1% (w/w) gum solution	20% (w/w) gum solution
L	25.15 ± 0.04	25.36 ± 0.13
a	−0.217 ± 0.12	−0.29 ± 0.08
b	0.5 ± 0.06	0.65 ± 0.11
ΔE	25.16 ± 0.03	25.35 ± 0.13

average molecular weight with mean values of 164 ml g^{-1} and 9.0×10^5 Da, respectively.

The physico-chemical and functional properties of GA are illustrated in the following section as reported by different researchers.

Physico-chemical properties. *Colour.* GA is found in colours varying from orange-brown to pale white. The tears of GA have a matt surface texture. The gum acquires a paler hue, and the glassy appearance in broken or of kibbled form. GA is also available commercially in the form of granules, flakes and powders.

Food colour is an essential factor which makes it appealing. The result of GA addition on colour change in food items can be determined by colour analysis of gum at different concentrations (Bertuzzi *et al.*, 2012). Bhushette and Annapure (2017) determined the colour values (luminescence L*, greenness -a* and yellowness b*) of GA solutions as presented in Table 3.3. With the increase in the gum concentration in an aqueous solution, the colour of the solution changes to brownish. With the increase in concentration, greenness and luminescence decrease whereas yellowness increases.

Carbohydrate analysis and monosaccharides composition. Most exudate gums possess >90% (by dry weight) carbohydrates. GA mainly comprises monosaccharides like arabinose, galactose, glucuronic acid and rhamnose (Phillips *et al.*, 2008). Table 3.4 shows the physico-monosaccharides present in GA and chemical properties of GA compared to gum *Acacia nilotica* and gum ghatti (Lopez-Torrez *et al.*, 2015). GA's functional and thermal properties may vary due to compositional differences in the polysaccharide chain.

Functional properties. *Solubility, water absorption capacity (WAC) and oil absorption capacity (OAC).* GA is easily soluble in water. It is odourless, tasteless, translucent and does not interact with a chemical compound. GA is designated as hard gum. The WAC and OAC of GA determined by (Bhushette and Annapure, 2017) are summarized in Table 3.4. As GA is easily soluble in water, it has a low WAC (approx. 45%). The result was similar to the findings of (Prinyawiwatkul *et al.*, 1997) that WAC decreases with an increase in water solubility. OAC and WAC depend on the protein and carbohydrate content of a gum (Chandra and Samsher, 2013).

Rheological properties. The viscosity of GA solution is an important property which influences the organoleptic properties of the food products in which GA finds application (Glicksman, 1982).

GA exhibits Newtonian behaviour at lower concentration (1–5% w/w). However, at higher concentration (10–20% w/w) and with an increase in shear rate, it exhibited non-Newtonian behaviour and showed shear thinning behaviour (Bhushette and Annapure, 2017). Similar behaviour was reported for *Albizia zygia* gum (Eddy *et al.*, 2013), brea plant gum (Bertuzzi *et al.*, 2012), *Acanthophyllum bracteatum* root gum (Jahanbin *et al.*, 2012) and gum *Kondagogu* (Vinod *et al.*, 2008). At higher concentrations, there was an increase in viscosity. This may be due to an increase in inter-molecular attractive force with an increasing number of molecules per unit volume.

The presence of electrolytes and higher and lower pH of solution leads to a decrease in viscosity of the GA solution. Heating of a GA solution leads to an irreversible decrease in viscosity, which could be attributed to the denaturation

Table 3.4. Comparison of physico-chemical properties of GA with *Acacia nilotica* gum and gum ghatti. Author's own table.

Physico-chemical properties	GA (Karamalla *et al.*, 1998)	*Acacia nilotica* gum (Bhushette and Annapure, 2017)	Gum ghatti (Kang *et al.*, 2011b)
Moisture (%)	10.75	6.49 ± 0.18	13.42 ± 0.12
Ash (%)	3.77	1.81 ± 0.06	1.14 ± 0.01
Nitrogen (%)	0.328		0.327 (Thombare *et al.*, 2018)
Protein (%)		2.25 ± 0.00	1.18 ± 0.08
Specific rotation (°)	–31.31		–30 to –40 (Deshmukh *et al.*, 2012)
pH	4.66	4.76 ± 0.02 (1% gum solution)	4.9 5.1 (Thombare *et al.*, 2018)
Equivalent weight	1436	–	–
Total uronic acid content (%)	13.71	12.782 ± 0.476	12.83 ± 0.23
Galactose (%)	44	26.44	21.97 ± 0.14
Arabinose (%)	27	39.62	37.07 ± 0.26
Rhamnose (%)	13	10.15	1.09 ± 0.03
	GA (Bhushette and Annapure, 2017)	*Acacia nilotica* gum (Bhushette and Annapure, 2017)	
Bulk density (g/cc)	0.405 ± 0.01	0.667 ± 0.00	–
Tap density (g/cc)	0.521 ± 0.01	0.883 ± 0.02	–
Hausner's index	1.287 ± 0.01	1.324 ± 0.013	–
Compressibility (%)	28.71 ± 0.44	23.32 ± 0.00	–
Solubility (%)	93.40 ± 0.01	92.09 ± 2.11	87.75 (Thombare *et al.*, 2018)
Water absorption capacity (%)	44.63 ± 0.62	46.35 ± 4.13	–
Oil absorption capacity (g/g)	1.216 ± 0.01	1.216 ± 0.02	–

and precipitation of protein-rich components possessing high molecular mass (Chikamai and Banks, 1993).

Elamin *et al.* (2013) studied the effect of soil types (sand, clay and sandy clay) and rainfall (low, medium and high) on the viscosity of GA in the gum belt of North and South Kordofan and the Blue Nile States in Sudan. The analysis showed high average viscosity for GA obtained from low rainfall areas in sandy and clayey soils in medium rainfall areas. Sandy soil in medium rainfall areas and clayey soil in medium and high rainfall areas produced GA of low average viscosity, having significant differences between them. The study's results revealed no significant differences in pH for GA from different rainfall areas and soil types.

Emulsification properties. GA is used as an emulsifier for encapsulation of flavour materials such as essential oils used in dried food mixtures like soups and cakes or the preparation of soft drinks. Chikamai *et al.* (1996) found that heating solutions at 100°C for >6 hour resulted in a significant reduction in emulsifying properties, while heating to 65°C for 24 hour had little effect on emulsion stability.

Dickinson *et al.*, 1988, 1989, 1991a, b; Phillips *et al.*, 1990), reporting the interfacial and emulsification properties of GA and gums of other acacia species, did not find any correlation between emulsification behaviour and protein content. Studying the emulsification behaviour of GA in beverages and concentrated emulsions, Ray *et al.* (1995) found that an increase in GA emulsion stability was associated with an increase in protein content and molecular weight, as also reported by Randall *et al.* (1988).

Gum ghatti

Chemistry

Composition. Gum ghatti (GG) is a natural polymer of several monosaccharides present in the form of their calcium and magnesium salt (Al-Assaf *et al.*, 2009, 2021; Thombare *et al.*, 2018). Due to the unique compositional property of GG being only partially soluble in water, it possesses water-soluble as well as water-insoluble fractions. Most of the literature available discusses the structure or composition of refined GG obtained after precipitation or filtration, where the risk of losing the insoluble portion is high. According to Kang and co-workers (2011a), the refined GG consists of about 78% total sugars, 4% protein, 1% ash and 13–15% moisture (Kang *et al.*, 2011a), and traces of tannin (<0.1%) were also reported (Sakai *et al.*, 2013). Though monosaccharide composition from refined GG observed by various authors differs to some extent, consistently, arabinose emerges to be the dominant component with 34–38%, followed by galactose (21–25%), glucuronic acid (15–19%), mannose (5–8%), xylose (0.8–1.2%), rhamnose (0.8–1.2%) (Sakai *et al.*, 2013) and traces of glucose (< 1%) (Kang *et al.*, 2011a). This is why GG is often called arabinogalactan. The abundant presence of arabinose along with xylose and mannose can serve as criteria for the detection of GG (Al-Assaf *et al.*, 2009). The GG contains 4–9% protein, which is higher than the protein content of gum arabic (approx. 2%), which makes GG a rich source of protein (Barak *et al.*, 2020).

Structure. A series of fundamental works elucidating the structure of GG was carried out from 1955 to 1965, mainly by Aspinall *et al.*, (1955, 1958, 1965). As reported by them, the structure of GG, through stepwise fragmentation, methylation and Smith degradation, was identified as a backbone of repeating D-mannopyranose and D-glucuronic acid units, to which other sugars like galactopyranose, arabinofuranose and arabinopyranose are attached as side chains.

Kang and co-workers reported a more precise structure in 2011 and 2012 using advanced analytical techniques like methylation-GC-MS, NMR spectra, 1D (^{1}H, ^{13}C) and 2D (COSY (Correlation Spectroscopy), TOCSY (Total Correlation Spectroscopy), NOESY (Nuclear Overhauser Effect Spectroscopy) and HSQC (Heteronuclear Single Quantum Coherence)). For this, GG was separated into different fractions by serial precipitation using various proportions of ethanol in aqueous GG solution (F50, F65, F80 as 50%, 65% and 80% ethanol in water precipitate) and FS (supernatant of F80). The structure analysis studies of different fractions indicated almost similar structures for all three precipitated fractions (F50, F65, F80) and a distinct structure and composition for the FS (Kang *et al.*, 2011c), which can be considered as the highly water-soluble fraction. The structure of GG, which precipitated after adding 80% ethanol in water (F80), is presented in Fig. 3.4. The structure is mainly composed of a backbone made up of 1,6-linked galactopyranosyl (Galp) residues which are substituted at O-3 and O-4. On the other hand, side chains comprise of →2)-Araf-(1→4)-GlcpA-(1→6)-Galp-(1→6)-Galp-(1→ which are often terminated by arabinofuranosyl units and seldom by rhamnopyranosyl, arabinopyranosyl, galacto and glucuronopyranosyl units (Kang *et al.*, 2011c).

The structure of another fraction of GG, which remained in solution after 80% ethanol addition, was found to be globular, highly branched and have minor acetyl substitutions. The structure, shown in Fig. 3.5 (Kang *et al.*, 2011c), was reported to be made up of arabinose, galactose, rhamnose and glucose with

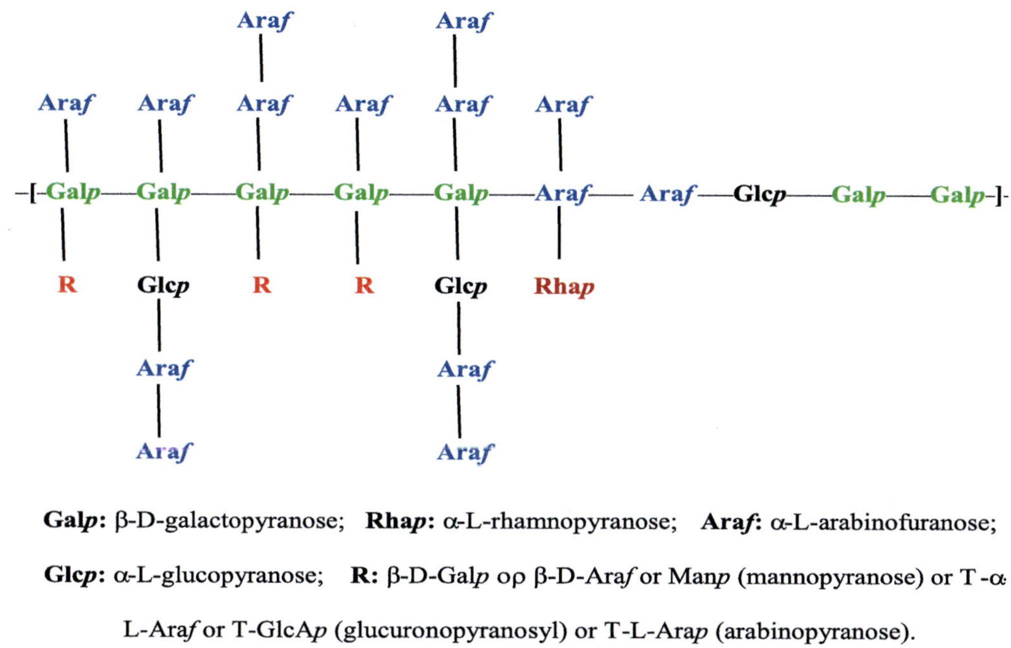

Fig. 3.4. Typical chemical structure of an ethanolic precipitate fraction from gum ghatti polysaccharides. Author's own image.

4–5% protein and about 2% uronic acids. The backbone comprises of 1,6-linked galactopyranosyl chain, having side chains attached at O-3 and O-4 positions by different component sugar residues, namely →6)-β-D-Galp-(1→, →2)-α-L-Araf-(1→, →5)-a-L-Araf-(1→, t-α-L-Araf and →2,3,5)-a-L-Araf-(1→.

Kang and co-workers also examined the glycoprotein structure of GG, confirming for the first time the covalent attachment of protein to polysaccharides (Kang *et al.*, 2012). The glycoprotein was reported to comprise the backbone of galactose linked through 1,6 linkage (Kang *et al.*, 2012). This protein component of GG provides hydrophobicity, which is confirmed by its higher hydrophobic score (566) compared to gum arabic (241). The distribution of these hydrophobic amino acids and hydrophilic polysaccharides in GG makes it an excellent emulsifier for oil in water-type colloidal suspensions (Kang *et al.*, 2014).

Properties

GG is composed of magnesium and calcium salts comprising soluble and insoluble portions. GG is partly soluble in water at 100°C. The insoluble portion contains calcium ions and forms gel in water, which may be dissolved by the maceration process (Jefferies *et al.*, 1977). Gel formation takes place due to calcium ions and intermolecular linkages. The viscosity of the insoluble gel fraction was 10–30 times higher than the soluble fraction of GG (Jefferies *et al.*, 1978, 1982).

The polysaccharide chain length and degree of branching are responsible for the difference in solubility and viscosity of the two fractions. The low solubility and higher viscosity of the gel fraction may be due to the accumulation of anionic groups and bivalent Ca^{2+} ions on different molecules, which is also found in alginates (Williams, 1982). GG has a rod-like

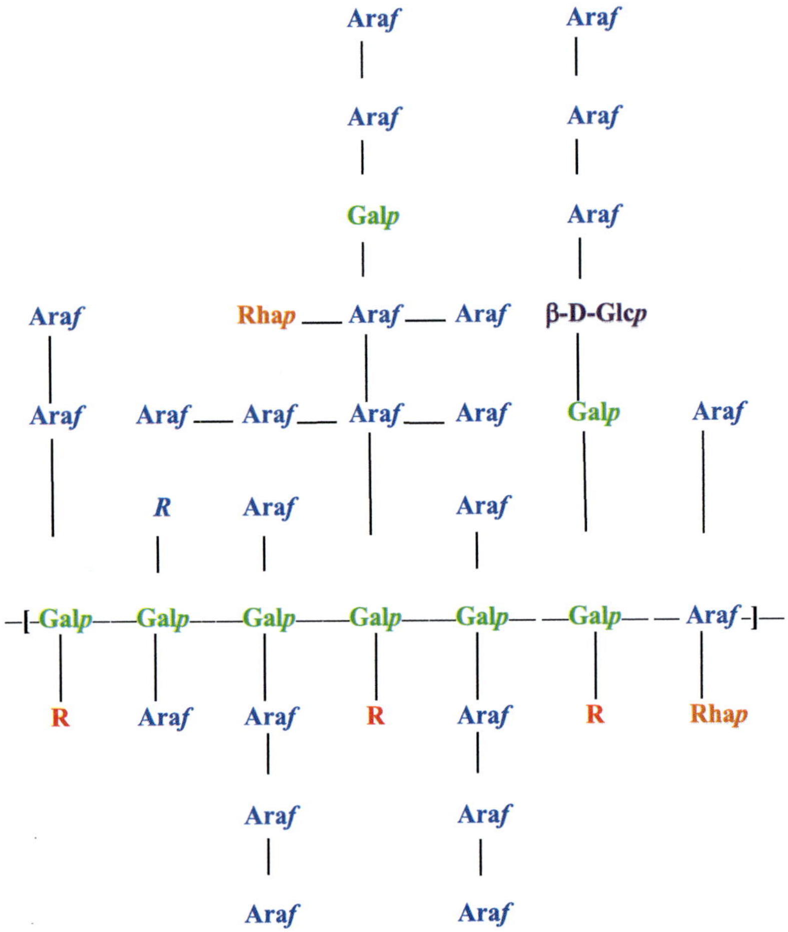

Galp: β-D-galactopyranose; **Araf:** α-L-arabinofuranose; **Rhap:** α-L-rhamnopyranose; **Glcp:** β-D -glucopyranose; **R:** mannopyronose or Glcp or Gal*p*, Terminal (T) α-L-Araf and (T)-Gal*p* and T-L-Ara*p*.

Fig. 3.5. Typical chemical structure of a water-soluble fraction from gum ghatti polysaccharides. Author's own image.

Table 3.5. Physico-chemical properties of GG studied by different research groups. Author's own table.

(Sarkar *et al.*, 2018)		(Kang *et al.*, 2015)		(Thombare *et al.*, 2018)	
Properties	Value	Properties	Value	Properties	Value
Bulk density	1.13±0	Moisture (%)	13.42±0.12	Moisture (%) (Volatile matter)	12.66
True density	1.66	Ash (%)	1.14±0.01	Ash (%) Acid insoluble ash (%)	1.97 0.15
Compressibility index (%)	31.91	Protein (%)	2.18±0.08	Nitrogen (%)	0.327
Solubility index (%)	71.56±00.02	Total uronic acid content (%)	12.83±0.23	Solubility (%)	87.75
Water holding capacity (g water absorbed /100g gum)	304.33±0 00.5	Galactose (%)	21.97±0.14	Bark and foreign organic matter	0.39
Oil holding capacity (g of oil absorbed/ 100g gum)	159.66±00.5	Arabinose (%)	37.07±0.26	Viscosity (cP)	1176
Angle of repose	38.13±0	Rhamnose (%)	1.09±0.03	pH	4.9–5.1

molecular structure that causes tangles, thus contributing to structural viscosity and leading to pseudoplasticity at lower concentrations (Meer *et al.*, 1973; Meer, 1980; Glicksman, 1983).

Rheological properties. In their rheological study on GG, Al-Assaf *et al.* (2008) found that GG forms a viscous solution at a concentration of 300 g/kg. As with other gums, the viscosity increased with the concentration of the GG solution. At a low shear rate, GG exhibited shear thinning properties, but at a higher shear rate, it exhibited Newtonian behaviour. The hydrated GG solution at ambient temperature exhibited the highest viscosity at pH 7.8 and drops on both sides. Above 8.0 pH, the solution became stringy. The emulsion stability of GG was found to be higher at high temperatures (60°C), indicating its resistance to heat. Kaur *et al.* (2009) also investigated the effect of the gum concentration on the viscosity of gatifolia SD (a commercial product of spray-dried gum ghatti) solution and found that the viscosity increased with increasing concentration.

Surface tension. Huang *et al.* (2001) measured the surface tension for different gum concentrations (0.01–1.5% w/w). According to this, the surface tension decreased with increasing gum concentration from 0.01 to 0.50% and then reached a plateau at a gum concentration of more than 1%. In this range, the surface tension was 37.9 dyne/cm. The surface tension appears to be related to the protein content of the gum and its structure.

Physico-chemical properties. Physico-chemical properties of GG studied by various researchers are summarized in Table 3.5 (Kang *et al.*, 2015; Sarkar *et al.*, 2018; Thombare *et al.*, 2018). The authors have determined different physico-chemical properties by taking GG from different sources. In this context, no regional variations in the physico-chemical properties could be found. Thus, there is the scope of studying regional variation in properties of gum ghatti influenced by soil and climatic variation.

A comparison of the physico-chemical properties of GG with gum karaya and gum arabic is presented in Table 3.6. It can be seen that GG has a lower ash content than gum

Table 3.6. Comparison of physico-chemical properties of GG with *Acacia senegal* gum and *Sterculia urens* gum. Author's own table.

Physico-chemical properties	Gum *Acacia senegal* (Karamalla *et al.*, 1998)	Gum ghatti (Kang *et al.*, 2011c)	Gum *Sterculia urens* (Sahu *et al.*, 2019)
Ash (%)	3.77	1.14 ± 0.01	4.62
Nitrogen (%)	0.328	0.327	0.16
Protein (%)	–	2.18 ± 0.08	1.06
Specific rotation	–31.31°	–30° to –40°	–
pH	4.66	4.9–5.1	4.26
Total uronic acid content (%)	13.71	12.83 ± 0.23	40
Galactose (%)	44	21.97 ± 0.14	26
Arabinose (%)	27	37.07 ± 0.26	–
Rhamnose (%)	13	1.09 ± 0.03	15–30
Bulk density (g/cc)	0.405 ± 0.01	0.99	0.79 ± 0.02
Tap density (g/cc)	0.521 ± 0.01	–	0.90 ± 0.04
Hausness index	1.287 ± 0.01	–	1.14 ± 0.06
Compressibility (%)	28.71 ± 0.44	20.66 ± 0.46	12.31 ± 4.19
Solubility (%)	93.40 ± 0.01	87.75	60.06 ± 0.1
Water absorption capacity (%)	44.63 ± 62	–	–
Oil absorption capacity (g/g)	1.216 ± 0.01	–	–
Coefficient of friction (Sarkar *et al.*, 2018)			
Glass	0.60	0.62	0.55
Plywood	0.62	0.56	0.62
Steel	0.60	0.54	0.74
Aluminium	0.58	0.52	0.55
Sunmica	0.63	0.55	0.56
Sunmica laminated	0.59	0.57	0.57
Angle of repose	39.48	31.13	38.73

karaya and gum arabic. It has a higher protein content than gum karaya and is comparable to gum arabic in this respect. The uronic acid content of GG is lower than that of gum karaya, and the rhamnose content of GG is significantly lower compared to gum karaya and gum arabic. Other physico-chemical properties of GG were comparable to those of gum karaya and gum arabic. With medium solubility compared to gum karaya and gum arabic, GG is a unique gum that can be used for a wider range of applications.

Genotoxicity of GG. Based on the evaluation of the genotoxic potential of GG, Hobbs *et al.* (2012) found that there was no toxicity or mutagenicity at doses up to 6 mg/plate, the maximum dose recommended by OECD. Maronpot *et al.* (2013) performed the toxicity study of GG according to the OECD guidelines on female and male rats, administering different doses of GG (0.0, 0.5, 1.5 and 5%) in the base diet. The 5% GG dose had no observable adverse effect level (NOAEL).

Guar gum

Chemistry

Various analytical techniques have been used to determine the structure and behaviour of guar gum (Chudzikowski, 1971; Whistler and

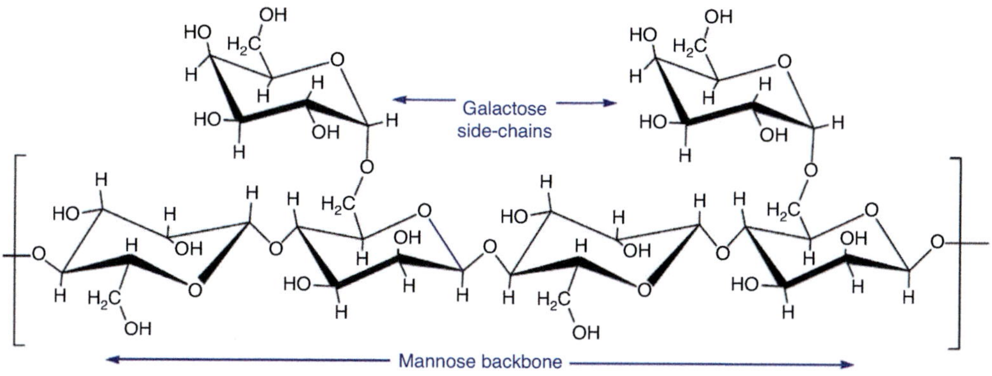

Fig. 3.6. Chemical structure of guar gum polysaccharides. Author's own image.

Hymowitz, 1979; Tripathy and Das, 2013). Accordingly, it consists of linear backbone chains of (1→4)-β-D-mannopyranosyl units with branch points of α-D-galactopyranosyl units attached by (1→6) linkages (Fig. 3.6). Many research studies revealed the ratio of mannose to galactose units in guar gum in the range of 1.6:1–1.8:1 (Grasdalen and Painter, 1980). This ratio has also been approximated to 2:1 in other studies (Parija *et al.*, 2001). The absence of any uronic acid makes guar gum different from the great majority of plant gums and mucilages. Guar has one of the highest molecular weights of all naturally occurring water-soluble polysaccharides. The results obtained with advanced techniques such as size exclusion chromatography as well as low angle laser light scattering, show the average molecular weight in the range of 10^6 to 2×10^6 Da (Vijayendran and Bone, 1984).

Properties

Galactomannans are insoluble in organic solvents like hydrocarbons, alcohols, esters, ketones, etc., with the only exception of formamide. Water is the most important solvent for galactomannans, as it not only hydrates but also forms colloidal solutions of unusually high viscosity, even at very low concentrations. On contact with water, the galactose side chains attached to the mannose backbone interact with the surrounding water molecules, resulting in intermolecular chain entanglement that increases viscosity. As the concentration of guar gum in water increases, this entanglement increases further, leading to gelation or thickening. Even 1% aqueous dispersion of high-quality guar gum may have a viscosity of up to 10,000 cP (Parija *et al.*, 2001). It is also interesting that, although guar gum is hydrophilic, it is not hygroscopic. Water vapour in air merely changes the moisture equilibrium of galactomannans.

Guar gum attains its full viscosity potential in cold water, while most other natural gums, including locust bean galactomannan, require prolonged cooking time. As with most hydrocolloids, the viscosity of a guar gum solution is highly dependent on the shear rate and can better be described as an apparent viscosity. As the viscosity decreases with increasing shear rate, guar gum solutions exhibit shear-thinning properties with non-Newtonian, pseudoplastic behaviour (Tripathy and Das, 2013). Temperature also has an inverse relationship with the viscosity of the guar gum solution. As the temperature of the guar gum solution rises, the water molecules around the guar gum molecules lose their order, causing the conformation to be disrupted and the viscosity to decrease (Srichamroen, 2007). Because guar gum is non-ionic, it is stable and has a consistent viscosity over a wide pH range. The practical range of stability is between pH 4 and 10.5. Maximum viscosity can be obtained at pH values of 6–9, and the lowest at approximately 3.5. Except for values above 10.5 and below 4, the viscosity potential is not greatly affected by pH (Smith and Montgomery, 1959). Due to the numerous hydroxyl groups in the chain, guar

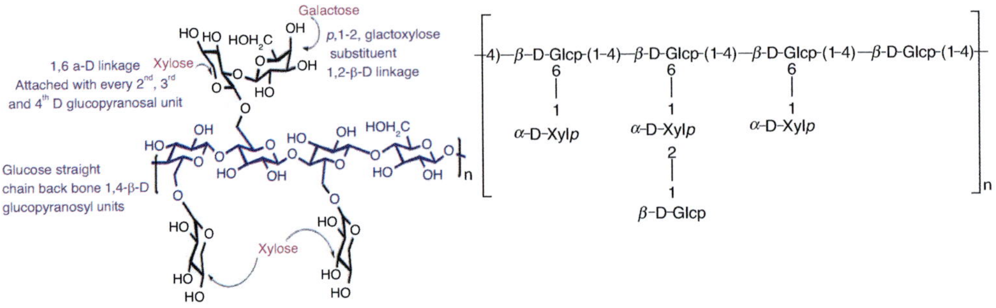

Fig. 3.7. Schematic illustration of the molecular structure of tamarind seed gum. Author's own image.

gum forms hydrogen bonds in aqueous solution, which explains much of its behaviour.

The basic mannose chain structure of galactomannan along with the single-unit galactose branches, further increases the number of exposed hydroxyl groups. This results in a product that exhibits an unusual effect on other hydrated colloidal systems through hydrogen bonding. As guar gum can form hydrogen bonding with both hydrated mineral and organic surfaces, it can influence most of the systems by its action.

Hence, the addition of even small amounts of guar gum can significantly alter the electrokinetic properties of the treated system. Therefore, guar gum can act as a dispersant for organic systems, especially those with functional hydroxyl or carboxylic groups, and as a coagulant for inorganic ones, in particular for those with clay characteristics. Since guar gum solution is biodegradable, it tends to microbial degradation. The non-preserved guar gum solution should be used within 24 hour or must be preserved with some preservatives.

Tamarind seed gum

Chemistry

Tamarind seed gum, a crude extract of tamarind seeds, is rich in polysaccharides (approx. 65–72%) (Kumar and Bhattacharya, 2008) and contains glucose, xylose and galactose units, in a molecular ratio of approximately 3:2:1 (Wang *et al.*, 1997; Freitas *et al.*, 2005) (Fig. 3.7). Its structure is based on a β(1→4)-D-glucan backbone, substituted at position 6 of the glucopyranosyl units mainly by single α-D-xylopyranosyl residues as well as by disaccharide side chains composed of β-D-galactopyranosyl-(1→2)-α-D-xylopyranosyl residues. In addition, tamarind seed gum is a high-molecular-weight polysaccharide (720–880 kDa) (Wang *et al.*, 1997; Sims *et al.*, 1998; Freitas *et al.*, 2005), which forms viscous solutions when dissolved in water like many polysaccharide gums extracted from plant materials. It has potential for commercial applications in the pharmaceutical industry for controlling drug release (Sumathi and Alok, 2002) and in textile printing as a thickener (Abo-Shosha *et al.*, 2008). Particularly in the food industry in Japan, refined tamarind seed gum is used as an approved food additive to modify texture as a thickening, stabilizing and gelling agent (Nishinari *et al.*, 2000).

Composition of tamarind seed. The composition of tamarind seeds varies greatly depending on the variety and degree of ripeness. The seeds make up about 30% of the total fruit and consist of the seed coat or testa (20–30%) and the kernel or endosperm (70–80%). The composition of the kernel is shown in Table 3.7 (Khounvilay and Sittikijyothin, 2012).

The amino acid analysis of tamarind seed protein conducted by Bhattacharya *et al.* (1994) shows that it is rich in glutamic acid, aspartic acid, glycine and leucine but deficient in sulfur-containing amino acids (methionine and cysteine). The proportions of hydrophobic

Table 3.7. Proximate composition of tamarind seed kernel (%). Author's own table.

Moisture: 11.4–22.7
Protein: 15.0–20.9
Fat/oil: 3.9–16.2
Crude fibre: 2.5–8.2
Carbohydrate: 65.1–72.2
Total ash: 2.4–4.2

amino acids (alanine, valine, leucine, isoleucine and phenylalanine) and hydrophilic amino acids (lysine, histidine, aspartic acid, glutamic acid and arginine) are 28.2 and 42.3%, respectively (Bhattacharya *et al.*, 1994).

The edible seed kernel is rich in phosphorus, potassium and magnesium (Calcium: 120.0, Magnesium: 180.0, Potassium: 1020.0, Sodium: 210.0, Iron: 80.0, Zinc: 100.0, Palmitic acid: 14–20, Stearic acid: 6–7, Oleic acid: 15–27, Linoleic acid: 36–49, Arachidic acid: 2–4, Behenic acid: 3–5, Lignoceric acid: 3–8, Beta-sitosterol: 66–72, Campesterol: 16–19, Stigmasterol: 11–14) (Gunasena and Hughes, 2000). However, the calcium content is low in comparison to that of other cultivated legumes (Bhattacharya *et al.*, 1994).

The seed kernel is also a rich source of fatty oil. The major fatty acids are palmitic, oleic and linoleic acid (Andriamanantena *et al.*, 1983). The lipids contain a relatively large proportion of unsaturated fatty acids, with linoleic acid (36–49%) in highest concentration. Other major fatty acids are oleic (15–27%) and palmitic acid (14–20%). Several hydrocarbons such as β-amyrin, campesterol and β-sitosterol have been identified in the unsaponifiable matter of the seeds.

Properties

Rheological behaviour of tamarind kernel powder (TKP) dispersion. The flow behaviour of TKP dispersion in water is an important functional property in characterizing the material and the product development, and for the design of the processing units. The rheological behaviour of TKP dispersion was studied by Prabhanjan (1989), Bhattacharya *et al.* (1991), Prabhanjan and Ali (1995), Yamanaka *et al.* (2000) and Salazar-Montoya *et al.* (2002). The effects of concentration (2, 4, 6, 8 and 10% TKP powder) on the flow behaviour index, the consistency index and the apparent viscosity were investigated. They behave like non-Newtonian, shear-thinning fluids (Fig. 3.8) with yield stresses. The yield stress values are low at low concentrations of TKP but become prominent at high concentrations. An increase in TKP concentration increases the non-Newtonian behaviour, i.e. the flow behaviour index diminishes, whereas the consistency index, the yield stress and the apparent viscosity increase.

Cassia tora seed gum

Chemistry

Cassia tora gum (CTG) is derived from the seeds of *Cassia tora*, a common herbaceous annually occurring weed throughout India. The pods are 15–22.5 cm long and up to 0.625 cm in diameter, containing flattened dark seeds. *Cassia tora* seed is composed of hull (27%), endosperm (32%) and germ (41%) (Soni and Pal, 1996). Gum obtained from the seeds of *Cassia tora* is known as 'Cassia tora seed gum'. Chemically, it is a neutral heteropolysaccharide of galactose and mannose (i.e. galactomannans). Methylation studies showed that the backbone of the polysaccharide consists of 1→4 linked D-mannopyranose and D-glucopyranose units. The pH of this gum mucilage is approximately 7 (Joshi and Verma, 1964). The protein content of *Cassia tora* seeds is about 23.2% and they are rich in all essential amino acids, particularly methionine and tryptophan (Pawar and D'mello, 2011).

Proximate analysis of cassia tora seed. The result of the proximate analysis of *Cassia tora* seed is summarized in Table 3.8 (Sahu *et al.*, 2023). The *Cassia tora* seed sample at 10.82% (on wet basis) moisture content contained on average 4.44% ash, 9.25% fat, 28.13% fibre, 23.44% protein and 25.68% carbohydrate. The results are the mean of three determinations. Ash content was observed to be nearly equal to the values reported for *Cassia tora* (Adamu *et al.*, 2013) and *Cassia alata* seed (Ukhun and Ifebigh, 1988).

Mineral content of cassia tora seed. The major mineral contents in *Cassia tora* seed were

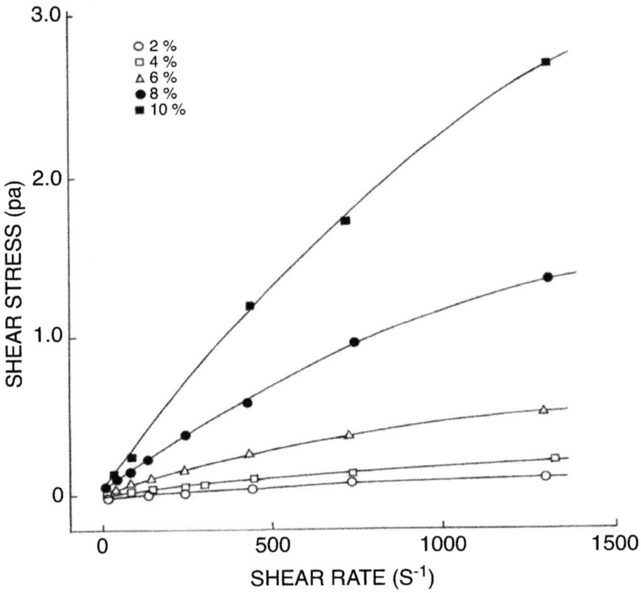

Fig. 3.8. Rheogram of TKP dispersions at different concentrations. Author's own image.

Table 3.8. Proximate composition of *Cassia tora* seed (%). Author's own table.

Moisture content (on wet basis): 10.82 ± 0.10
Fat content: 9.25 ± 1.57
Carbohydrate: 25.68 ± 0.22
Ash content: 4.44 ± 0.30
Fibre: 28.13 ± 1.88
Protein: 23.44 ± 1.87

determined by Sahu *et al.* (2023). *Cassia tora* seed was found to be rich in calcium (630.0 mg/kg), phosphorus (240.0 mg/kg) and iron (37.1 mg/kg) compared to the normal levels of most other seeds. Vadivel and Janardhanan (2000) observed lower calcium, iron and phosphorus levels in the seeds of *Cassia hirsuta* L.. In contrast, *Cassia obtusifolia* contains higher calcium (4100 and 6590 mg/kg) for varieties FSY-1013 and FSY-1034, respectively (Crawford *et al.*, 1990). Ukhun and Ifebigh (1988) also observed lower calcium contents in *Cassia alata* seed compared to *Cassia tora* seed.

The calcium and iron content of *Cassia tora* seed were recorded by Adamu *et al.* (2013) to be comparatively lower than the values determined in the present study. Iron and phosphorus levels in *Cassia tora* seed were found to be higher than those analysed in *Cassia floribunda* (Vadivel and Janardhanan, 2000). Siddhuraja *et al.* (1995) also reported lower values for mineral content in *Cassia laevigata* when compared to the observed values.

Gum-Resins

Asafoetida

Chemistry

Asafoetida consists of three main fractions, including resin (40–64%), gum (approx. 25%) and essential oil (10–17%) (Takeoka, 2001). The resin fraction contains ferulic acid and its esters, coumarins, sesquiterpene coumarins and other

Table 3.9. Chemical composition of asafoetida gum-resin presented as % (w/w). Author's own table.

Component	% (w/w)
Dry matter content	98.6 ± 0
Total sugar	67.1 ± 1.1
Rhamnose	10.8
Arabinose	25.3
Galactose	58.8
Glucuronic acid	5.1
Total protein	6.8 ± 0
Water-soluble protein	4.9 ± 0.1

Table 3.10. Emulsifying properties of asafoetida gum-resin. Author's own table.

Property	Value
Emulsion activity index (EAI)	38.6 ± 01 (m^2/g)
Emulsion stability index (ESI)	149.6 ± 0.5 (min)
Emulsion capacity (EC), 0.5% w/v	881.0 ± 2.0 (mL oil/g powder)
Emulsion capacity (EC), 1.0% w/v	477.5 ± 7.7 (mL oil/g powder)
Emulsion stability (ES), [a] (Treatment A)	75.0 ± 0.7 (%)
Emulsion stability (ES), [b] (Treatment B)	76.2 ± 0.3 (%)

[a]Keeping emulsion at 20°C, 24 hour.
[b]Emulsion pre-treatment (80°C, 30 min), then keeping at 20°C.

terpenoids. The gum includes glucose, galactose, L-arabinose, rhamnose, glucuronic acid, polysaccharides and glycoproteins including arabinogalactan protein (AGP), and the volatile fraction contains sulfur-containing compounds, monoterpenes and other volatile terpenoids. The resin and volatile compounds of asafoetida are traditionally used to treat asthma, intestinal parasites and influenza (Iranshahy and Iranshahi, 2011). AGP remaining from ethanolic extraction of resins and volatile oil is generally considered as a by-product. Polysaccharides of asafoetida contain units of (1 → 3)-β-D-galactan as the main chain and terminal-α-L-Araf, terminal-α-L-Rhap, (1 → 3)-α-L-Araf, (1 → 5)-α-L-Araf, terminal-β-D-Galp, β-D-GlcA and (1 → 4)-β-D-GlcA in the side chains. The chemical composition of asafoetida is presented Table 3.9, also providing the monosaccharide composition of the gum fraction of the gum-resin. The major constituents of *Ferula asafoetida* and chemical structures of important sesquiterpene coumarins and sulfur-containing compounds present in the plant have been reviewed and compiled by Amalraj and Gopi (2017).

The study of surface and interfacial tension of asafoetida gum revealed that surface tension decreased from 60.3 ± 0.2 mN/m at a concentration of 0.2% to 54.5 ± 1.5 mN/m at a concentration of 1.5% (w/w) in solutions. Similar results were achieved for the interfacial tension. The minimum interfacial tension observed for asafoetida was 17.6 mN/m.

Properties

Table 3.10 gives the emulsifying properties of asafoetida gum-resin. The foaming ability (FA) and foaming stability of the gum-resin were found to be 53 ± 2.7% and 1440 ± 60 min respectively.

Asafoetida gum has Newtonian flow behaviour, which is characterized by a linear relationship between shear stress and shear rate. The natural pH of asafoetida is around 6. In the alkaline pH range, no effects on the rheological behaviour of the solution are observed, whereas in the acidic range, the rheology can be altered by adding a shear-thinning agent. While this apparent viscosity at high shear rates usually increases with increasing pH, this parameter decreases in the gum-resin solution with increasing temperature (Saeidy *et al.*, 2018).

Guggul

Chemistry

The oleo-gum resin of the guggul tree is a complex mixture of gum, minerals, essential oils, terpenoids, sterols, ferrulates, flavanones and sterones and several other, not yet identified substances (Hanus *et al.*, 2005). The resin yields two fractions upon ethyl acetate extraction. The ethyl acetate-soluble fraction comprises 45% of the gum-resin whereas the insoluble fraction consists of the carbohydrate gum, which is about 55% of the gum-resin. Most of the bioactive components were found in the ethyl acetate-soluble fraction, while the insoluble carbohydrate fraction has no hypolipidemic effects (Nityanand and Kapoor, 1975). The ethyl acetate-soluble fraction consists mainly of di- and triterpenoids, steroids, lignans and various fatty acid esters (Dev, 1989; Hanus *et al.*, 2005).

Further pH gradient fractionation of the ethyl acetate-soluble fraction of guggul yields 4% acidic fraction, 1% basic fraction and 95% neutral fraction. The neutral fraction contains the bioactive components of guggul. Additional fractionation of the neutral fraction led to isolation of a major non-ketonic fraction (88%) and a small ketonic fraction (12%). The hypolipidemic activity was found to be associated with the ketonic preparation, which contains a number of steroids, including the two isomers *E*- and *Z*-guggulsterone (*cis*- and *trans*-4,17(20)-pregnadien-3,16-dione). These compounds, which constitute approximately 2% of gum guggul and 5% of guggulipid, by weight, are pregnane derivatives. They are reportedly devoid of any estrogenic, antiestrogenic or progestational activity, and recent studies confirm that they do not activate the conventional steroid receptors. Pharmacological studies revealed that the pure guggulsterone isomers had pronounced hypolipidemic activity (Satyavati, 1988).

The synthesis and stereochemistry of these compounds were already reported long before their actual isolation (Benn and Dodson, 1964). The compounds isolated from gum guggul were found to be identical to the synthesized compounds in all respects. The chemical synthesis method was used to produce the *E* and *Z* forms of guggulsterone so that the crystal structures of these steroids could be determined (Sarkhel *et al.*, 2001; Gupta *et al.*, 2006).

Besides the guggulsterones, sterols have also been isolated from guggul and evaluated for their pharmacological properties. Some of these properties, such as the antidiabetic and anti-inflammatory activities of specific fractions, have been patented (Bosely *et al.*, 2004; Pratap *et al.*, 2005).

Several analytical methods such as thin-layer chromatography (TLC), high-performance thin-layer chromatography (HPTLC) and HPLC applied for the separation and estimation of guggulsterones in oleo-gum resin and small tissue samples have been described (Ramawat *et al.*, 2008).

Several metabolites of guggulsterone have also been identified in various fungi, such as *Aspergillus niger* and *Cephalosporium aphidicola*. The hydroxylated metabolites, *cis*- and *trans*-7β-hydroxyl-4,17(20)-pregnadien-3,16-dione and 6β,11α-dihydroxyl-4,17(20)-pregnadien-3,16-dione, show antibacterial activity, whereas the 11α-hydroxylated metabolite has a free radical-scavenging activity (Atta-ur-Rahman *et al.*, 1998; Choudhary *et al.*, 2005).

Properties

Siddiqui *et al.* (2013) described the most important physico-chemical parameters of essential oils received from guggul exudates collected from Morena, Madhya Pradesh, India. The authors observed that the yield of the essential oil, obtained by hydro-distillation of the crushed exudates of guggul samples, was approximately 0.86% and its colour was yellowish to brown. The acid value of the essential oil obtained from freshly harvested guggul exudates was analysed as 5.24 (mg/KOH/g) and the saponification value was found to be 60.89 mg/KOH/g for the oil.

Myrrh

Chemistry

Commiphora myrrha (Syn. *C. molmol*) from the family of Burseraceae, is an endemic plant of north-east Africa (Dolara *et al.*, 1996) and Arabia (Etman *et al.*, 2011). The shrubs of

Commiphora myrrha secrete a natural reddish-brown oleo-gum resin named myrrh. The term myrrh is derived from the Arabic and Hebrew word 'murr', which means 'bitter' (El Ashry *et al.*, 2003). Myrrh contains a water-soluble gum portion and an oleoresin part, which is soluble in alcohol (Dolara *et al.*, 1996). Terpenoids, steroids, flavonoids, lignans, carbohydrates and long-chain aliphatic alcohol derivatives have been isolated from different species of the genus *Commiphora* (Su *et al.*, 2011). Myrrh contains 23–40% resin, 40–60% gum and 2–8% essential oil. The resin part of myrrh contains α-, β- and γ-commiphoric acid, commiphorinic acid, heeraboresene, α- and β-heerabomyrrhols, commiferin, kertosteroids, compesterol, β-sitosterol, cholesterol, α-amyrone and 3-epi-α-amyrin (Lemenih and Teketay, 2005). The gum consists of 18% protein and 64% carbohydrates. The carbohydrate portion of the gum has a chemical structure similar to acacia gum (El Ashry *et al.*, 2003). Hydrolysis of the gum results in arabinose, galactose, xylose and 4–0-mythylglucuronic acid (Lemenih and Teketay, 2005). Myrrh essential oil is pale yellow in colour (El Ashry *et al.*, 2003) and contains about 20 different furano-sesquiterpenoids (Su *et al.*, 2012) that are mainly responsible for the characteristic odour of myrrh (Abdul-Ghani *et al.*, 2009). The presence of three components that makes myrrh an oleo-gum resin, and their use in semi-solid formulations of Traditional Persian Medicine (TPM), suggest that it might have the potential to function as a multipurpose excipient in pharmaceutical dosage forms (Taghipour and Abadi, 2023).

Properties

The smell of myrrh is described as warm and balsamic, sweet and somewhat spicy, sharp and pungent when fresh. For this reason, perfumers have used myrrh as an absolute, oil or resinoid for a long time in oriental spicy bases, chypre accords, woody bases, forest notes, pine scents, etc. (Tucker, 1986). Myrrh has also been used in a number of modern perfumes, such as 'Fidji' by Guy Laroche, 'Onna' by Gay Farn and 'Volcan d'Amour' by Diane von Furstenburg (Fragrance Foundation, 1983). Another noteworthy homage to myrrh is 'Imperial Myrrh' by Giorgio Armani, in which the gum-resin plays a prominent role in the fragrance and which, with its captivating blend of lemon, spices, myrrh and frankincense, creates an oriental atmosphere.

The US Food and Drug Administration approved myrrh for use in foodstuffs at an early stage, while the Council of Europe did not include myrrh in the list of plants and plant parts approved for use in foodstuffs until 1981 (Ford *et al.*, 1992). Typical uses include: adhesive thickener, thickener, stabilizer, flavouring, fixative and emulsifier in food, clarification in beverages and release agent for rubber products.

Myrrh has also been used for some time as a bioinsecticide, particularly as a termite repellent and as a mosquito repellent when mixed as incense (Farah, 1994). Some components of myrrh resin (*Commiphora rostrata*) have been found to repel the corn rootworm, while gum-resin from *Commiphora africana* caused morphological deformities in adult beetles and pupae, significantly reducing the number of adults and increasing mortality rates (Karamalla, 1997).

The oleo-gum resin of *Commiphora myrrha* is used extensively in Arabic and Persian medicine for the treatment of some inflammatory disorders, various infections and bronchial complaints. It is also used as a mouthwash and to cure various stomach problems, including stomach cancer (Angeel *et al.*, 1987; Evans, 1989; Iwu, 1990; Brown, 1995; Rao *et al.*, 2001). Myrrh is also used in Arabia to treat tumours of the spleen, liver, stomach, breast, head, nose and eyes (Hartwell, 1982; Angeel *et al.*, 1987; Evans, 1989). Nowadays, untreated myrrh resin is used throughout East Africa and Saudi Arabia as an anti-inflammatory and antirheumatic medicine (Iwu, 1990). Two of the compounds found in myrrh (furanoeudesma-1,3-diene and curzarene) have been found to possess a pronounced analgesic effect (Archaeology, 1996).

In Europe, myrrh and myrrh tincture are officially listed in the European Pharmacopoeia and approved in the monographs of Commission E (Wichtl and Bisset, 1994; European Pharmacopoeia, 2023). The tincture is used as a monopreparation and also as an ingredient in various dental products, mouthwashes and toothpaste.

Frankincense (*Salai guggul*)

Chemistry

The composition of the essential oil distilled from frankincense and other components of the gum-resin obtained from various *Boswellia* species varies depending on the individual species and on the climate, harvesting conditions and geographical location (Ammon, 2006; Al-Harrasi and Al-Saidi, 2008). Frankincense is reported to contain 60–85% resin (mixture of various terpenoids), 6–30% gums (mixture of polysaccharides) and 5–9% essential oil (Rijkers *et al.*, 2006). The resin fraction is composed of pentacyclic triterpenes, of which boswellic acids form the most important active functional group (Goyal *et al.*, 2011). In contrast, the gum portion consists of pentose and hexose sugars with some oxidizing and digestive enzymes. The essential oil is a mixture of mono-, di-, and sesquiterpenes (Goyal *et al.*, 2011). In a study, it was reported that the resinous part of *Boswellia serrata* (so-called 'Indian frankincense') contains monoterpenes such as α-thujone, macrocyclic diterpenoids such as incensole, incensole oxide and iso-incensole oxide, the diterpene alcohol serratol, triterpenes such as α- and β-amyrins, pentacyclic triterpenic acids such as boswellic acids and tetracyclic triterpenic acids such as tirucall-8,24-dien-21-oic acids (Siddiqui, 2011). The four major boswellic acids found in frankincense are β-boswellic acid (BA), acetyl-β-boswellic acid (ABA), 11-keto-β-boswellic acid (KBA) and 3-O-acetyl-11-keto-β-boswellic acid (AKBA), which have been shown to be mainly responsible for inhibiting pro-inflammatory enzymes (Siddiqui, 2011). Among these four boswellic acids, AKBA is the most important inhibitor of 5-lipoxygenase, which is known to trigger inflammation (Siddiqui, 2011). AKBA has been found to be effective against a large number of inflammatory diseases such as arthritis, bronchial asthma, chronic colitis, ulcerative colitis, Crohn's disease and cancer (Raja *et al.*, 2011). The mechanism of action involves AKBA binding to 5-lipoxygenase in a calcium-dependent and reversible manner, acting as a non-redox, non-competitive inhibitor (Qurishi *et al.*, 2010).

Properties

Frankincense essential oil has anti-inflammatory and antibacterial properties for all skin types, including conditions such as acne and aging and dry skin, and is therefore particularly beneficial for acne-prone skin (Alraddadi and Shin, 2022). Furthermore, the essential oil promotes the formation of new cells, maintains skin elasticity and soothes dry and chapped skin (Hamidpour *et al.*, 2013). It reduces or prevents the appearance of sun spots and age spots and thus produces an even skin tone (Mikhaeil *et al.*, 2003). It also reduces skin redness and irritation (Eyre *et al.*, 2003) and has been successfully applied as a skin healing agent for bruises and infected wounds (Michie and Cooper, 1991). It is an effective remedy for wrinkles and skin aging (Kaur *et al.*, 2020) and can also be used to treat psoriasis, eczema (Togni *et al.*, 2014) and other skin diseases.

In addition, frankincense has been and is used in perfumery, in the food and beverage industry, for religious rituals and for various other local purposes (Ford *et al.*, 1992; Farah, 1994; FAO, 1995; Lemenih *et al.*, 2003). There is particular potential in the development of new natural products in the pharmaceutical, fragrance and flavouring sectors.

As already mentioned, the pharmaceutical efficacy is primarily due to the boswellic acids contained in frankincense (Corsano and Iavarone, 1964; Safayhi *et al.*, 1992; Ammon *et al.*, 1993; Huang *et al.*, 2000).

Pharmacological applications have shown that frankincense also has anti-tumour and anticancer properties (Huang *et al.*, 2000) and shows anti-inflammatory (Shao *et al.*, 1998; Safayhi *et al.*, 2000; Krieglstein *et al.*, 2001) and anti-proliferative activity (Glaser *et al.*, 1999; Hoernlein *et al.*, 1999). Furthermore, boswellic acids have been shown to be effective against chronic colitis (Gupta *et al.*, 2001), bronchial asthma (Gupta *et al.*, 1998; Safayhi *et al.*, 2000) and human leukaemia HL-60 cells (Shao *et al.*, 1998). Controlled double-blind studies have also shown that *Boswellia* extracts can be successfully used to treat ulcerative colitis (Singh and Atal, 1986). Due to the anti-inflammatory effect of boswellia extract or AKBA, results were achieved that were comparable to those of conventional treatments with synthetic standard medications such as prednisolone and sulfasalazine (Yamada *et al.*, 1993). However, treatment with boswellia extract was associated with significantly fewer side effects than treatment with the two synthetic steroids mentioned above (Gupta *et al.*, 1997).

References

Abdul-Ghani, R.A., Loutfy, N. and Hassan, A. (2009) Myrrh and trematodoses in Egypt: An overview of safety, efficacy and effectiveness profiles. *Parasitology International* 58, 210–214.

Abo-Shosha, M.H., Ibrahim, N.A., Allam, E. and El-Zairy, E. (2008) Preparation and characterization of polyacrylic acid/karaya gum and polyacrylic acid/tamarind seed gum adducts and utilization in textile printing. *Carbohydrate Polymers* 74, 241–249.

Adamu, H.M., Ushie, O.A. and Elisha, B. (2013) Chemical and nutrient analysis of raw and fermented seeds of *Cassia tora*. *Journal of Physical Science and Innovation* 5, 125–138.

Akiyama, Y., Eda, S. and Kato, K. (1984) Gum arabic is a kind of arabinogalactan-protein. *Agricultural and Biological Chemistry* 48, 235–237. DOI: 10.1080/00021369.1984.10866126.

Al-Assaf, S., Amar, V. and Philips, G. (2008) Characterization of gum ghatti and comparison with gum arabic. In: Williams, P. (ed.) *Gums and Stablizers for Food Industry*. Royal Society of Chemistry, Cambridge, UK, pp. 280–290.

Al-Assaf, S., Phillips, G. and Amar, V. (2009) Gum ghatti. In: Phillips, G. and Williams, P. (eds) *Handbook of Hydrocolloids*. CRC Press, Boca Raton, Florida, pp. 477–494.

Al-Assaf, S., Phillips, G.O. and Amar, V. (2021) Gum ghatti. In: *Handbook of Hydrocolloids*, 3rd edn. Woodhead Publishing, Sawston, UK, pp. 653–672.

Al-Harrasi, A. and Al-Saidi, S. (2008) Phytochemical analysis of the essential oil from botanically certified oleogum resin of *Boswellia sacra* (Omni Luban). *Molecules* 13, 2191–2189.

Ali, K.S.E. and Daffalla, H.M. (2018) Physicochemical and functional properties of the gum arabic from *Acacia senegal*. *Annals Food Science and Technology* 27, 27–34.

Ali, B.H., Ziada, A. and Blunden, G. (2009) Biological effects of gum arabic: A review of some recent research. *Food and Chemical Toxicology* 47, 1–8. DOI: 10.1016/j.fct.2008.07.001.

Alraddadi, B.G. and Shin, H.-J. (2022) Biochemical properties and cosmetic uses of *Commiphora myrrha* and *Boswellia serrata*. *Cosmetics* 9, 119. DOI: 10.3390/cosmetics9060119.

Amalraj, A. and Gopi, S. (2017) Biological activities and medicinal properties of asafoetida: A review. *Journal of Traditional and Complementary Medicine* 7, 347–359. DOI: 10.1016/j.jtcme.2016.11.004.

Ammon, H.P. (2006) Boswellic acids in chronic inflammatory diseases. *Planta Medica* 72, 1100–1116.

Ammon, H.P.T., Safayhi, T., Mark, T. and Sabieraj, J. (1993) Mechanism of anti-inflammatory actions of curcumine and boswellic acids. *Journal of Ethnopharamacology* 38, 113–119.

Anderson, D. and Stoddart, J. (1966) Studies on uronic acid materials: Part XV. The use of molecular-sieve chromatography in studies on *Acacia senegal* gum (gum arabic). *Carbohydrate Research* 2, 104–114.

Anderson, D.M.W. and Bridgeman, M.M.E. (1985) The composition of the proteinaceous polysaccharides exuded by *Astragalus microcephalus*, *A. gummifer* and *A. kurdicus* - The sources of Turkish gum tragacanth. *Phytochemistry* 24, 2301–2304. DOI: 10.1016/S0031-9422(00)83031-9.

Anderson, D., Hirst, S.E., Rahman, S. and Stainsby, G. (1967) Studies, on uronic acid materials: Part XVIII. Light-scattering studies on some molecular-weight fractions from *Acacia senegal* gum. *Carbohydrate Research* 3, 308–317.

Anderson, D.M.W., Dea, I.C.M., Karamalla, K.A. and Smith, LF. (1968) Studies on uronic acid materials: Part XXIV. An analytical study of different forms of the gum from *Acacia senegal* willd. *Carbohydrate Polymers* 6, 97–103. DOI: 10.1016/S0008-6215(00)80062-2.

Anderson, D.M.W., McNab, C.G.A., Anderson, C.G., Brown, P.M. and Pringuer, M.A. (1983) Studies of uronic acid materials, part 58*: Gum exudates from the genus *Sterculia* (gum karaya). *International Tree Crops Journal* 2, 147–154. DOI: 10.1080/01435698.1983.9752749.

Andriamanantena, R.W., Artaud, J., Gaydou, E.M., Iatrides, M.C. and Chevalier, J.L. (1983) Fatty acid and sterol compositions of malagasy tamarind kernel oils. *Journal of the American Oil Chemists' Society* 60(7), 1318–1321.

Angeel, A.M., Mossa, J.S., Tariq, M., Al-Yahya, M.A. and Al-Said, M.S. (1987) *Plants Used in Saudi Folk Medicine*. King Saud University Press, Riyadh.

Archaeology (1996) Medicinal myrrh. *Newsbrief* 49, 3. Archaeological Institute of America. Available at: https://archive.archaeology.org/9605/newsbriefs/myrrh.html (accessed 30 April 2025).

Aspinall, G., Hirst, E. and Wickstrom, A. (1955) Gum ghatti (Indian gum). The composition of the gum and the structure of two aldobiouronic acids derived from it. *Journal of the Chemical Society* 1160. Available at: https://doi.org/10.1039/JR9550001160

Aspinall, G., Auret, B. and Hirst, E. (1958) Gum ghatti (Indian gum). Part II. The hydrolysis products obtained from the methylated degraded gum and the methylated gum. *Journal of the Chemical Society* 221–230. Available at: https://doi.org/10.1039/JR9580000221

Aspinall, G., Bhavanandan, V. and Christensen, T. (1965) Gum ghatti (Indian gum). Part V. Degradation of the periodate-oxidised gum. *Journal of Chemical Society* 2677. Available at: https://doi.org/10.1039/JR9650002677

Atta-ur-Rahman, Choudhary, M.I., Shaheen, F., Ashraf, M. and Jahan, S. (1998) Microbial transformations of hypolipemic E-guggulsterone. *Journal of Natural Products* 61, 428–431.

Azouka, A., Huggett, R. and Harrison, A. (1993) The production of shellac and its general and dental uses: A review. *Journal of Oral Rehabilitation* 20, 393–400.

Banerjee, P.K., Srivastava, B.C. and Kumar, S. (1982) Cohesive-energy density of shellac. *Polymer* 23, 417–421. DOI: 10.1016/0032-3861(82)90345-7.

Bar, H. and Bianco-Peled, H. (2021) The unique nanostructure of shellac films. *Progress in Organic Coatings* 157, 106328. DOI: 10.1016/J.PORGCOAT.2021.106328.

Barak, S., Mudgil, D. and Taneja, S. (2020) Exudate gums: Chemistry, properties and food applications – A review. *Journal of the Science of Food and Agriculture* 100, 2828–2835. DOI: 10.1002/jsfa.10302.

Benn, W. and Dodson, R. (1964) The synthesis and stereochemistry of isomeric 16-hydroxy-17(20)-pregnenes. *Journal of Organic Chemistry* 29, 1142–1148.

Bertuzzi, M.A., Slavutsky, A.M. and Armada, M. (2012) Physicochemical characterisation of the hydrocolloid from Brea tree (*Cercidium praecox*). *International Journal of Food Science & Technology* 47, 768–775. DOI: 10.1111/j.1365-2621.2011.02907.x.

Bhattacharya, G. (1940) A note on the refractive index of shellac. *Indian Journal of Physics* 14, 237–246.

Bhattacharya, S., Bal, S., Mukherjee, R.K. and Bhattacharya, S. (1991) Rheological behaviour of tamarind (*Tamarindus indica*) kernel powder suspension. *Journal of Food Engineering* 13, 151–158.

Bhattacharya, S., Bal, S., Mukherjee, R.K. and Bhattacharya, S. (1994) Functional and nutritional properties of tamarind (*Tamarindus indica*) kernel protein. *Food Chemistry* 49, 1–9. DOI: 10.1016/0308-8146(94)90224-0.

Bhushette, P.R. and Annapure, U.S. (2017) Comparative study of *Acacia nilotica* exudate gum and acacia gum. *International Journal of Biological Macromolecules* 102, 266–271. DOI: 10.1016/j.ijbiomac.2017.03.178.

Bosely, J.A., AL, B. and Rogers, J.S. (2004) Food composition for reducing insulin resistance. US Patent no.6737442B2.

Bose, P.K. (1963) Chemistry of lac. *Journal of the Institution of Chemists (India)* 25, 83–98.

Brown, D. (1995) *Encyclopedia of Herbs and Their Uses*. Dorling Kindersley, New York.

Chandru, S. and Samsher, S. (2013) Assessment of functional properties of different flours. *African Journal of Agricultural Research* 8, 4849–4852. DOI: 10.5897/AJAR2013.6905.

Chauhan, G., Verma, A., Doley, A. and Ojha, K. (2019) Rheological and breaking characteristics of Zr-crosslinked gum karaya gels for high-temperature hydraulic fracturing application. *Journal of Petroleum Science and Engineering* 172, 327–339. DOI: 10.1016/j.petrol.2018.09.043.

Chikamai, B.N. and Banks, W.B. (1993) Gum arabic from *Acacia senegal* (L) Willd. in Kenya. *Food Hydrocolloids* 7, 521–534. DOI: 10.1016/S0268-005X(09)80246-4.

Chikamai, B.N., Banks, W.B., Anderson, D.M.W. and Weiping, W. (1996) Processing of gum arabic and some new opportunities. *Food Hydrocolloids* 10, 309–316. DOI: 10.1016/S0268-005X(96)80006-3.

Choudhary, M.I., Shah, S.A., Sami, A., Ajaz, A. and Shaheen, F. (2005) Fungal metabolites of (E)-guggulsterone and their antibacterial and radical-scavenging activities. *Chemistry & Biodiversity* 2, 516–524.

Chudzikowski, R.J.C. (1971) Guar gum and its application. *Journal of the Society of Cosmetic Chemists* 22, 43–60.

Coelho, C., Nanabala, R., Ménager, M., Commereuc, S. and Verney, V. (2012) Molecular changes during natural biopolymer ageing – The case of shellac. *Polymer Degradation and Stability* 97, 936–940. DOI: 10.1016/J.POLYMDEGRADSTAB.2012.03.024.

Connolly, S., Fenyo, J. and Vandevelde, M. (1988) Effect of a proteinase on the macromolecular distribution of *Acacia senegal* gum. *Carbohydrate Polymers* 8, 23–32. DOI: 10.1016/0144-8617(88)90033-1.

Corsano, S. and lavarone, C. (1964) Isolamento dall'incenso dell'acido 3-acetil-11-ossi-β-bos- vellico. *Gazzetta Chimica Italiana* 94, 328–339.

Crawford, L., McDonald, G.M. and Friedman, M. (1990) Composition of sickle pod (*Cassia obtusifolia*) toxic weed seeds. *Journal of Agricultural and Food Chemistry* 38, 2169–2175.

Dauqan, E. and Abdullah, A. (2013) Utilization of gum arabic for industries and human health. *American Journal of Applied Sciences* 10, 1270–1279. DOI: 10.3844/ajassp.2013.1270.1279.

Deshmukh, A.S., Setty, C.M., Badiger, A.M. and Muralikrishna, K.S. (2012) Gum ghatti: A promising polysaccharide for pharmaceutical applications. *Carbohydrate Polymers* 87, 980–986. DOI: 10.1016/j.carbpol.2011.08.099.

Dev, S. (1989) Chemistry of *Commiphora mukul* and development of a hypolipidemic drug. In: Rehman, A. (ed.) *Studies in Natural Product Chemistry*. Elsevier, Amsterdam, pp. 695–719.

Dickinson, E., Murray, B.S., Stainsby, G. and Anderson, D.M.W. (1988) Surface activity and emulsifying behaviour of some Acacia gums. *Food Hydrocolloids* 2, 477–490. DOI: 10.1016/S0268-005X(88)80047-X.

Dickinson, E., Elverson, D.J. and Murray, B.S. (1989) On the film-forming and emulsion-stabilizing properties of gum arabic: Dilution and flocculation aspects. *Food Hydrocolloids* 3, 101–114. DOI: 10.1016/S0268-005X(89)80020-7.

Dickinson, E., Galazka, V.B. and Anderson, D.M.W. (1991a) Emulsifying behaviour of gum arabic. Part 2: Effect of the gum molecular weight on the emulsion droplet-size distribution. *Carbohydrate Polymers* 14, 385–392. DOI: 10.1016/0144-8617(91)90004-V.

Dickinson, E., Galazka, V.B. and Anderson, D.M.W. (1991b) Emulsifying behaviour of gum Arabic. Part 1: Effect of the nature of the oil phase on the emulsion droplet-size distribution. *Carbohydrate Polymers* 14, 373–383. DOI: 10.1016/0144-8617(91)90003-U.

Dolara, P., Luceri, C., Ghelardini, C., Monserrat, C., Aiolli, S. *et al*. (1996) Analgesic effects of myrrh. *Nature* 379, 29.

Dror, Y., Yachin Cohen, Y. and Yerushalmi-Rozen, R. (2006) Structure of gum arabic in aqueous solution. *Wiley Online Library* 44, 3265–3271. DOI: 10.1002/polb.20970.

Duvallet, S., Fenyo, IC. and Vandevelde, M. (1993) The characterization of gum arabic from an experimental field of ferlo (North Senegal). *Food Hydrocolloids* 7, 319.

Dziezak, J.D. (1991) A focus on gums. *Food Technology (Chicago)* 45, 116–132.

Eddy, N.O., Abechi, S.E., Ameh, P. and Ebenso, E.E. (2013) GCMS, FTIR, SEM, physiochemical and rheological studies on *Albizia zygia* gum. *Walailak Journal of Science and Technology* 10, 247–265.

El Ashry, E.S.H., Rashed, N., Salama, O.M. and Saleh, A. (2003) Components, therapeutic value and uses of myrrh. *Die Pharmazie-An International Journal of Pharmaceutical Sciences* 58, 163–168.

Elamin, E.E., Balall, M.E. and Mahmoud, A.E. (2013) Effect of soil type and rainfall on *Acacia senegal* (L.) Willd. *Gum Viscosity* 6456, 6–9.

Elmanan, M., Al-Assaf, S., Phillips, G.O. and Williams, P.A. (2008) Studies on acacia exudate gums: part VI. interfacial rheology of *Acacia senegal* and *Acacia seyal*. *Food Hydrocolloids* 22, 682–689. DOI: 10.1016/j.foodhyd.2007.02.008.

Etman, M., Amin, M., Nada, A.H., Shams-Eldin, M. and Salama, O. (2011) Emulsions and rectal formulations containing myrrh essential oil for better patient compliance. *Drug Discoveries & Therapeutics* 5, 150–156.

European Pharmacopoeia (2023) *European Pharmacopoeia*, 11th edn. Deutscher Apothekerverlag, Stuttgart, Germany, p. 11.0-11.2.

Evans, W.C. (1989) *Trease and Evans' Pharmacognosy*, 13th edn. Bailliere Tindall, London, pp. 474–475.

Eyre, H., Hills, M.J. and Watkins, S.D. (2003) Compositions containing Boswellia extracts. US Patent no.6,589,516.

FAO (1995) *Flavours and Fragrances of Plant Origin. Non-Wood Forest Products 1*. Food and Agricultural Organization (FAO), Rome.

Farag, Y. and Leopold, C. (2009) Physicochemical properties of various shellac types. *Dissolution Technologies* 16, 33–39. DOI: 10.14227/DT160209P33.
Farah, A.Y. (1994) *The Milk of the Boswellia Forests. Frankincense Production among the Pastoral Somali.* EPOS, Environmental Policy and Society, Uppsala, Sweden.
Ford, R.A., Api, A.M. and Letizia, C.S. (1992) Monographs on fragrance raw materials. *Food and Chemical Toxicology* 30(Suppl), 1–138.
Fragrance Foundation (1983) *The Fragrance Foundation Reference Giude.* Fragrance Foundation, New York.
Freitas, R.A., Martin, S., Santos, G.L., Valenga, F., Buckeridge, M.S. *et al.* (2005) Physico-chemical properties of seed xyloglucans from different sources. *Carbohydrate Polymers* 60, 507–514.
Gardner, W. and Whitmore, W. (1929) Nature and constitution of shellac I – preliminary investigation of the action of organic solvents. *Industrial & Engineering Chemistry* 21, 226–229.
Glaser, T., Winter, S., Groscurth, P., Safayhi, H., Sailer, E.R. *et al.* (1999) Boswellic acids and malignant glioma: Induction of apoptosis but no modulation of drug sensitivity. *British Journal of Cancer* 80, 756–765.
Glicksman, M. (1982) *Food Hydrocolloids.* CRC Press, Boca Raton, Florida.
Glicksman, M. (ed.) (1983) *Gum Ghatti (Indian Gum).* CRC Press, Boca Raton, Florida.
Goodrum, L.J., Patel, A., Leykam, J.F. and Kieliszewski, M.J. (2000) Gum arabic glycoprotein contains glycomodules of both extensin and arabinogalactan-glycoproteins. *Phytochemistry* 54, 99–106. DOI: 10.1016/S0031-9422(00)00043-1.
Gören, A.C., Bilsel, G., Öztürk, A.H. and Topçu, G. (2010) Chemical composition of natural colophony from *Pinus brutia* and comparison with synthetic colophony. *Natural Product Communications* 5, 1729–1732.
Goyal, S., Sharma, P., Ramchandani, U., Shivastanva, S.K. and Dubey, P.K. (2011) Novel anti-inflammatory topical herbal gels containing *Withania somnifera* and *Boswellia serrata*. *International Journal of Pharmaceutical and Biological Archive* 2, 94.
Grasdalen, H. and Painter, T. (1980) NMR studies of composition and sequence in legume-seed galactomannans. *Carbohydrate Research* 81, 59–66.
Gunasena, H.P.M. and Hughes, A. (2000) *Tamarind: Tamarindus indica L.* International Centre for Underutilised Crops, University of Southampton, Southampton, UK.
Gupta, I., Parihar, A. and Malhotra, P. (1997) Effects of *Boswellia serrata* gum resin in patients with ulcerative colitis. *European Journal of Medical Research* 2, 37–43.
Gupta, I., Gupta, V., Parihar, A., Gupta, S., Ludtike, R. *et al.* (1998) Effects of *Boswellia serrata* gum resin in patients with bronchial Asthma: Results of a double-blind, placebo-controlled 6-week clinical study. *European Journal of Medical Research* 3, 1–5.
Gupta, I., Parihar, A., Malhotra, P., Gupta, S., Ludtke, R. *et al.* (2001) Effects of gum resin of *Boswellia serrata* in patients with chronic colitis. *Planta Medica* 67, 391–395.
Gupta, V.K., Bandhoria, P., Gupta, B.D. and Gupta, K.K. (2006) Crystal structure of guggulsterone Z. *Crystallography Report* 51, 265–270.
Hamidpour, R., Hamidpour, S., Hamidpour, M. and Shahlari, M. (2013) Frankincense (乳香 rǔ xiāng; boswellia species): From the selection of traditional applications to the novel phytotherapy for the prevention and treatment of serious diseases. *Journal of Traditional and Complementary Medicine* 3, 221–226.
Hanus, L.O., Rezanka, T., Dembitsky, V.M. and Moussaieff, A. (2005) Myrrh – commiphora chemistry. *Biomedical Papers of the Medical Faculty of the University Palacky, Olomouc, Czech Republic* 149, 3–27.
Hartwell, R. (1982) *Plants Used Against Cancer. A Survey*, Vol. 2. Quaterman Publication, Lawrence, Massachusetts, pp. 88–93.
Hobbs, C.A., Swartz, C., Maronpot, R., Davis, J., Recio, L. *et al.* (2012) Evaluation of the genotoxicity of the food additive, gum ghatti. *Food and Chemical Toxicology* 50, 854–860. DOI: 10.1016/j.fct.2011.11.021.
Hoernlein, R.F., Orlikowsky, T., Zehrer, C., Niethammer, D., Sailer, E.R. *et al.* (1999) Acetyl-11-keto-beta-boswellic acid induces apoptosis in HL-60 and CCRFCEM cells and inhibits topoisomerase. *Journal of Pharmacology and Experimental Therapeutics* 288, 613–619.
Huang, M.-T., Badmaev, V., Ding, Y., Liu, Y., Xie, J.-G. *et al.* (2000) Anti-tumor and anti-carcinogenic activities of triterpenoid, β-boswellic acid. *Biofactors* 13, 225–230.

Huang, X., Kakuda, Y. and Cui, W. (2001) Hydrocolloids in emulsions: Particle size distribution and interfacial activity. *Food Hydrocolloids* 15, 533–542. DOI: 10.1016/S0268-005X(01)00091-1.

Idris, O.H.M., Williams, P.A. and Phillips, G.O. (1998) Characterisation of gum from *Acacia senegal* trees of different age and location using multidetection gel permeation chromatography. *Food Hydrocolloids* 12, 379–388. DOI: 10.1016/S0268-005X(98)00058-7.

Imeson, A. (2010) *Food stabilisers, Thickeners and Gelling Agents*, Vol. 3. Wiley-Blackwell, Chichester, UK.

Imeson, A. (ed.) (1992) *Thickening and Gelling Agents for Food*. Springer Science Business Media, Hong Kong.

Iranshahy, M. and Iranshahi, M. (2011) Traditional uses, phytochemistry and pharmacology of asafoetida (Ferula assa-foetida oleo-gum-resin) - A review. *Journal of Ethnopharmacology* 134, 1–10. DOI: 10.1016/j.jep.2010.11.067.

Islam, A.M., Phillips, G.O., Sljivo, A., Snowden, M.J. and Williams, P.A. (1997) A review of recent developments on the regulatory, structural and functional aspects of gum arabic. *Food Hydrocolloids* 11, 493–505. DOI: 10.1016/S0268-005X(97)80048-3.

Iwu, M.M. (1990) *Handbook of African Medicinal Plants*. CRC Press, Boca Raton, Florida, pp. 160–161.

Jahanbin, K., Moini, S., Gohari, A.R., Emam-Djomeh, Z. and Masi, P. (2012) Isolation, purification and characterization of a new gum from *Acanthophyllum bracteatum* roots. *Food Hydrocolloids* 27, 14–21.

Jefferies, M., Pass, G. and Phillips, G.O. (1977) Viscosity of aqueous solutions of gum ghatti. *Journal of the Science of Food and Agriculture* 28, 173–179. DOI: 10.1002/JSFA.2740280211.

Jefferies, M., Pass, G., Phillips, G.O. and Zakaria, M.B. (1978) The effect of metal ion content on the viscosity of gum ghatti. *Journal of the Science of Food and Agriculture* 29, 193–200. DOI: 10.1002/JSFA.2740290218.

Jefferies, M., Konadu, E.Y. and Pass, G. (1982) Cation effects on the viscosity of gum ghatti. *Journal of the Science of Food and Agriculture* 33, 1152–1159. DOI: 10.1002/JSFA.2740331113.

Jermyn, M.A. (1962) Fungal cellulases X. further purification of the?-glucosidase of *Stachybotrys atra*. *Australian Journal of Bological Sciences* 15, 769–786.

Joshi, S.B. and Verma, K.C. (1964) Panwar gum as suspending and emulsifying agent (*Cassia tora*). *Indian Journal of Pharmacology* 26, 175–177.

Kamath, N.H. and Mainkar, V.B. (1950) Iodometric determination of acid value of lac. *Analytical Chemistry* 22, 724–726. DOI: 10.1021/AC60041A035.

Kang, J., Cui, S.W., Chen, J., Phillips, G.O., Wu, Y. *et al.* (2011a) New studies on gum ghatti (*Anogeissus latifolia*) part I: Fractionation, chemical and physical characterization of the gum. *Food Hydrocolloids* 25, 1984–1990. DOI: 10.1016/J.FOODHYD.2010.12.011.

Kang, J., Cui, S.W., Phillips, G.O., Chen, J., Guo, Q. *et al.* (2011b) New studies on gum ghatti (*Anogeissus latifolia*) part III: Structure characterization of a globular polysaccharide fraction by 1D, 2D NMR spectroscopy and methylation analysis. *Food Hydrocolloids* 25, 1999–2007. DOI: 10.1016/J.FOODHYD.2010.11.020.

Kang, J., Cui, S.W., Phillips, G.O., Chen, J., Guo, Q. *et al.* (2011c) New studies on gum ghatti (*Anogeissus latifolia*) part II: Structure characterization of an arabinogalactan from the gum by 1D, 2D NMR spectroscopy and methylation analysis. *Food Hydrocolloids* 25, 1991–1998. DOI: 10.1016/J.FOODHYD.2010.11.021.

Kang, J., Cui, S.W., Guo, Q., Chen, J., Wang, Q. *et al.* (2012) Structural investigation of a glycoprotein from gum ghatti. *Carbohydrate Polymers* 89, 749–758. DOI: 10.1016/j.carbpol.2012.04.004.

Kang, J., Guo, Q., Phillips, G.O. and Cui, S.W. (2014) Understanding the structure-emulsification relationship of gum ghatti - A review of recent advances. *Food Hydrocolloids* 42, 187–195. DOI: 10.1016/j.foodhyd.2014.04.010.

Kang, J., Guo, Q., Wang, Q., Phillips, G.O. and Cui, S.W. (2015) New studies on gum ghatti (*Anogeissus latifolia*) part 6: Physicochemical characteristics of the protein moiety of gum ghatti. *Food Hydrocolloids* 44, 237–243. DOI: 10.1016/j.foodhyd.2014.09.013.

Karamalla, A. (1997) The chemical characterization of myrrh and frankincense and opportunities for commercial utilization. In: Mugah, J.O., Chikamai, B.N., Mbiru, S.S. and Casadei, E. (eds) *Conservation, Management and Utilization of Plant Gums, Resins and Essential Oils. Proceedings of a Regional Conference for Africa held in Nairobi, Kenya, 6-10 October*, pp. 75–78.

Karamalla, K.A., Siddig, N. and Osman, M. (1998) Analytical data for *Acacia senegal* var. *Senegal* gum samples collected between 1993 and 1995 from Sudan. *Food Hydrocolloids* 12, 373–378. DOI: 10.1016/S0268-005X(98)00005-8.

Kaur, A., Singh, T.G., Dhiman, S., Arora, S. and Babbar, R. (2020) Novel herbs used in cosmetics for skin and hair care: A review. *Plant Archives* 20, 3784–3793.

Kaur, L., Singh, J. and Singh, H. (2009) Characterization of gum ghatti (*Anogeissus latifolia*): A structural and rheological approach. *Journal of Food Science* 74, E328–E332. DOI: 10.1111/J.1750-3841.2009.01244.X.

Khounvilay, K. and Sittikijyothin, W. (2012) Rheological behaviour of tamarind seed gum in aqueous solutions. *Food Hydrocolloids* 26, 334–338. DOI: 10.1016/J.FOODHYD.2011.03.019.

Khurana, R.G., Singh, A.N., Upadhye, A.B., Mhaskar, V.V. and Dev, S. (1970) Chemistry of lac resin–III. Lac acids–3: An integrated procedure for their isolation from hard resin; chromatography characteristics and quantitative determination. *Tetrahedron* 26, 4167–4175. DOI: 10.1016/S0040-4020(01)93058-3.

Krieglstein, F.C., Anthoni, K., Rijcken, E.J.M., Laukötter, M., Spiegel, H.-U. *et al.* (2001) Acetyl-11-keto-β-boswellic acid, a constituent of a herbal medicine from *Boswellia serrata* resin, attenuates experimental ileitis. *International Journal of Colorectal Disease* 16, 88–95. DOI: 10.1007/s003840100292.

Kumar, C.S. and Bhattacharya, S. (2008) Tamarind seed: Properties, processing and utilization. *Critical Reviews in Food Science and Nutrition* 48(1), 1–20.

Kumbhare, V. and Bhargava, A. (1999) Studies on the nutritional composition of *Sterculia* species. *Journal of Food Science and Technology* 36, 542–544.

Kunkel, M.E. and Seo, A. (1994) *In vitro* digestibility of selected polymers. *Journal of Environmental Polymer Degradation* 2, 245–251. DOI: 10.1007/BF02071972.

Langenheim, J.H. (2003) *Plant Resins: Chemistry, Evolution, Ecology and Ethnobotany*. Timbers Press, Portland, Oregon.

Le Cerf, D., Irinei, F. and Muller, G. (1990) Solution properties of gum exudates from *Sterculia urens* (karaya gum). *Carbohydrate Polymers* 13, 375–386. DOI: 10.1016/0144-8617(90)90037-S.

Lemenih, M. and Teketay, D. (2005) Frankincense and myrrh resources of Ethiopia: II. Medicinal and industrial uses. *Ethiopian Journal of Science* 26, 161–172.

Lemenih, M., Abebe, T. and Olsson, M. (2003) Gum and resin resources from some Acacia, Boswellia and Commiphora species and their economic contributions in ILiban, South-East Ethiopia. *Journal of Arid Environments* 55(3), 465–482. DOI: 10.1016/S0140-1963(03)00053-3.

Lopez-Torrez, L., Nigen, M., Williams, P., Doco, T. and Sanchez, C. (2015) *Acacia senegal* vs. *Acacia seyal* gums – part 1: Composition and structure of hyperbranched plant exudates. *Food Hydrocolloids* 51, 41–53. DOI: 10.1016/J.FOODHYD.2015.04.019.

Mahendran, T., Williams, P.A., Phillips, G.O., Al-Assaf, S. and Baldwin, T.C. (2008) New insights into the structural characteristics of the arabinogalactan-protein (AGP) fraction of gum arabic. *Journal of Agricultural and Food Chemistry* 56, 9269–9276. DOI: 10.1021/jf800849a.

Maronpot, R.R., Davis, J., Moser, G., Giri, D.K. and Hayashi, S.-M. (2013) Evaluation of 90-day oral rat toxicity studies on the food additive, gum ghatti. *Food and Chemical Toxicology* 51, 215–224. DOI: 10.1016/j.fct.2012.09.037.

Marvelys, L., Maritza, M., Lilian, S., Pinto Gladys, L. and Julio, H. (2006) Structural elucidation of the polysaccharide from *Sterculia apetala* gum by a combination of chemical methods and NMR spectroscopy. *Food Hydrocolloids* 20, 908–913. DOI: 10.1016/j.foodhyd.2005.09.005.

Meer, G. (1980) Gum ghatti. In: Davidson, R.L. (ed.) *Handbook of Water-Soluble Gums and Resins*. McGraw Hill, New York, ch.9, p. 1.

Meer, G., Meer, W.A. and Gerard, A. (1973) Gum ghatti. In: Whistler, R.L. and BeMiller, J.N. (eds) *Industrial Gums*. Academic Press, New York, pp. 265–271.

Michie, C.A. and Cooper, E. (1991) Frankincense and myrr as remedies in children. *Journal of the Royal Society of Medicine* 84, 602–605.

Mikhaeil, B.R., Maatooq, G.T., Badria, F.A. and Amer, M.M. (2003) Chemistry and immunomodulatory activity of frankincense oil. *Zeitschrift Für Naturwissenschaft C* 58, 230–238.

Murthy, H.N. (2022) Chemical constituents and applications of gums, resins, and latexes of plant origin. In: Murthy, H.N. (ed.) *Gums, Resins and Latexes of Plant Origin*. Reference Series in Phytochemistry, Springer, Cham, Switzerland.

Nie, S.P., Wang, C., Cui, S.W., Wang, Q., Xie, M.Y. *et al*. (2013) A further amendment to the classical core structure of gum arabic (*Acacia senegal*). *Food Hydrocolloids* 31, 42–48. DOI: 10.1016/J.FOODHYD.2012.09.014.

Nishinari, K., Yamatoya, K. and Shirakawa, M. (2000) Xyloglucan. In: Phillips, G.O. and Williams, P.A. (eds) *Handbook of Hydrocolloids*. Woodhead Publishing, Sawston, UK, pp. 247–267.

Nityanand, S. and Kapoor, N. (1975) Hypolipidaemic effect of ethyl acetate fraction of *Commiphora mukul* (guggul) in rats. *Indian Journal of Pharmacology* 7, 106.

Panda, H. (2004) *The Complete Technology Book on Natural Products: Forest Based*. Gyan Books, New Delhi.

Parija, S., Misra, M. and Mohanty, A.K. (2001) Studies of natural gum adhesive extracts: An overview. *Journal of Macromolecular Science, Part C: Polymer Reviews* 41, 175–197.

Pawar, H.A. and D'mello, P.M. (2011) *Cassia tora* Linn.: An overview. *International Journal of Pharmaceutical Sciences and Research* 5, 2286–2291. Available at: http://dx.doi.org/10.13040/IJPSR.0975-8232.2(9).2286-91

Pérez-Mosqueda, L.M., Ramírez, P., Alfaro, M.C., Rincón, F. and Muñoz, J. (2013) Surface properties and bulk rheology of *Sterculia apetala* gum exudate dispersions. *Food Hydrocolloids* 32, 440–446. DOI: 10.1016/j.foodhyd.2013.02.007.

Phillips, G.O., Ogasawara, T. and Ushida, K. (2008) The regulatory and scientific approach to defining gum arabic (*Acacia senegal* and *Acacia seyal*) as a dietary fibre. *Food Hydrocolloids* 22, 24–35. DOI: 10.1016/J.FOODHYD.2006.12.016.

Phillips, G.O., Wedlock, D.J. and Williams, P.A. (eds) (1990) *Gums and Stabilisers for the Food Industry*. Oxford University Press, Oxford, UK.

Prabhanjan, H. (1989) Studies on modified tamarind kernel powder. I. Preparation and physicochemical properties of sodium salt of carboxymethyl derivatives. *Starch/Staerke* 41, 409–414.

Prabhanjan, H. and Ali, S.Z. (1995) Studies on rheological properties of tamarind kernel powder, its derivatives and their blends with maize starch. *Carbohydrate Polymers* 28, 245–253.

Pratap, R., Pal, R., Singh, S., Shankar, G., Nath, C. *et al*. (2005) Method of treating a cognitive memory dysfunction using guggulipid. US Patent No. 6,896,901 B2.

Prinyawiwatkul, W., McWatters, K., Beuchat, L. and Phillips, R.D. (1997) Functional characteristics of cowpea (*Vigna unguiculata*) flour and starch as affected by soaking, boiling, and fungal fermentation before milling. *Food Chemistry* 58, 361–723.

Qurishi, Y., Hamid, A., Zargar, M.A., Singh, S.K. and Saxena, A.K. (2010) Potential role of natural molecules in health and disease: Importance of boswellic acid. *Journal of Medicinal Plants Research* 4(25), 2778–2785.

Raj, V., Lee, J.H., Shim, J.J. and Lee, J. (2021) Recent findings and future directions of grafted gum karaya polysaccharides and their various applications: A review. *Carbohydrate Polymers* 258, 117687. DOI: 10.1016/j.carbpol.2021.117687.

Raja, A.F., Ali, F., Khan, I.A., Shawl, A.S. and Arora, D.S. (2011) Acetyl-11-keto-β-boswellic (AKBA): Targeting oral cavity pathogens. *BMC Research Notes* 4, 406. DOI: 10.1186/1756-0500-4-406.

Ramawat, K.G., Marthur, M., Dass, S. and Suthar, S. (2008) Guggulsterone: A potent natural hypolipidemic agent from *Commiphora wightii* – problems, perseverance, and prospects. In: Ramawat, K.G. and Merillon, J.M. (eds) *Bioactive Molecules and Medicinal Plants*. Springer, Berlin, Germany, pp. 101–121.

Randall, R.C., Phillips, G.O. and Williams, P.A. (1988) The role of the proteinaceous component on the emulsifying properties of gum arabic. *Food Hydrocolloids* 2, 131–140.

Rao, R.M., Khan, Z.A. and Shah, A.H. (2001) Toxicity studies in mice of *Commiphora molmol* oleo-gum resin. *Journal of Ethnopharamacognosy* 76, 151–154. DOI: 10.1016/s0378-8741(01)00189-1.

Rasmussen, S.C. (2017) The early history of polyaniline: Discovery and origins. *Substantia* 1, 99–109. DOI: 10.13128/substantia-30.

Ray, A.K., Bird, P.B., Iacobucci, G.A. and Clark, B.C. (1995) Functionality of gum arabic. Fractionation, characterization and evaluation of gum fractions in citrus oil emulsions and model beverages. *Topics in Catalysis* 9, 123–131. DOI: 10.1016/S0268-005X(09)80274-9.

Rijkers, T., Ogbazghi, W., Wessel, M. and Bongers, F. (2006) The effect of tapping for frankincense on sexual reproduction in *Boswellia papyrifera*. *Journal of Applied Ecology* 43(6), 1188–1195.

Saeidy, S., Nasirpour, A., Djelveh, G., Ursu, A.V., Marcati, A. *et al*. (2018) Rheological and functional properties of asafoetida gum. *International Journal of Biological Macromolecules* 118, 1168–1173. DOI: 10.1016/j.ijbiomac.2018.06.177.

Safayhi, H., Mack, T. and Saieraj, J. (1992) Boswellic acids: Novel, specific, nonredox inhibitors of 5-lipoxygenase. *Journal of Pharmacology and Experimental Therapeutics* 261, 1143–1146.

Safayhi, H., Boden, S.E., Schweizer, S. and Ammon, H.T. (2000) Concentration-dependent potentiating and inhibitory effects of Boswellia extracts on 5-Lipoxygenase product fomation in stimulated PMNL. *Planta Medica* 66, 110–113.

Sahu, P., Pisalkar, P.S., Patel, S. and Katiyar, P. (2019) Physico-chemical and rheological properties of karaya gum (*Sterculia urens* Roxb.). *International Journal of Current Microbiology and Applied Sciences* 8, 672–681. DOI: 10.20546/ijcmas.2019.804.072.

Sahu, P., Patel, S., Pisalkar, P.S., Katiyar, P. and Khokhar, D. (2023) Nutritional, physical and thermal properties of *Cassia tora* (Charota) seeds. *The Pharma Innovation Journal* 12, 2291–2299.

Sakai, E., Katayama, T., Ogasawara, T. and Mizuno, M. (2013) Identification of *Anogeissus latifolia* wallich and analysis of refined gum ghatti. *Journal of Natural Medicines* 67, 276–280. DOI: 10.1007/S11418-012-0678-3.

Salazar-Montoya, J.A., Ramos-Ramirez, E.G. and Delgado-Reyes, V.A. (2002) Changes of the dynamic properties of tamarind (*Tamarindus indica*) gel with different saccharose and polysaccharide concentrations. *Carbohydrate Polymers* 49, 387–391.

Sarkar, P.C. and Shrivastava, A.K. (1997) FTIR spectroscopy of lac resin and its derivatives. *Pigment and Resin Technology* 26, 370–377. DOI: 10.1108/EUM0000000004464/FULL/HTML.

Sarkar, P., Sahu, U., Binsi, P., Nayak, N., Ninoa, G. *et al.* (2018) Studies on physico-chemical and functional properties of some natural Indian gums. *Asian Journal of Dairy and Food Research* 37, 126–131.

Sarkhel, S., Yadava, U., Prakas, P., Jain, G.K., Singh, S. *et al.* (2001) Guggulsterone E, a lipid-lowering agent from *Commiphora mukul*. *Acta Cryallography* E57, 285–286.

Satyavati, G.V. (1988) Gum guggul (*Commiphora mukul*) – The success story of an ancient insight leading to a modern discovery. *Indian Journal of Medical Research* 87, 327–335.

Shao, Y., Ho, C.-T., Chin, C.-K., Badmaev, V., Ma, W. *et al.* (1998) Inhibitory activity of boswellic acids from *Boswellia serrata* against human leukaemia HL-60 cells in culture. *Planta Medica* 64, 328–331.

Sharma, K.K. (2017) Lac insects and host plants. In: Omkar (ed.) *Industrial Entomology*. Springer, Singapore, pp. 157–180.

Sharma, K.K., Monobrullah, M.d., Mohanasundaram, A. and Ramani, R. (2016) *Beneficial Insect Farming*. ICAR-Indian Institute of Natural Resins and Gums, Ranchi, India.

Sharma, S., Shukla, S. and Vaid, D. (1983) Shellac-structure, characteristics & modification. *Defence Science Journal* 33, 261–271.

Siddhuraja, P., Vijayakumari, K. and Janardhanan, K. (1995) Nutritional and antinutritional properties of the underexploited legumes *Cassia laevigata* Willd. and *Tamarindus indica* L.. *Journal of Food Composition and Analysis* 8, 351–362.

Siddiqui, M.Z. (2011) *Boswellia serrata*, a potential antiinflammatory agent: An overview. *Indian Journal of Pharmaceutical Sciences* 73, 255–261.

Siddiqui, M.Z., Thomas, M. and Prasad, N. (2013) Physicochemical characterization and antioxidant activity of essential oils of guggul (*Commiphora wightii*) collected from Madhya Pradesh. *Indian Journal of Pharmaceutical Sciences* 75(3), 368–372.

Silva, S.C.C.C., Araujo Braz, E.M. de, Amorim Carvalho, F.A. de, Sousa Brito, C.A.R. de, Brito, L.M. *et al.* (2020) Antibacterial and cytotoxic properties from esterified *Sterculia* gum. *International Journal of Biological Macromolecules* 164, 606–615. DOI: 10.1016/j.ijbiomac.2020.07.031.

Sims, I.M., Gane, A.M., Dunstan, D., Allan, G.C., Boger, D.V. *et al.* (1998) Rheological properties of xyloglucans from different plant species. *Carbohydrate Polymers* 37, 61–69.

Singh, G.B. and Atal, C.K. (1986) Pharmacology of an extract of salai guggal ex-*Boswellia serrata*, a new non-steroidal anti-inflammatory agent. *Agents Actions* 18, 407–412.

Singh, A.N., Upadhye, A.B., Mhaskar, V.V., Dev, S., Pol, A.V. *et al.* (1974) Chemistry of lac resin–VII: Pure lac resin–3: Structure. *Tetrahedron* 30, 3689–3693.

Smith, F. and Montgomery, R. (1959) *The chemistry of plant gums and mucilages*. Reinhold, New York.

Soni, P.L. and Pal, R. (1996) Industrial gum from *Cassia tora* seeds. *Trends in Carbohydrate Chemistry* 2, 33–44.

Srichamroen, A. (2007) Influence of temperature and salt on viscosity property of guar gum. *Naresuan University Journal: Science and Technology* 15, 55–62.

Stephen, A. (ed.) (1995) *Food Polysaccharides and Their Applications*. CRC Press, London.

Street, C.A. and Anderson, D.M.W. (1983) Refinement of structures previously proposed for gum arabic and other acacia gum exudates. *Talanta* 30, 887–893. DOI: 10.1016/0039-9140(83)80206-9.

Su, S., Wang, T., Duan, J.-A., Zhou, W., Hua, Y.-Q. *et al.* (2011) Anti-inflammatory and analgesic activity of different extracts of *Commiphora myrrha*. *Journal of Ethnopharmacology* 134, 251–258.

Su, S., Hua, Y., Wang, Y., Gu, W., Zhou, W. *et al.* (2012) Evaluation of the anti-inflammatory and analgesic properties of individual and combined extracts from *Commiphora myrrha*, and *Boswellia carterii*. *Journal of Ethnopharmacology* 139, 649–656.

Sumathi, S. and Alok, R. (2002) Release behavior of drugs from tamarind seed polysaccharide tablets. *Journal of Pharmacy and Pharmaceutical Sciences* 5, 12–18.

Taghipour, Z. and Abadi, H.H. (2023) Herbal medicine for cardiovascular desease in view of Avicenna. In: Schulz, H., Emami, S.A. and Nadjafi, F. (eds) *Medicinal Plants Used in Traditional Persian Medicine*. CAB International, Wallingford, UK, pp. 230–280.

Tahir, A., Elkheir, M. and Yagoub, A. (2007) Effect of tree and nodule age on some physicochemical properties of gum from *Acacia senegal* (L.) Wild., Sudan. *Research Journal of Agricultural and Biological Sciences* 3, 866–870.

Takeoka, G. (2001) Volatile constituents of asafoetida. In: Takeoka, G.R., Guntert, M. and Engel, K.-H. (eds) *Aroma Active Compounds in Foods*. American Chemical Society, Washington, DC.

Tamburini, D., Dyer, J. and Bonaduce, I. (2017) The characterisation of shellac resin by flow injection and liquid chromatography coupled with electrospray ionisation and mass spectrometry. *Scientific Reports* 7, 1–15. DOI: 10.1038/s41598-017-14907-7.

Thombare, N., Mate, C.J., Tamilarashi, K., Chowdhury, A.R. and Srivastava, S. (2018) Physico-chemical characterization and microbiological evaluation of gum ghatti as potential food additive. *Multilogic in Science* VIII, 316–319.

Togni, S., Maramaldi, G., Pierro, F. and Biondi, M. (2014) A cosmeceutical formulation based on boswellic acids for the treatment of erythematous eczema and psoriasis. *Clinical, Cosmetic and Investigational Dermatology* 7, 321–327.

Tripathy, S. and Das, M.K. (2013) Guar gum: Present status and applications. *Journal of Pharmaceutical and Scientific Innovation* 2, 24–28.

Tucker, A.O. (1986) Frankincense and myrrh. *Economic Botany* 40(4), 425–433.

Ukhun, M.E. and Ifebigh, E.O. (1988) Compositional chemistry of *Cassia alata* seeds. *Food Chemistry* 30, 205–210.

Vadivel, V. and Janardhanan, K. (2000) Chemical composition of underutilized legume *Cassia hirsute* L. *Plant Foods for Human Nutrition* 55, 369–381.

van den Berg, K.J., Boon, J.J., Pastorova, I. and Spetter, L.F. (2000) Mass spectrometric methodology for the analysis of highly oxidized diterpenoid acids in old master paintings. *Journal of Mass Spectrometry* 35, 512–533.

van den Werf, I.D., Berg, K.J., Schmitt, S. and Boon, J.J. (2000) Molecular characterization of copaiba balsam as used in painting techniques and restoration procedures. *Studies in Conversation* 45, 1.

Vijayendran, B.R. and Bone, T. (1984) Absolute molecular weight and molecular weight distribution of guar by size exclusion chromatography and low-angle laser light scattering. *Carbohydrate Polymers* 4, 299–313.

Vinod, V.T.P., Sashidhar, R.B., Sarma, V.U.M. and Vijaya Saradhi, U.V.R. (2008) Compositional analysis and rheological properties of gum kondagogu (*Cochlospermum gossypium*): A tree gum from India. *Journal of Agricultural and Food Chemistry* 56, 2199–2207. DOI: 10.1021/JF072766P.

Vinod, V.T.P., Sashidhar, R.B., Sarma, V.U.M. and Raju, S.S. (2010) Comparative amino acid and fatty acid compositions of edible gums kondagogu (*Cochlospermum gossypium*) and karaya (*Sterculia urens*). *Food Chemistry* 123, 57–62. DOI: 10.1016/j.foodchem.2010.03.127.

Wadia, M.S., Khurana, R., Mhaskar, V.V. and Dev, S. (1969) Chemistry of lac resin—I. Lac acids (part 1): Butolic, jalaric and laksholic acids*. *Tetrahedron* 25, 3841–3854. DOI: 10.1016/S0040-4020(01)82916-1.

Walde, S.G., Balaswamy, K., Shivaswamy, R., Chakkaravarthi, A. and Rao, D.G. (1997) Microwave drying and grinding characteristics of gum karaya (*Sterculia urens*). *Journal of Food Engineering* 31, 305–313. DOI: 10.1016/S0260-8774(96)00082-9.

Wang, Q., Ellis, P.R., Ross-Murphy, S.B. and Burchard, W. (1997) Solution characteristics of the xyloglucan extracted from *Detarium senegalense* Gmelin. *Carbohydrate Polymers* 33, 115 124.

Wang, Q., Burchard, W., Cui, S.W., Huang, X. and Phillips, G.O. (2008) Solution properties of conventional GUM arabic and a matured GUM arabic (Acacia (sen) SUPER GUM). *Biomacromolecules* 9, 1163–1169. DOI: 10.1021/bm7011696.

Weinberger, H. and Gardner, W.H. (1933) Nature and constitution of shellac. *Industrial and Engineering Chemistry - Analytical Edition* 5, 267–270. DOI: 10.1021/AC50084A024.

Weiping, W. and Branwell, A. (2000) Tragacanth and karaya. In: Philips, G.O. and Williams, P.A. (eds) *Handbook of Hydrocolloids*, Woodhead Publishing, Sawston, UK, pp. 231–246.

Whistler, R.L. (1993) Exudate gums. In: *Industrial Gums: Polysaccharides and Their Derivatives*, 3rd edn. Elsevier Inc, Amsterdam, pp. 309–339.

Whistler, R.L. and Hymowitz, T. (1979) *Guar: Agronomy, Production, Industrial Use, and Nutrition*. Purdue University Press, West Lafayette, Indiana.

Wichtl, M. and Bisset, N.G. (eds) (1994) *Herbal Drugs and Phytopharmaceuticals*. Medpharm Scientific Publishers, Stuttgart, Germany.

Williams, P.A. (1982) Adsorption of polyelectrolytes onto barium sulphate. PhD thesis, University of Salford, Salford, UK.

Yadav, H. and Karthikeyan, C. (2019) Natural polysaccharides: Structural features and properties. *Polysaccharide Carriers for Drug Delivery* 1–17. DOI: 10.1016/B978-0-08-102553-6.00001-5.

Yamada, T., Deitch, E., Specian, R.D., Perry, M.A., Sartor, R.B. *et al.* (1993) Mechanisms of acute and chronic intestinal inflammation induced by indomethacin. *Inflammation* 17, 641–662.

Yamanaka, S., Yuguchi, Y., Urakawa, H., Kajiwara, K., Shirakawa, M. *et al.* (2000) Gelation of tamarind seed polysaccharide xyloglucan in the presence of ethanol. *Food Hydrocolloids* 14, 125–128.

4 Harvesting and Trade with Special Focus on Sustainability Aspects

Satish Chandra Sharma[1] and Niranjan Prasad[2]*
[1]*ICAR-Indian Institute of Natural Resins and Gums, Namkum,* Ranchi, *India;* [2]*ICAR-National Institute of Secondary Agriculture, Namkum, Ranchi, India*

Abstract
Natural resins and gums (NRGs) are low-volume and high-value products that are the result of a healing mechanism in selected tree species as a response to natural or man-made injury. NRGs are widely used in daily life and have great importance as non-timber forest products (NWFPs). Harvesting of NRGs for domestic consumption and sale is very common among the forest dwelling communities, particularly tribal communities in India. The natural resins and gums are non-toxic, biodegradable and eco-friendly in nature for use in various industries. A recent trend towards the use of plant-based and natural products demands the replacement of synthetic additives with natural ones. Today, the whole world is increasingly interested in natural drugs and excipients. Therefore, there is tremendous potential to develop the sector further. NRGs are tapped by exposing the resin and gum ducts by making suitable incisions on the stem of the respective trees. Most of the natural resins and gums are collected in small quantities by forest dwellers adopting traditional tapping methods. Benefits mainly depend on the quality of the produce. The prevalent techniques and devices of resin and gum tapping are traditional, location specific, less efficient and time consuming. A reduction in the drudgery of NRG tappers/collectors and an increase in the efficiency of tapping techniques and devices with concomitant increases in yield could be realized with mechanization. Similarly, mechanized tapping techniques and devices will minimize injury to the trees and help in sustainable production of NRGs. The sustainability of the livelihoods of rural and tribal people might be improved with the adoption of mechanization of the NRG sector.

Introduction

Natural resins and gums (NGRs) are produced from selected plant species in response to artificial or man-made injury. The NGRs help the wound recover and heal, protecting it from insect infections and fungal and microbial invasion. This phenomenon of plant healing helps the NRG tappers produce gum/resin/gum-resin through the blazing method or by injecting plant growth regulator into the stem of the plant to induce NRG production. The plant produces such products as natural exudate from the stem or root.

Non-timber forest products (NTFPs) account for 70% of export profits from the Indian forest sector, 50% of overall forest income and 55% of employment in the sector (Tiwari and Campbell, 1995). The gross value of NTFPs harvested in India is US$0.79 billion, with the national average value being US$31.49 per ha. In a similar vein, India's NTFP-based companies employ 1.6 million man days annually (Behera, 2009). An estimate of the average gross value

*Corresponding author: niranjan_ilri_1999@yahoo.com

 Natural Gums and Resins: Botany and Sustainable Uses in Medicine, Nutrition, Perfumery and Cosmetics (eds Hartwig Schulz *et al.*)
DOI: 10.1079/9781800626690.0004

of products and services supplied by the forestry industry is US$4961.8 million, or around 2.37% of GDP.

In addition to being a significant employer, the NRG industry provides one of the most significant means of subsistence support for more than 50 million people who live in forest and sub-forest areas. With over 45,000 plant species, including roughly 120 that provide gum and resin, India is a great centre of plant biodiversity (MacKinnon and MacKinnon, 1986). In 2019/20, the average yearly exports of natural resins, gums and gum-resins were 7662.21 t, 272,594.79 t and 1182.68 t worth US$482.3 million, US$3304.0 million and US$76.5 million, respectively (Yogi *et al.*, 2021a).

Human beings have known about NRGs since ancient times and they have a prominent position among NWFPs. Aside from goods that are directly consumed as food, fuel or medication, these are perhaps the most used and traded NWFPs. NRG tapping entails making the appropriate incisions on the stems of the corresponding trees to expose the gum and resin ducts. Many plant species have unique passageways or chambers where NRGs are secreted. They develop in specialized structures known as ducts. When exposed to air, NRGs tend to harden out on the bark of the trees (Giri *et al.*, 2008). The age of the trees, their genetic characters and the surrounding environment all affect the size, frequency of occurrence and distribution pattern of resin canals and gum ducts (Negi, 1992). Commonly utilized in daily life, non-timber forest products are becoming increasingly important. The NRGs may be used in a variety of sectors since they are environmentally benign, biodegradable and non-toxic. Global food sectors are becoming more and more visible as prospective NRG users at present.

Generally, the gathering of NTFPs involves the weakest segments of society, especially indigenous people. They gather these products for their livelihood support. According to Rawat and Vaneet (2006), more than 5 million tribal people in India rely on NTFPs for both work and household income. Tribal communities with diverse cultural and socio-economic origins live in varied densities across the nation in various ecological and geo-climatic circumstances (hilly, forested, desert, etc.). In the past, the majority of tribes relied on hunting and gathering or subsistence farming. A significant proportion of remote indigenous communities still rely on the forests for their livelihoods. Basic needs like food, fuel and building materials are provided in the forest-based tribal economy by the forest's production. To get sufficiently remunerative prices for the products, scientific methods must be followed for the tapping and collection of NRGs. The gum industry can provide employment and a steady additional income to rural people and thereby stop their migration into the towns and cities. This chapter covers harvesting and collection techniques for commercially important NRGs as well as variation in and factors affecting the yield of NRGs.

Resins

Resins are more or less solid products with a complex chemical composition. When heated, they will often soften and melt (Negi, 1992). Resins are very often associated with volatile oils (oleoresins), gums (gum-resins) and oil and gum (oleo-gum resins). No clear distinction can be made between these different groups, as many products can be considered as belonging to more than one group. In India, lac, pine rosin and dammars are commercially important for various applications and product development in different industries.

Lac

As already mentioned in Chapter 2, lac is not a resin in the sense of the International Organization for Standardization (ISO) definition, since it is produced by breeding lac insects, which belong mainly to various species of the genus *Kerria*. Nevertheless, the production and postharvest processes of lac will be briefly discussed here.

Lac is produced by culturing lac insects belonging to a few species mainly of the genus *Kerria*. The lac insect can thrive only on specific plants called lac hosts. In India, commonly used lac host species are: ber (*Ziziphus mauritiana*), palas (*Butea monosperma*) and kusum (*Schleichera oleosa*). The quick-growing bushy plants *Flemingia semialata* and *Calliandra calothyrsus* are gaining popularity for systematic lac production on plantation scale (Sharma, 2017). Rain tree (*Albizia saman*) is an important lac host in coastal regions and southern parts of India. Earlier, most lac production came from naturally occurring plants in the forest and sub-forest areas, but nowadays, it is gaining importance as a major agroforestry component for land use diversification and increasing the income of farmers, especially in rain-fed areas.

Lac is mainly produced in India, Thailand, Indonesia, parts of China, Myanmar, Philippines, Vietnam and Cambodia. In India, lac is mainly produced in Jharkhand, Chattishgarh, Madhya Pradesh, West Bengal, Maharashtra, Odisha and parts of Utter Pradesh, Andhra Pradesh, Gujarat and North-Eastern Hill region (Sharma *et al.*, 2020).

Under lac production three main operations are performed, namely, raising of lac host plantation (nursery raising and planting), management of lac host plantation (irrigation, weed management, gap filling, nutrient management, intercropping, plant protection, canopy management and pruning) and major cultural operations (pruning, inoculation, *phunki* removal, pesticide spraying and harvesting). These three major operations under lac cultivation are detailed as below.

Raising of lac host plantation

Nursery raising and planting are the major activities considered under raising of lac host plantation. Seedlings are raised either in polythene bags or in nursery beds. Raising of seedlings in polythene bags is preferred due to easy handling and transportation, ease of establishment and because they are economical. Seedlings raised in polythene bags or in a nursery are transplanted to the main field during monsoon season at an age of 4–12 months. The age at which seedlings are transplanted varies with species, depending upon growth and development.

Management of lac host plantation

To manage planted lac host in the main field, cultural operations like irrigation, weed management, gap filling, nutrient management, intercropping, plant protection, canopy management and pruning are performed.

Major cultural operations

Under major cultural operations, operations such as pruning, inoculation, *phunki* removal, pesticide spraying and harvesting of matured lac crops are carried out to maintain the lac host plantation, as detailed below with their aims.

1. **Pruning:** This includes removal (cutting) of unsuitable shoots of the host plant to facilitate the growth of new shoots.
2. **Inoculation:** Bundles of *broodlac* (lac sticks with mature gravid female insects) are placed on the branches of host trees, allowing young lac larvae/crawlers to come out of their mother cells and settle on the host plant.
3. ***Phunki* Removal:** Used-up *broodlac* after the complete emergence of lac larvae from female cells is called *phunki*. Usually, it is recommended to remove *phunki* from host trees 3 weeks after placing the *broodlac* on the trees.
4. **Pest and Disease Control:** Insecticides are applied to control insects and fungicides to control fungi.
5. **Harvesting:** This is the cutting of mature or immature (*ari*) lac crops from the host tree.

Sticklac

Lac growers harvest mature lac crops by cutting the branches of host trees with lac encrustation. These harvested lac-bearing branches are known as lac stick, which is scraped manually to separate the lac encrustations. After removal of mature lac encrustations from the lac stick, it is commercially known as sticklac, which mainly contains three useful products: resin, dye and wax.

Production and trade

India is the leading producer of lac in the world. A survey conducted by the Indian Council of Agricultural Research-Indian Institute of Natural Resins and Gums (ICAR-IINRG) indicated that the production of sticklac in India during 2019/20 was 18,944 t. In India, Jharkhand ranks first among lac growing states, followed by Chhattisgarh, Madhya Pradesh, West Bengal, Maharashtra and Odisha (Sharma *et al.*, 2006). These six states contribute >98% of national lac production (Yogi *et al.*, 2021a). Jharkhand contributed 54.6%, followed by Chhattisgarh (18.3%), Madhya Pradesh (13%), West Bengal (5.1%), Maharashtra (3.9%) and Odisha (3.4%). The production of lac from lac crops is highly dependent on biotic (threat from enemy insects) and abiotic (temperature, humidity, rainfall) factors. It is not only influenced by the climatic factors affecting the insect directly but also indirectly by those affecting the host tree. The risk of crop failure is high in a season with an adverse climate, as observed in 2010/11, when lac production in India was at a minimum, as shown in Fig. 4.1. During the past two decades, the average production of lac in India was 18,385 t, with the highest production, of 23,239 t, recorded in 2006/07. Recently, due to the adoption by the farmers of lac production on cultivable, bushy hosts, i.e. *Flemingia semialata*, a trend of increasing production has been observed from 2017/18 onwards (Fig. 4.1). Being a small-sized host plant, *F. semialata* can be cultivated in fields like other crops and becomes ready for lac cultivation in 1 year, whereas for a large tree host, one has to wait for 5–10 years. Also, due to the small size of *F. semialata*, lac culture operations like pruning, inoculation, spraying and harvesting can be performed even by women farmers, as climbing a tree is not required.

Lac is produced mostly in Asian countries but is being consumed and imported almost all over the world. Major global producers and importing countries of lac and its value-added products are depicted in Fig. 4.2. Export of lac and lac-based products from India and their export value during 2007–2020 are depicted in Fig. 4.3. The export of lac and its value-added products (shellac, aleuritic acid, seedlac, dewaxed shellac, bleached lac, shellac wax, button lac and gasket lac) from India

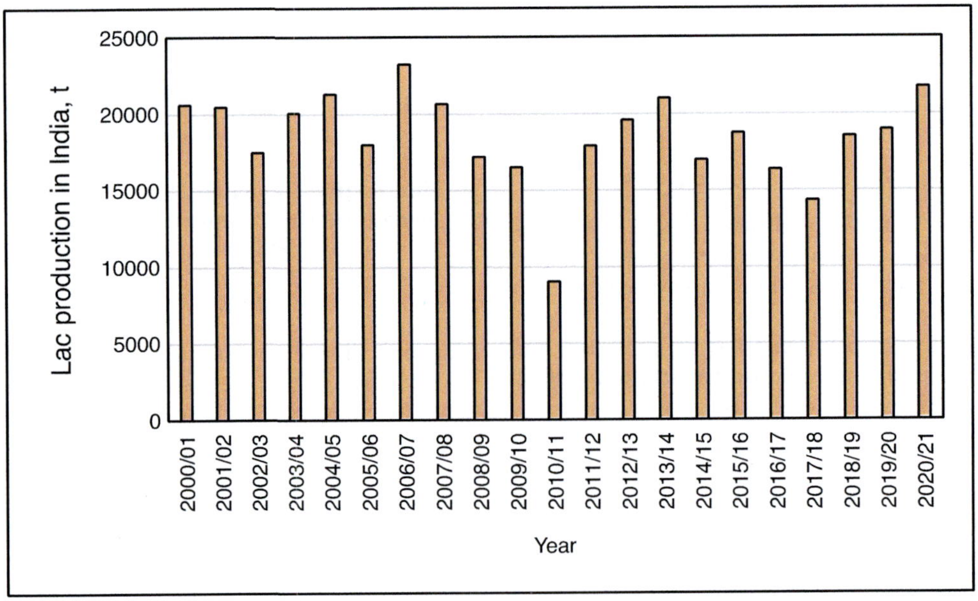

Fig. 4.1. Trends in lac production in India. (By Thombare *et al.*, 2022.) Used with permission.

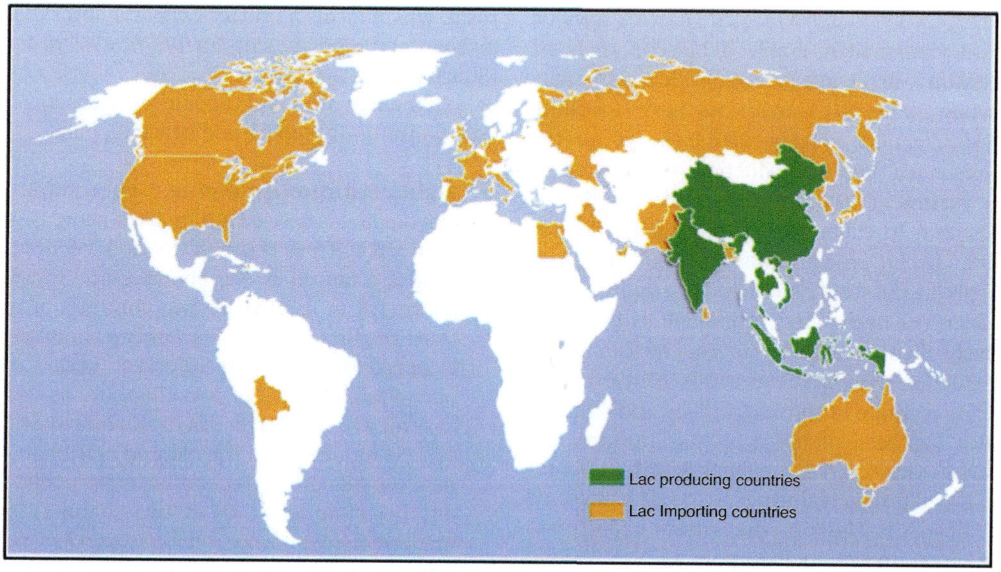

Fig. 4.2. Major lac producing and importing countries of the world. (By Thombare *et al.*, 2022.) Used with permission.

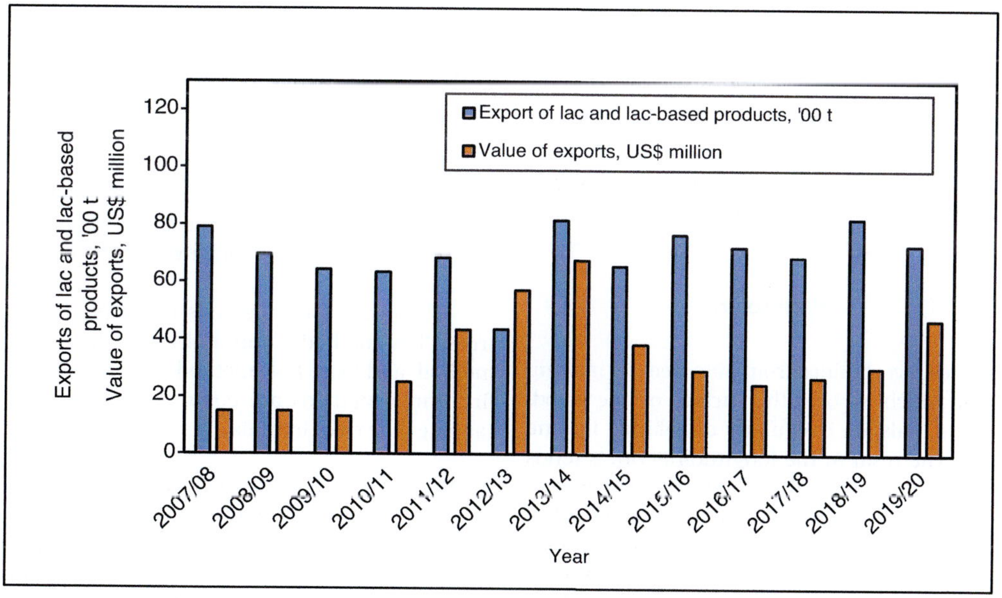

Fig. 4.3. Export of lac and lac-based products and their export value. (By Thombare *et al.*, 2022.) Used with permission.

during 2019–2020 was 7293t, with a value of >US$48 million (Table 4.1). The roots of instability in export values of lac, especially during 2011–2014, with its peak touching US$67.7 million during 2013/14, go to the unpredicted drop in production, almost half of the average, in the year 2010/11. The drastic reduction in exports during 2012/13 created panic in lac-based industry about its short supply in the future from India, resulting in a sudden rise in the price of lac and its products, which then stabilized due to consistency in production from 2013 onwards. Since then, a steady increase in the demand for and value of NRGs has been observed on the international market. The export share value-wise of different lac-based products during 2019/20 is depicted in Table 4.1. These products were exported to various countries, including the USA (37.25%), Bangladesh (9.54%), Germany (7.34%), the Republic of Korea (5.53%), Afghanistan (5.45%), Iraq (4.84%), the UK (2.63%), Egypt (2.47%), Pakistan (2.31%), Spain (2.01%), China (1.97%), Japan (1.93%), Indonesia (1.89%), France (1.84%), Thailand (1.56%), Italy (1.54%), Sri Lanka (1.12%), Australia (1.10%) and Iran (1.02 %). Globally, Indian lac is exported to 76 countries in Europe, America, Asia, Africa and Australasia (Yogi *et al.*, 2021b). The future prediction of the lac sector revealed that about 7500–8000t of lac and its value-added products would be required to meet overseas demand, worth US$80–90 million.

Pine rosin

Pine resin is obtained from pine trees by making a cut which exposes the surface of the wood. Resin canals are irregularly distributed in pine trees. The resin in the longitudinal ducts tends to reach the surface through transverse ones in the rays. Accordingly, for tapping the trees only a shallow incision is necessary as the flow of resin from the transverse ducts stimulates secretion of resin in longitudinal ones. The maximum flow of resin is from the top of the incision, where both the longitudinal and transverse ducts are cut. On exposure to air, the volatile oil in the oleoresin (turpentine oil) gradually evaporates leaving a clear, hard glossy substance called rosin, which forms a protective coating over the incision and tends to stop further flow of resin (Sharma *et al.*, 2018a).

In India two kinds of tapping are in practice: light continuous tapping and heavy tapping.

1. **Light continuous tapping**: Light continuous tapping is carried out on trees above 0.9 m in girth. Trees between 0.9 and 1.8 m in girth are tapped in one channel, and those above 1.8 m in girth in two channels at a time. The first channel or set of two channels is tapped for 5 years. At the end of 5 years, a new channel or set of two channels, as the case may be, is started, leaving an interspace of 100 mm between the old and new channels. At the end of second five-year period, another channel or set of two channels is made, leaving another interspace, and so on until tapping has progressed right round the tree. This is the common mode of tapping being practised in India, and the trees are regularly tapped without any rest.
2. **Heavy tapping**: In heavy tapping (also known as tapping to death), maximum possible production of the resin is derived before the trees are due for felling. It is achieved by cutting on the tree as many channels as it can bear, with a minimum interspace of 100 mm between the successive channels. It is generally started 5 years in advance of main felling in prospective regeneration areas and 2 years in advance of thinning in areas marked for thinning. The lowest girth prescribed for heavy tapping is 600 mm.

There are four methods of tapping pine trees: 'box method', 'cup and lip method', 'rill method' and 'bore hole method'. At present, the first three methods are commonly applied for tapping pine trees in India.

1. **Box method**: This is the oldest method of resin tapping. A cavity or box of 100 mm × 100 mm and up to 120 mm deep is chopped at the base of the trees (Fig. 4.4). It is meant to collect the resin as it exudes from the blaze or incision that is made just below the box by chipping away the bark and outer layer of the sapwood. The resin oozes out of the blaze and is collected in the box. However, this method is very damaging and the trees tend

Table 4.1. Export of lac and its products from India during 2019/20. Author's own table.

Lac products	Export (t)	Value (US$ million)
Shellac	1600.13	10.53904
Aleuritic acid	81.02	2.476179
Seedlac	3577.14	22.29796
Dewaxed shellac	308.5	2.932012
Bleached lac	216.02	2.47356
Shellac wax	3.15	0.042738
Button lac	1507.51	7.513524
Lac dye	0.1	0.000821
Total	7293.57	48.27583

to die within a few years of resin tapping being started.

2. **Cup and lip method:** In this method, the outer bark of the tree is scraped off with an adze to a reasonably smooth surface 600 mm long, 150 mm wide and 250 mm above the point where the lip is fixed. In light tapping, channels are initially located on the south or south-west face of the tree as better yields are obtained from the warmer aspects. Subsequent channels are made in an anticlockwise direction. A cut about 150 mm broad and slightly slanting outwards is made with a curved chisel and mallet about 250 mm below the lower edge of the blaze. The lip, a rectangular piece of galvanized iron (150 mm × 50 mm), is driven into the cut to collect resin into a pot kept below. The pot is partially covered to prevent pieces of bark, dirt and so on from falling into it, and also to minimize the evaporation of the resins that accumulate in it (Fig. 4.5.). In order to open up the clogged resin ducts and aid in the smooth and continuous flow of resin, the channels are freshened at definite intervals. The cup and lip method of tapping has a number of disadvantages. Even though a channel depth of 25 mm has been prescribed for the blazes, very often the depth exceeds the prescribed limit. The inherent hacking action involved in using the adze makes it very difficult to control the depth of the blazes. Also, the tapper may make much deeper blazes in the hope of getting more resin. Deep cuts around the hole result in loss of timber and make the trees less resistant to wind storms. Moreover, after covering the circumference of the tree, a second cycle of tapping is not possible because of the slow healing of deeper blazes. This generally results in an abrupt fall in resin production.
3. **Rill method:** This is an improved method, standardized at Forest Research Institute in Dehradun, India, to overcome the problems of the box and cup and lip methods. For freshening of the blaze, the tapper stands near the tree on one side of the blaze and holds the freshening knife at the lowest point of the control groove. The knife is then pulled up by the tapper along the blaze line marked on the tree. The depth of the rill is about 2 mm into the wood. After freshening both arms of the blaze, a 1:1 mixture of dilute sulfuric acid (20%) and dilute nitric acid (20%) is sprayed on the freshly cut rill with the help of a spray bottle (Fig. 4.6). Exudation of oleoresin starts soon after the rills are made (Fig. 4.7). The pot containing the oleoresin is emptied into a collection can. The resin adhering to the pot is removed with the help of a scraper. The control groove is also increased to avoid accumulation of resin in it.
4. **Bore hole method**: The Forest Research Institute in Dehradun, India, has developed a new tapping technique known as the bore hole technique of resin extraction from *Pinus roxburghii*. In this method, with the help of a machine, holes are made near the ground level into the tree's sapwood to open the resin ducts, and the exudating resin is collected in closed containers (Fig. 4.8). The hole in each tree is made approximately 100 mm above the ground. It has been found from studies that a hole of 150 mm depth and 25 mm diameter is suitable for obtaining maximum resin yield. The holes are drilled straight into the tree stem with a slight slope towards the opening so that the resin drains freely. Immediately after making the hole, stimulants/chemicals (a mixture of sulfuric acid and 2-chloroethylphosphonic acid) are sprayed inside it using a plastic bottle/spray bottle. A spray volume of 1–2 ml is applied to each hole. Chemical treatment is done once

Fig. 4.4. Box method. (By B. Dutt, Solan, India.) Used with permission.

Fig. 4.5. Cup and lip method. (By B. Dutt, Solan, India.) Used with permission.

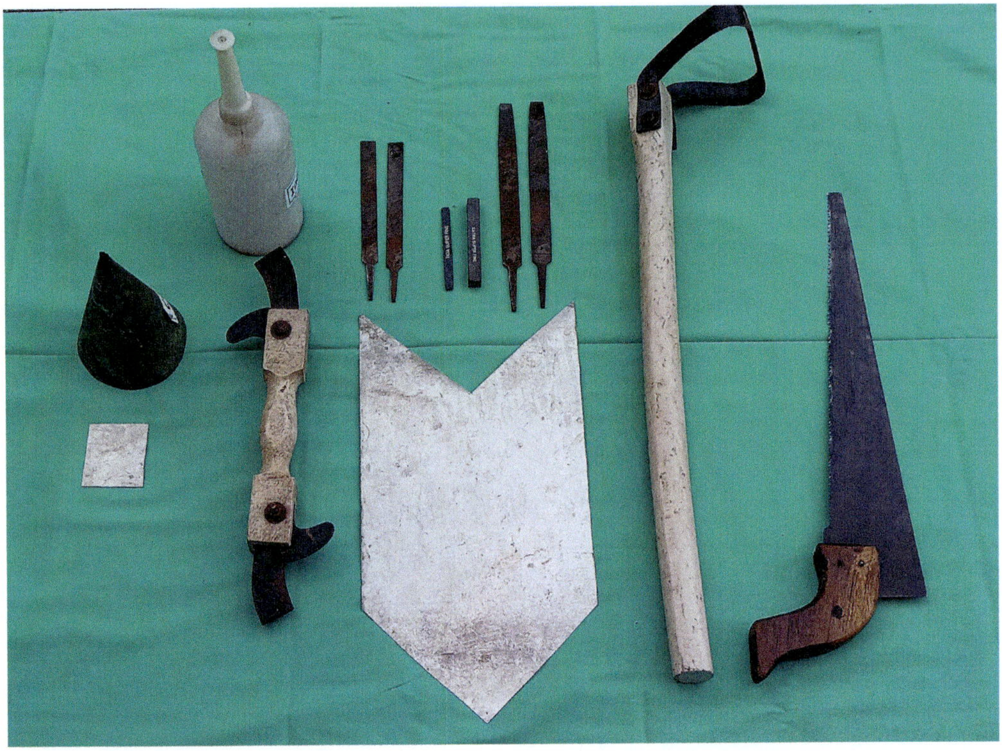

Fig. 4.6. Equipment used for rill method. (By S.C. Sharma, ICAR-IINRG, Ranchi, India.) Used with permission.

only. After treatment, a spout is installed inside the hole by gently hammering with a small mallet or pushing with the palm of the hand to achieve compression fitting in the hole. The spout is joined to a collection container (a polybag made of high-density polyethylene, HDPE, 353 × 120 mm) tightly fixed to each spout (Fig. 4.9).

Once the polybag is filled with resin, it is removed from the tree and the resin is poured into a collection can (*balti*). A new polybag is immediately tied for further collection of resin. Some of the advantages of the bore hole method of resin extraction are as follows:

- There is less stress due to the small size (25 mm diameter) of the hole. The hole heals quickly.
- The technique is very suitable for the protection of the tree against fire, insects, pests and diseases.
- Prolonged resin flow can be obtained from the bore hole for a period of several months without wounding the stem.
- The holes are drilled using a machine, which is why the labour productivity of this technique is several times higher than that of other methods.
- The technique could be applied very effectively in conservation and management of pine resources.

Production and trade

Total worldwide production of rosin, which is made up of resin extraction from pine, was about 1.2 million t per annum in 1995 (Coppen and Hone, 1995). Global trade of pine resin ranged from 300,000 to 400,000 t only in 2017. China and Indonesia are the leading producers of pine resin, followed by India. A major share of resin production in India comes from states of Himachal Pradesh and Uttarakhand.

Fig. 4.7. Rill method. (By Niranjan Prasad, ICAR-IINRG, Ranchi, India.) Used with permission.

In India, resin yield varies between 4 and 7 kg/tree per year. Resin production for India from 2010/11 to 2019/20 is depicted in Fig. 4.10. According to FAO reports (Coppen and Hone, 1995), India's raw resin production peaked at about 75,000 t in 1975/76 and has been declining steadily since then. Production in 1990/91 was less than 25,000 t. The main reason for the decline was the loss of trees through deforestation, either because many of them had reached the end of their productive lifespan and there were no new pine stands to replace them, or because the damage to the trees caused by the use of inefficient, incorrectly applied methods of tapping led to a ban on tapping by the Forest Department.

Details of rosin trade between India and trading partners around the world by quantity and value for the year 2023/24 are presented in Table 4.2.

Fig. 4.8. Bore hole method of rosin extraction. (By Niranjan Prasad, ICAR-IINRG, Ranchi, India.) Used with permission.

Dammar

Most dammar is produced by tapping trees. However, some of the resin is also collected in fossilized form from the ground. Generally, dammars are produced from the tree species *Shorea robusta*, *Vateria indica* and *Canarium strictum*, known as sal, white and black dammars, respectively.

White resin obtained from the wood of *Vateria indica* (a critically endangered species) is known as white dammar. This tree is very popular due to its value as timber and for medicinal uses. In Kerala, white dammar

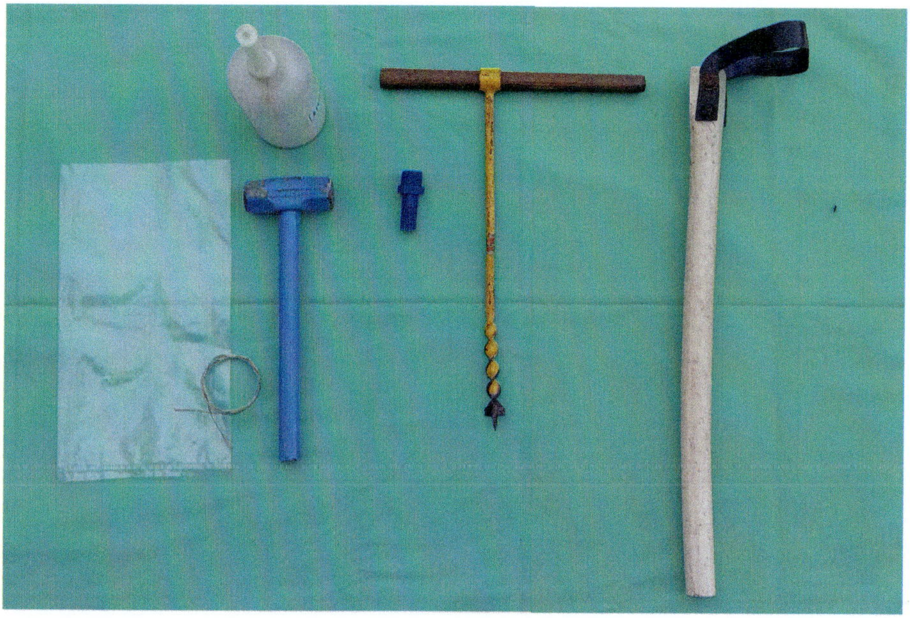

Fig. 4.9. Equipment used for bore hole method. (By S.C. Sharma, ICAR-IINRG, Ranchi, India.) Used with permission.

trees are abundant and mainly seen in semi-evergreen and wet evergreen forests. These evergreen trees attain a maximum height of 30m and are found in south-western India from North Kanara in Karnataka to Tirunelveli in Tamil Nadu through the whole of Kerala. White dammar production in Kerala is very low due to the difficulties in extraction. Tribal communities reported that very few trees of *Vateria indica* produce dammar, hence no artificial resin tapping method is employed and resins are collected from natural incisions only (ICAR-IINRG, 2016). The black resin obtained from the bark of *Canarium strictum* is known as black dammar or *dhoopa*. This medicinal evergreen tree species is highly appreciated for its aromatic resin. This gigantic tree occurs in the evergreen forests of Western Ghats and Eastern Himalayas in India and attains a height up to 40 m. In Kerala it is seen in all evergreen forests, and collection of black dammar is localized in different areas (ICAR-IINRG, 2016). Presently, black and white dammar are collected from forests throughout the state of Kerala, with the exception of core areas of wildlife sanctuaries. Various scheduled tribal cooperative societies of Kerala are involved in dammar harvesting and collection, mainly confined to the *Canarium strictum* species. Approximately, 17 t of black dammar and 460 kg of white dammar are produced annually in Kerala (ICAR-IINRG, 2016).

In India, 3–5 narrow strips of bark are removed at a height of 900 1200 mm above the ground when tapping sal dammar. In about 12 days, the grooves become filled with the resin, which oozes out as a whitish liquid that becomes soon brown on drying. The cavities of the grooves are freshened, after which the exudation continues and the resin is collected as before. The process is repeated several times in a year. When extracting white dammar, semicircular incisions are made on the trunk of the tree up to the surface of the sapwood. Blazes or cuts are so spaced as to cause least damage to the tree. The resin starts oozing from the incisions in 3–4 days and continues for 60–90 days. The resin is also exuded when the bark is scorched by lighting fires around the base of the tree. This method gives a high yield of resin but damages the timber and may even kill the tree. Black dammar resin is obtained by making vertical incisions in the bark in a belt about

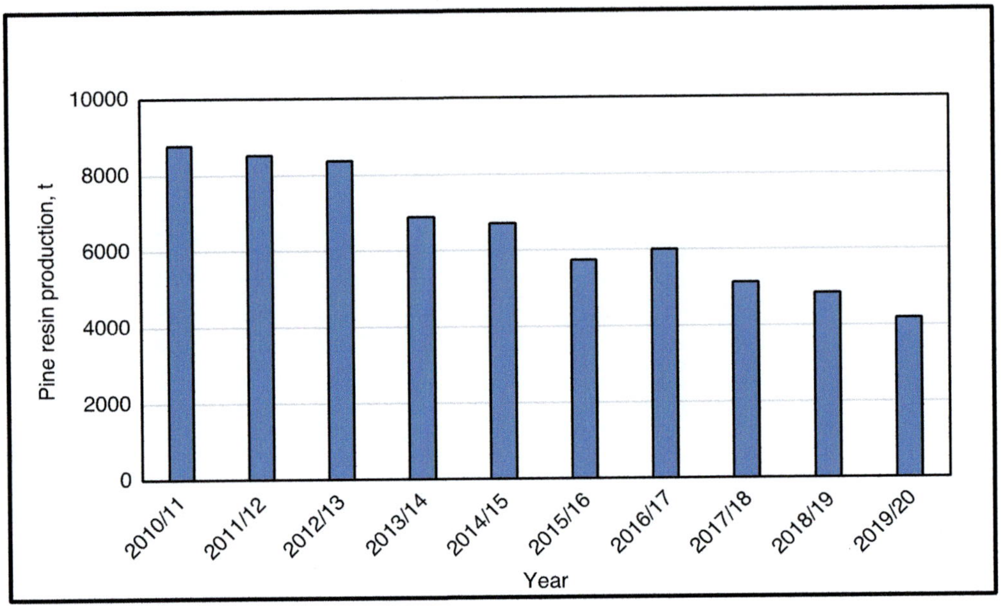

Fig. 4.10. Pine resin production in India between 2010/11 and 2019/20. (By Yogi *et al.*, 2021a.) Used with permission.

1.8 m from the bottom and then lighting a fire around the base of the tree. The resin flows out after 2 years and continues for about 10 years. The flow lasts for about 6 months every year from November to April. The viscous resin that oozes out hardens into a somewhat translucent mass of a bright shining colour that is collected manually.

Sal dammar harvesting from Shorea robusta *trees*

Scientists at the Indira Gandhi Agricultural University (IGAU) in India have conducted extensive studies to develop harvesting times and effective methods for harvesting sal dammar from *Shorea robusta* trees. To harvest exudate from sal trees, different tapping methodologies were adopted and it was observed that the summer was the best season for harvesting in the region. It was also observed during the study that a combined application of 2-chloroethylphosphonic acid and sulfuric acid were most effective for production of the sal dammar biopolymer (ICAR-IINRG, 2019).

White dammar harvesting from Vateria indica *trees*

Two methods – box method and deep cut method – were experimented with at Kerala Agricultural University, under the Network Project on Harvesting, Processing and Value Addition of Natural Resins and Gums, for white dammar harvesting and collection from *Vateria indica* trees, and three grades of dammar resin were obtained from selected tree diameter classes (<50 cm, 50–100 cm, 100–150 cm and >150 cm). During the experimentation, it was observed that white dammar exudes from the sap wood of the respective trees (ICAR-IINRG, 2017). The detailed methodologies adopted for the aforementioned dammar harvesting are detailed in the respective sections.

Box method. The box method was adopted for white dammar harvesting and collection from the *Vateria indica* tree species in the Malakkapara and Peechi areas of the Kerala region of India by the scientists of Karala Agricultural University. In this method, box type blazes of 20 × 20 cm

Table 4.2. Indian exports and imports of rosin for 2023/24. Author's own table.

Export destination countries	Quantity (t)	Value (US$ million)
UAE	63.84	0.08
Cote D'Ivoire	60.40	0.11
Israel	41.8	0.05
Kenya	40.25	0.07
Iran	8.0	0.01
Nepal	7.91	0.01
Djibouti	6.05	0.01
Sri Lanka	3.6	0.01
USA	3.13	0.01
Others (13)	7.76	0.03
Total	242.74	0.39
Import source countries		
Indonesia	32,029.78	35.18
Brazil	5250.00	6.23
Argentina	1998.00	1.99
Vietnam	1058.65	1.20
Nepal	817.35	1.35
Singapore	652.83	0.76
China	558.08	0.85
Malaysia	76.80	0.08
USA	39.56	0.04
Honduras	24.00	0.05
Others (5)	36.73	0.09
Total	42,541.78	47.82

were developed on the tree trunk of the trees at a suitable height from ground level. On the basis of experiments, it was concluded that the box method for white dammar resin extraction from *Vateria indica* is not suitable as a very small amount of dammar resin was collected. However, the trees with a girth >150 cm produced a higher quantity of resin (ICAR-IINRG, 2017).

Deep cut method. Experimentation on white dammar resin extraction from the *Vateria indica* tree species naturally growing in the Malakkapara and Peechi areas of the Kerala region of India was conducted by scientists of Kerala Agricultural University, under the Network Project on Harvesting, Processing and Value Addition of Natural Resins and Gums, adopting the deep cut method. In this method, deep cuts with a depth of 2–3 cm were made on the tree trunk of *Vateria indica* trees at a suitable height from ground level (Fig. 4.11). Based on experiments, it was concluded that the deep cut method yielded a relatively higher amount of white dammar compared with the box method, which increases with enhancement in tree girth. It was also concluded that white dammar resin production was higher in the tree girth class range of 100–150 cm. The deep cut method adopted for white dammar resin harvesting from *Vateria indica* tree species at Malakkapara and Peechi showed a significant increase in the production of first-quality and third-quality resins (ICAR-IINRG, 2017).

Based on the study, it was observed that *Vateria indica* trees exude more resin from natural cracks and fissures than from blazes made by the resin producers, while *Canarium strictum* trees exude nearly the same quantity of resin from both natural and artificial wounds (ICAR-IINRG, 2017).

Black dammar harvesting from Canarium strictum *trees*

Three different harvesting methods – box method, bore hole method and strip cut method – were adopted for black dammar resin extraction from *Canarium strictum* trees growing abundantly in the Malakkapara and Nelliamapthy areas of the Kerala region. Trees of diameter classes < 50 cm, 50–100 cm, 100–150 cm and >150 cm were selected and blazes and bores were developed on the tree trunk at a suitable height from ground level. The maximum quantity of black dammar resin was harvested from trees having a girth class >150 cm (ICAR-IINRG, 2017).

Box method. The box method was adopted for black dammar harvesting and collection from *Canarium strictum* trees in the Malakkapara area of the Kerala region by scientists of Kerala

Fig. 4.11. Deep cut method of resin collection from *Vateria indica*. (By S. Gopakumar, Thrissur, India.) Used with permission.

Fig. 4.12. Collection of black dammar on plastic sheet. (By S. Gopakumar, Thrissur, India.) Used with permission.

Agricultural University, under the Network Project on Harvesting, Processing and Value Addition of Natural Resins and Gums. To harvest black dammar resins from selected tree species, blazes of 20 × 20 cm were developed removing the bark only on the tree trunk of the respective trees at a suitable height. The exuded black dammar resins from blazed trees were collected 1 month after blaze development. From the experimentation, it was observed that the blazes in the box shape were developed easily without any problem, but bark removal and exposure of a large area adversely affected the tree growth (ICAR-IINRG, 2017).

Bore hole method. A study of black dammar resin tapping from the tree species *Canarium strictum* using the bore hole technique was conducted at two sites in the Kerala region, Malakkapara and Nelliamapthy, by researchers from Karala Agricultural University. To harvest black dammar resin utilizing the bore hole methodology, initially a strip of bark was removed from the tree trunk at suitable height and a bore hole of 2.54 cm diameter was made using a hand-operated drill in an outward slanting position. A plastic pipe 5.0 cm in length was inserted at the bottom portion of the developed strip. A polythene bag of suitable size was tied with the pipe so that black dammar resin could be collected in a closed system to obtain dammar free from impurities. The cut portion of the tree trunk was covered with polythene sheet to prolong resin production. Based on the study, not only has it been concluded that the bore hole method used for black dammar resin production from *Canarium strictum* produced maximum resin yield, but it was also observed that the majority of the collected resin was of first quality in nature. In addition, during the study it was observed that exuded resin solidified on the top of the collection container when exposed to the environment and clogged the top opening of the collection container. Thus, prolonged resin collection in the closed container was not possible (ICAR-IINRG, 2017).

Strip cut method. The applied strip cut method was a modified technique of the bore hole method in which a small portion of bark was removed from the tree trunk of the *Canarium strictum* tree and a polythene sheet was fixed underneath the removed portion of bark. Resin oozing from the strip flowed directly onto the fixed polythene sheet and was collected (Fig. 4.12). In this method, injury to the tree was minimal, with relatively higher resin production compared with other methods adopted for black dammar resin harvesting from respective tree species. The strip cut method adopted for black dammar resin production from *Canarium strictum* tree species in the region produced the maximum quantity (90%) of first-grade black dammar resin and was found to be highly efficient (ICAR-IINRG, 2017). The highest production of resin was observed in trees having a girth >150 cm. The maximum quantity of first-quality resin was observed in trees of the same diameter. In this method, there was a gradual increase in the production of resin towards monsoon season and it continued up to December. In the strip cut method, the blaze healed naturally within 2–3 months, whereas, in traditional method, it took 6–10 months to produce new cells for blaze healing (ICAR-IINRG, 2018). The dammars produced from different tree species are shown in Fig. 4.13.

Period of harvesting/collection

Sal dammar collection is done several times during the year. Altogether, three lots are obtained: the first in June–July, the second in October and the third in January. Generally, the resin obtained in June–July is the best in terms of quantity and quality. Black dammar is collected mostly during December–April. If a fully productive tree is tapped once a month in the manner described above, it will yield about 4–5 kg of dammar resin at each tapping, i.e. about 48 kg/year. However, the yield varies from genotype to genotype and some trees may only be tapped every 3 months due to their low yields. Furthermore, it has been observed that resin production decreases significantly at the time of flowering and fruiting of the trees.

Production and trade

Dammar is produced from *Shorea robusta*, *Canarium strictum* and *Vateria indica*. Annual production of dammar in India is about 80–100t and used as painting and incense material.

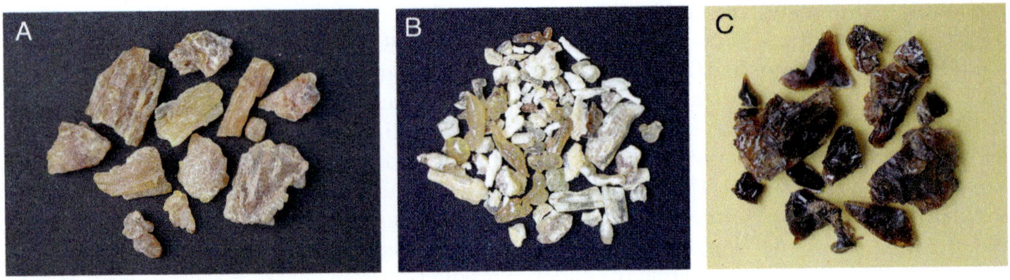

Fig. 4.13. Dammars obtained from different tree species: A. sal dammar; B. white dammar; C. black dammar. (By Nandkishore Thombare, ICAR-IINRG, Ranchi, India.) Used with permission.

Table 4.3. Indian imports and exports of dammar for 2023/24. Author's own table.

Export destination countries	Quantity (t)	Value (US$ million)
Sri Lanka	141.69	0.09
Nepal	14.77	0.02
Germany	1.21	0.01
Other (8)	5.66	0.02
Total	163.33	0.14
Import source countries		
Indonesia	17,718.29	10.15
Thailand	3612.68	9.28
Vietnam	1142.28	2.57
Lao	176.92	0.36
Singapore	16.31	0.01
Total	22,666.48	22.37
Gums		

Indian imports and exports of dammar during 2023/24 are given in Table 4.3. As can be seen there, more than 90% of dammar was imported to India from Indonesia and Thailand.

Gums

The most important gum trees are *Acacia senegal* (arabic), *Sterculia urens* (karaya), *Anogeissus latifolia* (ghatti), *Moringa oliefera* (moringa), *Buchnania lanzan* (chirounji) and *Acacia nilotica* (babul), as reported by Prasad *et al.* (2012).

Exudate gums

Exudate gums are solids obtained from plants that consist of mixtures of polysaccharides that are either water-soluble or absorb water and swell to form a gel or jelly when they come into contact with water. Exudation is often interpreted as a pathological reaction to injury to the plant, whether intentional ('tapping') or as a result of accidental injury or damage by insect pests.

Gum arabic

Gum formation in *Acacia senegal* trees takes place in the cambial zone of the stem or branch. It is promoted by stress conditions like drought, heat, insect attack or natural or artificial injury (Joseleau and Ullmann, 1990). Among different gum-producing *Acacia* species, systematic planting for gum production is done only for *Acacia senegal*. The gum production taking place due to artificial injury is known as tapping, which is required for *Acacia senegal*. In other *Acacia* species, gum formation occurs by natural exudation or natural injury (Islam *et al.*, 1997). These gum modules appear like tears of the tree. The gum produced by the tree is in liquid form, which solidifies on exposure to air in the injured area of the tree and works as a protective cover (Hassan, 2000; Verbeken *et al.*, 2003; Sanchez *et al.*, 2018), protecting the tree from attack by microorganisms, supporting wound healing and working as an anti-desiccant preventing water loss from the tree (Karamalla *et al.*, 1998).

The place of gum formation in a tree, whether at the point of injury or somewhere else in the tree, is uncertain (AwadElkarim, 1994; Hassan, 2000). About 90% of the gum comes from *Acacia senegal* trees (Awouda, 1988). The maximum average yield per season per tree from *Acacia senegal* may be several kilograms (Dione and Vassal, 1993). Maximum gum production from *Acacia senegal* occurs in trees aged 7–15 years; after 20 years, gum production begins to decline (Dione and Vassal, 1993).

Effect of *Acacia senegal* stand management on gum yield. The effect of stand management on gum yield per tree from *Acacia senegal* was studied at two locations in Sudan for 3 years. The study showed that the yield in the farmer's stands was 47–60% lower than in the research stands, for both wild and cultivated species (Ballal *et al.*, 2005). Inoculation of the soil with rhizobia at the beginning of the rainy season led to a significant increase in the percentage of trees producing gum arabic (GA), gum production per tree, fluorescein diacetate and acid phosphatase activity and soil microbial biomass. The C and N content in soil remained unaffected. The rhizobial inoculation had a positive effect on soil microbial function and gum production (Fall *et al.*, 2016). The study found that GA yield decreased with increased water use efficiency, and no biological nitrogen fixation occurred during the dry season (Gray *et al.*, 2013).

Effect of tapping on gum yield. The date of tapping plays an essential role in gum arabic yield. The highest yield in North Kordofan, Sudan, was observed in tapping carried out from 1 October to 1 November in natural and planted stands. In the study, a 40% reduction in yield was observed for late, 15 November to 15 December, tapping. In a research plantation, the average yield was 274.2 g/tree compared with 137.7 g/tree in a farmer's field in Sudan (Ballal *et al.*, 2005). The results of a study done by Tadese *et al.* (2016) indicated an increase in yield with an increase in the intensity of tapping. Gum arabic yield was observed to be greater using developed tapping practices compared with collection of naturally produced gum from *Acacia senegal* trees. Further, the yield decreased with increasing tree size (diameter) (Tadese *et al.*, 2016).

The ecological effects of understorey vegetation (50 and 100% grass cover, bare and burnt) and tapping date (1 October, 15 October and 1 November) on gum yield of *Acacia senegal* in Sudan have been described by Adam *et al.* (2013). In their study, gum picking started 40 days after tapping and continued for seven pickings at 15 day intervals. There was a significant effect of grass cover on the gum yield.

Effect of rainfall on gum yield. The study carried out at Demokeya, Sudan, concluded that gum yield per tree generally correlates with the average annual rainfall. The highest yield was 145.3 kg/ha in 1994, with the highest average annual rainfall of 463 mm. Ballal *et al.* (2005) concluded that the yield positively correlated with rainfall, minimum and maximum temperature during gum collection and intensity of tapping. Research carried out by Jahun and Salami (2020) in the Kwanar-duhuwam and Kadoma plantations of Jigawa State, Nigeria, revealed that gum arabic tapping done after 3 weeks of the rainy season resulted in the highest yield, of 163.9 g/picking and 172.25 g/picking, respectively. Harmand *et al.* (2012) conducted a study in Cameroon of *Acacia senegal* growth in an area with annual rainfall of 650–1250 mm. Based on the study, they concluded that the optimum date for tapping was 10 October for 650 mm isohyet and 25 November for 1250 mm isohyet. In the area with an average annual rainfall of 650–800 mm, the gum production was 100–500 g/tree, which corresponds to 50–250 kg/ha for 500 trees/ha. In an area with more than 1000 mm rainfall, gum production was lower and uncertain. It was found that the mean gum production did not vary significantly between soil types. Furthermore, annual gum production on sandy soils was more dependent on climatic conditions and varied more than on clay soils.

Traditional method of blazing. In the traditional method of gum arabic production, the wound is made in the form of a cut in the tree trunk with an axe. The cuts are made at a distance of about 300 mm and a depth of about 10 mm on the tree trunk (Fig. 4.14A). The width of the cut depends on the axe's width, which varies from place to place as they are manufactured by

Fig. 4.14. Traditional blazing: A. use of an axe for making blaze; B. gum exudation from blaze made by an axe. (By J.C. Tewari, Jodhpur, India.) Used with permission.

local blacksmiths. Exudation of the gum from the wound occurs in the form of tears after about a week. The gum is produced in a highly viscous form with more than 30% of volatile substances, including water. The gum produced starts drying after exudation and takes the shape of tears, as shown in Fig. 4.14B. The dried tears are collected by gum pickers either by hand or removing them with the help of a local tool with a metal hook mounted on a wooden pole. The tears thus detached from the tree fall on a cloth spread on the ground. Gum tears are then cleaned by removing foreign matter like bark, insects, leaves, etc. After cleaning, the gum tears are graded according to their size and colour. Usually, gum produced following the traditional method is darker in colour and sold at a lower price.

Rectangular blaze method. *Acacia senegal* trees are also tapped by making rectangular incisions of 600 mm × 50 mm in the branches. The gum production from the incision starts within 3–8 weeks, depending on weather conditions. The effect of different parameters (length of blaze, width and depth including blaze height and tree crown diameter) on gum production at an 8-year-old *Acacia senegal* plantation in Chad was investigated by Ngaryo *et al.* (2011), who found that tree crown diameter and tapping height significantly and positively affected gum production, while depth showed a significantly negative effect. Maximum gum production was observed with a 6 mm-deep blaze in October. Moderate gum production was recorded in September, with minimum production in December (Ngaryo *et al.*, 2011).

Gum inducer method. In the traditional and rectangular blaze method, trees were injured, and sometimes they died due to excessive injury made by gum pickers. Hall and Smith (1995) reported that induced stress increases ethylene biosynthesis and activates plants for gum exudation. Gum production also takes place with the application of gum inducers (plant growth regulators or inorganic acids) in the stem of the *Acacia senegal* tree. Usually, plant growth regulators like 2-chloroethylphosphonic acid and inorganic acids like sulfuric acid are used in gum production. Both substances contribute to creating stress conditions in the tree, leading to increased gum exudation. Several researchers have worked on applying chemical stimulants and hormones to induce gum production in *Acacia senegal*.

In order to enhance gum production of *Acacia senegal*, a dose of 2-chloroethylphosphonic acid was standardized by researchers from ICAR-Central Arid Zone Research Institute (CAZRI), Jodhpur, India (Tewari *et al.*, 2017). In this method, a hole of 15 mm in diameter and 50 mm deep is drilled in the tree's main trunk at the height of about 500 mm from ground level using either a hand-driven or battery-operated drill. The hole is drilled slanting inward at an angle of more than 30° so that the injected material does not flow out during the application. The required quantity of hormone is injected with the help of a syringe, and the hole is plugged with mud. In response to hormone treatment, the gum is produced in tears on the branches of treated trees. Generally, it is observed that mostly gum production takes place where cracks or fissures occur on branches. The gum pickers collect the harvested gum either in a cloth/jute bag or in a small basket made of bamboo or palm tree leaves.

In a study, the correlation between various girth classes and patterns of gum arabic production was explored. Mean maximum gum production of 375 g/tree/season was recorded from the diameter at breast height (DBH) group of 510–600 mm, followed by the 410–500 mm DBH group with mean gum production of 322.5 g/tree/season. In the 20–30 cm DBH group, mean gum production was found to be 210 g/tree/season. The study concluded that gum production was positively correlated with tree girth. Adoption of ICAR-CAZRI gum inducer technology by farmers for *Acacia senegal* may be a source of income for millions of farmers in arid and semi-arid regions in India (Pareek, *et al.*, 2017). Abib *et al.* (2013) found that the application of ethephon increased gum production in northern Cameroon. The production per tree was found at 77, 313 and 214 g at semi-arid sites and 30, 186 and 114 g at sub-humid sites when 0, 40 and 12 mg 2-chloroethylphosphonic acid per tree was applied, respectively. Generally, gum production was 400–600% higher in the chemically treated provenances than in the untreated trees at sub-humid sites.

Gum production mechanization. Adam *et al.* (2017) conducted a study in Kordofan, Sudan, on gum tapping and reported that most gum producers (40%) used a traditional small axe, followed by *sunki* (15%) and *makmak* (3%), to make incisions in the tree trunk for gum production, while the remaining producers (42%) used other tools. The study revealed that many intermediate types of equipment were developed, but their adoption was poor. The authors further reported that the participation of rural women in the gum production chain (cultivation, planting, collection and refining through primary processing) was about 85%. Pareek *et al.* (2017) reported the use of a manual or electrical drill to make a hole for plant growth or gum inducer application in the gum inducer method of gum tapping. Based on the research carried out, Aziz *et al.* (2018) concluded that there were considerable differences (p>0.05) in gum production, the number of trees blazed/day and difficulty of gum tappers in handling the tree using the tapping tool *sunki* and the axe (traditional tool) in the gum garden of En Nahud, Sudan. Since this review of gum mechanization, very little work has been done in this area and there is an opportunity to develop high-quality and efficient equipment in the near future to mechanize gum tapping and reduce the tedious work of gum-producing farmers.

Harvesting. Gum arabic production occurs during the dry season (November–May). Gum is also obtained from September to October, but the gum collected is of low viscosity and dark in colour. No gum production was observed during the rainy season as the trees were in full bloom. On average, a single tree may produce 250 g of gum annually, but the yield can vary from a few grams to a maximum of 10 kg (Giri *et al.*, 2008).

Gum production occurs from fissures formed on the tree bark due to stress conditions (heat, dryness, wounds and diseases). The gum exudes naturally from the tree's bark in thick form or as a foamy liquid, solidifies quickly into tears 20–75 mm in diameter, gradually dries and hardens when exposed to air. Gum produced in tear form on the blaze or cracks is removed manually utilizing a wooden pole (Fig. 4.15A), by hand picking or with a collection tool (Fig. 4.15B) (Giri *et al.*, 2008).

Trade. The major gum arabic exporting and importing countries in the world are depicted in Fig. 4.16. The estimated world crude gum arabic export was about 113,000 t in 2016. The estimated average export by different exporting countries from 1992 to 2016 is presented in Fig. 4.17. Sudan, Chad and Nigeria were the major gum arabic producers and exporters, contributing more than 90% of total world exports from 2014 to 2016. France and India are the two largest importers of the gum, accounting for about 75% of world imports from 2014 to 2016. The trend of gum arabic imports to India from

Fig. 4.15. Gum arabic harvesting: A. gum harvesting using a wooden pole (By J.C. Tewari, Jodhpur, India); B. using a locally made traditional device (By Alland and Robert, 2019).

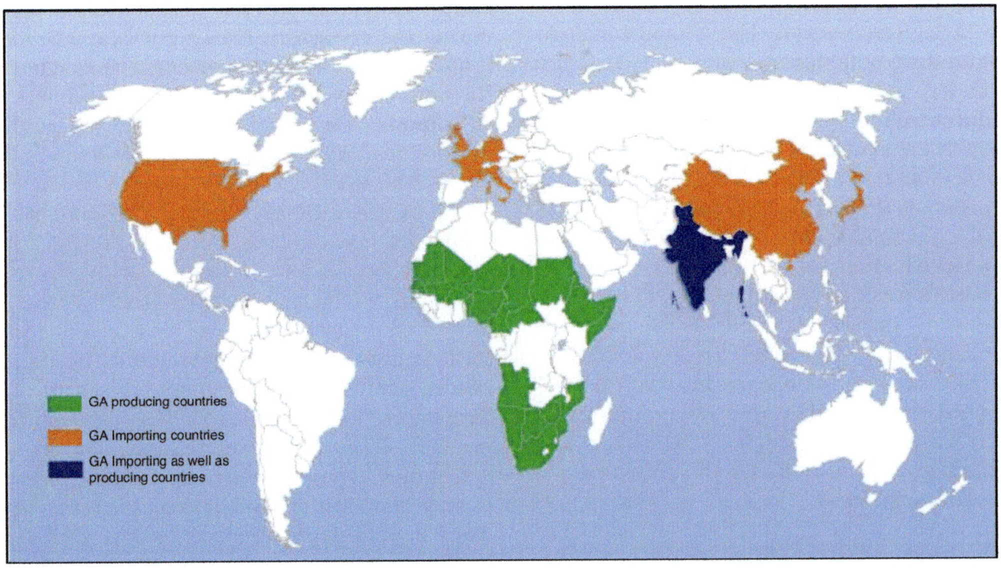

Fig. 4.16. GA exporting and importing countries in the world. (By Prasad *et al.*, 2022.) Used with permission.

1992 to 2019 is presented in Fig. 4.18 (Jales, 2018; Yogi *et al.*, 2021a, b, c). Other importers like Japan, the United States and the rest of Europe combined imported 23% of the gum from 2014 to 2016 (Jales, 2018). The average imports of gum arabic from various countries from 1992 to 2016 are presented in Fig. 4.19. European countries, Japan and the United States generally imported sorted, graded and purified raw material. India, on the other hand, mainly imported bulk, unsorted and uncleaned gum arabic, rejects and sieved material.

Gum karaya

Gum karaya is produced from various *Sterculia* tree species in the form of dried tears. It is formed as a by-product of metabolic pathways in response to biotic and abiotic stresses, i.e. injury, drought, heat, high temperature and infection. Natural exudation takes place either from cracks in the tree trunk or sometimes from roots in termite-infested sites. Gum karaya production takes place by tapping *Sterculia* tree species using various tapping methods. In India, gum

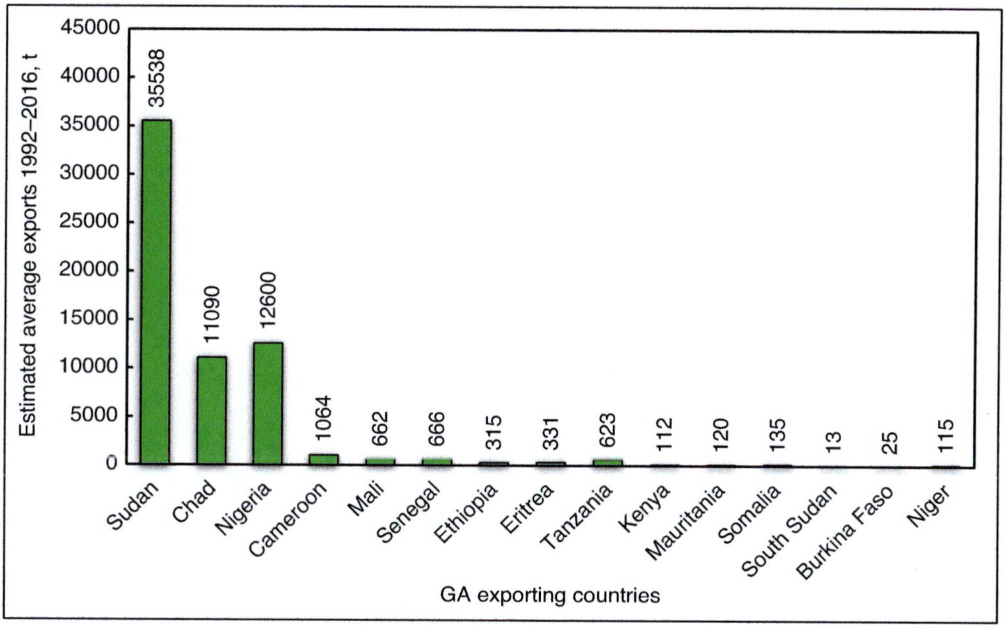

Fig. 4.17. Estimated average gum arabic exports by country, 1992–2016. (By Jales, 2018.) Used with permission.

karaya is mainly obtained from *Sterculia urens* trees, while in African countries it is exclusively obtained from *Sterculia setigera*.

Commercial production of gum karaya from *Sterculia* species started in India in the 1960s. The gum exudate was collected and moulded into the form of a disk for drying in the fire pits on open charcoal or a cooker. This practice of collecting gum continued until 1992, and later an improved method was developed and popularized. Tribal communities in Andhra Pradesh (India) were trained to adopt the improved method by the Grijan Cooperative Corporation Ltd., Vishakhapatnam. The method involved using a sickle/blaze knife for making blaze instead of an axe (Coppen, 1995).

In India, the best time for gum karaya tapping is from April to June, when ambient temperatures are high and humidity is low. During this period, maximum quantity of excretion is produced (Vasishth and Guleria, 2017). During the rainy season, a large amount of the gum produced is washed away and the remaining gum is of low quality and has lower viscosity (Verbeken *et al.*, 2003). The gum yield varies from tree to tree due to their individual genetic background and climatic conditions. The yield can vary from 20 to 30 g per blaze in off-season to 100 g per blaze during peak season. Sometimes trees with a high gum yield can produce 500–750 g of gum per blaze in the peak season.

The colour of gum karaya may vary from white to grey and this, along with foreign matter present in it, acts as a factor for classification into different grades. Gum karaya is commonly classified into five grades, namely 'hand-picked', 'superior No.1', 'superior No.2', 'fare average' and 'shifting' (Davidson, 1980; López-Franco *et al.*, 2009).

Various blazing and gum inducer methods have been developed for gum karaya production/tapping. A review of these methods is presented below.

Blazing method. As already mentioned, superior plants produce resin/gum/gum-resin in response to various injuries as a defence mechanism, which helps in healing the wound and protecting it from microbial and fungal infection (Lujan-Medina *et al.*, 2013).

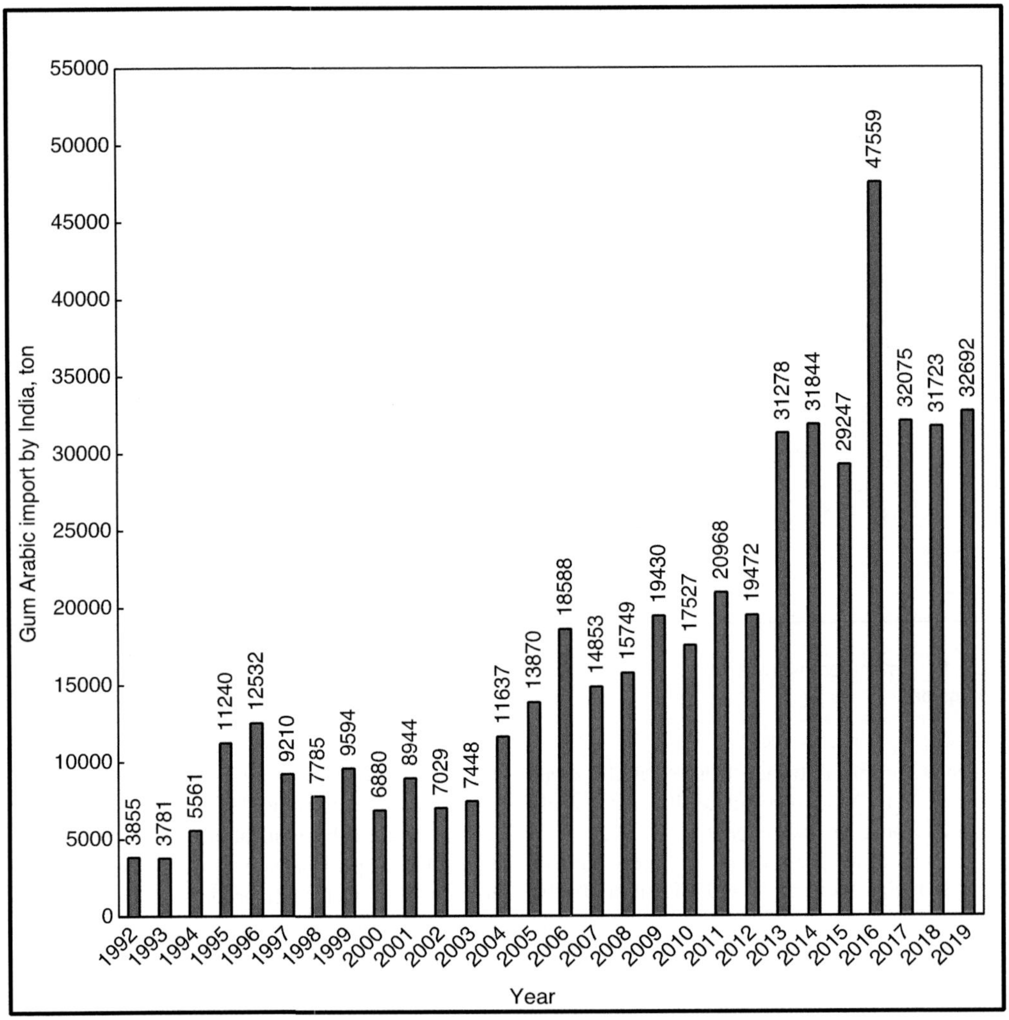

Fig. 4.18. Imports of gum arabic by India, 1992–2016. (By Jales, 2018; Yogi *et al.*, 2021a, b, c.) Used with permission.

Humans take advantage of this phenomenon by producing these natural raw materials, either by making blazes or by introducing some chemical inducer into the plant. Karaya trees produce gum of commercial importance which is mainly used in food and various medical applications. Different methods of extracting gum karaya have been developed and used in practice.

Gum karaya is mainly produced from the stem and the root (Kuruwanshi *et al.*, 2017). The tree sap is the source of gum karaya, which has good adhesive property (Upadhyay, 2017). Artificial gum tapping is done either by making blazes, by peeling and making deep cuts in the stem or by introducing gum inducers into the tree. Various conventional, rectangular and semicircular methods of gum tapping are presented below.

Conventional blazing of irregular shape. In the conventional blazing method, an irregularly shaped wound is cut into the trunk of the tree using an axe (Sharma *et al.*, 2016). The gum exudate on the wound is collected by hand after drying. The large lumps collected are broken and

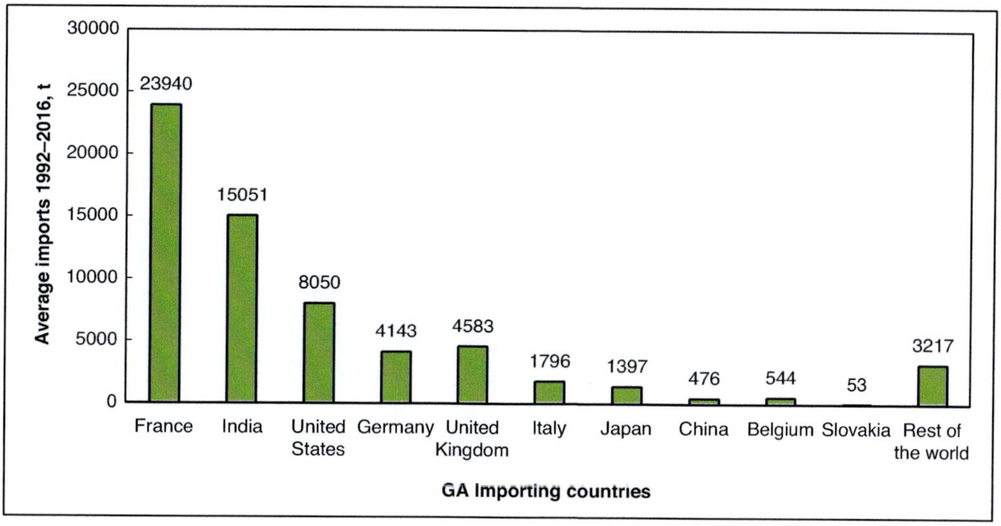

Fig. 4.19. Average gum arabic imports by country, 1992–2016. (By Jales, 2018.) Used with permission.

foreign matter like bark, insects, leaves, etc. are separated by hand picking. The gum produced by this method is darker in colour and fetches a lower price on the national and international market (Bhattacharya *et al.*, 2003).

Rectangular blazing. In the conventional method of gum tapping, large and irregularly shaped wounds are made which are injurious to the tree. To overcome the drawbacks of the conventional method, the forest departments in Damoh and Jabalpur in the state of Madhya Pradesh (India) and the Forest Research Institute, Dehradun, India (Forest Research Institute, 1973) have introduced an improved method of gum tapping. According to the recommendations of this method, tapping should be done on trees with a girth of more than 900 mm, and at the beginning of tapping the blaze size needs to be 150 mm tall, 100 mm wide and 25 mm deep. One blaze may be added for every 500 mm girth above 900 mm. It was recommended to collect gum 16 times in one tapping season at 2 week intervals. After each collection, removal of a 2-cm-high section above the previous one was recommended for freshening the blaze for further gum production. Following these recommendations and making blazes in each season allowing proper spacing between adjacent blazes, a rest period of 6 years should be provided before returning to the site of a previous blaze. Applying this method, an annual yield of 1.5–5.0 kg of gum per tree has been achieved.

A semicircular blazing method of gum karaya tapping (Fig. 4.20) was developed by the Kovel Foundation, Vishakapatnam, Andhra Pradesh (India), for continuous and sustainable tapping of gum of high quality (Sharma *et al.*, 2023). In this method, a sharp sickle is used for making a semicircular blaze (Fig. 4.21). A small piece of transparent square polythene sheet 15 cm × 15 cm is pinned on the blaze with a thorn for collection of the gum. This avoids contact of the gum with bark after exudation and with the human hand during gum collection. In this method, a bamboo basket is recommended for collecting the gum from the tree and tweezers for picking foreign matter from the gum during cleaning and grading. This method further recommends the use of a raised platform made from a wooden pole, bamboo mat and polythene sheeting to dry the gum. The drying platform is placed in an open, sunny space at waist height to allow the gum to dry quickly.

Gum inducer method. Apart from the blazing method of gum tapping, gum production

Fig. 4.20. Semicircular blazing method developed by the Kovel Foundation, Vishakhapatnam. (By S.C. Sharma, ICAR-IINRG, Ranchi, India.) Used with permission.

also takes place in karaya trees when gum inducers like plant growth regulators or inorganic acids are introduced into the tree. Several researchers have studied the effect of different gum inducers on gum karaya production, as described below.

Nair *et al.* (1995) carried out experiments to find the effect of 2-chloroethylphosphonic acid on gum yield and blaze healing. Ten holes, each of 5 mm in diameter and 20–35 mm deep, were made using a hand-drill on each tree around the stem. The holes were equally spaced and were drilled at a height of 1m from ground level. The holes were given an inward slope to avoid backflow of the inducer. Five consecutive holes were treated with the growth regulator and the other five (opposite to the treated holes) were kept as controls with distilled water.

The treatment resulted in a 20-fold increase in gum yield compared to the control. About 1.5 kg of high-quality gum was obtained from seven trees treated with 2-chloroethylphosphonic acid. Yield variation observed among trees was probably due to heterozygosity. There was no gum production from the control after 30 days, while the treated holes continued to secrete the gum. Histological examination showed that large gum ducts or cavities were formed in the secondary phloem and wood, covering a radius of 20–30 mm around each hole in the treated trees. Thick tissue developed at the hole site after 45 days and replaced the damaged tissues.

At Indira Gandhi Agriculture University, Raipur Centre, experiments on 10 gum karaya trees located at Kurrabhata village in Gariaband (Chattishgarh) were conducted by injecting 2-chloroethylphosphonic acid during May to June 2010 (Fig. 4.22A,B). The average gum yield was 320g per treated tree, while gum production in the control was negligible. The maximum gum production was found to be 650g on a tree

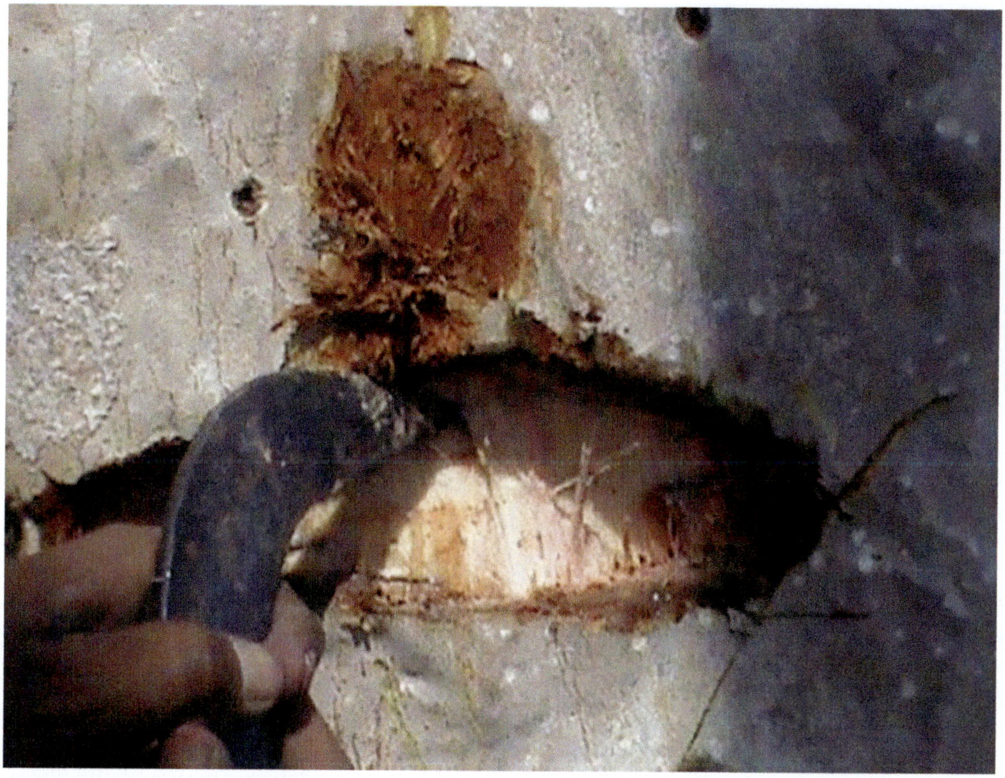

Fig. 4.21. Use of blazing knife for making arc-shaped blaze in a karaya tree for gum exudation. (By P. Katiyar, Raipur, India.) Used with permission.

with a girth of 500 cm while minimum production was 10 g on a tree with a girth of 76 cm. This method of gum tapping caused the least damage to the tree (ICAR-IINRG, 2012).

Vasishth and Guleria (2017) also carried out experiments for tapping gum karaya by applying the gum inducer method. They were able to show that the best results were achieved when 2-chloroethylphosphonic acid was applied in a quantity of 4 ml at a concentration of 300 mg/ml.

Sharma and Prasad (2013) conducted experiments for tapping gum karaya from *Sterculia urens* trees growing in Taimara Valley, India, using the standardized technique of tapping in a semicircular blaze, as developed by the Koval Foundation, Visakhapatnam, in collaboration with the Girijan Co-operative Corporation, Visakhapatnam (Andhra Pradesh).

After one month of treatment, exudated gum was collected and the blazes of each tree were refreshed by removing thin bark from the blaze to open the gum ducts for further gum exudation. The freshening process was continued at monthly intervals throughout the duration of the experiment.

The trees were found to excrete gum throughout the year, but the exudation was higher during hot weather conditions. The minimum and maximum gum yield per tree per year were 304.88 g and 1310.48 g, respectively, for trees tapped with a diameter range of 0.56–0.70 m. Physical damage to each tree was calculated to be 91.16 cm^2 on an area basis and 231.55 cm^3 on a volume basis resulting from one semicircular blaze made on each tree (Sharma and Prasad, 2013).

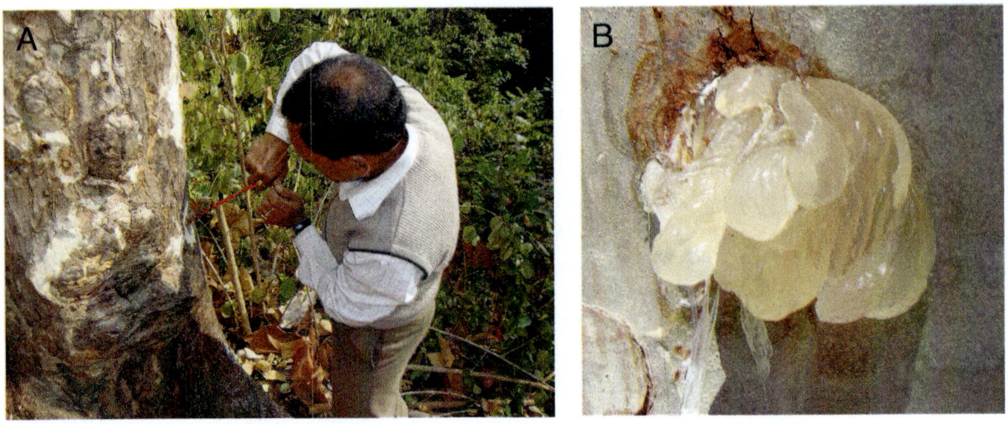

Fig. 4.22. Gum inducer method: A. use of a hand-drill for making holes in a karaya tree for injecting the gum inducer; B. exudation of gum karaya. (By S.C. Sharma, ICAR-IINRG, Ranchi, India.) Used with permission.

Effect of tapping technique on gum karaya yield. Sharma and Prasad (2013) conducted experiments on nine trees of *Sterculia urens* for gum karaya tapping. They observed that gum started oozing out after 4 hour in holes treated with the gum inducer technique and immediately in the case of the semicircular blazing technique. Mean gum karaya collection during the experimentation was found to be 753.68 g in cases using the semicircular blazing technique and 2307.16 g for cases using the inducer technology.

Nair *et al.* reported that 2-chloroethylphosphonic acid enhances gum formation and wound healing in *Sterculia urens* (Nair *et al.*, 1995). It was also found that gum production was higher using the inducer technology than using the semicircular blazing method in almost in all months of the study period. The inducer technology was found to be better for gum karaya tapping, with 17 times less damage to the *Sterculia urens* trees on a volume basis and about 200 times higher gum production compared to the semicircular blazing technique (Sharma *et al.*, 2018b).

Mechanization of gum karaya production/tapping. Prasad *et al.* (2012) reviewed the work carried out for mechanization of gum and resin tapping and reported the use of an axe and blazing knife for making cuts in the blazing method of tapping gum karaya. They further reported the use of hand- and battery-operated drills for making holes in the gum inducer method of gum tapping using different gum inducers. Kuruwanshi *et al.* (2017) developed a blazer for making blazes on the gum karaya tree for gum exudation. The use of the blazer controls the size and depth of the blaze for sustainable gum production and conservation of the gum-producing tree.

Harvesting/collection. The production of gum karaya starts immediately after making blazes on the tree. Maximum gum production takes place on the first day. The gum produced on the blaze is allowed to dry and removed in the form of tears with a sickle or directly by hand in the traditional method of gum collection (Fig. 4.23A). However, in the method developed by the Kovel Foundation, Vishakhapatnam, a polyethylene sheet is pinned below the blaze. The gum collected on the polythene sheet is removed along with the sheet and a new sheet is placed for subsequent gum collection without touching the gum (Fig. 4.23B).

An environment with low humidity and high temperatures is favourable for the production of the best gum quality. Gum karaya tapping in India is therefore carried out mainly from April to June. Sometimes, the gum is also collected in September and October, but the

Fig. 4.23. Gum karaya collection: A. using a sickle; B. on a polyethylene sheet. (By Krishna Rao, Visakhapatnam, India.) Used with permission.

natural product obtained during these months is of lower viscosity and darker in colour.

Trade. In India, gum karaya production has decreased from 6838 t in the years 1975 and 1976 to about 100 t in the years 2015 and 2016 (Yogi *et al.*, 2018). Before, India was the leading producer and exporter of gum karaya. During 1960–1980 the average gum karaya export from India was about 4000 to 6000 t (Coppen, 1995). The data on gum karaya export and import in India from 2008/09 to 2019/20 are presented in the Fig. 4.24. Imports have also risen continuously since 2007/2008 probably to meet the increased national demand.

The reason for the decline in production and export of this gum is attributed to the change in the Indian government's marketing policy. In the changed policy, the entire gum marketing from collection to export has been taken over by the government. Furthermore, the government has imposed a complete ban on the tapping of karaya trees for gum production in some potential gum karaya producing districts.

Some African countries, such as Senegal, Sudan, Ghana and Mali, have made great strides in the production and export of gum karaya and are competing with India. These countries have a large stock of karaya trees and the potential to become leading producers and exporters of gum karaya in the future with the introduction of improved production, collection and processing practices.

European countries like France and the United Kingdom, followed by Germany, Belgium, the Netherlands, Switzerland and Italy, are the leading importers of gum karaya. In addition to these European countries, the USA and Japan also import significant quantities of gum karaya from India and African countries (Coppen, 1995).

Gum ghatti

In *Anogeissus latifolia* trees, gum production occurs mainly in the summer or a dry winter and is caused by artificial or naturally occurring injuries. The dried exudate is harvested manually. Gum ghatti is mainly tapped from March to mid-June (Kramer and Kozlowski, 1979). The gum producers make artificial cuts in the tree bark to increase gum production. The cuts must be small and shallow to avoid injury to the plants and to guarantee a sustainable gum production.

Gum formation and yield. In *A. latifolia* bark or wood, there are no pre-formed gum-producing tissues. Thus, the trees do not produce gum on their own. However, *A. latifolia* has the potential

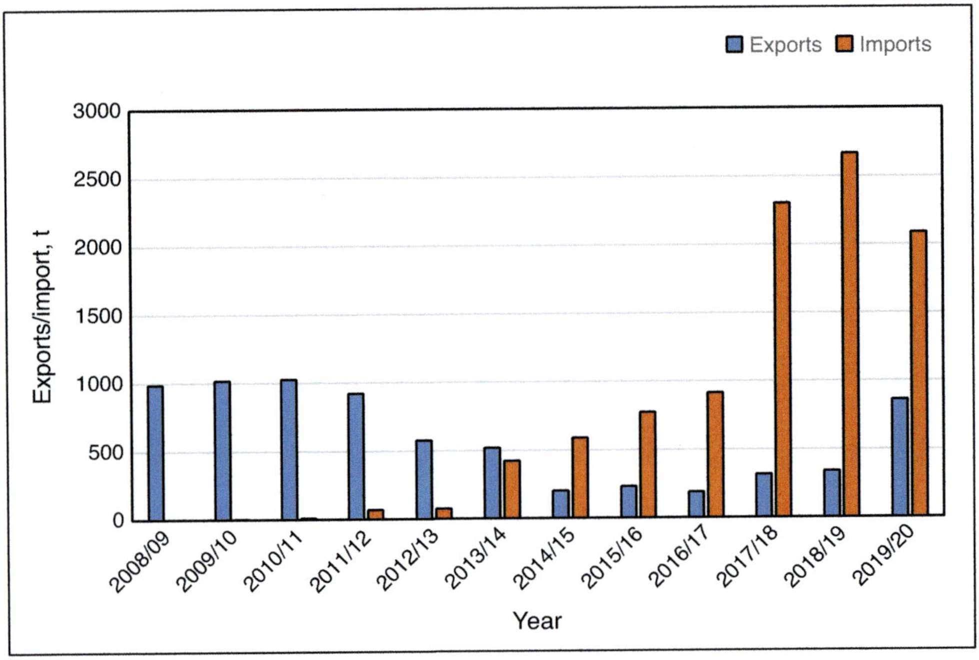

Fig. 4.24. Indian exports and imports of gum karaya from 2008/09 to 2019/20. (Directorate General of Commercial Intelligence and Statistics (2021).) Used with permission.

to produce gum when it is injured. Natural injuries (e.g. branch breaking by wind, injury caused by birds and animals) also lead to gum production. Ghosh and Purkayastha (1959) studied the anatomy of the gum cavities formed by injuries in the bark and cambial zone. Gum production depends on the height of the tree, climatic conditions and the tapping method. Gum ghatti is produced as tears or irregular masses and the production is usually higher in dry years (Ghosh and Purkayastha, 1959). On average, one tree produces approximately 1–2 kg of gum per tree per year.

In *A. latifolia* trees, bud formation/shoot growth and flowering/fruiting occur from June to July and September to January, respectively. In these periods, the trees utilize reserve metabolites for their growth. From April to May there is only minor growth activity in the tree, and there remains a larger reserve of metabolites in the wood parenchyma. This appears to be the reason for the higher gum production during April–May compared to other months. As gum yields during April–May were about 21 times higher compared to December–January using the gum inducer method, it was suggested to utilize the *A. latifolia* tree for gum tapping during April–May and leaving it to rest in the remaining period of the year (Bhatt, 1987).

The size and shape of the gum tears produced depend on the external and internal environment conditions of the tree, such as temperature, relative humidity, wind, shape/size of the fissure and internal pressure in the tree. The gum tears are mainly produced in two shapes, spiral (Fig. 4.25A) and round (Fig. 4.25B). Gum ghatti possesses a glassy appearance and, depending on the age of the exudate, its colour ranges from dark red to translucent white (Al-Assaf *et al.*, 2009).

Conventional method of gum ghatti tapping. The gum is harvested from wild trees by tribal people by making an incision in the tree's main stem with an axe. Gum production starts at the incision and the gum solidifies on drying. The gum collectors pick the dried gum by hand. To increase gum production, gum producers harvest the gum often, which injures the cambium, resulting in a shortening of the

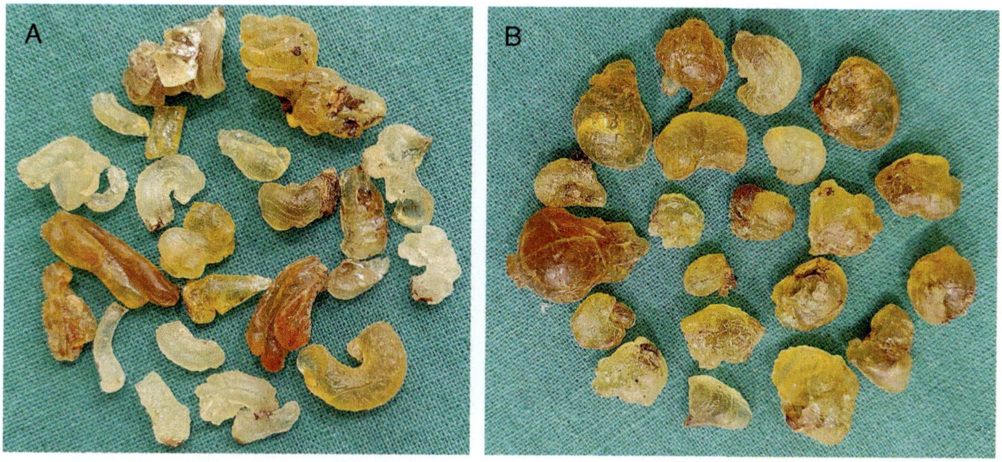

Fig. 4.25. Shape of gum ghatti tears: A. spiral; B. round. (By Nandkishore Thombare, ICAR-IINRG, Ranchi, India.) Used with permission.

tree's life if the injury does not heal (Ghosh and Purkayastha, 1959). Traditionally, trees are tapped to increase gum yield by using different tapping methods, i.e. blazing, bark stripping and making deep cuts into the tree with local tools like axes. This usually leads to deeper and larger blazes being made.

Further, due to the lack of guidelines for blaze depth, size and number per tree per season, gum tappers make deep and large blazes on the tree to get higher gum yield, which adversely affects tree health, and sometimes leads to the death of the tree. Thus, there is scope for developing a blazer that can make a blaze of desired depth and size. Moreover, there is a need to develop guidelines for blaze depth, size and number per tree per season for sustainable tapping of gum from *A. latifolia* trees and to maintain the health of the trees.

Inducer method for gum ghatti tapping. Like other gum trees, *A. latifolia* trees are also sometimes treated with stress hormones like 2-chloroethylphosphonic acid to increase the gum yield, which is directly related to the concentration of this plant hormone. Bhatt (1987) studied the effect of 2-chloroethylphosphonic acid application on gum exudation from *A. latifolia* trees. The plant hormone was applied in different concentrations (100, 200, 400, 800 and 1600 mg) of the active substance in a 4 ml solution. The study was carried out in April–May and December–January on *A. latifolia* trees free from visible defects and located in forests in Shivrajpur and Dang in the state of Gujarat in India. A significant increase in gum production took place in response to 1600 mg 2-chloroethylphosphonic acid treatment over other treatments in two seasons in three consecutive years. Gum yield was 21 times higher in April–May compared to December–January (Bhatt, 1987). Kuruwanshi *et al.* (2017) studied the effect of gum inducers on gum yield in *A. latifolia* trees in the Raipur district in the state of Chhattisgarh in India in 2015 (Fig. 4.26A,B,C,D). Maximum gum production was observed in May, followed by April, June and March (Kuruwanshi *et al.*, 2017).

An improved tapping method in which gummosis is increased significantly relies on the application of 2-chloroethylphosphonic acid (1600 mg) to enhance the gum ghatti yield (466 times), as reported by Bhatt (1987). Other commercial tapping approaches to obtain yields of gum ghatti utilize various methods like blazing, peeling or making deep cuts on the base of the bole or stem (Kuruwanshi *et al.*, 2018). Collection/picking of the soft gel-like gum from the tree bark is done manually at particular sites on a regular basis to provide a smooth and continuous flow of fresh gum. To increase the yield, the old gum is always harvested to make way for new gum (Amar *et al.*, 2006). The best time for

Fig. 4.26. Gum inducer method for *A. latifolia*: A. hole making with the hand-drill; B. introduction of the gum inducer with the syringe; C. covering the hole with moist clay; D. gum exudation. (By P. Katiyar, Raipur, India.) Used with permission.

gum collection before monsoon, and the largest amounts of gum are harvested throughout the late spring, for example between March and mid-June, as during this time, when the climate becomes warmer, the gum yield increases (Whistler, 1993; Giri *et al.*, 2008).

The main harvesting season is in April, when average yields approach approximately 1–2 kg/tree (Reddy *et al.*, 1964). Traditionally, solidified gum is hand-picked by tribal people living in the forest. They generally use axes or sharp-edged tools to separate the gum from the bark (Bhatt, 1987). This is the main reason for the impurities associated with freshly collected gums. Further contamination is due to the conventional methods used to store the gum. The world harvest of gum ghatti is relatively small and probably does not exceed 12,000 t (Bhatt, 1987; Kuruwanshi *et al.*, 2018).

India represents 30% of overall gum ghatti production worldwide, and 64% of it is produced in Maharashtra (Giri *et al.*, 2008; Yogi *et al.*, 2018). Apart from the Western Ghats of Maharashtra, the other regions where gum ghatti is harvested include the arid plateaus of Vindhyachal, Satpura, Madhya Pradesh, Jharkhand, Chhattisgarh, Bihar, Odisha, Gujarat and Andhra Pradesh (Srivastava and Ray, 2015; Kuruwanshi *et al.*, 2018).

Trade. The data for gum ghatti production in India and the quantity and value of Asian gum (*Acacia nilotica*) exports from India are presented in Fig. 4.27. The production of gum ghatti increased from 43.3 t in 2009 to 448.41 t in 2014 and later declined to 54.0 t in 2020 (Yogi *et al.*, 2021c). On average, during 2010–2021, India exported 510.45 t of Asian gum (Yogi *et al.*, 2021a) to different countries of the world. India is also an

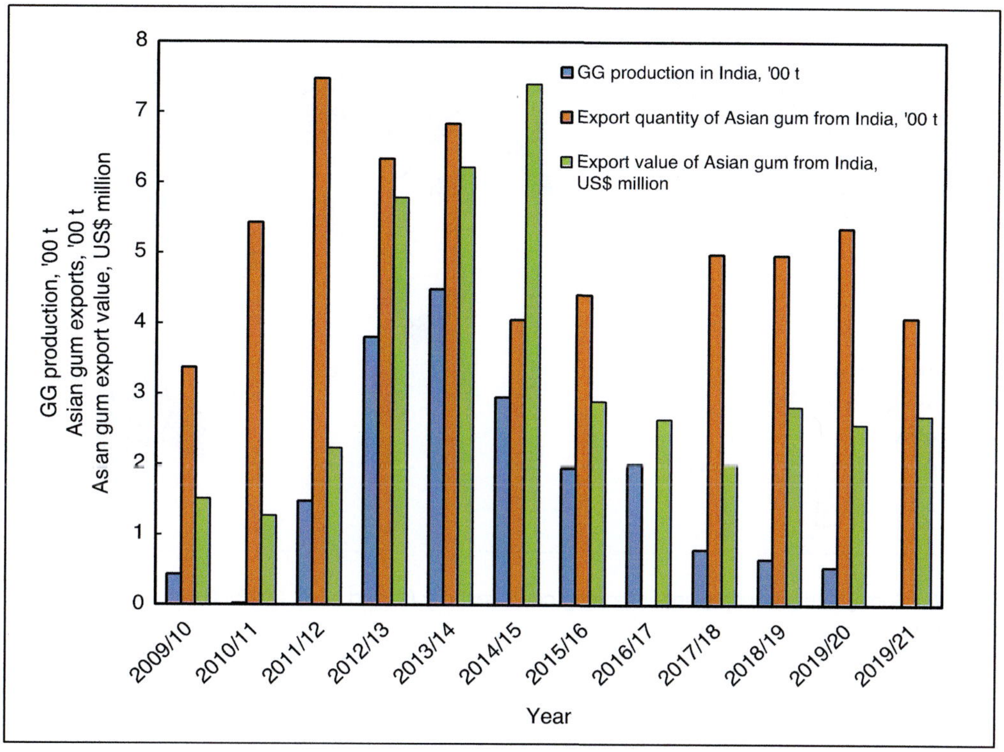

Fig. 4.27. Production of gum ghatti in India and exports of Asian gum (including gum ghatti) from India during 2009/10–2020/21. (By Prasad *et al*., 2023.) Used with permission.

importer of Asian gum, mainly imported from Indonesia and Singapore. Gum ghatti importing and exporting countries are indicated in Fig. 4.28.

Unfortunately, there is no official trade data available for Pakistan, Nepal, Sri Lanka, Myanmar and Bhutan. Therefore, it would be desirable for the future for the respective government agencies in these countries to compile the production and export data of gum ghatti, making it available to importing countries to enable them to meet their requirements for various applications. In this way, farmers and forest dwellers could also benefit as they could get reasonable prices for their produce.

*Moringa (*Moringa oleifera*) gum*

Moringa oleifera trees produce a natural gum (Fig. 4.29) which is harvested by hand from the moringa tree. This is the normal practice by local people of collecting moringa gum. After the gum tears have been collected, they are dried and stored for further use. As the yield of naturally obtained gum is comparatively low, attempts have been made to develop a scientific method of tapping the gum and to induce gummosis in moringa trees (ICAR-IINRG, 2023). Two methods, namely, the mechanical method and the drill method, were compared to find out their suitability for increasing gum production. It was concluded that the drill method, followed by the application of gum inducers, was the most suitable method for moringa gum production (Fig. 4.30). Maximum gum exudation was recorded in May, when temperatures rise sharply. Gum production using the mechanical method (Fig. 4.31) proved to be less efficient than the gum inducer method.

Experiments on extraction of moringa gum from *Moringa oleifera* trees in Tamil Nadu region were conducted by experts from the Forest College and Research Institute in collaboration

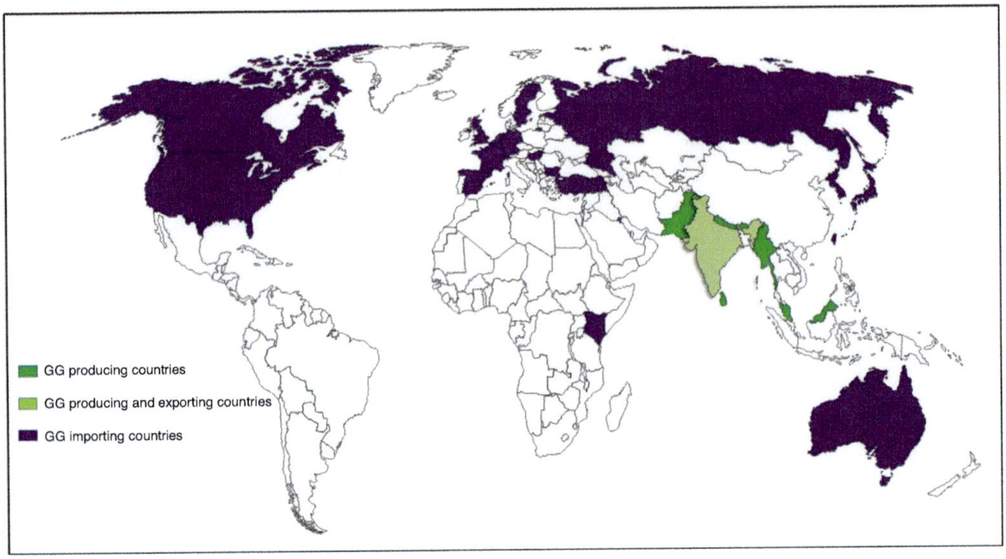

Fig. 4.28. Gum ghatti producing and importing countries. (By Prasad *et al.*, 2023.) Used with permission.

Fig. 4.29. Natural gum production. (By A.K. Mishra, Ranchi, India.) Used with permission.

Fig. 4.30. Gum production by inducer method. (By A.K. Mishra, Ranchi, India.) Used with permission.

Fig. 4.31. Gum production by mechanical method using blazer. (By A.K. Mishra, Ranchi, India.) Used with permission.

with other research institutes in India to optimize the yield in moringa resin production.

The trials showed that the maximum moringa gum yield (25.40 g/tree/month) was obtained with hand-operated drilling with 4 ml gum inducer application followed by hand-operated drilling with 2 ml inducer application for the diameter class of 90–120 cm. Minimum yield (1.73 g/tree/month) was obtained with hand-drilling without application of inducer in the diameter class of 30–60 cm (ICAR-IINRG, 2023).

To optimize the best tapping technique for enhanced moringa gum production with minimal injury to the *Moringa oleifera* trees, scientists of ICAR-National Institute of Secondary Agriculture conducted experiments with the mechanical method using a blazer to make the blaze and the gum inducer method using a hand-drill or a battery-operated drill to prepare the hole for the inducer injection (Fig. 4.32A,B,C) on suitable *Moringa oleifera* trees grown in the Kankebar area of Ramgarh district, Jharkhand.

Based on their experiments, the scientists came to the conclusion that the gum yield of trees treated with battery- and hand-operated drills with a drill diameter of 12.5 mm and 25 mm was approximately 15 times higher than without the use of gum inducer (2-chloroethylphosphonic acid). In contrast, it was only about four times higher for trees treated with hand-operated drills and a drill diameter of 25 mm.

In almost all treatments, a regular increase in moringa gum yield was observed with increasing ambient temperature, suggesting that moringa gum can be tapped throughout the year but that a higher quantity of gum can be collected during April–September. When using the gum tapping blazer, the blaze is produced on the tree trunk faster, more easily and with sharper edges (Fig. 4.32C) compared to the blaze produced using a chisel, and it oozes gum within a couple of hours. When gum was extracted from *Moringa oleifera* trees with a hand-drill, gum production was higher than with any other method (Sharma *et al.*, 2024).

Chirounji gum

The *Buchanania lanzan* trees are found in moist deciduous, dry deciduous and semi-evergreen forests, distributed throughout India, except the arid region of north-west India. It is a medium-sized tree having a dark grey or nearly black bark surface. The bark is rough with small quadrangular plates and deep and narrow fissures, resembling crocodile skin. Young branches are covered with silky hairs. The leaves are simple, broadly oblong, with a notch at the leaf apex (emarginate) and a leathery texture (coriaceous). Flowers are greenish-white and appear in terminal and axillary panicles without a peduncle. Fruits are initially green and turn black when ripe. The individual seed is surrounded by a hard seed coat and feels like a stone.

Gum production from this tree occurs naturally through cracking or natural injuries. Alternatively, farmers use locally available pruning knives to make blazes on the tree to induce gum production. Research has been conducted at the IGAU, Raipur, and the ICAR-IINRG, Ranchi, to develop sustainable methods for chirounji gum production.

The IINRG conducted experiments in tapping chirounji gum from *Buchanania lanzania* trees growing naturally in the village of Kewali in Ranchi district using different tapping techniques: bore hole making with battery-operated and hand-operated drills, both with a 12.5 mm drilling diameter, with a hand-operated drill with a 25 mm diameter, with and without the use of an inducer, a gum tapping blazer and a commercially available chisel to producing blazes of a defined size and shape. From the results obtained, it was concluded that the gum inducer technology was better for chirounji gum tapping from *Buchanania lanzania* trees using a hand-operated drill with gum inducer (2-chloroethylphosphonic acid) application, as gum production was higher compared to all other techniques (ICAR-IINRG, 2023).

Studies on gum production adopting the gum inducer and blazing methods for the *Buchanania lanzania* tree species in the Chhattisgarh region were conducted at IGAU, Raipur. The results indicated that the inducer method of gum extraction is significantly more effective for chirounji gum production during April–June compared with the blazing method.

Fig. 4.32. Bore hole/blaze development: A. using a battery-operated drill; B. using a hand-drill; C. using a gum tapping blazer. (By S.C. Sharma, ICAR-IINRG, Ranchi, India.) Used with permission.

*Babool (*Acacia nilotica*) gum*

Babool gum is collected from babool trees (Fig. 4.33) throughout the year except during the monsoon season. The best gum quality is obtained when the babool gum is white, the next best when it is red and the worst when it is black. The gumming process is strongly influenced by temperature and relative humidity, and microbes or insect infestation, e.g. termites, can also trigger gum production.

Traditionally, trees are tapped by blazing, stripping off the bark or making deep cuts with an axe. This damages the plants in various places, which often leads to the death of the tapped trees. Therefore, scientific and sustainable tapping methods have been developed to increase the gum yield in commercial tapping while ensuring the survival of the tapped trees. In this context, the gum inducer method of tapping, in which the trees are treated with stress hormone, has been proven to be a safe and scientific technique for stainable gum production.

The best time to tap the trees is at the beginning of the dry season, when the relative humidity is low. As a general rule, the higher the average temperature, the higher the gum yield. Natural factors that affect the vitality of the trees, such as hot weather, poor soil, lack of moisture, etc. have a negative effect on the yield. The mechanical method of gum tapping is not very effective in babool.

Trade. There is high demand for Asian gum (*Acacia nilotica*) both from and in India. The share of exports and imports of Asian gum by destination is depicted in Table 4.4. During 2023/24, about 381.6 t of Asian gum was exported from India to various countries in the world, of which Germany, Russia, Japan and China were the major importers. During the same period, India imported 530.35 t of Asian gum, mainly from Bangladesh, Singapore and Afghanistan.

Seed gum

Seed gums are derived from the endosperm of the seeds of various plant species, e.g. guar (*Cyamopsistetra gonoloba*), tamarind (*Tamarindus indica* L.), cassia (*Cassia tora*), and psyllium (*Plantago ovata*), and are used for various commercial purposes. Details on the production of gum from seeds of various plants are explained in the following sections.

Guar gum

Cultivation. Guar or cluster bean (*Cyamopsis tetragonolobus*) is an annual agricultural crop belonging to the Leguminosae family. It is grown in arid zones of west and north-west India, Pakistan, Sudan and parts of the USA. Guar gum, also known as cyamopsis gum, guaran, guyan, guarina or glucotard, is a natural non-ionic, water-soluble polysaccharide obtained from the guar plant. The guar plant is about 0.6m high and resembles the soybean plant in its general appearance and the arrangement of the pods on the vertical stem. The pods are 50–125 mm in length and contain an average of 5–6 light brown seeds (Chudzikowski, 1971).

Fig. 4.33. Babool gum exudate on tree. (By Niranjan Prasad, ICAR-IINRG, Ranchi, India.) Used with permission.

Guar gum is extracted from guar seeds, which are dicotyledonous plants and accordingly contain two halves of endosperm. The colour of the husk varies from variety to variety and is usually light to dark brown. The germ and the husk of guar seed are known as guar meal, which is very rich in protein. The embryo is the innermost part, which contributes about 43–47% of the seed weight. The endosperms surrounding the germ comprise 35–42% of the weight, while the outer husk makes up 14–17% (Seaman, 1980).

The prevailing monsoons on the Indian subcontinent provide an ideal climate for guar cultivation. India and Pakistan together grow around 90% of world's guar production, with India accounting for 80% of the total (Mudgil *et al.*, 2014). The total production of guar seeds in India in 2013/14 was 3,388,000.4 t, of which about 84.5% was produced in the state of Rajasthan alone (Directorate of Economics and Statistics, 2015). The major quantity of guar seeds produced are processed into guar split or guar gum for export and domestic consumption while some of the guar seeds are exported as such. India is the leading exporter of guar seeds as well as guar gum (Yogi *et al.*, 2015). In 2013/14, around 650,000 t of guar gum were produced in India, of which 601,000t were exported, more than 50% of which to the USA (Directorate of Economics and Statistics, 2015).

Guar or bush bean has been cultivated in India since ancient times. Guar is a drought-tolerant, annual legume that grows best in sandy soils and needs moderate, intermittent rainfall with plenty of sunshine. Guar is an upright, coarse legume with a deep taproot that reaches for moisture deep below the soil surface. Its roots also develop well laterally. Some varieties have small hairs on all parts of the plant, while others have glabrous (smooth, hairless) leaves, stems and pods. The leaves are alternate trifoliate and sit on long petioles. The plants are single-stemmed, finely branched or basally branched and grow to a height of 45–100 cm. The flowers are small and white. The pods are elongated and 5–10 cm long. The pods usually

Table 4.4. Exports and imports of Asian gum during 2023/24 from and to India, respectively. Author's own table.

Export destination countries	Quantity (t)	Value (US$ million)
Germany	68.74	0.38
Russia	50.00	0.22
Japan	43.55	0.35
China	42.51	0.15
Belgium	25.00	0.10
Malaysia	20.00	0.07
USA	17.59	0.07
UK	17.20	0.05
France	16.58	0.05
South Africa	15.44	0.07
Others (32)	64.99	0.42
Total	381.60	1.93
Import source countries		
Bangladesh	520.00	0.13
Singapore	6.83	0.01
Afghanistan	3.52	0.02
Total	530.35	0.16

contain 5–12 oval or cube-shaped seeds of varying size and colour.

For optimum plant growth, guar requires rain before sowing and again to induce the ripening of the seeds. Too much precipitation can cause the plant to become leafier, thereby reducing the number of pods and/or the number of seeds per pod, which affects the size and yield of the seeds.

Guar is a rainfed crop and is generally sown after the monsoon rainfall between the second half of July and early August. It is harvested from late October to November. The average yield of guar seed in India is around 500–700 kg/ha.

Guar gum extraction. Taking advantage of the difference in hardness of the various seed components, the endosperm is separated from the husk and embryo following a multistage grinding and sieving processes. These operations are also combined with other physical treatments to crack the seed and separate the parts. The germ of guar seed consists mainly of protein and the endosperm of galactomannan. The separated endosperm, commonly known as guar split, is commercially available and ground into powder for obtaining marketable guar gum. Sabahelkheir Murwan *et al.* (2012) have reported the detailed protocol for separating guar endosperms. To avoid any possible enzymatic activity during the soaking of guar seeds in water, the method was modified (Liyanage *et al.*, 2015) by performing an initial heat treatment at 100°C for 30 min to deactivate the enzymes and the germ. The endosperm obtained using standard commercial purification techniques is not completely pure. Traces of hull and germ left with the endosperm contribute to impurities. Powdered guar gum comprises 75–86% water-soluble galactomannan, 8–14% moisture, 5–6% protein, 2–3% fibre and 0.5–1% ash (Chudzikowski, 1971; John *et al.*, 1976).

However, since all seed components, including the husk and germ, are edible, their role as impurities is minimized. These impurities do not affect the suitability of guar gum as a food additive. As guar gum is of natural origin, its solubility and viscosity vary slightly from batch to batch. Major application areas of guar gum and its derivatives are in the textile, paper, food, cosmetic, explosive and mining industries (Seaman, 1980). Based on the percentage of impurities, the viscosity of the gum solution and various other parameters, guar gum is categorized into food grade and industrial grade.

For guar gum, there exist standard specifications of the Joint Expert Committee on Food Additives (JECFA), the Food Codex and the Bureau of Indian Standards (BIS) for grading and quality control.

Guar gum powder is obtained after removing the hull and germ from the seeds and grinding the endosperm into a fine powder, which is achieved by different processing methods. The proportion of endosperm ranges from 32 to 42%, depending on the variety and degree of maturity of the plant. So, the most important steps in the processing of guar seeds are cleaning, grading, dehusking, splitting and separation of endosperm, grinding and purification of the powder. On arrival at the processing plant, the seeds are sieved to remove dirt, stones, sand, metal residues, chaff and broken seeds. Commercially available vibrators, electromagnets and pushers are used for cleaning.

The dehulling and splitting of the seeds is done in two processes, namely dry grinding and wet grinding. In some industries, the husk is also carbonized by flaming and then separated. The yield of wet processing is about 8–10% higher than from dry processing. However, the quality of the gum produced in this way is not very high. Therefore, most industries prefer dry processing or carbonization of the seeds.

Burr mills, pin mills and modified hammer mills are usually used for splitting of guar seeds into two halves so as to separate the germ and endosperm. The splits are then heated in ovens and passed through the dehusking machines, usually consisting of a two-tiered chamber, each with a rotating saw-toothed blade. Splits stripped of their hull pass to sifters that separate the clean endosperm pieces on a 20-mesh screen.

Guar gum powder is produced from endosperm splits simply by grinding in attrition mills, hammer mills, ultra-fine grinders or other size-reduction equipment. However, guar gum with the best thickening power and fastest hydration rate is produced when the splits are first soaked in water and then flaked, extruded or ground.

The guar gum powder is usually chemically modified to give it new properties for a wider range of applications. For industrial applications, many guar gum products are formulated with additives that control the hydration rate, enzyme resistance, dispersibility, flowability or other special properties. The most common commercial derivatives of guar gum are hydroxyl-propyl guar, carboxy methyl guar and 2-hydroxy-3 (tri-methyl ammonium chloride) propyl guar.

Production and trade. India's share in the world trade of guar gum is 75.31%, followed by Pakistan (6.59%), Spain (5.77%), Italy (3.03%), Germany (1.56%), the USA (1.42%) and China (1.2%). The production of guar seed, potential of guar gum production, export of guar gum and value of export of guar gum from India for the period 2010/11–2019/20 are presented in Table 4.5.

Tamarind seed gum

Matured tamarind pods are harvested from the tamarind tree. The harvested pods are then broken or opened to separate the seeds from the ripe/soft tamarind pulp, fibre and epicarp. The seeds are then sorted by removing rotten seeds, debris, stones, etc. from the seed mass.

Manufacture of tamarind kernel powder. The tamarind seed kernel is covered by the seed coat (testa), which is difficult to remove from the kernel. For this reason, the healthy seeds are

Table 4.5. Production of guar seed, potential of guar gum production, export of guar gum and value of export of guar gum from India for the period 2010/11 to 2019/20. Author's own table.

Year	Guar seed production (t)	Potential of guar gum production (t)	Export of guar gum (t)	Value (US$ million)
2010/11	1965.3	655	364	631.65
2011/12	2217.6	739	627	3330.74
2012/13	2457.6	819	331	3893.53
2013/14	3585.3	1195	469	1931.60
2014/15	3275.9	1092	521	1498.54
2015/16	2824.9	942	256	462.61
2016/17	1849.5	617	318	408.16
2017/18	1624.3	541	367	576.35
2018/19	1312.1	437	362	586.23
2019/20	1448.0	483	236	373.27
Average	2256.05	752	385	1369.27

roasted after the sorting process, which makes the seed coat brittle and allows it to be removed from the kernel with little effort. The de-hulled tamarind seeds are crushed using a pulverizer to produce a creamy white tamarind kernel powder (TKP) with a yield of 55–60%. The TKP is further sieved to produce a fine powder that is used to isolate tamarind gum.

Extraction and isolation of tamarind gum. Tamarind seeds represent the raw material used in the manufacture of tamarind seed gum (xyloglucan). The tamarind gum is extracted either from whole tamarind seed or from TKP.

To obtain tamarind gum from the seeds, they are first soaked and then boiled in double distilled water for 5 hour to remove the dark outer layer. After this is removed, the inner white part is boiled again in sufficient distilled water with constant stirring to obtain a slurry. The resulting solution is cooled in a refrigerator to facilitate sedimentation of the undissolved portion.

The supernatant liquid is separated either by centrifugation (500 rpm for 20 min) or by simple decantation. Then, the solution is concentrated in a water bath at 60°C to one-third of the initial volume. This concentrated solution is cooled and mixed with acetone with continuous stirring. Precipitation occurs and the precipitate obtained is washed with acetone followed by vacuum drying (50–60°C), grinding, sieving and storage. The resulting powder is known as tamarind seed gum (Rao and Srivastava, 1973). In the extraction of gum from TKP, the only difference in the process is that TKP is mixed directly with cold water to obtain a slurry, which is then boiled for 20 min. The remaining steps are the same as described above, with some modifications in the operating conditions. It is recommended to store TKP in a dry place in moisture-proof containers, as the powder tends to deteriorate during storage under humid conditions. If TKP is stored inadequately, it turns rancid and brown due to enzymatic degradation. Hence, TKP should be defatted before storage or mixed with 0.5% of sodium bisulfite before packaging. The general characteristic of a good TKP is that it should have the characteristic taste when dissolved in water and be free from any burnt or other undesirable flavours. It should also have a good shelf life and be free from any insect pests, fungal infestation or foreign substances (El-Siddig *et al.*, 2006).

Table 4.6. Indian exports of tamarind flour meal and powder, 2023/24. Author's own table.

Country	Quantity (t)	Value (US$ million)
Australia	31.24	0.14
Sri Lanka	21.00	0.01
UAE	13.57	0.03
Maldives	7.34	0.01
Nepal	7.25	0.01
USA	4.95	0.01
UK	4.47	0.01
Others (22)	14.99	0.03
Total	104.81	0.25

Trade. Tamarind gum powder is used in the paper, textile, animal feed, pet food, mining and pharmaceutical industries. Exports of tamarind flour meal and powder from India for 2023/24 are shown in Table 4.6.

Cassia gum

The raw material used for the production of cassia gum usually contains some *Cassia obtusifolia* along with *Cassia tora* seeds. Both species grow at the same site and are difficult to distinguish visually, hence they are harvested together. Researchers have reported some detectable chemical differences in the amino acid and anthraquinone content of both seeds (Upadhyaya and Singh, 1986), but the close similarity in seed chemistry, morphology and ecological requirements between the two plant species is far more significant than the relatively minor differences between them. The proportion of *Cassia tora* seeds in the raw material ranges from 52.7–81.5%, while the remaining seed proportion is from *Cassia obtusifolia*.

The cassia pods are harvested together when the plants are ripe. The matured seeds are gathered and then dried in the sun. The seeds obtained from the pods undergo a cleaning process that includes mechanical steps to

Table 4.7. Indian exports and imports of *Cassia tora* seeds, 2023/24. Author's own table.

Export destination countries	Quantity (t)	Value (US$ million)
China	16,514.96	7.75
Taiwan	3837.48	1.86
Vietnam	1010.00	0.52
Japan	737.45	0.41
UK	255.71	0.44
Korea RP	246.04	0.14
Chile	24.70	0.05
Others (11)	2.17	0.01
Total	22,628.51	11.18
Import source countries		
Nigeria	20,422.29	8.29
China	200.00	0.10
Korea RP	10.51	0.01
Total	20,632.80	8.39

remove foreign matter (stones, dirt, agricultural waste, plant residues, etc.) and seeds other than *Cassia tora and Cassia obtusifolia*. Firstly, dust and coarse impurities are sieved out, and then seeds are passed through a gravity separator to remove stones and seeds with a similar size but different weight compared to *Cassia tora* or *Cassia obtusifolia* seeds. In this way, underdeveloped seeds are removed and the proportion of *Cassia occidentalis* seeds is reduced to a maximum of 0.05%.

After the seeds have been cleaned, they are subjected to a dehusking and splitting process in which the husk and germ are separated from the endosperm. The seeds are roasted under such conditions that the endosperm (split) remains intact and flexible, while the husk and germ, which are more sensitive to heat, become brittle. The physical stress crushes the husk and germ into a powder, which is removed from the intact split by sieving. Brushing and polishing steps finally remove last traces of husk and germ from the endosperm.

Finally, the splits are ground into a uniform powder with a small particle size. To extract the gum from cassia powder, the cassia flour is soaked in water, causing the gum to swell. The swollen gum is added to a mixture of water and isopropanol while stirring. The wet gum is then separated from the alcohol/water solution by filtration, dried and packaged. The residual isopropanol is expelled during the drying process.

Most of the anthraquinone content in cassia seeds is located in the outer husk. The anthrachinone concentration in whole seeds is approximately 10,000 mg/kg (1%). After processing cassia seeds by various operations (splitting, dehusking-degerming, polishing and brushing), the concentration of anthraquinones in the split can be reduced to an amount below 250 mg/kg. Mechanical means cannot reduce the anthraquinone content in the splits any further. Therefore, the anthraquinone content is further reduced from 250 mg/kg to below 0.5 mg/kg by subsequent extraction with isopropanol.

After drying of the product, the product batches are transferred to a blender where they are mixed, and finally, cassia gum powder is packed into 20 kg bags. The residual isopropanol and water are evaporated during the drying process so that only traces of the alcohol can be detected. The extraction step allows the isolation and purification of cassia gum, resulting in refined cassia gum. Refined cassia gum is used in food formulations, while semi-refined cassia gum has been used in pet foods (canned meat for cats and dogs) as a thickening and gelling agent since the 1990s (Mahungu and Meyland, 2010).

Trade. *Cassia tora* seeds are another commodity with high exports from India to various countries (Table 4.7).

*Psyllium (*Plantago *sp.)*

Psyllium is indigenous to the Mediterranean region and West Asia and is distributed across the Canary Islands, southern Spain, North Africa, the Middle East and north-western Asia. *Plantago ovata* is native to India and commercially cultivated in north Gujarat, western Rajasthan, Punjab, Hariyana and Uttar Pradesh.

There are about 250 species of the genus *Plantago* found worldwide. Psyllium is an annual herb that reaches a height of 12–18 cm. The plant is herbaceous with linear, lanceolate,

Fig. 4.34. Psyllium plant. (By Nandkishore Thombare, ICAR-IINRG, Ranchi, India.) Used with permission.

dentate and hairy leaves. The white flowers are borne in cylindrical panicles. The sepals are characterized by distinct midribs extending from the base to the tip. Petal lobes are oval with a notch at the tip. The seeds are oval, clearly carinate, light grey-pink, with a brown line running along their convex side. Commercial planting is done mostly during the winter season (November–December) and the plant flowers occur about 60 days after planting. It also grows in the wild, known as 'Indian desert wheat'.

Psyllium is one of the most important medicinal plants (Fig. 4.34), cultivated mainly for its husk. A quarter or more of the total seed yield consists of mucilage.

India is the world's leading producer and supplier of psyllium products, accounting for 98% of global production. The most important states in which this medicinal plant is cultivated are Rajasthan, Gujarat, Madhya Pradesh and Haryana.

Production and trade. Psyllium is originally native to Persia (now Iran). Indian pharmaceutical companies use psyllium husk for the production of various herbal drugs. The major consumers of psyllium husk (also known as isobgul) from India are the USA, Germany, Iran, Sweden, Bangladesh and the UK, which account for more than 75% of the total exports from

Fig. 4.35. Guggul tapping: A. use of blazer for making shallow cut; B. guggul in liquid form produced from cut made on the branch of guggul tree. (By Niranjan Prasad, ICAR-IINRG, Ranchi, India). Used with permission.

India. Psyllium husk exports in 2019/20 were 1695.79 t worth US$ 3.93 million (Yogi *et al.*, 2021a).

Gum-Resin

Guggul or Indian myrrh (*Commiphora mukul*)

The *Commiphora mukul* tree exudes a yellowish gum-resin called guggul. The gum-resin residues in the ducts are located in the soft bark of the tree and are obtained by tapping. The resin ducts occur in the bark portion near the cambial layer. Plants attaining a diameter of 75 mm are suitable for tapping. Usually, circular incisions 1.5 mm deep are made on the main stem, which should not extend beyond the thickness of the bark (Fig. 4.35). Guggul oozes out from these incisions as a pale yellow, aromatic liquid that quickly solidifies to form a golden-brown or reddish-brown agglomerate of tears or stalactites.

The gum-resin is collected manually or with a spear. It is scraped off the wound with a knife. The collection is done at intervals of 10–15 days. The trees are tapped from November to January, and the collection continues until May or June. A healthy tree yields 250–500 g of gum-resin in one season, and guggul plants typically begin producing resin after 5 years. From the sixth year onwards, gum-resin yield increases from 200 to 400 g per plant.

Production and trade

In 2019/20 India exported 1.44 t of Indian myrrh worth US$10 millon, mainly to the UK, Malaysia, Uruguay, Israel, the USA and Australia. India also imported 22.09 t of myrrh from Somalia, Kenya and the United Arab Emirates worth US$120 millions during 2019/20 (Yogi *et al.*, 2021a).

Salai gum or Indian olibanum (*Boswellia serrata*)

Salai gum, commonly known as Indian olibanum, is the gum-resin collected from the tree of *Boswellia serrata*. The gum-resin is collected in the form of chunks. Tapping and collection of salai gum generally begins in November and lasts until the end of May. Usually, trees of 900 mm girth and more are tapped at breast height to extract the gum-resin. Trees of lower girth classes should not be tapped since any damage can impair their growth. A thin strip about 300 mm in length and 200 mm wide is peeled from the trunk of the tree at a height of about 0.75m from the ground. Thus, the reddish phloem in which the resin canal and ducts lie is exposed. The blaze is freshened after every fourth or fifth day. The first collection of the gum-resin is carried out after 2 weeks. Collection is done using a scraper, keeping a tray with a semicircular edge around the blazed surface. Freshening of the blaze is done

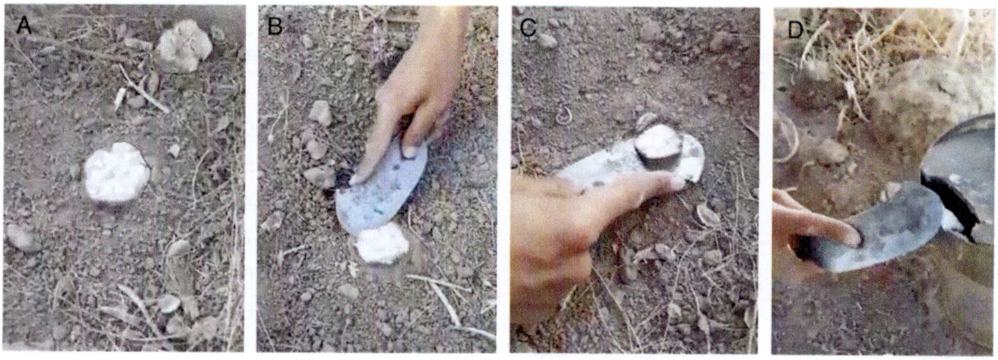

Fig. 4.36. Asafoetida (A. exudation from cut stem, B. exudate collection, C. freshening of the stem after exudate collection, and D. exudate collection in a tin can). (By Kumar, 2020.)

from time to time and the original blaze is slowly widened.

Average annual yield of salai gum per tree is estimated at about 1 kg, though a well-grown tree yields up to 2–2.5 kg of gum-resin per year. The yield of salai gum depends on the season, the location and the size of the tree. Generally, the larger the girth of a tree, the greater the yield. Yield has generally been found to be poor in moist regions and from trees which are old with black bark or dwarfed and suppressed. Medium-aged, healthy and vigorous trees produce comparatively better yields.

Production and trade

The gum resin of various *Boswellia* species is produced in Africa, West Asia, Oman and South Arabia. In India, *Boswellia serrata* (salai) is widely found in the Western Himalayas, Madhya Pradesh, Gujarat, Odisha, Rajasthan, Jharkhand and Maharashtra. In 2019/20 India imported 172.55 t of frankincense mainly from Somalia and Kenya. At the same time, it also exported 97.39 t of the gum-resin worth US$360million in 2019/20, mainly to Germany, Mexico, the UK, Trinidad, France and Guatemala (Yogi *et al.*, 2021a).

Asafoetida (*Ferula* sp.)

Asafoetida, often known as devil's dung or gods' food, is the dried aromatic gum-resin that exudes from the living rhizome, rootstock or taproot of various plant species of the genus *Ferula*, caused by cutting into the roots and rhizomes of the plants (Fig. 4.36A). Usually, the plants, which are about 4–5 years old, develop very thick and fleshy, carrot-shaped roots. A milky juice exudes from the cut surface which soon coagulates when exposed to air. After some days, the exudate gum-resin is scraped off (Fig. 4.36B) and a fresh slice of the root cut (Fig. 4.36C). Tapping is usually done in March and April, just before the plants flower. The scraped-off gum-resin is then collected in tin containers (Fig. 4.36D) for storage and transport from the collection centre.

Production and trade

Asafoetida is a popular spice that has become particularly important in Asian and North African cuisine (Kandziora and Melzig, 2015). In India, asafoetida was mainly supplied by Afghanistan, Iran and Uzbekistan in 2023/24. On the other hand, India also exported processed asafoetida to the United Arab Emirates, the USA, Saudi Arabia, Malaysia, Singapore, the UK, Canada, Myanmar, Australia and Qatar in 2023/24. Indian exports and imports of asafoetida are shown in Table 4.8.

Conclusion

NRGs are usually non-toxic, biodegradable and environmentally friendly and are therefore used

Table 4.8. Indian exports and imports of asafoetida, 2023/24. Author's own table.

Export destination countries	Quantity (t)	Value (US$ million)
United Arab Emirates	239.87	2.42
USA	139.90	2.02
Saudi Arabia	83.49	0.59
Malaysia	81.67	0.41
Singapore	69.35	0.74
UK	68.81	0.64
Canada	46.69	0.66
Myanmar	42.59	0.17
Australia	39.15	0.55
Qatar	37.78	0.41
Others (87)	310.35	2.36
Total	1159.65	10.97
Import source countries		
Afghanistan	1024.84	128.66
Iran	195.12	6.25
Uzbekistan	170.13	11.77
Kazakhstan	23.60	1.40
United Arab Emirates	23.49	1.23
Kyrgyzstan	16.50	0.39
Turkmenistan	2.11	0.13
Oman	0.78	0.10
Total	1456.57	149.93

in various industries. Commercially important NRGs are produced from various tree species and represent high-value products with low volumes. NRGs are tapped by exposing the resin and gum ducts of trees and are collected in small quantities by forest dwellers using prevalent standard tapping techniques. Improved efficiency of NRG harvesting could be achieved by reducing the labour input of NRG harvesters through the introduction of mechanical tapping techniques and devices. In the future, production systems will have to be made more efficient to ensure a sustainable use of the resources. As the availability of manpower is likely to decrease within the next years, new techniques for harvesting and collecting NRGs need to be developed with lower manpower inputs and higher yields to achieve competitive prices for the products.

Mechanized tapping techniques and devices will minimize tree damage and generally contribute to the sustainable production of NRGs. Improved gum production will support the gum industries in developing a variety of products and adding value to NRGs, thus allowing related industries to provide employment and a steady additional income for the rural population.

References

Abib, C.F., Ntoupka, M., Peltier, R., Harmand, J.M. and Thaler, P. (2013) Ethephon: A tool to boost gum arabic production from *Acacia senegal* and to enhance gummosis processes. *Agroforestry Systems* 87, 427–438. Available at: https://doi.org/10.1007/s10457-012-9564-y

Adam, H.E., Eltahir, M.E.S., Elhaja, M.E., Ibrahim, A.H., Abdelkareem, O.E.A. *et al.* (2017) Management of gum arabic production potentialities in the gum belt in Kordofan, Sudan. *International Journal of Environmental Planning and Management* 3, 1–9.

Adam, I.M., Ballal, M.E. and Fadl, K.E.M. (2013) Effect of grass density and date of tapping on *Acacia senegal* gum yield in North Kordofan state, Sudan. *Journal of Forestry Research* 24, 169–172. Available at: https://doi.org/10.1007/s11676-012-0268-y

Al-Assaf, S., Phillips, G. and Amar, V. (2009) Gum ghatti. In: Phillips, G. and Williams, P. (eds) *Handbook of Hydrocolloids*. CRC Press, Boca Raton, Florida, pp. 477–494.

Alland and Robert (2019) Sourcing and Supply Chain in Africa. Available at: https://www.youtube.com/watch?v=bwTqBRzs3RQ (accessed 24 March 2025).

Amar, V., Al-Assaf, S. and Phillips, G.O. (2006) An introduction to gum ghatti: Another proteinaceous gum. *Foods and Food Ingredients Journal of Japan* 211, 275–279.

AwadElkarim, M. (1994) Analytical studies on some crude and processed gum arabic samples with regard to quality aspects. MSc thesis, University of Khartoum, Sudan.

Awouda, E. (1988) Outlook from gum arabic production and supply in gum and stabilizers for the food industry. In: *Proceedings of 4th International Conference, Wrexham, July, 1987*. IRL Press, Oxford, UK, pp. 425–434.

Aziz, A., Hussien, E., Babiker, A.A., Elkarim, A., Osman, S. *et al.* (2018) A comparative study between tapping *Acacia senegal* by sunki and axe in West Kordofan, Sudan. *American Journal of Engineering, Technology and Society* 5, 1–4.

Ballal, M.E., El Siddig, E.A., Elfadl, M.A. and Luukkanen, O. (2005) Gum arabic yield in differently managed *Acacia senegal* stands in Western Sudan. *Agroforestry Systems* 63, 237–245. Available at: https://doi.org/10.1007/s10457-005-4162-x

Behera, B. (2009) Explaining the performance of state–community joint forest management in India. Ecological Economics 69, 177–185. https://doi.org/10.1016/j.ecolecon.2009.08.015

Bhatt, J.R. (1987) Gum tapping in *Anogeissus latifolia* (Combretaceae) using ethephon. *Current Science* 56, 936–940.

Bhattacharya, P., Joshi, B. and Hayat, S.F. (2003) An improved method of tapping gum from kullu *Sterculia urens*. *Forests Trees and Livelihoods* 13, 187–196. https://doi.org/10.1080/14728028.2003.9752454

Chudzikowski, R.J.C. (1971) Guar gum and its application. *Journal of the Society of Cosmetic Chemists* 22, 43–60.

Coppen, J.J.W. (1995) *Gums, Resins and Latexes of Plant Origin*. FAO Inter-Departmental Working Group, Rome.

Coppen, J.J.W. and Hone, G.A. (1995) *Gum Naval Stores: Turpentine and Rosin from Pine Resin*, Non-Wood Forest Products series. Food and Agriculture Organization, Rome, p. 2.

Davidson, R.L. (1980) *Handbook of Water-Soluble Gums and Resins*. McGraw-Hill, New York.

Dione, M. and Vassal, J. (1993) Experimental study on the means of production of trees of *Acacia senegal*: Looking back on the gum tree development programmes in the Senegalese Sahel. In: *Proceedings of 5th Sahel Workshop. Natural Resources and Social Conflicts in the Sahel, 4–6 January, 1993*. Arhus University, Arhus, Denmark, pp. 22–41.

Directorate General of Commercial Intelligence and Statistics (2021) Ministry of Commerce and Industry, Government of India. Available at: http://www.dgciskol.gov.in/ (accessed 15 March 2021).

Directorate of Economics and Statistics (2015) *Agricultural Statistics at a Glance 2015*. Directorate of Economics and Statistics, Ministry of Agriculture, Government of India.

El-Siddig, K., Gunasena, H.P., Prasad, B.A., Pushpakumara, D.K., Ramana, K.V. *et al.* (2006) *Tamarind: Tamarindus indica L.* Southampton Centre for Underutilised Crops, Southampton, UK.

Fall, D., Bakhoum, N., Sall, S.N., Zoubeirou, A.M., Sylla, S.N. *et al.* (2016) Rhizobial inoculation increases soil microbial functioning and gum arabic production of 13-year-old *Senegalia senegal* (L.) Britton, trees in the north part of Senegal. *Frontiers in Plant Science* 7. Available at: https://doi.org/10.3389/fpls.2016.01355

Forest Research Institute (1973) *Karaya Gum from Sterculia urens Roxb*, Industrial Series no.7. Forest Research Institute, Dehradun, India.

Ghosh, S.S. and Purkayastha, S.K. (1959) An anatomical study of wood and bark of *Anogeissus latifolia* tree with special reference to gum exudation. *Indian Forester* 88, 92–99.

Giri, S., Prasad, N., Pandey, S., Prasad, M. and Baboo, B. (2008) *Natural Resins and Gums of Commercial Importance - At a Glance*. ICAR-Indian Institute of Natural Resins and Gums, Ranchi, India.

Gray, A., Odee, D., Cavers, S., Wilson, J., Telford, A. *et al*. (2013) Does geographic origin dictate ecological strategies in *Acacia senegal* (L.) willd.? Evidence from carbon and nitrogen stable isotopes. *Plant and Soil* 369, 479–496. Available at: https://doi.org/10.1007/s11104-013-1593-4

Hall, M.A. and Smith, A.R. (1995) Ethylene and the responses of plants to stress. *Bulgarian Journal of Plant Physiology* 21, 71–79.

Harmand, J.M., Ntoupka, M., Mathieu, B., Njiti, C.F., Tapsou, J.M. *et al*. (2012) Gum arabic production in *Acacia senegal* plantations in the Sudanian zone of Cameroon: Effects of climate, soil, tapping date and tree provenance. *Bois et Forets Des Tropiques* 66, 21–33. Available at: https://doi.org/10.19182/bft2012.311.a20507

Hassan, E. (2000) Characterization and fractionation of *Acacia seyal* gum. MSc. thesis, University of Khartoum, Sudan.

ICAR-IINRG (2012) *Annual Report 2012*. Indian Institute of Natural Resins and Gums, Ranchi (India).

ICAR-IINRG (2016) *Annual report 2015–16*. ICAR-Indian Institute of Natural Resins and Gums, Ranchi (India).

ICAR-IINRG (2017) *Annual Report 2016–17*. ICAR-Indian Istitute of Naural Resins and Gums, Ranchi (India).

ICAR-IINRG (2018) *Annual Report 2017–18*. ICAR-Indian Institute of Natural Resins and Gums, Ranchi (India).

ICAR-IINRG (2019) *Annual Report 2018–19*. ICAR-Indian Institute of Natural Resins and Gums, Ranchi (India).

ICAR-IINRG (2023) *Annual Report 2022–23*. National Institute of Secondary Agriculture, Ranchi (India).

Islam, A.M., Phillips, G.O., Sljivo, A., Snowden, M.J. and Williams, P.A. (1997) A review of recent developments on the regulatory, structural and functional aspects of gum arabic. *Food Hydrocolloids* 11, 493–505. Available at: https://doi.org/10.1016/S0268-005X(97)80048-3

Jahun, I.K. and Salami, K.D. (2020) Influence of tapping dates on the yield of *Acacia senegal* (L.) wild at two different locations in Jigawa State, Nigeria. *FUDMA Journal of Sciences* 4, 246–249.

Jales, M. (2018) *Commodities at a Glance: Special Issue on Gum Arabic*. UNCTAD, Geneva, Switzerland.

John, H.M., William, G.N. and Herman, F.W. (1976) The role of the endosperm in the germination of legumes: Gallactomannas, nitrogen, phosphorous change in the germination of guar (*Cyamopsis tetragonolobus* L.). *American Journal of Botany* 63, 790–797.

Joseleau, J. and Ullmann, G. (1990) Biochemical evidence for the site of formation of gum arabic in *Acacia senegal*. *Phytochemistry* 20, 3401–3405.

Kandziora, T. and Melzig, M.F. (2015) Asafoetida. *Journal of Phytotherapy* 36, 37–43. Available at: https://doi.org/10.1055/s-0034-1395851

Karamalla, K.A., Siddig, N. and Osman, M. (1998) Analytical data for *Acacia senegal* var. *Senegal* gum samples collected between 1993 and 1995 from Sudan. *Food Hydrocolloids* 12, 373–378. Available at: https://doi.org/10.1016/S0268-005X(98)00005-8

Kramer, P.J. and Kozlowski, T.K. (1979) *Physiology of Woody Plants*. Academic Press, New York.

Kumar, B. (2020) Asafoetida (Raw): Original Collection at Tajakistan II. Available at: https://www.youtube.com/watch?v=6OwmYM7Elwo (accessed 24 March 2025).

Kuruwanshi, V.B., Katiyar, P. and Khan, S. (2017) Scientific approaches of gum tapping in gum karaya (*Sterculia urens* Roxb.) for high gum production. *International Journal of Current Microbiology and Applied Sciences* 6, 3366–3374. Available at: https://doi.org/10.20546/ijcmas.2017.608.402

Kuruwanshi, V.B., Katiyar, P. and Khan, S. (2018) Gum tapping technique and anatomical study of ethylene induced gum duct formation in Dhawda (*Anogeissus latifolia*). *Journal of Pharmacognosy and Phytochemistry* 1, 124–128.

Liyanage, S., Abidi, N., Auld, D. and Moussa, H. (2015) Chemical and physical characterization of galactomannan extracted from guar cultivars (*Cyamopsis tetragonolobus* L.). *Industrial Crops and Products* 74, 388–396.

López-Franco, Y., Higuera-Ciapara, I., Goycoolea, F.M. and Wang, W. (2009) Other exudates: Tragancanth, karaya, mesquite gum and larchwood arabinogalactans. In: Phillips, G.O. and Williams, P.A. (eds) *Handbook of Hydrocolloids*, 2nd edn. Woodhead Publishing, Sawston, UK, pp. 495–534. Available at: https://doi.org/10.1533/9781845695873.495

Lujan-Medina, G.A., Ventura, J., Lara-Ceniceros, A.C., Ascacio-Valdés, J., Boone-Villa, V.D. *et al.* (2013) Karaya gum: General topics and applications. *Macromolecules: An Indian Journal* 9.

MacKinnon, J. and MacKinnon, K. (1986) *Review of the Protected Areas System in the Indo-Malayan Realm*. The International Union for Conservation of Nature (IUCN), Gland, Switzerland.

Mahungu, S.M. and Meyland, L. (2010) *Cassia Gum: Chemical and Technical Assessment (CTA)*. FAO/WHO Report, Rome.

Mudgil, D., Barak, S. and Khatkar, B.S. (2014) Guar gum: Processing, properties and food applications—A review. *Journal of Food Science and Technology* 51, 409–418.

Nair, M.N.B., Shivanna, K.R. and Mohan Ram, H.Y. (1995) Ethephon enhances karaya gum yield and wound healing response: A preliminary report. *Current Science* 69, 809–810.

Negi, S. (1992) *Himalayan Wildlife Habitat and Conservation*. Indus Publishing, New Delhi.

Ngaryo, F.T., Goudiaby, V.A., Dagbenonbakin, G.D., Agbangba, E.C., Diatta, S. *et al.* (2011) Modeling the production of arabic gum (*Acacia senegal* (L.) willd) based on the proper tapping characteristics in the semi-arid Sahel of Chad. *International Journal of Science and Advanced Technology* 1, 24–30.

Pareek, K., Tewari, J.C., K, S., Gaur, M.K., Sharma, A. *et al.* (2017) Effect of different girth classes on gum arabic production from *Acacia senegal* in arid western Rajasthan. *IOSR Journal of Environmental Science, Toxicology and Food Technology* 11, 2319–2399. Available at: https://doi.org/10.9790/2402-1109021216

Prasad, N., Sharma, K., Tiwari, J., Gupra, R. and Thomas, M. (2012) Mechanization of natural gum and resin tapping. *Journal of Non-Timber Forest Products* 19, 279–289.

Prasad, N., Thombare, N., Sharma, S.C. and Kumar, S. (2022) Gum arabic – a versatile natural gum: A review on production, processing, properties and applications. *Industrial Crops and Products* 187, 115304. Available at: https://doi.org/10.1016/j.indcrop.2022.115304

Prasad, N., Thombare, N., Sharma, S. and Kumar, S. (2023) Natural exudate gum from ghatti tree (*Anogeissus latifolia*): A review on production, processing and marketing. *Journal of Non-Timber Forest Products* 29, 153–161. Available at: https://doi.org/10.54207/BSMPS2000-2023-Y019B3

Rao, P.S. and Srivastava, H.C. (1973) Tamarind in industrial gums. In: Whistler, R.L. (ed.) *Industrial Gums*. Academic Press, New York, pp. 369–411.

Rawat, R.S. and Vaneet, J. (2006) Non-timber forest produces from Western Himalayan forests. *Journal of Non-Timber Forest Products* 13, 161–165.

Reddy, K., Rajadurai, S., Sastry, K. and Nayudamma, Y. (1964) Studies on dhava tannins. I. The isolation and constitution of a gallotannin from dhava (*Anogeissus latifolia*). *Australian Journal of Chemistry* 17, 238.

Sabahelkheir Murwan, K., Abdalla Abdelwahab, H. and Nouri Sulafa, H. (2012) Quality assessment of guar gum (endosperm) of guar (*Cyamopsis tetragonoloba*). *ISCA Journal of Biological Sciences* 1(1), 67–70.

Sanchez, C., Nigen, M., Mejia Tamayo, V., Doco, T. and Williams, P. (2018) Acacia gum: History of the future. *Food Hydrocolloids* 78, 140–160. Available at: https://doi.org/10.1016/j.foodhyd.2017.04.008. hal-01602791

Seaman, J. (1980) Guar gum. In: Davidson, R.L. (ed.) *Handbook of Water-Soluble Gums and Resins*. McGraw-Hill, New York, ch. 6, pp. 1–19.

Sharma, K.K. (2017) Lac insects and host plants. In: Omkar, O. (ed.) *Industrial Entomology*. Springer, Singapore, pp. 157–180. Available at: https://doi.org/10.1007/978-981-10-3304-9_6

Sharma, S. and Prasad, N. (2013) Comparison of gum tapping techniques for gum karaya – a gum of commercial importance. *Journal of Non-Timber Forest Products* 20, 165–170.

Sharma, K.K., Jaiswal, A.K. and Kumar, K.K. (2006) Role of lac culture in biodiversity conservation: Issues at stake and conservation strategy. *Current Science* 91, 894–898.

Sharma, S., Prasad, N., Pandey, S. and Giri, S. (2016) A review on status of gum tapping and scope for improvement. *Agricultural Mechanization in Asia, Africa and Latin America* 47, 68–75.

Sharma, S., Prasad, N., Pandey, S. and Giri, S. (2018a) Status of resin tapping and scope of improvement - a review Agricul. *Agricultural Mechanization in Asia, Africa and Latin America* 49, 16–26.

Sharma, S., Prasad, N., Pandey, S. and Singh, R. (2018b) Mechanization status in commercially important NRGs production - a review. *Multilogic in Science* 8, 353–357.

Sharma, S., Prasad, N., Pandey, S. and Bhargava, V. (2020) Development of integrated small scale lac processing unit. *Agricultural Mechanization in Asia, Africa and Latin America* 51, 58–66.

Sharma, S.C., Pandey, S.K. and Prasad, N. (2023) Mechanization status in karaya gum tapping and scope for improvement - a review. *Agricultural Mechanization in Asia, Africa and Latin America* 54, 18–29.

Sharma, S.C., Pandey, S.K., Prasad, N. and Raj, H. (2024) Unveiling the potential of *Moringa oleifera* Lam. gums: An investigation into optimal tapping strategies. *Industrial Crops and Products* 214, 118511. Available at: https://doi.org/10.1016/J.INDCROP.2024.118511

Srivastava, S. and Ray, D.P. (2015) Natural gum resources in India and their commercial importance. *International Journal of Bioresource Science* 2, 151–155.

Tadese, S., Lemenih, M., Melese, B., Dika, G. and Dara, P.K. (2016) Effect of tapping intensity and tree diameter on gum Arabic yield of *Acacia senegal* (L.) wild in southern Ethiopia. *World Journal of Pharmaceutical and Life Sciences* 2, 229–238.

Tewari, J., Pareek, K., Shiran, K. and Prasad, N. (2017) On exudation of gum arabic through advance technology. *International Journal of Environmental Sciences and Natural Resources* 2, 555596. Available at: http://dx.doi.org/10.19080/IJESNR.2017.02.555596

Thombare, N., Kumar, S., Kumari, U., Sakare, P., Yogi, R.K. *et al.* (2022) Shellac as a multifunctional biopolymer: A review on properties, applications and future potential. *International Journal of Biological Macromolecules* 215, 203–223. Available at: https://doi.org/10.1016/J.IJBIOMAC.2022.06.090

Tiwari, D.D. and Campbell, J.Y. (1995) Developing and sustaining non-timber forest products: Some policy issues and concerns with special reference to India. *Journal of Sustainable Forestry* 31, 53–77.

Upadhyay, R.K. (2017) Nutritional, therapeutic, and pharmaceutical potential of plant gums: A review. *International Journal of Green Pharmacy* 11, S30–S48.

Upadhyaya, S.K. and Singh, V. (1986) Phytochemical evaluation of *Cassia obtusifolia* and *Cassia tora*. In: *Proceedings of the Indian Academy of Science: Plant Sciences*, Vol. 96(4). pp. 321–326.

Vasishth, A. and Guleria, V. (2017) Standardized gum tapping techniques to maximize yield from high-value Indian tree, *Sterculia urens*. *Journal of Forestry Research* 28, 615–619. Available at: https://doi.org/10.1007/s11676-016-0315-1

Verbeken, D., Dierckx, S. and Dewettinck, K. (2003) Exudate gums: Occurrence, production, and applications. *Applied Microbiology and Biotechnology* 63, 10–21. Available at: https://doi.org/10.1007/s00253-003-1354-z

Whistler, R.L. (1993) Exudate gums. In: Whistler, R.L. and BeMiller, J.N. (eds) *Industrial Gums: Polysaccharides and Their Derivatives*, 3rd edn. Academic Press, New York, pp. 309–339. Available at: https://doi.org/10.1016/B978-0-08-092654-4.50016-4

Yogi, R.K., Bhattacharya, A., Jaiswal, A.K. and Kumar, A. (2015) *Lac, Plant Resins and Gums Statistics 2014: At a Glance*. ICAR-Indian Institute of Natural Resins and Gums, Ranchi India.

Yogi, R.K., Kumar, A. and Singh, A.K. (2018) *Lac, Plant Resins and Gums Statistics 2016: At a Glance.* ICAR-Indian Institute of Natural Resins and Gums Namkum, Ranchi, India.

Yogi, R.K., Kumar, N. and Sharma, K. (2021a) *Lac, Plant Resin and Gums Statistics 2020: At a Glance.* ICAR-Indian Institute of Natural Resins and Gums, Ranchi, India.

Yogi, R.K., Kumar, N. and Sharma, K. (2021b) *Lac, Plant Resins and Gums Statistics 2018: At a Glance.* ICAR- Indian Institute of Natural Resins and Gums, Ranchi, India.

Yogi, R.K., Kumar, N. and Sharma, K. (2021c) *Lac, Plant Resins and Gums Statistics 2019: At a Glance.* ICAR-Indian Institute of Natural Resins and Gums, Ranchi, India.

5 Postharvest and Storage Management

Niranjan Prasad*, Satish Chandra Sharma, Sanjay Kumar Pandey and Sakharam Kale
ICAR-National Institute of Secondary Agriculture, Namkum, Ranchi, India

Abstract
Natural resins and gums (NRGs) are important non-timber forest products produced from a wide range of plants. With the exception of lac and guar gum, the majority of NRGs are processed by manual grading and cleaning before export. Most NRGs are exported in raw or primary processed form. Therefore, the NRG sector needs mechanization with modern equipment and machinery for processing and value-addition to maximize returns on investment and minimize labour cost, time and drudgery. Consequently, to produce high-quality processed products to earn higher foreign exchange on the international market, NRG postharvest processing requires mechanization. Improvement in processing, reduction in manpower requirement and drudgery and increased efficiency could be possible with the adoption of such mechanization of these plant-based raw materials. Generally, the NRG industry can provide a source of additional income to rural and tribal people to support their sustainable livelihood, minimizing migration to town and cities of people searching for employment, such is the enormous potential that exists for the NRG industry to grow further.

Introduction

Postharvest management of natural resins and gums (NRGs) involves one or more of the following: purification to obtain the desired component(s), separation of different constituents for different applications and derivatization of specific constituents. Postharvest management techniques are normally restricted to physical measures so as to allow minimal disruption to the gum or resin, especially when they are meant for applications related to human consumption and contact.

Resins

Many plant species have unique ducts or chambers where they secrete resin. Trees exude resins through their bark, and these harden when exposed to air. All natural resins come from plants, with the exception of lac, which is produced by the lac insect (*Kerria lacca* Kerr.) as reported by Sharma and Prasad (2013). These are soluble in diethyl ether, turpentine and various alcohols but insoluble in water. Natural resins such as rosin, copal, dammar, amber, sandarac and manila are particularly important for furniture coatings. Resins possess a distinct gloss and might be amorphous, brittle, translucent or semi-transparent. In addition, they often ignite with a smoky flame and are fusible.

Lac

As already mentioned in Chapters 2 and 4, 'lac' is not strictly a vegetable gum-resin, as it

*Corresponding author: niranjan_ilri_1999@yahoo.com

 Natural Gums and Resins: Botany and Sustainable Uses in Medicine, Nutrition, Perfumery and Cosmetics (eds Hartwig Schulz *et al.*)
DOI: 10.1079/9781800626690.0005

is obtained from lac insects. Nevertheless, some aspects related to the postharvest and storage of this natural product will be briefly discussed here. On maturity, the lac crop is harvested by cutting the branches containing the lac encrustations and scraping manually, mostly using a pruning knife. The scraped-off lac is known in the trade as sticklac. This product contains, in addition to the lac resin and its associated dye (erythrolaccin), a water-soluble lac dye (laccaic acid), wax and some impurities such as insect remains, sand and wood material. The extensive removal of lac dye and impurities in primary processing results in the semi-refined product known as 'seedlac', which is used to produce shellac and other value-added products (Thombare *et al.*, 2022). The impurity content in seedlac is about 3–7%, on average 5% (Bureau of Indian Standards, 1973). Seedlac is produced in the form of grains with a mesh size of 2–2.5 mm or smaller and is yellow or reddish-brown, depending on the host tree and the place from which the sticklac has been collected. In general, the shelf life of sticklac is improved by conversion to seedlac, which requires the following five processing steps: crushing, washing, drying, separating and sorting.

Equipment for primary processing

The Indian Council of Agricultural Research-Indian Institute of Natural Resins and Gums (ICAR-IINRG), Ranchi (India), has developed machines for establishing small-scale lac processing units with a capacity of 100 kg sticklac/day to promote primary processing at village level. An integrated small lac processing unit (100 kg/day) has also been developed at ICAR-IINRG to reduce manpower requirements and remove the drudgery in primary lac processing (Sharma *et al.*, 2024b). In order to make seedlac from sticklac at village level, researchers from ICAR-IINRG designed and developed a small-scale lac processing unit with a capacity of 100 kg sticklac/day. This unit consists of four distinct machines: a lac scraper-cum-crusher, a lac washing machine, a lac winnower and a lac grader (Sharma *et al.*, 2022). The unit is easy to operate and may be operated manually or by an electric motor.

In addition, an integrated small-scale processing unit has been developed by the same institute that includes the five primary units (sticklac handling, size reduction, grading, soaking and washing) (Fig. 5.1).

When the system's mechanisms were tested, they were found to work as required. The system's performance evaluation showed an average yield of seedlac as 70.84% of the weight of sticklac, with an average degree of contamination of 2.04% in the seedlac. The mean yield of seedlac produced utilizing the integrated small-scale lac processing unit was comparable to that of a non-integrated small-scale lac processing unit (Sharma *et al.*, 2020).

The quality parameters of seedlac prepared with the newly developed unit were determined in the Quality Evaluation Laboratory of the ICAR-National Institute of Secondary Agriculture (ICAR-IINRG), Ranchi, using standard methods. The following quality attributes of seedlac prepared from sticklac were evaluated and found to be within the specified ranges: flow (fluidity), life under heat (heat polymerization time), colour index, insolubility in hot alcohol (impurity), acid value, moisture content and wax content, as reported by Sharma *et al.* (2020).

Shellac

Seedlac is refined to obtain a pure natural product by one of the two following processes: (i) heat process: manual or mechanized, (ii) solvent process. In the heat process, the seedlac is subjected to melting and hot filtration. It is melted in steam-heated hydraulic press (Fig. 5.2) and squeezed through layers of cotton cloth to remove particulate impurities. The filtered molten lac is then transferred to steam-heated vessel from where it is pressed through hot rollers (Fig. 5.3) into a continuous sheet. The sheet solidifies on cooling and is broken to obtain flakes of resin called shellac (Fig. 5.4).

Small quantities of shellac are also produced through a manual process and this is known as 'handmade shellac'. This product is prepared manually from seedlac using a *bhatta* (earthen oven fired by wood charcoal) process (Fig. 5.5). The same technique used for handmade shellac preparation is followed for preparation of button lac, with the difference that the molten lac is moulded (Bhowmik, 1962; Sharma *et al.*, 2022). In this method, the seedlac is filled into a sausage-shaped cotton bag 9m long with a diameter of around 50 mm. The bag is then gradually passed in front of

Fig. 5.1. Integrated small-scale lac processing unit. (By R.P. Srivastava, ICAR-IINRG, Ranchi, India.) Used with permission.

a charcoal-fired hearth to melt the lac. The molten lac is then forced out through the fabric (which acts as a sieve) by twisting the bag. After being filtered, the molten material is either stretched using a glazed ceramic cylinder to create sheets that are around 1–2 mm thick, referred to as handmade shellac (Fig. 5.6), or it is moulded into the form of discs, referred to as button lac (Fig. 5.7). The residue left inside the cloth bag contains lac, sand and insect debris and is referred to as *kiri* lac. The whole operation is done by a team of three specialist workers.

Mechanization in button lac and handmade shellac preparation. Some effort was made to mechanize the traditional method of button lac manufacture and a machine was developed for preparation of button lac (ICAR-ILRI, 2003) to minimize the labour requirement and the drudgery associated with the *bhatta* process.

Fig. 5.2. Steam-heated hydraulic press. (By S.C. Sharma, ICAR-IINRG, Ranchi, India.)

However, this first attempt to mechanize the process needs to be further developed as there is a manual aspect to the machine that is labour intensive and requires trained manpower to prepare button lac or handmade shellac.

A commercially available extruder (Fig. 5.8) was modified to function as a substitute for the traditional button lac/shellac preparation method in order to mechanize the preparation process (ICAR-IINRG, 2012; Prasad, 2014). In this context, it was observed that while lac filtration occurred initially and filtered materials came out of the filter unit, later, raw material melted inside the material handling system and restricted the flow of materials from the hopper to the filtration unit.

Mechanization of machine-made shellac. Machine-made shellac is prepared from seedlac and produced commercially using a steam-heated hydraulic press (Fig. 5.2) and flaking roller (Fig. 5.3). Pandey *et al.* (2016) reported that there are two types of machine-made shellac: heat-processed, which is manufactured with seedlac as the raw material; and solvent-processed, which is manufactured using sticklac, seedlac or *kiri* (a by-product of the lac industry). Solvent-processed shellac includes garnet lac, dewaxed decolorized lac and shellac. In the mechanized process of making shellac, the required amount of seedlac is placed over the filter cloth within the steam-heated hydraulic press, which contains steam-heated grids. Within the steam-heated hydraulic press, the hydraulically operated up-stroke pressing ram develops high pressure, which filters the melted seedlac through filter cloth. The filtered lac comes out from the bottom of the steam-heated filter press and is collected in trays. The melted lac thus produced is transferred to a collecting pan with a steam-heating facility. The molten lac is then fed into the sheeting/flaking rollers (Fig. 5.3), which transform it into a thin sheet of uniform thickness. The molten lac that comes out of the sheeting rollers is either manually or mechanically stretched on a conveyer. When the uniformly thick sheet is exposed to the atmosphere, it hardens and becomes breakable by hand into flakes, popularly known as

Fig. 5.3. Flaking roller. (By S.C. Sharma, ICAR-IINRG, Ranchi, India.)

Fig. 5.4. Shellac flakes. (By Niranjan Prasad, ICAR-IINRG, Ranchi, India.)

machine-made shellac (Karim, 1962; Sharma *et al.*, 2014; Sharma *et al.*, 2022).

Lac dye

Lac dye, a by-product of the lac industry, can be recovered from wash water obtained during primary processing of lac in which the basic raw material sticklac is converted into a semi-refined product, seedlac. Apart from lac resin, the sticklac also contains water-soluble lac dye (laccaic acid, up to 1%). Recovery of by-product lac dye also helps in hygienic disposal of effluent. Lac dye is a mixture of at least five closely related compounds, all being anthraquinone derivatives which have been assigned the names laccaic acid A, B, C, D and E. Since lac dye is non-toxic, apart from dyeing of wool and silk, it is also used as food colouring material. The use of this dye has the advantage of being natural over synthetic dyes.

Pine rosin

The solid form of the resin that is derived from pines and some other plants, mostly conifers, is known as rosin, or Greek pitch. It is semi-transparent and comes in shades ranging from black to yellow. Rosin is brittle at room temperature, but it melts at stovetop temperatures. Due to the pressing demand for oleoresins during World War II, resins received greater attention.

Fig. 5.5. Bhatta for making handmade shellac. (By S.J. Kale, ICAR-IINRG, Ranchi, India.)

Their chemical characteristics and related applications, such as the production of varnishes, adhesives and food glazing agents, make them valuable. Pine tree species (*Pinus roxburghii*, *Pinus caribaea*, *Pinus wallichiana* A.B. Jackson, *Pinus keysia*) naturally produce rosin, which is abundantly available and is an inexpensive hydrocarbon biomass. The economically significant oleoresin produced by *chir* pine is used as a raw ingredient in India's turpentine and rosin industries. Specifically, the north-eastern regions, Himachal Pradesh, Uttarakhand and Jammu and Kashmir's highlands, are the major sources of *chir* pine resin. Between 8000 and 9000t of raw rosin are produced annually from pine trees in the northern hill states, and a sizable amount of resin is collected in Uttarakhand and Himachal Pradesh. The rosin and turpentine oil factories process over 85% of the total resin produced in the states.

Pine resin processing in India

Resin is collected in tins and transported to key locations before being transported to the factories. The cities of Nahan and Bilaspur in Himachal Pradesh are the locations of two turpentine and resin oil factories in India. Turpentine oil and rosin are produced by distillation of the resin obtained from *chir* pine trees (Fig. 5.9A). When crude resin arrives at the processing industry, it is quite impure and contains bark, needles, dirt, insects, etc. In order to collect clean resin, there is a need to adopt alternative tapping methods like the bore-hole method where the chances of impurities getting mixed in is less. After being transported to the processing factories, the tins are kept in an appropriate shed. Prior to processing, the top of the tin is cut and any floating dirt on the resin's surface is manually removed. Then, turpentine oil is added to the resin to soften it as it can be challenging to deal with in, especially in the winter. The screw

Fig. 5.6. Handmade shellac. (By Niranjan Prasad, ICAR-IINRG, Ranchi, India.)

elevator raises the resin and the impurities to the mixing vats, which are sizeable mild steel tanks. The V-shaped mild steel tanks are equipped with steam jackets to heat the resin. An indirect heating mechanism is used to melt the resin, with steam fed into the mild steel vessels via copper coils for 4–6 hour and kept at a temperature of 95°C. After being mechanically agitated with iron arms to remove larger impurities like pine needles and bark fragments, the molten resin is then pumped into a rest vat using a centrifugal pump. The resin is left to rest in the rest vat for 18–20 hour. If there is any dust or other impurities, it settles by gravity during this rest period. Through a bottom valve, the dust and impurities are removed, and the lower-grade rosin is then transferred for processing separately. From the rest vat to the compression chamber and ultimately to the distillation kettle, the resin is transported by steam pressure (Fig. 5.9B). In the distillation kettle, the resin is cooked by indirect steam for 1–2 hour. In the process, turpentine oil and resin moisture evaporate and collect in the condenser and a separator is used to obtain the turpentine oil. Giri *et al.* (2008) reported that, depending on the resin quality, the rosin in the distillation kettle is withdrawn at 165°C, collected in carts and mixed with 100–200g of oxalic acid to make it more glossy and clear (Fig. 5.9C).

Both turpentine and rosin oil are widely used in various industries for different

Fig. 5.7. Button lac. (By Niranjan Prasad, ICAR-IINRG, Ranchi, India.)

applications. For commercial application, they are categorized into several grades. Table 5.1 lists the 12 rosin grades that are now in demand in the trade sector and that the Indian Standards Institute has recognized.

Storage

Rosin is packed in tin barrels of 200 kg capacity and stored in cool conditions in a shed.

White dammar

The white dammar produced on the trees is allowed to dry before collection. The primary processing of white dammar includes drying of collected resin either in the mild sun or in a shed. Dried resin is cleaned by sieving and hand-picking to remove foreign matter, then packed in sacks to be transported to sale outlets, either nationally or abroad.

Dammar is sometimes dewaxed to improve the quality of the varnishes made from it. The dewaxed dammar is prepared by dissolving dammar in a hydrocarbon solvent and precipitating and removing a high-melting, resinous fraction. The remaining soluble fraction is more compatible with the cellulose component of cellulose lacquers. In India, an essential oil is distilled from dammar resin which is used in the fragrance and pharmaceutical industry.

Gums

Certain plants and bushes secrete a viscous substance called 'gum', which becomes solid when dried but dissolves in water. There are

Fig. 5.8. Extruder for mechanization of button lac/shellac manufacturing. (By S.C. Sharma, ICAR-IINRG, Ranchi, India.)

primarily two types of gums: seed gums and exudate gums. The gums found in the seeds of several plants, such as guar, tamarind and *cassia tora*, are known as seed gums. Exudate gums, or gums derived from plants, are solids containing various combinations of polysaccharides that either dissolve in water or absorb it and expand to form a gel or jelly when submerged in water (Sharma *et al*., 2024a). *Acacia nilotica* (babul), *Acacia senegal* (kumat), *Anogeissus latifolia* (dhawra/ghatti), *Lannea coromandelica* (jhingan), *Boswellia serrata* (Indian frankincense) and *Commiphora wightii* (guggul) are the most significant gum-producing plants (Sharma *et al*., 2018). Exudate gums are produced often from a tree's stem but can occasionally also be obtained from the roots. Exudation is frequently interpreted as a pathological reaction to plant injury, whether intentional ('tapping') or the result of accidental injury or damage by insect borers. Although several plant groups produce gums, only a few tree species are commercially used for the marketing of this non-wood forestry product.

Exudate gums

Gum arabic

Freshly collected gum arabic (GA) contains approximately 10–15% moisture and some impurities (bark, leaves, soil, sand, etc.). At the time of exudation, GA is white to light brown in colour, but turns brown when it comes in contact with bark and foreign matter. Before marketing, the gum manufacturer carries out primary processing of the GA, which may involve drying, cleaning and grading. In African gum belt countries (Kenya, Somalia,

Fig. 5.9. Stages of the production process of rosin from pine resin: a. pine resin collection; b. steam-distillation process (by M.F. Ansari, ICAR-IINRG, Ranchi, India); c. purified rosin (by Nandkishore Thombare, ICAR-IINRG, Ranchi, India). Used with permission.

Ethyopia, Sudan, etc.), the level of GA processing is minimal, with the exception of Sudan, where gum processing has become increasingly essential. Nigeria and Senegal also have facilities for GA processing. In other countries, gum processing is only performed at the primary level, such as cleaning, sorting, grading and kibbling. Enhanced processing infrastructure for gum processing in Sudan has boosted the competition in crude gum production and helped local traders and producers get enhanced prices for GA. In 2003, the Sudanese government awarded licences to 12 gum-processing companies, enabling them to get four to five times higher prices for GA than during 1998–2002 (Couteaudier, 2007).

Despite modern developments in some GA-producing African countries, high-quality processing still takes place outside Africa, particularly in Europe and North America, and more than 90% of processed GA was exported from these regions from 2014 to 2016. High-quality GA processors can be identified in two types: spray-dried gum processors and processors producing customized products as per individual client requirements (Jales, 2018).

Primary processing. *Drying*. In GA processing, drying is the most important unit operation. While the freshly formed GA

Table 5.1. Quality criteria for different grades of rosin. Author's own table

Rosin	Grade	Full name	Lovibond colour		
			Red	Yellow	Blue
Pale	X	Extra white	1.35	13	–
	WW	Water white	1.85	19.5	–
	WG	Window glass	2.6	30	–
	N	Nancy	3.6	41	–
Medium	M	Mary	4.9	51	–
	K	Kitty	6.2	60	–
	I	–	8.8	80	–
	H	Harry	12	100	–
Dark	G	–	17	100	–
	F	–	27	120	0.1
	E	–	47	139	1.2
	D	Dark	75	160	3.2

contains about 30% moisture, the gum dries naturally on the tree until harvest, so that the harvested GA finally contains about 10–15% moisture. As per climatic conditions, the collected GA is sun-dried for a further 5–15 days to lower the moisture content below 10%. Furthermore, drying in the sun helps to lighten the colour of the gum, thus improving its quality (Whistler, 1993). Alternatively, a solar dryer can be used for faster drying of GA while avoiding mixing foreign matter during the process. If the GA has a moisture content of more than 10% (on a dry weight basis), it is susceptible not only to colour changes (from white to dark brown) but also to fungal infestation. Thus, the moisture content in GA should be less than 10% before storage. Generally, gums with fungal infestation and dark brown colour fetch lower prices because there is less demand for them in the market. Thus, there is scope for developing driers and recommending suitable driers for GA drying at the producer level.

Cleaning and grading. The cleaning and grading of GA are usually carried out manually or using machines in GA-producing countries. No primary processing of GA takes place at the gum producers' level. After the GA has been received by the processing industry, foreign matter (bark, leaves, soil, sand, etc.) adhered to the GA is removed. Bark and leaves (of larger size) are separated manually by winnowing or hand picking, whereas sieving removes finer objects (soil and sand). After the removal of foreign material, the GA is graded based on size, colour and the amount of impurity remaining in the gum. GA of the highest quality has a bigger tear size and lighter colour with less foreign matter compared to gums of lower quality. The commonly available grades of GA in the market, along with their characteristics, are listed in Table 5.2.

Secondary processing. In some cases, GA is further processed after primary processing and before sale. This may include mechanical grinding of the GA; the process, popularly called 'kibbling', converts the gum to nodules of different sizes which are then graded by sieving. Alternatively, secondary processing may include the removal of impurities from gum by dissolution and filtration, followed by drying with a spray-dryer or by drum drying (Islam *et al.*, 1997). Various research projects have looked at the secondary processing of GA, as summarized below:

Processing by heating gum solution. GA produced from *Acacia senegal* was processed by heating the gum in solution at 65 and 100°C.

Table 5.2. GA products with a short description. Author's own table.

Type or grade	Description
Wet gum	the fresh gum collected from the stem or branches of Acacia senegal trees
Natural gum	partially-dried gum obtained after three weeks of collection
Cleaned	whole or broken lumps, 10–20 mm in diameter
Cleaned amber and sifted (CAS)	whole or broken tubers of more than 20 mm in diameter with a pale to dark amber colour, including dust and sieved material
Hand-picked selected (HPS)	whole or broken nodules of more than 30 mm diameter, light colour with manually removed dust and shiftings
Kibbled gum	mechanically size reduced and sieved CAS or HPS
Mechanical powder	mechanically powdered HPS or CAS
Spray-dried powder	spray-dried powdered HPS, CAS or kibbled gum
Instant soluble gum	granulated gum of enhanced solubility
Specialty gum	proprietary and customized gum

Heating the gum solution for 6 hour at 100°C showed loss of emulsification characteristics and significant protein degradation. However, the desired level of viscosity and gelling reduction was achieved. A reduction in gelling without a significant reduction in emulsification characteristics was observed by heating at 65°C within 24 hour (Chikamai *et al.*, 1996).

Processing using enzyme biotechnology. Chikamai and co-workers processed GA by taking three different enzymes (pronase, viscozyme and f3-O-galactosidase). The processed GA showed reduced viscosity and gelling with varied effects on other properties of the gum. It was concluded that pronase is unsuitable for GA processing, where emulsification and stabilization are essential properties. The desired viscosity and gelling reduction level for industrial application were achieved within 24 hour using viscozyme and f3-O-galactosidase. The use of f3-O-galactosidase improved properties like emulsification and protein content and could prove to be the processing technology of the future (Chikamai *et al.*, 1996). Tinnelly and Wenzhi (2007) developed a methodology for manufacturing GA powder, which included different unit operations like removal of impurities, grinding, hydrolysis, clarification and decolouration, centrifugation, ceramic membrane purification, spray-drying and cooling. The advantages of this technology are that it involves a reasonable technology route and is a labour- and time-saving process combined with lower production costs and increased production efficiency. The products manufactured using this technology are characterized by high purity, good quality and stable molecular weight (Tinnelly and Wenzhi, 2007).

Transportation and storage. GA in crude form is stored in burlap or jute sacks and exported, whereas graded GA is packed in previously unused heavy-strength jute sacks of about 80 kg each. Processed and semi-processed (kibbled, granules and powdered) GA is exported in drums, polyethene-lined, multi-layered craft paper bags or cardboard boxes. GA stored in a dry place under cooled conditions (21–24°C) has a long shelf life (Jales, 2018).

Value chain of GA. An overview of the GA value chain with a simplified schematic is depicted in Fig. 5.10. This describes the activities associated with gum production from producer to consumer, highlighting value addition at each consecutive stage. This includes tree raising, gum tapping, reducing moisture content (drying), final cleaning, sorting and grading, transportation, storage, processing/value addition, marketing and industrial utilization. Before the final consumer receives the GA, various activities such as sorting and grading, transportation and processing will take place at various stages and at various locations, involving numerous players in the individual GA value

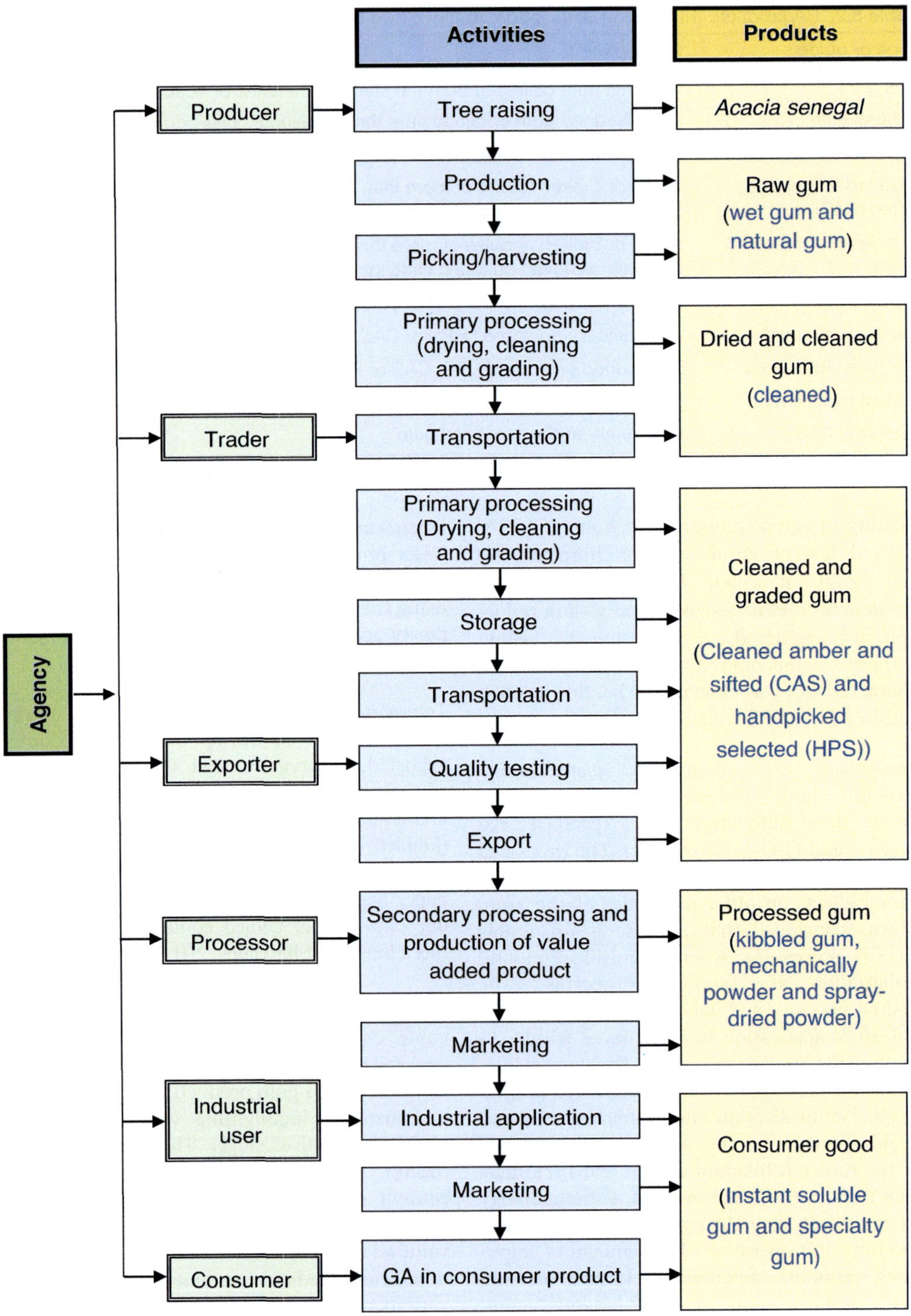

Fig. 5.10. Simplified value chain of gum arabic. (Prasad *et al*., 2022.) Used with permission.

chains. Most stakeholders in the value chain are producers in developing countries, including tappers and collectors. Other associated people in the GA value chain include professionals like input suppliers, local business people, traders, processors, persons involved in storage and transportation, exporters and importers, and processors and users of different industries and government sectors. The GA products are summarized with a short description in Table 5.2.

Gum karaya

Freshly harvested gum karaya may contain about 20% moisture. It also contains impurities like bark, leaves, soil, sand, etc. At time of exudation, gum karaya is white in colour. But as soon as it comes in contact with bark or foreign matter, it gains colour. To avoid the colour change, it is recommended to collect exudating gum on a plastic sheet. Though gum production takes place during the rainy season, it is not collected, as it is of very poor quality and usually infected by moths and fungus. On average, one mature karaya tree produces 1–5 kg of gum per season.

In Senegal, gum karaya production and collection is carried out from September to January and from March to July. The gum collected is processed manually and mechanically and divided into two grades, hand-picked and standard. Gum karaya collected in Senegal is of inferior quality compared to that produced in India. Based on colour and impurity present in the gum, it is marketed in five grades in India: hand-picked, Superior No.1, Superior No.2, fair average and sifting. Due to awareness created amongst gum pickers in India of improved methods of gum tapping, collection, processing and storage, the quality of gum has increased over a period of time.

Drying. At the time of exudation, gum karaya contains about 30–35% moisture. Natural drying of the gum takes place on the tree after its exudation until collection. Freshly collected gum contains about 20% moisture. The collected gum is dried in the sun for 5–15 days to reduce the moisture content to about 10%. At the same time, the colour of the gum is lightened. The Kovel Foundation in Vishakhapatnam recommended sun drying of gum karaya on an elevated platform made of wooden poles, bamboo mats and polythene sheets to prevent pets and children from disturbing the drying process. In 2013, a solar dryer was also successfully tested at ICAR -IINRG in Ranchi to speed up the drying process and to prevent foreign objects from mixing with the gum.

Apart from cleaning and grading, drying is the most important primary processing unit operation in gum karaya processing. Several researchers have studied the drying characteristics and kinetics of gum karaya in order to identify suitable drying techniques. Singh and Singh (2014) studied the drying kinetics of three grades of gum karaya (grades I, II and III) using a cabinet-type convective dryer. The experimental data of the study fitted best to Page's Model for prediction of drying kinetics. In the study it was found that the drying process and drying rate increased with a rise in air temperature. It was also observed that the drying rate constant for a given temperature was minimal for Grade I and maximal for Grade III.

Sahu *et al.* (2017) studied drying characteristics and kinetics of gum karaya grits of geometric mean diameter in the range 4.5–99 mm using a hot air dryer. The study was carried out for five temperature levels in the range 40–80°C at a constant air flow rate of 1.5 m/s.

The experimental data revealed that complete drying took place in the falling rate period and that the constant rate period was absent at all temperature levels. The initial moisture removal rate was fast and later gradually decreased, becoming very low, unlike other biological materials, probably due to the hardening that occurs in gum karaya.

Microwave drying and grinding characteristics of gum karaya were studied by Walde *et al.* (1997) using a domestic microwave oven and domestic dry grinder, respectively. The study indicated that the gum karaya sample dried in the microwave oven was crispy and consumed less energy in grinding without affecting its properties.

Cleaning and grading. In most countries where gum karaya is produced, cleaning and grading is done manually or mechanically in processing units. Hardly any primary processing is done at gum picker level. After gum karaya is received in the processing unit, it is subjected to cleaning by removing foreign matter (bark, leaves, soil, sand, etc.). Larger-sized foreign matter like bark and leaves are removed by hand-picking or winnowing, whereas finer objects like soil and sand are removed by sieving. After most of foreign matter has been removed, the gum karaya is separated into different grades based on colour and impurity still remaining in the gum. The higher-quality gum is lighter in colour and contains fewer impurities compared with inferior quality gum.

Storage. Gum karaya containing more than 10% moisture (dry basis) tends to change its colour from white to dark brown and becomes infested with fungi. The use of gum karaya in various industries depends on its ability to form a viscous and colloidal solution, which is mainly due to the presence of acetyl groups in the gum. However, the gum easily loses its acetyl groups when stored in airy containers in a hot and humid environment; this loss is particularly significant in powdered gum karaya. To avoid this loss of acetyl groups, gum karaya is usually stored in airtight containers at a storage temperature below 25°C.

Studies were carried out to control deacetylation of gum karaya during storage, for quality retention. For this, the samples of gum karaya were packed in low-density polyethylene (LDPE), high-density polyethylene (HDPE), aluminium-coated plastic (ACP) and polypropylene (PP) films filled with air or nitrogen and kept at ambient and cold conditions (domestic refrigerator). Quality parameters such as drying loss, swelling, volatile acid content and water absorption were studied in stored samples for a period of 1 year. The highest swelling index of gum was observed in samples that had been stored and refrigerated in an HDPE bag filled with nitrogen (ICAR-IINRG, 2016).

Gum ghatti

Freshly exuded gum ghatti has a high moisture content (sometimes even higher than 30%) and is very soft in nature. Due to the high moisture content, there is an increased risk of contamination and microbiological spoilage at this stage. Thus, to reduce the moisture content for safe storage (10–13%) and further processing, the collected exudates are dried in the sun for a few days or mechanically dried in a few hours. Dried harvested gum nodules acquire a rounded tear-like shape after drying and are suitable for further processing (Sakai *et al.*, 2013). Contaminations with derivatives like bark, sand, dust, and so on may be between 1 and 3%, while ash content may be in the range between 3 and 5%, depending on the grade of the material. Gum ghatti can be categorized into three grades based on the colour: white, yellow and red. Grade 1 is off-white to buff coloured, grade two is light amber to brown and grade three is dark brown. The colour and hence the grades are dependent on the contamination level present in gum ghatti. Generally, appropriate knowledge of gum ghatti tree plantations and work steps like cleaning, winnowing, and so on yield the best quality of gum ghatti. Gum ghatti processing varies according to the individual needs of the gum processing and manufacturing industries.

Processing. The collected and stored gum ghatti can be further processed to improve its quality for direct use and to achieve a higher profit in its processed form in food and pharmaceutical products. Gums with varying levels of impurities and colour need primary processing such as cleaning, sorting and grading (Amar *et al.*, 2006). Gums with large bark pieces may be cleaned by hand in primary processing centres. Minor impurities on the bark that adhere to the gum ghatti samples may be removed with a pulverizer machine. The powdered gum ghatti is then separated by sieving through screens of different mesh sizes according to particle size, and the bark residues are removed in this way. Gum ghatti free from bark can be further sorted by hand to remove particles that are not suitable for storage and marketing and can then be divided into grades 1–3, according to impurity levels and colours (Sakai *et al.*, 2013). Gum ghatti tears can also be further processed by pulverizing them into a fine powder. Meshes of different sizes can be used for the separation of the gum ghatti powder based on particle sizes, as per the requirement of the users and consumers. During the pulverizing process, the particles may split, which leads to an optimal separation of adhered impurities from the gum, providing high-quality purified clear gum ghatti powder.

Such preparations are partially dissolved in water, filtered and sterilized. The final product is either prepared directly as the gummy, lumpy form after sterilization or as a powder by drum drying or spray-drying.

The gum ghatti powder produced by such spray-drying techniques (e.g. commercially available Gatifolia SD preparation) is completely soluble in water and of consistent colour (Al-Assaf *et al.*, 2021). Further, the viscosity of aqueous solution of spray-dried gum ghatti falls between gum arabic and gum karaya, and the gum has a high degree of solubility and emulsification properties. As such, these preparations are extensively used as emulsifiers in several emulsion formulations that are not stabilized by gum arabic and other gums (Barak *et al.*, 2020).

Storage. The journey of gum ghatti from the tree to the end consumer can take between 3 and 12 months (Amar *et al.*, 2006). Therefore, proper storage and packaging are required to maintain its quality parameters. Primary processed gum ghatti, which is categorized into different grades, can be conveniently packed in jute bags of up to 50 kg capacity for bulk storage and transportation.

When stored in a warehouse, large-scale storage of gum ghatti should be carried out under cool and dry conditions to avoid the formation of agglomerates and lumps and to enhance the shelf life. When gum ghatti is stored in large quantities, it has a slightly sweet taste and a very faint odour (Sakai *et al.*, 2013).

Gum moringa

The stem of a wounded drumstick tree (*Moringa oleifera*) produces a mucilaginous gum with a mild taste. This gum is originally white, but when it comes into contact with bark and is exposed to the environment, it turns into reddish brown or brownish black. The gum exudate is allowed to dry on the tree for easy collection. The dried gum is then collected and further dried to reduce moisture, pulverized and sieved.

Chirounji gum

Chirounji gum exudes in the form of long tears from *Buchanania lanzan* trees and gets solidified when it comes in contact with the environment. The long tears of the gum are broken into small pieces for further processing. The broken gum is cleaned manually using a suitable sieve and sorted according to the impurities present in the gum.

Babool gum

Babool gum, is obtained from the *Acacia nilotica* tree and usually has a moisture content of 13%. The gum's colour, which varies from very pale yellowish-brown to dark reddish-brown, is determined by the amount of tannins present in the product. The lighter, more valued gums are very viscous and soluble in water, whereas the tannins in the darker gum reduce their solubility. A similar processing technique is used for babool gum as for gum arabic.

Seed gums

Gums derived from the endosperm portion of the seeds of different plant species, guar (*Cyamopsis tetragonoloba*), tamarind (*Tamarindus indica*), cassia (*Cassia tora*) and psyllium (*Plantago ovata*), are commercially used. The details of how to process gums from seeds of these plants are briefly explained in the respective sections below.

Guar gum

Guar seeds are composed of the hull, endosperm and germ, typically in the proportion of 15, 40 and 45%, respectively. The germ consists predominantly of protein, while the endosperm mainly contains the sugar compound galactomannan. Germ and hull are used as animal feed after proper treatment (Vishwakarma *et al.*, 2012).

During processing, the three components are separated from each other. The guar gum powder is obtained after removing the hull and germ from the seed and the endosperm has been ground into a fine powder. The main unit operations involved in the processing of guar seeds are cleaning, grading, dehusking, splitting and separation of endosperm, and grinding and purification of the powder. Guar gum is traded mainly as guar splits, guar gum and guar gum derivatives, with different levels of processing (Mudgil *et al.*, 2014).

On arrival at the processing plant, seeds are screened for removal of dirt, stones, sand, metal debris, chaff and broken seeds. Standard seed-cleaning vibrators, electromagnets and sifters are used for cleaning. Dehulling and splitting of the seeds are performed by two processes, dry grinding (Fig. 5.11) and wet grinding (Fig. 5.12) (Vishwakarma *et al.*, 2009). In some industries the shell is charred by flame treatment and then separated. The yield of wet processing is 8–10% higher than that of dry processing, but since the quality of the gum is not good (Gunjal and Kadam, 1991), the industry generally uses dry processing or charring of the seeds (Vishwakarma *et al.*, 2016).

Burr mills, pin mills and modified hammer mills are usually used for splitting guar seeds into the germ and the endosperm. The splits are then heated in ovens and passed through dehusking machines usually consisting of a two-tiered chamber, each with a rotating saw-toothed blade. The dehusked spillings are passed through sieving machines that separate the clean endosperm pieces on a 20-mesh sieve. Guar gum powder is produced from the endosperm splits by grinding in attrition mills, hammer mills, ultra grinders or other crushing equipment. However, guar gum with the best thickening power and fastest hydration rate is produced when the splits are first soaked in water and then flaked and ground (Vishwakarma *et al.*, 2009).

An industrial flow process and layout of commercial guar processing plant along with their various sections (seed splitting and germ separation, split heating, dehulling) suggested by researchers are depicted in Figs 5.13–5.14 (Vishwakarma *et al.*, 2009).

Natural guar gum is usually derivatized to enhance certain properties for specific applications. For industrial applications, many guar gum products (hydroxypropyl guar, carboxymethyl guar and 2-hydroxy-3(trimethylammonium chloride) propyl guar) are formulated with additives that control the rate of hydration, enzyme resistance and dispersibility (Thombare *et al.*, 2016).

Storage and handling. Guar gum powders are generally packed only in sound, clean, dry polyethylene bags which are stowed in jute bags or multi-layer kraft paper bags (Fig. 5.15). Guar gum powder and its derivatives are stable in dry form. It has a long storage life in its dry form, provided that it is warehoused properly. The properties of guar gum remain unchanged for 12–18 months. However, when exposed to humid conditions, guar gum absorbs moisture, which results in microbiological degradation, fermentation and lumping of the powder, and the properties of the gum are adversely affected. Therefore, guar gum should be packed in moisture-proof packets/containers and stored in a cool, dry place away from direct sunlight. Furthermore, it is recommended to consume guar gum within a reasonable time period once the bag is opened. The shelf life of guar gum may be extended by adding suitable preservatives.

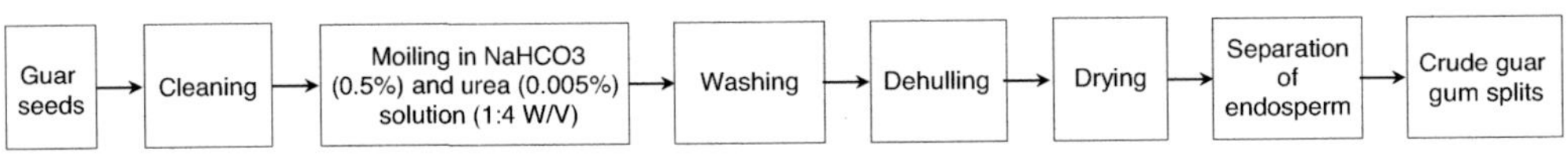

Fig. 5.11. Wet milling process of guar gum splits. (Vishwakarma *et al.*, 2009.) Used with permission.

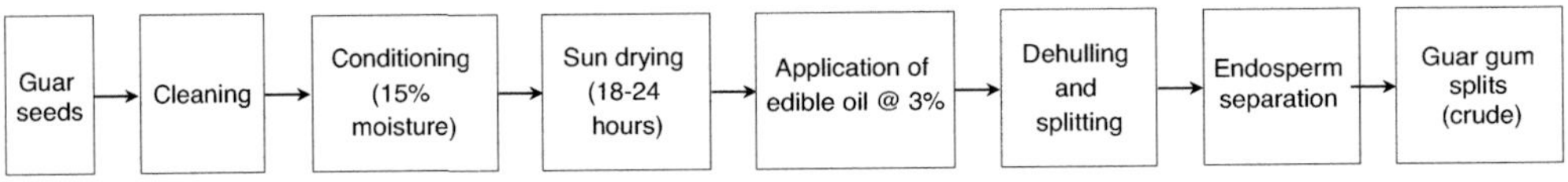

Fig. 5.12. Dry milling process of guar gum splits. (Vishwakarma *et al.*, 2009.) Used with permission.

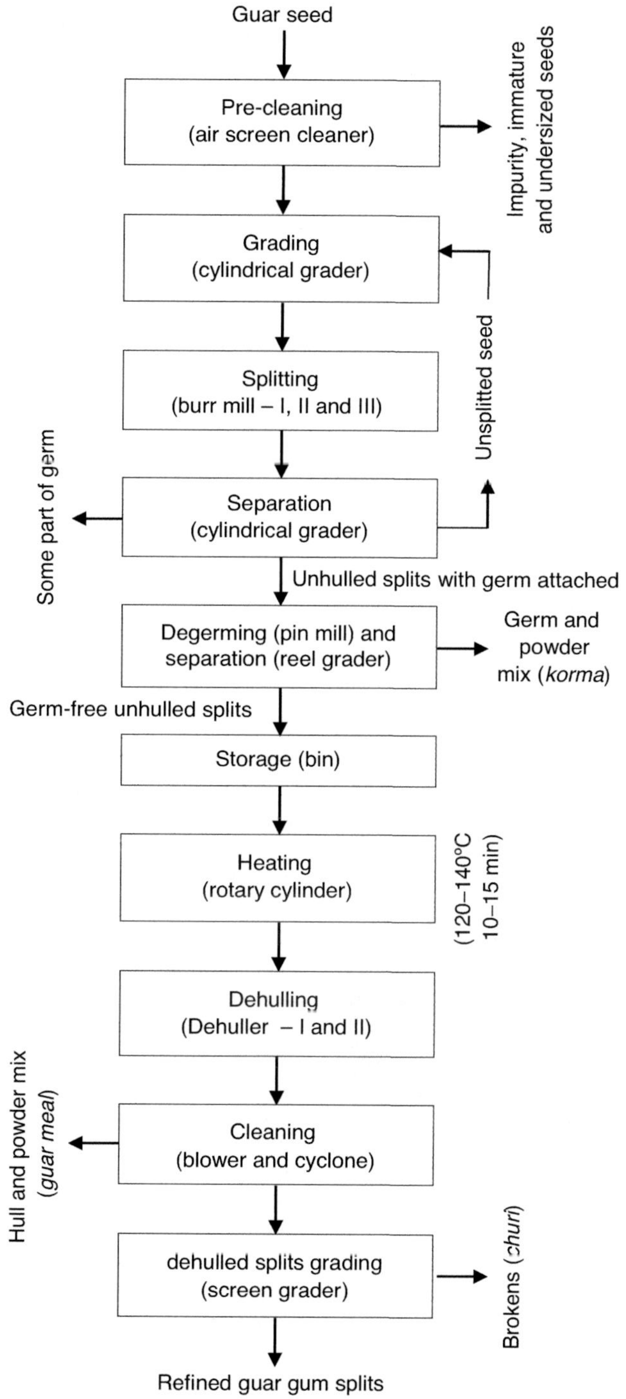

Fig. 5.13. Flow process of guar processing for commercial production of guar gum. (Vishwakarma *et al.*, 2009) Used with permission.

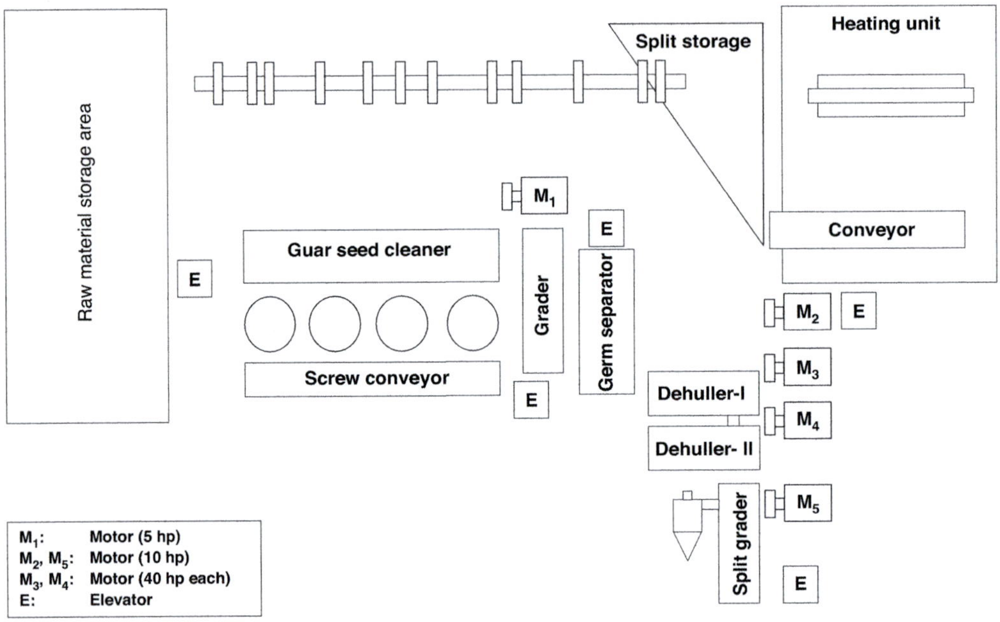

Fig. 5.14. Layout of commercial guar processing plant. (Vishwakarma *et al.*, 2009) Used with permission.

Fig. 5.15. Storage of guar gum. (By Giri *et al.*, 2008.) Used with permission.

Tamarind seed gum

Tamarind seeds are obtained as a by-product of the tamarind pulp industries. The endosperm component of tamarind seeds contains gum which is used for making tamarind kernel powder (TKP). The stages of TKP production from tamarind seed is depicted in Fig. 5.16. TKP finds use in the paper, food and textile sectors, among other industries (Abo-Shosha *et al.*, 2008). Since tamarind gum and tamarind kernel powder are essentially the same raw materials, they are used interchangeably (Thombare *et al.*, 2014). This odourless, natural, non-toxic and bio-degradable material is mainly used as an adhesive in food products. In addition to its binding properties, tamarind kernel powder also has a high concentration of carbohydrates.

On arrival, tamarind pods are sun-dried over 5–7 days in households. The fresh fruits are also dried using small-scale dehydrators. It is simple to crack open the dried, ripe pods and extract the pulp and fibres from the shattered shells. Either hand-kneading or washing in water is used to remove the seeds. The seeds are then cleaned to get rid of any pulp and dried for 2 days in the shade. Finally, seeds are kept in sealed containers in a cool, shaded area. These laborious methods require a lot of manpower and are not economical.

The process of making tamarind kernel powder involves a number of unit operations, including sorting, roasting, separating, stripping, grinding, screening and packing (Whistler, 1993). A specific sorting procedure

Fig. 5.16. Stages of tamarind gum preparation: a. tamarind pods (by A. Balasubramanium, TNAU, Coimbatore, India); b. tamarind seeds (by Nandkishore Thombare, ICAR-IINRG, Ranchi); c. tamarind kernels (by A. Balasubramanium, TNAU, Coimbatore, India); d. tamarind kernel powder (by Nandkishore Thombare, ICAR-IINRG, Ranchi). Used with permission.

is required to remove any form of dirt in order to make the kernel clean. The screen of the cleaner machine is always quivering and aspirating, discarding smaller particles of dirt rather than larger ones. In order to lower moisture content and soften the shell, roasting is a crucial step. This makes it easier to separate the shell from the endosperm. As the shell and endosperm are tightly bound, separating them is a challenging task. It is noticed that roasting at 150°C for 15 min makes the kernels brittle, so that they can be easily removed from the shell. Control over the roasting process is required, especially when roasting temperature and timing are taken into account. The purpose of stripping is to remove the unwanted components. It is a simple stage in which the shell is less firm and dry than the kernel. The kernel must be separated from the shell, and this is done by winnower. After the kernel has been separated, it is ground into the appropriate powder consistency. A thick liquid without clodding is preferable when tamarind kernel powder is employed as a thickener. Before heating the powder to a liquid state, it is important to make sure that it has been well separated (Pulungan, *et al.*, 2001). To obtain better results, whole tamarind powder should pass through a 60-mesh sieve. The TKP produced in this way is packaged in waterproof bags and stored in dry, well-ventilated storage rooms to ensure a long shelf life.

Mechanization in TKP manufacturing. The machine-assisted pulp extraction method was developed under the Post-Harvest Technology Scheme (PHTS) of ICAR at the University of

Agricultural Sciences (UAS) in Bangalore, India. A tamarind dehuller with a capacity of 85 kg/h and a dehulling efficiency of 89.09% was developed by researchers from Tamil Nadu Agricultural University, Coimbatore, India. The main components of the dehuller are sharp L-shaped pegs made up of 15 × 3 mm steel rods mounted on the focal shaft encased with a 20 × 5 mm mild steel oblong sieve. To remove the seeds from tamarind pods, the researchers from that organization also developed a tamarind deseeder that achieves a deseeding efficiency of 89.15% (Kumar *et al.*, 2015; Jansi *et al.*, 2016).

Cassia gum

The endosperm of *Cassia obtusifolia*, commonly known as *Cassia tora*, which is a wild plant and thrives as a weed throughout India, is the source of cassia gum (Joshi and Biyani, 2015). In addition to its use as a food additive, cassia gum is well known for its thickening and gelling properties. After the seeds have been processed into cassia gum in India, it is exported to various countries, including France, Belgium and Australia.

Gum processing. On arrival at the processing plant, the seeds undergo various mechanical cleaning stages to remove impurities. The subsequent stages include dehusking and splitting the seeds. Seeds are heat-treated to facilitate removal of the husk and germ mechanically to obtain splits. Later, the seeds are degermed by milling and screening the endosperm. After milling and screening, the endosperm is ground to obtain a uniform powder with a small grain size.

Psyllium gum

One of the most important medicinal plants cultivated for its husk is the psyllium plant, also known as isabgol. Psyllium husks are available in various grades according to mesh size, purity, etc. They are mainly marketed as husks, husk powder and industrial powder. The dehusked seeds are used to feed dairy animals. Since the husks tend to absorb moisture, they are packed in plastic-coated bags and drums.

A quarter or more of the total seed yield consists of husk/mucilage. Psyllium seed mucilage contains a large number of proteins, which produce colloidal mucilage used in different medical systems such as Ayurveda, Unani and allopathy. India is the world's leading supplier and producer of psyllium seed and husk, accounting for 98% of worldwide production. Major psyllium herb growing states in the country are Rajasthan, Gujarat, Madhya Pradesh and Haryana. The main part of the world's supply of psyllium seed and husk is shipped to the USA, Western Europe and Korea. The seeds and shells are mainly used as a thickening agent in the food industry, as a base for cosmetics, for the production of dry toothpaste powder and genomic gels and as animal feed. Psyllium husks remain undigested after entering the digestive tract and have the ability to bind water, which has a positive effect on the treatment of diarrhoea, irritable bowel syndrome and constipation. They are also used as a daily dietary supplement to promote and maintain regular digestion. Regardless of other dietary factors or intestinal diseases, the inert mass of the husks aids in supplying a steady volume of solid food.

Gum processing. Processing of psyllium includes cleaning the seeds at cleaning units. The seeds are then passed through grinding mills eight times, which leads to separation of the husks. The husks are further purified by gravity separators before packaging. Psyllium husks are available in various grades according to mesh size, purity, etc.

On arrival at the processing plant, the psyllium grains are passed through various unit operations (pre-cleaning, separation of metallic and stone particles, gravity separation, sorting on colour basis, dehusking) for psyllium gum production. After passing through various conveyance systems, the raw material is cleaned using a drum sieve-type vibratory pre-cleaner to remove impurities (stones, paper, pieces of wood, etc.) along with other contaminants. To separate the metallic and stone particles from the pre-cleaned grains of psyllium, a magnetic separator and a destoner are used. After removal of metallic and stone particles from the pre-cleaned psyllium grains, products of the same size but with different specific weights are separated utilizing a gravity separation machine. A colour sorter machine is also utilized to sort the psyllium grains of different colour, which vary depending on the variety. The grains, sorted by colour, are then dehusked. The husk is separated from the non-husk part of the *psyllium* seed without substantial disruption. The husk is then passed through gravity separators before customized packaging is performed (NIFTEM, 2022).

Gum-Resins

The term 'gum-resin' refers to a resinous material that comprises of a natural mixture of resin and gum. Gum-resins contain traces of essential oil and possess characteristics of both gums and resins. These are often obtained from plants that thrive in arid and dry environments. Due to the presence of the gum content, which mainly consists of various polysaccharides, they are reasonably water-soluble. Usually, their flavour and smell are distinct and pungent.

Guggul or Indian myrrh

Guggul (*Commiphora mukul*), which forms on the branches of the tree, dries as hard drops on the tree itself before it is collected. Then the resin is cleaned by sieving and hand-picking to remove foreign matter and packed in sacks for transfer to the point of sale. The collected gum is graded according to its purity and from part of the harvest an essential oil is produced by applying steam-distillation. The solids that remain are removed by filtering. The remaining water is evaporated under vacuum, and the concentrated extracts are lyophilized to yield gum as a fine powder.

Salai

Salai gum, commonly known as 'Indian frankincense', is the gum-resin obtained from *Boswellia serrata*. This product contains gum, essential oil and non-volatile boswellic acids, which have gained particular importance in medicine. Salai gum is collected in the form of lumps and then processed accordingly.

Processing

After the collection of salai gum, the impurities (bark pieces and other debris) present in it are physically removed. The gum-resin is then placed in a bamboo basket for about one month. During storage, the essential oil fraction of the gum-resin separates. This is locally known as *ras*. *Ras* is utilized in paints and varnishes and accounts for about 8–10% of the raw material. The remaining semi-solid to solid fraction is thoroughly dried and occasionally treated with soapstone powders to make it brittle. After that, a wooden mallet or chopper is used to break it into small pieces. During this process, more bark and other impurities are physically removed. The gum-resin is then graded according to its colour and impurities (Warrier and Muthupandiyan, 2022). Generally, four grades are distinguished in the market: superfine and quality I, II and III, in decreasing order of quality.

Asafoetida

Asafoetida, often known as the devil's dung or the gods' food, is the dried aromatic gum-resin that exudes from the live rhizome, rootstock or taproot of many plant species in the genus *Ferula*. Ground-level pits where the tap root is located are used to collect the milky juice that is exuded from *Ferula asafoetida* plants (Fig. 5.17). The upper part of the *asafoetida* plant is freshened at regular intervals by reopening the clogged resin ducts for further exudation. The pits containing the tap roots of the *asafoetida* plant are covered with a suitable material such as sheet metal or locally available material after the stem has been cut or after tapping to collect the resinous gum. Asafoetida gum-resin is initially a paste, but after drying it becomes brown in colour and the milky juice exuded from the root turns into a dark resin-like substance. Natural asafoetida gum-resin is combined with flour and gum arabic since it has a very strong odour and taste and cannot be used directly in its pure form in cooking.

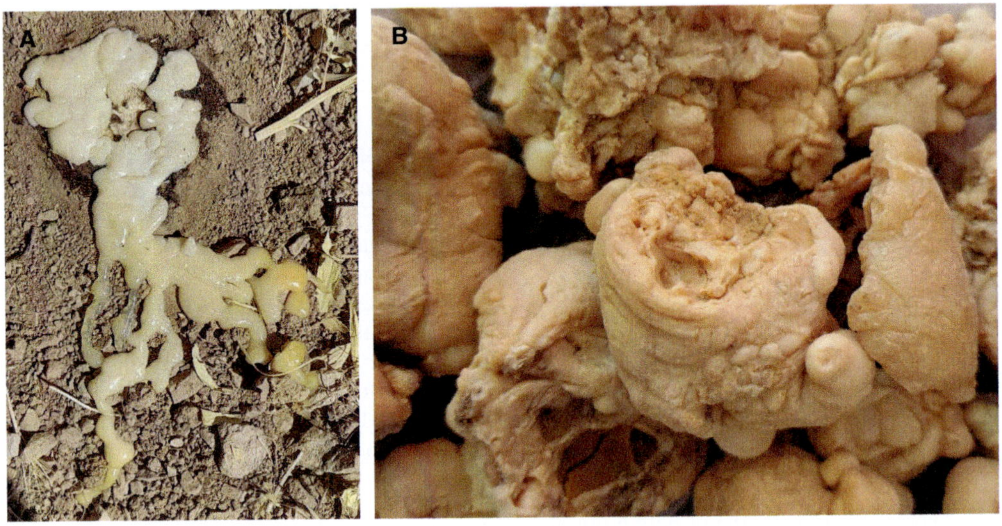

Fig. 5.17. Gum-resin obtained from Ferula asafoetida: a. milky juice exudating when the roots are cut; b. gum-resin that is formed when the milky juice dries in the air. (By Farsad Nadjafi.) Used with permission

The commercial version of the gum-resin, known as 'compounded asafoetida', is a fine powder that, in addition to 30% asafoetida gum-resin, includes gum arabic and rice flour to prevent the formation of lumps (Giri *et al.*, 2008). The special advantage of the compounded form is that it is easier to use. To obtain the essential oil from asafoetida gum-resin it is steam-distilled.

Conclusion

With the exception of lac and guar gum, most natural resins and gums are exported after processing. Other NRGs are exported after manual grading, cleaning and sizing in primary processed form. Therefore, postharvest processing of NRGs must be done utilizing contemporary equipment and procedures in order to produce high-purity processed products to earn maximum return from the export market. Additionally, there is an opportunity to develop derivatives and value-added products based on NRGs for the export market, which might fetch multiple times more foreign exchange. The duration of operation, manpower requirements and drudgery involved in processing of NRGs can be minimized by using machines and equipment developed for NRGs processing. In order to enhance productivity of the manpower in NRG processing and value addition, solar-powered machines and devices are used, particularly in isolated locations where the power supply is a problem. Application of suitable methods and machinery/equipment will certainly minimize the labour-intensiveness of NRG processing and value addition, leading to a reduction of processing costs and improvement in quality of processed products. Adoption of primary processing and value addition of NRGs at the village level will enhance the return to the producer/collector and generate employment for the rural and tribal populations involved in the production catchment.

References

Abo-Shosha, M.H., Ibrahim, N.A., Allam, E. and El-Zairy, E. (2008) Preparation and characterization of polyacrylic acid/karaya gum and polyacrylic acid/tamarind seed gum adducts and utilization in textile printing. *Carbohydrate Polymers* 74, 241–249. Available at: https://doi.org/10.1016/j.carbpol.2008.02.011

Al-Assaf, S., Phillips, G.O. and Amar, V. (2021) Gum ghatti. In: Phillips, G.O. and Williams, P.A. (eds) *Handbook of Hydrocolloid*, 3rd edn. Woodhead Publishing, Sawston, UK, pp. 653–672.

Amar, V., Al-Assaf, S. and Phillips, G.O. (2006) An introduction to gum ghatti: Another proteinaceous gum. *Foods and Food Ingredients Journal of Japan* 211, 275–279.

Barak, S., Mudgil, D. and Taneja, S. (2020) Exudate gums: Chemistry, properties and food applications – a review. *Journal of the Science of Food and Agriculture* 100, 2828–2835. Available at: https://doi.org/10.1002/jsfa.10302

Bhowmik, T. (1962) Technology of lac - (a) indigenous method of processing lac. In: Mukhopadhyay, B. and Muthana, M.S. (eds) *A Monograph on Lac. Indian Lac Research Institute*. Ranchi, India, pp. 156–166.

Bureau of Indian Standards (1973) *IS 15:1973. Specification for Seedlac (Second Revision)*. Bureau of Indian Standards, New Delhi.

Chikamai, B.N., Banks, W.B., Anderson, D.M.W. and Weiping, W. (1996) Processing of gum arabic and some new opportunities. *Food Hydrocolloids* 10, 309–316. Available at: https://doi.org/10.1016/S0268-005X(96)80006-3

Couteaudier, T. (2007) *Export Marketing of Sudanese Gum Arabic*. World Bank, Washington, DC.

Giri, S., Prasad, N., Pandey, S., Prasad, M. and Baboo, B. (2008) *Natural Resins and Gums of Commercial Importance - At a Glance*. ICAR-Indian Institute of Natural Resins and Gums, Ranchi, India.

Gunjal, B.B. and Kadam, S.S. (1991) *Handbook of World Food Legumes*. CRC Press, Boca Raton, Florida.

ICAR-IINRG (2012) *Annual Report 2011-12*. Indian Institute of Natural Resins and Gums, Ranchi, India.

ICAR-IINRG (2016) *Annual report 2015-16*. ICAR-Indian Institute of Natural Resins and Gums, Ranchi, India.

ICAR-ILRI (2003) *Annual Report 2002-03*. Indian Lac Research Institute, Ranchi, India.

Islam, A.M., Phillips, G.O., Sljivo, A., Snowden, M.J. and Williams, P.A. (1997) A review of recent developments on the regulatory, structural and functional aspects of gum arabic. *Food Hydrocolloids* 11, 493–505. Available at: https://doi.org/10.1016/S0268-005X(97)80048-3

Jales, M. (2018) *Commodities at a Glance: Special Issue on Gum Arabic*. UNCTAD, Geneva, Switzerland.

Jansi, S.R., Rajkumar, P. and Kailappan, R. (2016) Design and development of a power operated tamarind huller cum deseeder. *Agricultural Mechanization in Asia, Africa, and Latin America* 47, 41–46.

Joshi, U.M. and Biyani, K.R. (2015) Extraction of a novel seed gum from Cassia tora seeds and its characterization. *International Journal of Pharmacy and Pharmaceutical Research* 4, 243–251.

Karim, B. Mukhopadhyay, B. and Muthana, M.S. (eds). (1962) *Technology of lac – (b) machine-made shellac*. Indian Lac Research Institute, Ranchi, India, pp. 167–178.

Kumar, P.K., Pandian, N.K.S., Kumar, P.R., Kumar, A.S. and Balakrishnan, M. (2015) Development and evaluation of a continuous type tamarind deseeder. *Scientific Journal Agricultural Engineering* 2, 49–59.

Mudgil, D., Barak, S. and Khatkar, B.S. (2014) Guar gum: Processing, properties and food applications – a review. *Journal of Food Science and Technology* 51, 409–418.

NIFTEM (2022) *Detailed Project Report-Isabgol Processing Unit Under PMFME Scheme. National Institute of Food Technology Entrepreneurship and Management*. Ministry of Food Processing Industries, Sonipat, Haryana, India.

Pandey, S.K., Prasad, N. and Sharma, S.C. (2016) Lac processing-seedlac, shellac, bleached lac and lac dye: Skill and entrepreneurship development-oriented options. In: Yogi, R.K., Kumar, A., Singh, A.K., Yadav, S.K.S. and Jaiswal, A.K. (eds) *Natural Resins and Gums: Agribusiness Modules for Skill and Entrepreneurship Development. ICAR-Indian Institute of Natural Resins and Gums*. Ranchi, India, pp. 81–86.

Prasad, N. (2014) *Final Report of NAIP Sub-project on "A Value Chain on Lac and Lac based Products for Domestic and Export Markets"* (Final Project Report) Indian Institute of Natural Resins and Gums, Ranchi, India.

Prasad, N., Thombare, N., Sharma, S.C. and Kumar, S. (2022) Gum arabic – a versatile natural gum: A review on production, processing, properties and applications. *Industrial Crops and Products* 18, 1153047. Available at: https://doi.org/10.1016/j.indcrop.2022.115304

Pulungan, M.H., Sukardi, Sri Maryani, E.F., Atifah, N. and Sucipto, S. (2001) Tamarind kernel powder (*Tamarindus indica*) processing (reviews on roasting temperature and roasting duration). *Journal Technology Pertanian* 2, 150–158.

Sahu, P., Patel, S., Pisalkar, P.S. and Katiyar, P. (2017) Studies on drying characteristics of gum karaya (*Stericulia urens* Roxb.). *Trends in Biosciences* 10(3), 1114–1117.

Sakai, E., Katayama, T., Ogasawara, T. and Mizuno, M. (2013) Identification of *Anogeissus latifolia* Wallich and analysis of refined gum ghatti. *Journal of Natural Medicines* 67, 276–280. Available at: https://doi.org/10.1007/S11418-012-0678-3

Sharma, A., Sharma, L. and Goyal, R. (2018) A review on Himalayan pine species: Ethnopharmacological, phytochemical and pharmacological aspects. *Pharmacognosy Journal* 10, 611–619. Available at: https://doi.org/DOI:10.5530/pj.2018.4.100

Sharma, S.C. and Prasad, N. (2013) Comparison of gum tapping techniques for gum karaya – a gum of commercial importance. *Journal of Non-Timber Forest Products* 20, 165–170.

Sharma, S.C., Prasad, N., Pandey, S. and Bhargava, V. (2020) Development of integrated small scale lac processing unit. *Agricultural Mechanization in Asia, Africa and Latin America* 51, 58–66.

Sharma, S.C., Prasad, N. and Pandey, S.K. (2014) Effect of storage material and aging on quality parameters of shellac. *Journal of AgriSearch* 1, 92–95.

Sharma, S.C., Pandey, S.K. and Prasad, N. (2022) Equipments for manufacturing lac based value added products. *Journal of AgriSearch* 9, 249–254. Available at: https://doi.org/10.21921/jas.v9i03.11010

Sharma, S.C., Pandey, S.K., Prasad, N. and Raj, H. (2024a) Unveiling the potential of *Moringa oleifera* Lam. gums: An investigation into optimal tapping strategies. *Industrial Crops and Products* 214, 118511. Available at: https://doi.org/10.1016/J.INDCROP.2024.118511

Sharma, S.C., Prasad, N. and Pandey, S.K. (2024b) Seedlac manufacturing through integrated lac processing unit: An industrial product. *Indian Forester* 150, 267–274. Available at: https://doi.org/10.36808/if/2024/v150i3/169874

Singh, J. and Singh, B. (2014) Drying kinetics of gum karaya (*Sterculia urens*) using convective air dryer. *British Journal of Pharmaceutical Research* 4(13), 1647–1656.

Thombare, N., Srivastava, S. and Chowdhury, A.R. (2014) Multipurpose applications of tamarind seed and kernel powder. *Science Reporter* 51, 32–33.

Thombare, N., Jha, U., Mishra, S. and Siddiqui, M.Z. (2016) Guar gum as a promising starting material for diverse applications: A review. *International Journal of Biological Macromolecules* 88, 361–372. Available at: https://doi.org/10.1016/j.ijbiomac.2016.04.001

Thombare, N., Kumar, S., Kumari, U., Sakare, P., Yogi, R.K. *et al.* (2022) Shellac as a multifunctional biopolymer: A review on properties, applications and future potential. *International Journal of Biological Macromolecules* 215, 203–223. Available at: https://doi.org/10.1016/J.IJBIOMAC.2022.06.090

Tinnelly and Wenzhi, D (2007) Technique for producing arabic gum powder. CN101143995A.

Vishwakarma, R.K., Shivhare, U.S., Nanda, S.K. and Patil, R.T. (2009) Status of post harvest technology of guar (*Cyamopsis tetragonoloba*) in India. *Agricultural Mechanization in Asia, Africa and Latin America* 40, 65–72.

Vishwakarma, R.K., Shivhare, U.S. and Nanda, S.K. (2012) Physical properties of guar seeds. *Food and Bioprocess Technology* 5, 1364–1371. Available at: https://doi.org/10.1007/s11947-011-0514-x

Vishwakarma, R.K., Shivhare, U.S. and Nanda, S.K. (2016) An evaluation of wet processing method about steady state rheological properties of guar. *Journal of Advances in Food Science & Technology* 3, 203–212.

Walde, S.G., Balaswamy, K., Shivaswamy, R., Chakkaravarthi, A. and Rao, D.G. (1997) Microwave drying and grinding characteristics of gum karaya (*Sterculia urens*). *Journal of Food Engineering* 31, 305–313. Available at: https://doi.org/10.1016/S0260-8774(96)00082-9

Warrier, K.C.S. and Muthupandiyan, S. (2022) *Boswellia serrata* Roxb. ex Coleb.: A threatened tree. *International Journal of Plant & Soil Science* 34, 527–534. Available at: https://doi.org/DOI:10.9734/ijpss/2022/v34i242669

Whistler, R.L. (1993) Exudate gums. In: Whistler, R.L. and BeMiller, J.N. (eds) *Industrial Gums: Polysaccharides and Their Derivatives*, 3rd edn. Academic Press, New York, pp. 309–339. Available at: https://doi.org/10.1016/B978-0-08-092654-4.50016-4

6 Processing of Natural Gums and Resins

Niranjan Prasad*, Sanjay Kumar Pandey, Sakharam Kale and Satish Chandra Sharma

ICAR-National Institute of Secondary Agriculture, Namkum, Ranchi, India

Abstract

This chapter explores the processing and applications of natural resins, gums and gum-resins across pharmaceutical, cosmetic, perfumery and food industries. These natural polymers are valued for their biodegradability, non-toxicity and multifunctional properties, making them sustainable alternatives to synthetic materials. In the pharmaceutical industry, resins like shellac are used for drug encapsulation, coating and targeted delivery due to their pH-sensitive release properties. Gums such as guar gum and gum arabic serve as binders, disintegrants and stabilizers, while gum-resins like guggul and frankincense offer anti-inflammatory and antimicrobial benefits. The cosmetic and perfumery industries leverage resins as fragrance fixatives and texturizing agents, gums as thickeners and emulsifiers and gum-resins for their aromatic and skin-conditioning properties. Shellac, for instance, is used in nail polishes and hair sprays, while guar gum enhances the texture of lotions and shampoos. In the food industry, these natural polymers act as stabilizers, thickeners and glazing agents. Shellac coatings extend the shelf life of fruits, while gums like xanthan and tragacanth improve the texture and stability of dairy and bakery products. Gum-resins such as asafoetida and myrrh are used as flavouring agents and preservatives. Despite their advantages, challenges like variability in composition and stability issues exist. However, ongoing research into modifications and composite formulations aims to enhance their performance, ensuring their continued relevance in industrial applications. This chapter highlights the versatility and growing importance of natural resins, gums and gum-resins in meeting consumer demand for sustainable, functional ingredients.

Introduction

Due to their versatile properties, natural resins, gum-resins and gums are used in the manufacture of various products in the pharmaceutical, cosmetic, perfume and food industries. The respective properties of the plant-based raw materials will be discussed in more detail below.

Pharmaceutical industry

- **Natural gums:** Gums like guar gum, xanthan gum, and acacia gum are used as binders, disintegrants, and controlled release agents in tablet formulations. They help in creating gels, suspensions and emulsions, allowing better stability and controlled release of active pharmaceutical ingredients (APIs).

*Corresponding author: niranjan_ilri_1999@yahoo.com

DOI: 10.1079/9781800626690.0006

- **Natural resins:** Resins such as mastic and shellac are used as coating agents for tablets and capsules, providing a protective layer that controls the release of the drug and enhances shelf life.
- **Gum-resins:** Gum-resins such as myrrh and frankincense are used in wound healing, anti-inflammatory treatments and topical applications due to their antimicrobial and analgesic properties.

Cosmetics industry

- **Natural gums:** Gums like xanthan gum and carrageenan act as thickeners and stabilizers in lotions, creams and gels, improving texture and providing a smooth application.
- **Natural resins:** Resins such as benzoin and dammar are valued for their film-forming abilities and are often used in nail polishes, hair sprays and mascara to create a lasting hold. Some resins also serve as natural preservatives and antioxidants.
- **Gum-resins:** These are incorporated into skincare products for their anti-inflammatory and soothing properties. For example, frankincense and myrrh are used in anti-aging creams and serums, as they promote skin elasticity and have antibacterial benefits.

Perfumery industry

- **Natural gums:** Gums such as gum arabic are used as fixatives in perfumes to help bind and extend the release of volatile fragrances and improve longevity.
- **Natural resins:** Resins such as labdanum, benzoin and styrax are commonly used in perfumery for their rich, complex and long-lasting aroma profiles. They often serve as base notes, adding depth and warmth to perfumes.
- **Gum-resins:** Frankincense and myrrh are valued for their unique fragrances and are frequently used in incense, spiritual applications and as base notes in perfumes, providing a resinous, earthy scent.

Food industry

- **Natural gums:** Gums like guar gum, xanthan gum and locust bean gum are used as thickening, stabilizing and emulsifying agents in a variety of products, such as sauces, dressings, ice creams and baked goods.
- **Natural resins:** Resins like shellac are used as glazing agents for confectioneries, fruits and supplements, creating a shiny, protective coating that prolongs shelf life and improves appearance.
- **Gum-resins:** Myrrh and asafoetida are used as flavouring agents and possess preservative properties. Asafoetida, for example, is commonly used as a spice in certain cuisines, adding a unique flavour and aroma to various dishes.

Pharmaceutical Industries

The traditional use of excipients in drug formulations was to act as inert vehicles to provide necessary weight, consistency and volume for the correct administration of the active ingredient, but in modern pharmaceutical dosage forms they often fulfil multifunctional roles, such as modifying release, improvement of the stability and bioavailability of the active ingredient, enhancement of patient acceptability and ensuring ease of manufacture. New and improved excipients continue to be developed to meet the needs of advanced drug delivery systems (Raymond *et al.*, 2006; Patel *et al.*, 2007). Polymers have been successfully investigated and employed in the formulation of solid, liquid and semi-solid dosage forms and are specifically useful in the design of novel drug delivery systems. Both synthetic and natural polymers have been investigated extensively for this purpose (Varshosaz *et al.*, 2006). In most cases, the use of natural polymers for pharmaceutical applications is attractive because they are economical, readily available, non-toxic and capable of chemical modifications, potentially biodegradable and with few exceptions also biocompatible. A large number of plant-based pharmaceutical excipients are available today. Many researchers have explored the usefulness of plant-based materials as pharmaceutical excipients. Ability to produce a wide range of materials based on their properties and molecular weight, natural polymers became a thrust area in a majority of investigations in drug delivery systems (Banker and Anderson, 1987). Natural resins and gums can also be modified to meet the individual requirements of drug delivery systems and thus can compete with several synthetic excipients available on the market (Bhardwaj *et al.*, 2000).

Natural resins and gums have been studied for their application in different pharmaceutical formulation, such as matrix controlled systems, film coating agents, buccal films, microspheres, nanoparticles, viscous liquid formulations like ophthalmic solutions, suspensions and implants. Besides this, natural resins and gums have also been investigated in the design of novel drug delivery systems, such as those that target delivery of the drug to a specific region in the gastrointestinal tract or in response to external stimuli (Morkhade *et al.*, 2006). Their applicability and efficacy has been proven (Pandey and Khuller, 2004; Chamarthy and Pinal, 2008; Alonso-Sande *et al.*, 2009). They have also been utilized as viscosity enhancers, stabilizers, disintegrants, solubilizers, emulsifiers, suspending agents, gelling agents, bio-adhesives and binder forms (Guo *et al.*, 1998). Important natural gums and resins have been investigated, and some of those successfully used in different pharmaceutical applications are presented in the following section.

Resin

Shellac

Shellac, with its unique properties, finds different applications in pharmaceuticals, and this area has vast scope for research. Lac and its derivatives are used in making a number of pharmaceutical coatings. The development of shellac-coated sustained-release pellet formulation is one such example, given its recent application in pharmaceuticals (Farag, 2010). It is used for the microencapsulation of several medicines and vitamins, as the lac coating restricts the sudden release of active ingredients into the body and acts as a moisture barrier during storage. Lac is popular in Ayurvedic and Unani medicines used for treating hiccups, cough, osteitis, osteomyelitis, dropsy, obesity and inflammation. A shellac-alginate-based formulation was also found to be helpful for the encapsulation of peppermint essential oil (Foglio Bonda *et al.*, 2020). Quercetin-fortified self-assembly of novel almond gum-shellac nanoparticles revealed improved antioxidant activity (Sedaghat Doost *et al.*, 2019).

A study was performed to develop oil encapsulation systems using alginate and shellac as wall materials. The semi-spherical, smooth-surfaced beads had an encapsulation efficiency of 98.7% and total oil content of 38.6%. Beads release more oil in intestinal digestion than in gastric digestion (Morales *et al.*, 2017). The developed system also has the potential for making new products in the food industry due to its delivery capacity in oil encapsulation and oil-soluble compounds. Bioactive compounds taken orally have to reach the gastrointestinal tract by crossing the stomach's acidic environment barrier. It is, therefore, necessary to prevent the molecules from degrading in acidic pH in the delivery system and releasing it into the small intestine. The unique property of shellac being insoluble in water at neutral to acidic pH and soluble in alkaline pH makes it an ideal coating material for drugs intended for colon target release. When a particular drug is coated with shellac, it prevents or restricts the release of the active ingredient in the acidic pH of the stomach and releases the maximum portion of it in the intestinal alkaline environment, making it a pH-sensitive smart release material. Patel and co-workers developed pH-responsive colloidal particles of shellac to test the target-specific delivery of silibinin drugs. The drug encapsulated within the shellac particles could sustain the acidic pH environment and release a significant part of it (>90%) in an alkaline pH similar to gastrointestinal tract pH (Patel *et al.*, 2011). Fabrication of oat protein shellac beads at neutral pH was studied. The structure restricts swelling, which prevents premature riboflavin diffusion and achieves 85.5% survival and 80% amylase activity of *Lactobacillus acidophilus*. Accordingly, it was concluded that shellac beads are capable of being employed in future delivery systems of sensitive bioactive compounds in food and biomedical applications (Yang *et al.*, 2018).

Cinnamon bark extract containing shellac nanoparticles was synthesized and stabilized with xanthan gum to study the pH-responsive delivery system (Muhammad *et al.*, 2020). The particles were found to be stable in the acidic environment of the stomach, while >90% of the active ingredient (cinnamon polyphenol, which has antioxidant properties) was released in the alkaline pH, similar to the pH of the intestine. Besides this, custom-made shellac-based nanofibres prepared by the electrospinning process (Limmatvapirat *et al.*, 2011; Nawinda *et al.*, 2020) also exhibited excellent drug release properties.

It has been found that the coating of shellac on drug tablets improved the mechanical properties and efficiency of the shellac coated drug (Soradech *et al.*, 2012). An *in situ* forming gel containing bleached shellac and other shellac-based gels also showed desirable drug release properties for doxycycline hyclate and other drugs for treating ailments such as periodontitis (Phaechamud and Choncheewa, 2016; Phaechamud *et al.*, 2016, 2018, 2019; Phaechamud and Setthajindalert, 2018). Moisture-protective matrix tablets using shellac coatings were prepared, which were found to be successful in taste-masking and extended-release (Pearnchob *et al.*, 2004). Various matrices or formulations developed based on shellac have been reported to exhibit controlled release behaviour of different drugs such as metronidazole (Limmatvapirat *et al.*, 2008a), monolaurin (Chinatangkul *et al.*, 2018), doxycycline hyclate (Senarat *et al.*, 2021) and theophylline (Soradech *et al.*, 2012).

Shellac has various applications as a film in the pharmaceutical industry due to its low water permeability and controlled release properties. Nevertheless, due to its low stability, insolubility at neutral pH and need for an organic solvent to get dissolved, the use of other polymeric counterparts like cellulose derivatives and polyacrylates gained preference (Limmatvapirat *et al.*, 2004). Several efforts were made to improve the stability and properties of shellac films by partial hydrolysis (Limmatvapirat *et al.*, 2004), alkali treatments (Limmatvapirat *et al.*, 2008b), use of plasticizers like Jeffamine® (Bar and Bianco-Peled, 2020), gelatin, polyethene glycol and diethyl phthalate (Soradech *et al.*, 2013). Also, the effect of various plasticizers on film and other shellac properties is discussed in detail by Qussi and Suess (2006) and Luangtana-anan *et al.* (2007). An investigation was carried out into shellac's moisture-protection and taste-masking potential. It was found that tablets with shellac coating exhibited lower water absorption rates. The stability of the drug, acetylsalicylic acid, was also found to be greater in shellac-coated tablets. These tablets required a lower coating than with common cellulose derivatives with similar effects without changing drug-release properties. Coated tablets worked as a matrix for extended release of the drug, and the unpleasant flavour of acetaminophen was also suppressed (Pearnchob *et al.*, 2003). Multiple-layer coatings to tablets using calcium chloride, citric acid, etc. for the enteric sub-coat and shellac for the outer sub-coat were tried for the controlled release of theophylline. The coatings could prolong the release of the drug in the surrounding media. As different sub-coats have different interactions with shellac, this modifying sub-coat technique can be used to develop custom-made drug delivery systems.

It has been found that bleached shellac in solution phase, when it comes in contact with some non-compatible or aqueous solvent, converts itself to an opaque rigid gel which remains suspended in the resultant matrix. A higher concentration of bleached shellac increased the viscosity of the formulation, gel formation and duration of drug release but reduced the injectability (Phaechamud *et al.*, 2016). Efforts were made to improve the injectability, gel-forming properties and antimicrobial properties of *in situ* forming microparticle formulations by using different oils as the external phase of non-aqueous oil emulsion (Chuenbarn *et al.*, 2021), using solvents like *N*-methyl pyrrolidone (Phaechamud *et al.*, 2016), dimethyl sulfoxide, 2-pyrrolidone and eutectic (1:1 menthol:camphor) (Phaechamud and Setthajindalert, 2018). Various solvents have shown different viscosity and injectability for bleached shellac, and 2-pyrrolidone emerged as the most suitable solvent for bleached shellac-based *in situ* gel-forming formulations (Phaechamud and Setthajindalert, 2018).

Shellac has also been found helpful in tissue engineering applications. Jacoby *et al.* (2017) fabricated capillary-like structures of macro inlet and outlet vessels using a shellac blend and observed a thrombosis-free movement through the network imitating the arrangement of a true capillary.

Gums

Gums as natural polysaccharides possess a complex, branched polymeric structure due to which they exhibit high cohesive and adhesive properties which are highly useful in pharmaceutical preparations. They are ingredients in dental and other adhesives and in laxatives. These hydrophilic polymers are useful as tablet binders, disintegrants, emulsifiers, suspending agents, gelling agents, stabilizing agents, thickening agents, protective colloids in suspension and sustaining agents in tablets (Deore and Khadabadi, 2008). Pharmaceutical applications of some gums are used commercially as adjuvants in various pharmaceutical formulations, as summarized below.

Guar gum

Guar gum is a natural polysaccharide which finds application as a binder, as a disintegrating agent in tablet formulations and as a thickener, stabilizer, emulsifier and suspending agent in liquid formulations. Further, properties of guar gum can be modified to obtain variable solubility, swelling and film-forming ability. Due to these unique properties of guar gum and its derivatives, lots of research has been carried out to develop guar gum and its derivative-based formulations as colon-specific, antihypertensive, protein and transdermal drug delivery systems (Kaith *et al.*, 2014). Guar gum polysaccharides have the flexibility to obtain a desirable drug release profile as well as offering cost-effectiveness and broad regulatory acceptance, which make it suitable for use in targeted drug delivery in various forms, such as oral controlled drug delivery systems, coatings, tablets matrices, hydrogels and nanoparticles.

Guar gum is resistant to dissociation in the acidic environment in the stomach, possesses gelling properties to reduce release of drugs and is susceptible to enzymatic and microbial degradation in the large intestine. These characteristics of guar gum make it a convenient hydrophilic matrix for oral controlled delivery of drugs (Krishnaiah *et al.*, 1998a, b). Guar gum has been successfully studied as a matrix for the colon-specific controlled release of dexamethasone and other anti-inflammatory agents (Kenyon *et al.*, 1997; Wong *et al.*, 1997), indomethacin (RamaPrasad *et al.*, 1998), 5-amino salicylic acid (Krishnaiah *et al.*, 1999), albendazole (Krishnaiah *et al.*, 2001a), metronidazole (Krishnaiah *et al.*, 2002a) and mebendazole (Krishnaiah *et al.*, 2001b). The colon-targeted release of a guar gum-bound tablet was also confirmed in healthy human volunteers with a gamma scintigraphic study using a technetium 99m-DTPA as tracer (Krishnaiah *et al.*, 1998a; Krishnaiah *et al.*, 1998b), which showed maximum drug release in the colon as a consequence of enzymatic degradation. In compression-coated tablets, guar gum has proved to be an effective drug-release retardant at lower pH for the release of metronidazole (Krishnaiah *et al.*, 2002b), 5-fluorouracil (Krishnaiah *et al.*, 2002b; Sinha *et al.*, 2004, p. 101). The minimum amount of the drug was released in simulated gastrointestinal fluids, whereas in the presence of rat caecal content medium, the maximum amount of the drug was released.

Grafted guar gum-based products such as poly(acrylamide) grafted guar gum, when tested for *in vitro* release of 5-amino salicylic acid, revealed that increasing the percentage of grafting decreases the rate of drug release (Sen *et al.*, 2010). Cationic guar gum grafted with polyacrylic acid (Huang *et al.*, 2007) and amphiphilic guar gum grafted with polycaprolactone (Tiwari and Prabaharan, 2010) were synthesized and, when tested for release of ketoprofen, showed more drug release at alkaline pH. Cross-linked products of guar gum, such as glutaraldehyde cross-linked hydrogels for release of methotrexate (Rubinstein *et al.*, 1997; Chaurasia *et al.*, 2006) and phosphated cross-linked low-swelling hydrogels for the release of hydrocortisone (Gliko-Kabir *et al.*, 2000a, b), were substantiated for colon-targeted drug release and also showed positive results, retaining their degradation properties even after cross-linking. Introduction of guar gum has improved the thermal sensitivity of poly(*n*-isopropylacrylamide) hydrogel (Li *et al.*, 2008) and polyacrylic hydrogels (Li *et al.*, 2006), which are thermo-responsive gels with interpenetrating polymer networks.

Studies on the release of ketoprofen, nifedipine, diltiazem hydrochloride (Friend *et al.*, 1997; Altaf *et al.*, 1998; Al-Saidan *et al.*, 2005) and trimetazidine dihydrochloride (Krishnaiah *et al.*, 2002b) revealed that the dissolution rate of hydrophilic drugs from guar gum formulations is almost independent of stir speed under normal dissolution conditions. This proves guar gum formulations were stable under stressed conditions and establish it as a potential candidate for antihypertensive drug delivery. Cross-linked products of guar gum like glutaraldehyde cross-linked poly(acrylamide)-graft-guar gum microspheres (Soppimath *et al.*, 2001; Soppirnath and Aminabhavi, 2002) and micro gels of poly(acrylamide)-graft-guar gum (Toti and Aminabhavi, 2004), when tested for release of antihypertensive drugs such as verapamil hydrochloride and nifedipine, were found to be responsive to external stimuli like pH and the ionic strength of surrounding media. Specificity of guar gum-based products towards only alkaline pH was confirmed by the controlled delivery of protein drugs loaded in alginate-guar gum hydrogel cross-linked with glutaraldehyde (George and Abraham, 2007), which otherwise could be degraded in the acidic pH environment of the stomach. Due to the good film forming and slow release ability of the guar gum-based derivatives, carboxymethyl guar gum (Murthy *et al.*, 2004) and acryloyl grafted guar gum (Thakur *et al.*, 2009) showed potential to deliver the drug across the skin barriers in a controlled manner. Gelatin grafted carboxymethyl guar gum mixed with curcumin showed controlled release of curcumin in *in vitro* cell line studies and the antimicrobial studies displayed growth inhibition against gram positive and gram negative organisms at a considerable level (Manna *et al.*, 2015). Recently, microparticles based on guar gum and succinic anhydride, cross-linked with sodium trimetaphosphate, were prepared and evaluated as a carrier for drug delivery (Seeli and Prabaharan, 2016). Because of its pH-dependent drug release profile and no apparent cytotoxicity, the synthesized material was found to be an excellent carrier for colon-specific drug delivery.

Guar gum works as water-soluble fibre, which makes it suitable as a bulk-forming laxative, promoting regular bowel movements. It also relieves constipation and related chronic functional bowel ailments, like diverticulosis, Crohn's disease, colitis and irritable bowel syndrome (Takahashi *et al.*, 2009). Due to its gel-forming properties, guar gum has cholesterol- and glucose-lowering effects. Due to the gel-forming capacity of guar gum soluble fibre, an increased satiation is achieved because of slow gastric emptying. Diet with guar gum added decreased the appetite, hunger and desire for eating (Butt *et al.*, 2007), which helps in weight loss and obesity prevention (Heini *et al.*, 1998; Kovacs *et al.*, 2001). Guar gum and its derivatives are also useful against other diseases, like cholera and diarrhoea (Patel *et al.*, 2014). The mechanism through which guar gum lowers cholesterol involves increasing the excretion of bile acids in faeces and decreasing enterohepatic bile acid. This enhances the production of bile acids from cholesterol and thus hepatic-free cholesterol concentration is reduced (Rideout *et al.*, 2008). Guar gum and its derivatives are well established in controlling blood sugar level and hence suppressing diabetes. Studies on the anti-diabetic evaluation of guar gum (Jenkins *et al.*, 1977; Saeed *et al.*, 2012), partially hydrolyzed guar gum (Dall'Alba *et al.*, 2013), guar gum hydrolysate (Suzuki and Hara, 2004) and modified partially depolymerized guar meal (Gatenby *et al.*, 1996) have revealed that guar gum reduced the postprandial rise in blood glucose and insulin concentrations, thereby controlling diabetes. Partially hydrolyzed guar gum (PHGG) was found to be effective against cholera in adults (Alam *et al.*, 2008) and diarrhoea (Alam *et al.*, 2000). Guar gum derivatives improve gastrointestinal function (Lampe *et al.*, 1992) and hence prevent the risk of colorectal cancer by controlling functional constipation (Sakata and Shimbo, 2006; Belo *et al.*, 2008). PHGG was also found to be useful for treatment of irritable bowel syndrome (Parisi *et al.*, 2005). Kuo *et al.* tested partially hydrolyzed guar gum for prevention of $FeCl_3$-induced arterial thrombosis and hyperlipidemia in a high-fat diet fed to hamsters, with encouraging results (Kuo *et al.*, 2009). Another study indicated that granulated guar gum was found to be effective in long-term treatment of hypercholesterolemia. Furthermore sulfonated guar gum was reported to reduce cholesterol and fibrinogen effectively (Zhu *et al.*, 2008). Glycosylated guar gum and its sulfated derivatives were reported to possess chemo-preventive and anti-inflammatory properties. These derivatives also showed potency in preventing cancer by inhibiting carcinogen-activating enzymes and promoting the carcinogen detoxification enzyme glutathione-*S*-transferase (Eldeen *et al.*, 2006).

Guar gum contains soluble dietary fibres, which form a gel-like material when dissolved in water. Guar seed possesses 52–58% total dietary fibre (TDF) of the seed's dry weight, out of which the water-soluble endosperm contributes 26–32% soluble dietary fibre (SDF) (Kays *et al.*, 2006). When raw guar gum is consumed in large quantities, the dietary fibres dissolve in the digestive juices; this impairs nutrient absorption and decreases protein and lipid utilization.

The resulting problems can be easily overcome by using partially hydrolyzed guar gum. It is now being used commercially in various beverages, food products and medicinal foods as a functional dietary fibre all over the world with the trade name 'Sunfibre' (Gupta and Verma, 2014). Further, when PHGG was compared with gamma radiation depolymerized guar gum (RDGG) for their suitability as dietary fibre during model intestinal fermentation, better activities in *in vitro* assays were demonstrated by RDGG (Gupta *et al.*, 2015).

Gum arabic

Gum arabic (GA) was once extensively used in the pharmaceutical industry, but has now been replaced by celluloses and modified starches in many applications. It is still used as a suspending agent, emulsifier, adhesive and binder in tableting and in demulcent syrups (Whistler, 1993). In cosmetics, GA functions as a stabilizer in lotions and protective creams, where it increases viscosity, imparts spreading properties and provides a protective coating and a smooth feel. It is used as an adhesive agent in blusher and as a foam stabilizer in liquid soaps (Whistler, 1993). In the pharmaceutical industry, GA finds applications in controlling viscosity, improving density, keeping insoluble drugs in suspension and avoiding the precipitation of solid particles in solution (Goswami and Naik, 2014). Further, GA also finds applications as an emulsifying agent, binder in tablets, masking material for acrid testing material and demulcent for cough syrups. Uses of GA are also reported in preparing kaolin, calamine, cod liver oil emulsion, vitamin premixes, antiseptic preparation and food supplements (Saha *et al.*, 2018). The gum is sold in capsule form in some developed countries at health food stores and pharmacies (Jales, 2018).

In the UK and France, GA is approved as dietary fibre and is used as dietary fibre alone or in combination with food (Phillips *et al.*, 2008; Cherbut *et al.*, 2009; Phillips and Phillips, 2011). The product's compatibility with the diet for patients with diabetes mellitus helped reduce systolic blood pressure, which may lead to improved cardiovascular function and avoiding renal diseases.

Gelatin is widely used in biomedical applications as it is biodegradable and for its non-antigenicity, non-immunogenicity and biocompatibility. Scaffolds play a significant role in cell culture (Hutmacher, 2000). Biopolymer scaffolds are non-toxic and versatile (Peng *et al.*, 2016). Sarika *et al.* (2014) developed a cross-linked gelatin scaffold using GA for application in cell culture. The developed product was non-adherent and non-cytotoxic, the desired properties for spheroid cell culture.

Lamsen *et al.* (2020) encapsulated vitamin D3 with GA to improve its stability. The encapsulated vitamin D3 exhibited desired dispersion stability in the pH range of 2.0–7.4, indicating its potential for application in beverages. Tsai *et al.* (2014) prepared new solutions for polyelectrolyte complex (PEC) using chitosan/pectin/GA and membranes for biomedical applications. The use of PEC membranes for carriers of insulin led to steady and complete drug release for 6 hour, thus indicating its use in controlled drug release applications.

Skin is an essential organ in mammals and plays a role in protection from external injuries, self-healing and sensory activities (Zhong *et al.*, 2010). Different skin injuries such as burns and wounds cause skin loss, and its recovery takes time. Different scaffolds (sponges, films, mico/nanofibres and gel) have been prepared from synthetic and natural polymers. Among natural polymers, polysaccharides are recommended as tissue engineering scaffolds (Hadavi *et al.*, 2017). To simulate the skin's natural extracellular matrix, a scaffold was prepared using GA, corn protein (Zein) and polycaprolactone (PCL). The prepared GA/Zein/PCL scaffolds were found to have a porous structure. They exhibited high hydrophilic property, desirable porosity (~ 80%) and tensile strength of 1.36–3 MPa with an elongation of 19.13–44.06% needed for skin tissue engineering. Further, the scaffold containing GA possessed antibacterial properties, making it suitable for skin tissue engineering (Pedram Rad *et al.*, 2018).

Gum karaya

Most gum karaya is consumed in the pharmaceutical industry, where it is used in diverse applications. It functions as an adhesive in leak-proof sealing rings for post-surgical drainage pouches or ostomy bags (Boiteau and Blondeel, 1981). Coarse gum particles are very effective as bulk laxative as they absorb water and swell to 60–100 times their original volume. They are neither digested nor absorbed in the human gastrointestinal tract. Powdered gum karaya is widely applied on dental plates as an adhesive (Steinhardt and Goldwater, 1962). When brought in contact with the moist surfaces of the mouth, the gum does not dissolve but swells and provides a more comfortable and tighter fit of the plate. Furthermore, it is very resistant to bacterial and enzymatic degradation. In tampons, gum coatings form a gelatinous medium in contact with body fluids, preventing irritation of the mucous membranes and facilitating removal after use (Boiteau and Blondeel, 1981).

The unique properties of gum karaya can be further modified for wider application. A number of research works have been carried out to modify its physical and chemical properties by adding natural or synthetic polymers. It also fulfils the raw material requirement of several industries making products used in medical application. This section summarizes recent research and development work related to medical application of gum karaya.

Waters *et al.* (1987) reported the use of gum karaya in increasing solubility of nimodipine, which is used in treatment of vascular hemierania. The gum also finds application as a carrier of less soluble drugs by improving their solubility and regulated release in the human body (Kumar *et al.*, 2011; Chavan *et al.*, 2017). Deacetylation of gum karaya using hydroxides reduces its viscosity, which helps in increasing drug solubility and rate of diffusion (Babu *et al.*, 2002a, b; Prajapati *et al.*, 2013; Postulkova *et al.*, 2017). The developed beads had good buoyancy, mucoadhesion and a slow rate of drug release.

Freitas *et al.* (2020) investigated natural polysaccharides as carriers for administration of oral insulin. The developed formulation preserved approximately 64% of insulin dose in simulated gastric medium. Bera *et al.* (2015) developed floating mucoadhesive alginate beads entrapped in oil. The beads were coated with cross-linked alginate–*Sterculia* gum gel membrane for delivery of risperidone. The beads had 83.73 ± 0.81% drug entrapment efficiency and 70.84 ± 0.27% cumulative drug release after 8 h. Singh and Sharma (2011) synthesized hydrogels by functionalization of gum karaya with polyvinylpyrrolidone (PVP) for release of ornidazole used as the model antidiarrhoeal drug. Singh and Sharma (2008) developed novel hydrogels using gum karaya for anti-ulcer drug delivery. Singh and Vashishtha (2008) prepared gum karaya-based hydrogels using 2-hydroxyethylmethacrylate (HEMA) and acrylic acid by radiation-induced cross-linking polymerization for use in delivery of drugs.

Singh and Singh (2018) carried out radiation-induced graft-copolymerization of gum karaya polysaccharide in developing the hydrogels for drug delivery formulations. Gamma-rays-induced graft-copolymerization of gum karaya polysaccharide was carried out in the development of the hydrogels for drug delivery formulation (Singh *et al.*, 2011).

In floating drug delivery systems (FDDS), the density of drug formulation needs to be lower than the density of gastric juice to improve residence time and drug availability in the upper gastrointestinal tract. Such a system is useful in drug administration at specific sites for gastric diseases. Gum karaya is suitable for developing an FDDS system as it has solubility in an acidic environment, ability to swell and low density. Further, it is a biodegradable and biocompatible material. Modified gum karaya with a degree of deacetylation of 64% exhibited the desired density, swelling index, viscosity and solubility (Sarathchandiran and Kumar, 2013). Singh *et al.* (2010) synthesized an FDDS, using gum karaya and alginate for anti-ulcer drug delivery. They used $CaCl_2$ as a cross-linker for the synthesis of the FDDS.

Gum karaya-based microsphere formulation for site-specific drug delivery for treatment of ulcerative colitis was developed by both ionic gelation and emulsion-ionic gelation methods (Sirisha *et al.*, 2017). Gelatin/gum karaya microspheres were synthesized for controlled release of distigmine bromide following the emulsion cross-linking method. Results illustrated that the microspheres were sensitive to the pH of test mediums, indicating their possible use for intestinal drug delivery (Reddy *et al.*, 2019).

Metformin HCl microcapsules prepared by coating with alginate and gum karaya were found to be suitable for controlled release of Metformin HCl for a long time period (Kumar *et al.*, 2011). Gum karaya also exhibited good potential use as wall material for ellagic acid microencapsulation (Luján-Medina *et al.*, 2015). Further, gum karaya was used for coating ornidazole for selective release in treatment of diarrhoea (Singh and Sharma, 2009; Singh *et al.*, 2010; Upadhyay, 2017).

Coarse gum karaya particles/granules absorb water, swelling to 60–100 times their original volume, and prove to be an excellent bulk laxative. It is neither digested nor absorbed in the human gastrointestinal tract. Due to these properties, gum karaya finds use in oral suspension and super-disintegrant in medical applications (Munday and Cox, 2000).

Gum karaya powder is also used on dental plates (Steinhardt and Goldwater, 1962) as a denture fixative. Fine powder of gum karaya is dusted on the denture plate for comfortable and tight fit. Mild alkali blended with gum karaya powder exhibited improved adhesiveness of the denture powder (Jayaramudu *et al.*, 2013). It was found to be resistant to enzymatic and bacterial degradation. The gelatinous medium formed in contact with body fluids prevents irritation of the mucous membrane and facilitates easy after-use removal of tampons (Boiteau and Blondeel, 1981).

Padil and Černík (2013) synthesized copper oxide-based nanoparticles using gum karaya as a capping and reducing agent, which showed antimicrobial activity against *Staphylococcus aureus* and *Escherichia coli*. The synthesized nanoparticles were found to possess non-metagenic and non-teratogenic properties. They synthesized nanoparticles by polyelectrolyte complexation using polycationic chitosan and polyanionic polysaccharide obtained from *Sterculia striata* gum. Magalhães *et al.* (2016) prepared and studied *Sterculia striata* polysaccharides and chitosan-based nanocomplex for use as a drug release device for chloroquine (Magalhães *et al.*, 2016).

Singh and Pal (2008) modified gum karaya and developed a wound dressing for antimicrobial agent (tetracycline hydrochloride) delivery. It was found in the study that the polymer absorbed 8.3 ± 0.1 g of simulated wound fluid and released 0.820 ± 0.6 mg of the drug to the simulated fluid.

Gum karaya is also used as an adhesive in leakproof sealing rings for post-surgical drainage and ostomy bags (Marsan, 1967; Sanderson, 1996). Sparberg *et al.* (1966) used gum karaya in a solid state ring used in permanent and disposable ileostomy devices. The ring acts as an adherent and healing agent. Use of a disposable device soon after surgery prevented post-operative skin problems and patients became accustomed to the device immediately. The gum karaya ring is suitable for use as a permanent device instead of adhesive in patients with adhesive or solvent allergy.

From the above research work carried out on medical application of gum karaya, it has been established that the gum has been used extensively and has further potential. As this gum works also as a laxative, there is scope for developing gum karaya-based formulations in the form of syrup, capsules or tablets for treating constipation in both human beings and animals.

Gum ghatti

Medicinal uses of gum ghatti (GG) have been reported since ancient times in Ayurvedic and Unani therapies, and its unique properties also make it valuable for use in different food items. Table 6.1 summarizes the traditional medicinal uses of GG.

GG has excellent potential in the delivery of drugs, suspending agents, and disintegrants in pharmaceutical applications. It is also used as

Table 6.1. Traditional uses of *Anogeissus latifolia* gum. Author's own table.

Indigenous and remedial uses	References
Used with cup of water or milk for lactation	(Jagtap *et al.*, 2009)
Used to control dysentery	(Kala, 2016)
Used as tonic (postnatal) for women after child birth	(Dubey *et al.*, 2012)

a bulk laxative. In the modern pharmaceutical industry, it is frequently utilized as an emulsifier and may be a better emulsifying agent than GA (Ido *et al.*, 2008). Studies of different natural gums and their derivatives for various medical formulations indicated that GG could be applied as a potential vehicle for pharmaceuticals as a tablet binding agent. Due to its highly viscous nature in an aqueous solution, GG may be a superior stabilizer in the formulation of thick drug emulsions and suspensions (Lin *et al.*, 2018).

Modified GG finds extended applications in different pharmaceutical formulations. Researchers have carried out work to alter individually the physical and chemical properties of GG for a broader medical application in praxis. Some recent work carried out in the field of various medical applications of GG is presented below in Table 6.2.

Gum tragacanth

Gum tragacanth (GT) is an effective suspending agent and prevents insoluble materials, which are usually the active ingredients, from settling out in aqueous mixtures. Oil-soluble active components are solubilized in the dispersed oil phase of oil-in-water emulsions. By stabilizing the emulsion and keeping the oil droplets uniformly distributed throughout the product, the gum facilitates the absorption of the water-insoluble components, thus improving the action of the pharmaceutical (Whistler, 1993). GT is used as a stabilizer in dermatological creams and lotions, also contributing a protective coating and smooth handfeel (Leupold *et al.*, 1962). Combined with glycerin, it acts as a binding agent in the production of tablets (Weiping and Branwell, 2000). It is used as a basis for jelly lubricants and as a suspending agent in jellies and toothpastes, providing spreadability and a shiny, creamy appearance (Grossmith, 1956; Nebergall, 1956). In the glazing process of ceramic materials, the gum is used to produce glazes of increased hardness and resistance to handling (Lloyd and Read, 1964).

GT hydrogels coupled with organic monomers and cross-linking agents have been investigated for drug delivery applications because of their biocompatibility. GT-g-poly(acrylic acid) copolymer hydrogel synthesized using N,N′-methylenebisacrylamide as cross-linking agent has been experimentally investigated for release of the antibiotic amoxicillin to the colon (Singh and Sharma, 2014). Controlled release strategy is utilized in cancer treatment to improve the delivery of anticancer therapeutics. GT-based carriers have been adopted as drug delivery vehicles for their specific properties and application (Deb and Vimala, 2018).

Nano-carriers are excellent systems for improving drug loading and release. Nano-carrier-based GT was prepared by using GT or modified GT and an enhancing agent. Lecithin-GT mucoadhesive hybrid nanocomposite was used as a nano-carrier for enhancing amphotericin B oral bioavailability (Jabri *et al.*, 2018). Others have developed GT-based drug carriers for the release of various drugs (diclofenac, methotrexate, quetiapine, fluorouracil, silymarin, naringin, ibuprofen, theophylline and loperamide) (Ahmadi *et al.*, 2018; Shahiwala and Zarar, 2018).

GT, in the form of cream, hydrogels and composites, has been studied for expediting wound healing. Creams were prepared from GT (6 and 9 wt%) using a mixture of water and glycerin (4:1 wt%) as the vehicle for wound healing in rabbits. The wound-healing profile of the group treated with the cream containing GT was significantly better compared with the untreated control. The best healing effect was observed with the cream containing 6% GT (Moghbel *et al.*, 2005).

Apart from GT blends and creams, composites of GT with other polymers such as PCL have

Table 6.2. Medical applications of GG. Author's own table.

Specific medical applications of GG	Reference
Stabilizer in syrup formulation	(Deshmukh *et al.*, 2012)
Stabilizer and emulsifier in pharmaceutical emulsion and suspension of powdered and oil-soluble vitamins	(Cushman and Schick, 1975)
Binder in tablets	(Jain and Dixit, 1988) (Ravi *et al.*, 2013)
Release modifier in mucoadhesive drug delivery application	(Joshi *et al.*, 2010) (Puri *et al.*, 2021)
GG and polyvinyl alcohol-based biodegradable nanofibre mats for application in wound healing	(Dey *et al.*, 2020)
GG as a binder in making porous hydroxyapatite scaffolds for use in making spongy bone structures in bone tissue engineering	(Lett *et al.*, 2020)
GG and chitosan-based polyelectrolyte nanoparticles for antibacterial application	(Shelly *et al.*, 2013)
Drug delivery	
GG-based interpenetrating network of electrically active hydrogel for colon-specific drug delivery	(Sharma *et al.*, 2015)
Sodium carboxymethyl GG-based microparticulate system for the sustainable release of drug ropinirole hydrochloride	(Ray *et al.*, 2017)
Alginate-GG modified montmorillonite composite matrices for coating alginate-GA gel membrane for intragastric delivery of flurbiprofen (FLU) for an extended period	(Bera *et al.*, 2017)
Programmable GG-based hydrogel matrices for embedding anticancer therapeutic for controlled release in the harsh environment of the stomach	(Pal *et al.*, 2019)
Gellan gum and GG in making beads of polymeric polyelectrolyte complex synthesized from chitosan and gellan gum for sustainable delivery of diclofenac sodium in the intestine	(Verma *et al.*, 2021)
The GG-based pH-sensitive polymeric delivery system for antidiabetic therapy	(Bhosale *et al.*, 2021)
GG-hydroxyapatite matrix for drug delivery	(Padmanabhan *et al.*, 2021)

been studied. Nanofibres of curcumin-loaded PCL/GT were studied for their role in the healing of skin wounds in diabetic rats. Wounds covered with the nanofibres experienced faster wound closure with well-formed granulation tissues (Ranjbar-mohammadi and Bahrami, 2016).

Polysaccharide gum-biobased materials have attracted considerable attention in bone tissue engineering (Maia *et al.*, 2018). Adipose-derived mesenchymal stem cells were cultured on uncoated tissue culture plates and plates coated with collagen-based hydrogels and GT hydrogels (Haeri *et al.*, 2016). The result indicated that GT hydrogel may be used in orthopedic applications for enhancing the adhesion, proliferation and osteogenic differentiation of stem cells. Polymeric GT alginate beads were used to enhance the biological properties of calcium alginate, a common cell encapsulation system for bone tissue engineering (Kulanthaivel *et al.*, 2017).

GT is a suitable medium for enzyme immobilization and cell growth. Examination of the kinetic behaviour of immobilized glucose oxidase on GT carrier indicated that the optimum conditions for effective encapsulation of glucose oxidase were 1% w/v GT, 10% w/v $FeCl_3$, 1% w/v $AlCl_3$, and 2 hour of gelation time at 50°C. The kinetic parameters for both free and immobilized glucose oxidase were 9.00 and 12.375 mmol, respectively (Otadi and Mobayen, 2011). GT has also been used as a matrix for immobilization of whole cells. For example, GT was used for immobilization of *Escherichia coli* for conversion of benzyl penicillin to 6-aminopenicillanic acid (Otady and Vaziri, 2005). Experimental and computational studies of the interaction between human serum albumin and GT showed no adverse change in protein conformation during binding of the protein to GT. The results indicate that GT may be used as a matrix for immobilization of human proteins (Moradi *et al.*, 2018a, b).

GT has also been applied as a retarding agent in sustained-release drug delivery systems and as a membrane for delivery of therapeutic agents. GT/polylactic acid (PLA) nanofibrous scaffolds have been used experimentally for regeneration of damaged peripheral nerves (Ranjbar-mohammadi and Bahrami, 2016). The gum has potential application as an excipient for oral delivery of peptide/protein because of its ability to form polyelectrolyte complexes and hydrogels at a proper pH (Nur *et al.*, 2016). Further, GT is used in some countries as a topical agent to relieve rheumatoid arthritis, toothache and neck pain, or as an orally administered agent for treatment of stomach pain (Bagheri and Dashti-R, 2015).

Gum-resin

Guggul

Commiphora species are known for their medicinal properties and exhibit important biological activities such as anti-inflammatory, antibacterial, antimicrobial, antioxidant, antiarthritic, anticancer, anti-ulcer, hypocholesterolaemic and hypolipidaemic, heptatoprotective, muscle relaxing, antimalarial, antimycobacterial, antischistosomal, larvicidal and molluscidal (El Ashry *et al.*, 2003; Francis *et al.*, 2004; Al-Howiriny *et al.*, 2005). Several pharmacological active components have been identified in the plant, including *E*- and *Z*-guggulsterones and gugulipid, both found in the ethyl acetate extract of the plant (Nagarajan *et al.*, 2001). In 1987, guggulipid, the extract of guggul, was officially recognized in India as a lipid-lowering remedy and it is widely used for this indication (Szapary *et al.*, 2003). Guggulipid, a purified ketonic fraction, is presently used in India and Europe for hyperlidemia and hypocholesterolaemia (Singh *et al.*, 1994).

It has been found that guggulsterone reduces total serum cholesterol (Coon and Ernst, 2003) and stimulates thyroid activity (Tripathi *et al.*, 1984). The thyroid-stimulating activity of Z-guggulsterone facilitates increased iodine uptake and production of thyroid hormones, while it can also modulate cholesterol levels and cause weight loss. The increase of thyroid hormone synthesis resulting from the use of guggul was also associated with increased oxygen uptake in liver and muscle tissue (Tripathi *et al.*, 1988). This ability to increase the rate of metabolism is exploited to promote guggulipid as a product to reduce body weight.

Apart from its lipid-lowering (Satyavati *et al.*, 2013), anti-obesity (Francis *et al.*, 2004) and anti-inflammatory properties (Patil *et al.*, 1972), guggul has a potential role in cancer prevention (Cui *et al.*, 2003; Shishodia and Aggarwal, 2004; Ramawat *et al.*, 2008). Further, guggul is well known for its antidyslipidemic properties (Ulbricht *et al.*, 2005). The oleoresin of *Commiphora wightii* increases the number of leucocytes in the blood and stimulates phagocytosis, thus supporting the immune system in fighting diseases.

Guggul is also often used in toothpastes for its astringent activity on the mucous membrane of the mouth and throat. Sometimes a mouthwash is prepared with a combination of guggul, honey and rectified spirit dissolved in rose petal infusion for the treatment of stomatitis (inflammation of the mouth).

Frankincense

Olibanum is produced mainly by four *Boswellia* species from different regions, which are *Boswellia serrata* from India, *Boswellia carterii* from East Africa and China, *Boswellia frereana* from North-east Africa (Somalia) and *Boswellia sacra* from the Middle East (Chevrier *et al.*, 2005; Mathe *et al.*, 2007). The composition of the essential oil and other contents changes from species to species and differs depending on the climate, harvest conditions and geographic location (Ammon, 2006; Al-Harrasi and Al-Saidi,

2008). Frankincense is reported to contain 60–85% resins (mixtures of terpenes), 6–30% gums (mixture of polysaccharides) and 5–9% essential oil (Rijkers *et al.*, 2006). The resin portion is composed of pentacyclic triterpenes, in which boswellic acid is the active functional group (Goyal *et al.*, 2011). The gum portion consists of pentose and hexose sugars and some oxidizing and digestive enzymes. The essential oil is a mixture of monoterpenes, diterpenes and sesquiterpenes (Goyal *et al.*, 2011). Boswellic acids form the main active component of frankincense (Al-Harrasi and Al-Saidi, 2008). The four major boswellic acids (pentacyclic triterpenic acids) found in frankincense are: β-boswellic acid (BA), acetyl-β-boswellic acid (ABA), 11-keto-β-boswellic acid (KBA) and 3-O-acetyl-11-keto-β-boswellic acid (AKBA), which have been shown to be responsible for the inhibition of pro-inflammatory enzymes (Siddiqul, 2011). Among these four boswellic acids, AKBA is the most important inhibitor of an enzyme called 5-lipoxygenase, which is responsible for inflammation (Siddiqul, 2011). AKBA has shown to be effective against a large number of inflammatory diseases such as arthritis, bronchial asthma, chronic colitis, ulcerative colitis, Crohn's disease and cancer (Raja *et al.*, 2011). The mechanism of action is the binding of AKBA to 5-lipoxygenase in a calcium-dependent and reversible manner, and it acts as a non-redox type, non-competitive inhibitor (Qurishi *et al.*, 2010).

Asafoetida

In general, asafoetida consists of around 68% carbohydrates, 16% moisture, 4% protein, 1% fat, 7% minerals and 4% dietary fibre (Mahendra and Bisht, 2012). The gum-resin contains 40–64% resin, about 25% gum and 10–17% essential oil (Takeoka, 2001). The resin fraction contains ferulic acid and its esters, coumarins, sesquiterpene coumarins and other terpenoids. The gum includes glucose, galactose, 1-arabinose, rhamnose, glucuronic acid, polysaccharides and glycoproteins, and the volatile fraction contains sulfur-containing compounds, monoterpenes and other volatile terpenoids (Iranshahy and Iranshahi, 2011). Sulfur compounds in *Ferula asafoetida* resin show various biological activities and can be valuable in medicine (Iranshahi *et al.*, 2006). Three major sulfur constituents that have been identified include 2-butyl 1-propenyl disulfide, 1-(methyl thio) propyl 1-propenyl disulfide and 2-butyl 3-(methyl thio)−2-propenyl disulfide (Takeoka, 2001).

Asafoetida is traditionally used for the treatment of different conditions, such as whooping cough, asthma, ulcer, epilepsy, stomach ache, flatulence, bronchitis, intestinal parasites, weak digestion and influenza and is also used as an antispasmodic (Takeoka, 2001; Lee *et al.*, 2009; Al-Jafari *et al.*, 2012; Mahendra and Bisht, 2012). Asafoetida is an effective remedy for several diseases of the stomach. The digestive stimulant actions of asafoetida are the most commonly experienced beneficial physiological effect which works through enhanced secretion of saliva and activity of salivary amylase. It plays an important role in the digestion of dietary lipids by stimulating bile flow. It enhances bile acid secretion and also enhances the activities of digestive enzymes of the pancreas and small intestine. Moreover, it is used to address low acid levels in the stomach, stomach pressure, flatulence and loose stools. It is specially useful for women's ailments, being used in the treatment of problems such as unwanted abortion, sterility, difficult and excessive menstruation and leucorrhoea. Recent pharmacological and biological studies have also shown several activities, such as antioxidant (Dehpour *et al.*, 2009; Kavoosi and Rowshan, 2013), antimicrobial (Shrivastava *et al.*, 2012; Divya *et al.*, 2014; Padhy *et al.*, 2014; Patil *et al.*, 2015; Bhatnager *et al.*, 2015), antiviral (Mahendra and Bisht, 2012), antifungal (Kamble and Patil, 2008; Angelini *et al.*, 2009; Rani *et al.*, 2009; Mostafa *et al.*, 2013), cancer chemopreventive (Saleem *et al.*, 2001), antidiabetic (Abu-Zaiton, 2010), anticarcinogenic (Saleem *et al.*, 2001; Mallikarjuna *et al.*, 2003), antispasmodic and hypotensive (Fatehi *et al.*, 2004), relaxant (Bagheri *et al.*, 2014; Khazdair and Boskabady, 2015), neuroprotective (Tayeboon *et al.*, 2013; Moghaddam and Farhadi, 2015) and molluscicidal properties (Kumar and Singh, 2006).

Myrrh

Terpenoids, steroids, flavonoids, lignans, carbohydrates and long-chain aliphatic alcohol derivatives have been isolated from different *Commiphora* species (Su *et al.*, 2011). Myrrh contains 23–40% resin, 40–60% gum and 2–8% essential oil. The resin part of myrrh contains α-, β-, and γ-commiphoric acid, commiphorinic acid, heeraboresene, α- and β-heerabomyrrhols, commiferin, kertosteroids, compesterol, β-sitosterol, cholestrerol, α-amyrone and 3-epi-α-amyrin (Lemenith and Teketay, 2005). Furthermore, the gum consists of 18% protein and 64% carbohydrate. The carbohydrate fraction of the gum has a similar chemical structure to acacia gum (El Ashry *et al.*, 2003). Hydrolysis of the gum results in arabinose, galactose, xylose, and 4-O-mythylglucuronic acid (Lemenith and Teketay, 2005). Its essential oil has a pale yellow colour and contains about 20 different furanosesquiterpenoids (Su *et al.*, 2012), which are mainly responsible for its characteristic odour (Abdul-Ghani *et al.*, 2009). Studies have proved myrrh's antibacterial and antifungal effects against *Staphylococcus aureus*, *Escherichia coli*, *Pseudomonas aeruginosa* and *Candida albicans* (Dolara *et al.*, 2000).

Cosmetics and Perfumery Industries

Various natural resins and gums, known for their aromatic, adhesive, and stabilizing properties, are widely used in the cosmetics and perfumery industries. Some of these specific applications are listed below.

Benzoin, myrrh, and frankincense are natural fixatives that help stabilize and preserve fragrance compositions. These natural raw materials extend the longevity of fragrances by slowing down the evaporation of volatile essential oils, ensuring that perfumes last longer on the skin.

Furthermore, they provide the individual characteristic deep, earthy and warm notes in perfumes. Used as base notes in perfumes, resins like labdanum and opopanax add richness and depth, making them especially valuable in oriental and woody fragrances. They are also burned as incense for their fragrant smoke, widely used in aromatherapy.

Resins improve texture and adhesion in makeup and skincare products. In mascaras, eyeliners and lipsticks, resin compounds enhance the texture and improve adhesion, helping products stay on longer. They also contribute to glossiness in lip products.

Certain resins have protective, soothing and conditioning properties. Gum-resins like myrrh and frankincense, used in creams and balms, help protect the skin, reduce inflammation and have a calming effect. Some resins possess natural antimicrobial and antioxidant properties. Resins are often used to enhance the shelf life of natural cosmetics by inhibiting microbial growth.

Guar gum

Uses of guar gum and its derivatives in the cosmetics industry have been growing for past few decades. The properties of guar gum that contribute to its uses in cosmetics are cold-water solubility, thickening, solvent resistance, film forming, a wide pH range resistance, stability, non-toxic nature, safety and low price. A significant amount of guar gum is used in the manufacture of toothpaste, shaving foam and other cream formulations, where it ensures that the product is free-flowing. In emulsion systems like cream and lotions, guar gum offers phase separation, prevents syneresis, increases emulsion stability, prevents water loss and acts as a protective colloid. It also imparts spreadability and pleasant after-feel in lotions. In cosmetics to be stored in cold conditions, guar gum stabilizes the emulsion during the freeze-thaw cycle and prevents water phase from condensing out of the system (Chudzikowski, 1971; Srichamroen, 2007). Guar gum and its derivatives are extensively used in shampoos and cleansing preparations, where they act as inert and compatible fillers which also impart thickness and a stabilizing effect. The gum is used in aerosol dispensing aqueous liquid preparation where it causes reduction of migration of fog. Guar gum is used in hair colourants as a thickener and in dry face mask preparations as an emulsifying agent (Tripathy and Das, 2013).

Gum arabic

GA finds application in different cosmetics due to its unique properties. It is used as an emulsifier and stabilizer in lotion, as a stabilizer for foam in liquid soap, as a thickener in mascara and lipstick, as film-making material in peel-off masks and as a dental adhesive (Dauqan and Abdullah, 2013). It is also used as an adhesive in blusher (Viinanen *et al.*, 2011).

Gum tragacanth

One of the most important cosmetic applications of GT is its use as a natural product in hair gels to set hair into a particular style. These biobased hair gels are known as katira gels, the name of which originates from the local name of tragacanth in Iran. In addition, GT has been used extensively as an emulsifying and suspending agent in a variety of formulations such as creams and emulsions (Samavati *et al.*, 2014). When GT is mixed with water and glycerol, it produces a useful excipient for binding of pill masses in cosmetics formulations (Phillips and Williams, 2009).

Myrrh

Myrrh possesses antimicrobial and antifungal characteristics that have mainly been attributed to its essential oil (Alves *et al.*, 2015). The use of myrrh for its antibacterial properties goes back to the Sumerians in 1100 BC, who used myrrh to treat infected teeth. In addition, it has been used for treatment of skin candidiasis (De Rapper *et al.*, 2012). In ancient Egypt, myrrh was also used as an essential ingredient in mummification. The Egyptians stuffed the corpses with crushed myrrh so that the resin could develop its antiseptic properties during the mummification process. (Abdel-Maksoud and El-Amin, 2011).

Nutrition and Food Industry

The food industry faces many challenging scenarios in meeting consumers' desires. The industry wants to use ingredients that are natural and that fulfil their technological roles in foods. Among these ingredients, resins, gums and hydrocolloids are the compounds widely used as innovative agents in the food industry. Resins and gums are very versatile biopolymers, extensively used in the food sector as ingredients or additives, and fulfil several technological and, sometimes, nutritional functions. This versatility is intrinsically related to their molecular composition, which gives these polysaccharides certain properties, such as gelling, thickening, moisture retention, emulsification and stabilization. In the food industry, they are widely used in confectionery, as ice cream stabilizers, as food emulsions, in the microencapsulation of flavours and dyes, as clarifiers and as beverage stabilizers. This section of the chapter aims to discuss the knowledge of applications of natural gums and resins in the food industry.

Resins

Natural resins are used in the food industry for their unique properties, particularly in coatings, encapsulation, preservation and as glazing agents. Here are some of their main applications.

Shellac

Shellac has a distinct structure and properties which make it suitable for wide use in the food industry. The various possible applications of shellac include its use as a raw material for fabricating food waxes, food coatings and biodegradable films. It is also applied as a food foaming agent, oil-gelling agent and food emulsifier. Further, shellac is employed in the preparation of food delivery systems, such as microcapsules, coated carriers, nanofibre films, nanoparticles and microparticles.

In the food industry, shellac has long been used as a wax to enhance the postharvest preservation of fruits and vegetables. The use of protective wax on some fruits and vegetables is an effective postharvest technique that maintains their quality and extends their shelf life. In principle, the preservative effects of these waxes on fruits and vegetables are achieved by reducing the gas transmission rate, protecting against external mechanical damage and shielding them from microorganism attack. In this context, shellac is applied during the production of commercial food waxes, or in combination with other waxes to maximize their preservative potential

(Jo *et al.*, 2014). Studies on the effects of shellac wax applications have been carried out on fresh apples, revealing improved colour and reduced weight loss (Drake and Nelson, 1990). Similarly, Jo *et al.* (2014) found that a shellac-based wax covering on Fuji apples improved appearance and sensory values, prevented physical damage, reduced weight loss (from 7.22% to 4.82%) and maintained hardness. Moreover, it inhibited microorganism growth (from 2.67 log CFU/g to 1.30 log CFU/g for total aerobic bacteria and from 2.31 log CFU/g to 1.58 log CFU/g for yeast and moulds), thereby extending the shelf life of the apples (Jo *et al.*, 2014). A mixture of shellac and wax, used as a fruit preservative, also showed a preservation-extending effect on mangoes by reducing their respiration rate from about 160 to 80 ppm $kg^{-1}h^{-1}$. In addition, weight loss was reduced from approximately 16% to 8%, subsequently lowering the rate of decay from 85% to around zero (Fu *et al.*, 2013). Similarly, commercial wax (polyethylene–shellac) was found to decrease weight and firmness losses, modify the internal atmosphere and lower the respiratory rate of clementine mandarins (Contreras-Oliva *et al.*, 2012). Another research group found that the addition of antifungal agents to shellac-based wax improved the quality of citrus fruits (Kouassi *et al.*, 2012). During the processing of Spanish-style green olives (Manzanilla olives), the mechanically harvested fruits were stored under nitrogen, coated and then stored again under nitrogen. Commercial wax (mainly shellac) used in coating the olives suppressed the enzymatic hydrolysate oxidation of oleuropein under anoxic conditions, thereby decreasing the development of brown spots on the olives (Ramirez *et al.*, 2015). Although shellac as a food wax has been widely used in the preservation of fruits and vegetables, much work can still be carried out to develop more functional shellac-based food waxes. Previous studies on shellac food waxes mainly focused on its postharvest preservation effects. In fact, the uses of shellac-based waxes for preharvest preservation, reducing pests and diseases and increasing yield, have more obvious significance. The impressive barrier and biocompatibility characteristics of shellac could protect the fruits and vegetables without posing a health threat.

Generally, food coating provides preservative functions by controlling gas and water vapour exchange between the products and the surrounding environment. It also improves the mechanical properties or structural integrity of fresh food, consequently extending its shelf life (Chitravathi *et al.*, 2014). Shellac-based edible coatings effectively improved the quality of foods (such as fruits and vegetables), protecting them from the external environment and consequently extending their shelf life, as mentioned earlier. Nevertheless, it is worth noting that self-polymerization significantly decreases the mechanical properties and stability of pure shellac coatings. Incorporating amine-containing substances such as Jeffamine® (Bar and Bianco-Peled, 2020) can prevent self-polymerization through the amine-carboxyl reaction, improving the properties of shellac coatings. In addition, the incorporation of functional ingredients in the edible coatings can further enhance their preservative effects. The use of shellac-based emulsions loaded with active substances (such as essential oils) to fabricate shellac-based food coatings is expected to easily improve their preservative capacity. In addition, the current application of shellac as a food coating is mainly limited to the preservation of fruits and vegetables. However, the fact that other waxes (such as carnauba wax) have been widely used as a food coating in confectionaries (such as soft sweets and chocolate) and instant foods (such as bread and cake) (de Souza *et al.*, 2019) hints at the promising and extensive application for shellac-based food coatings in the future.

Shellac has excellent film-forming and binding properties that, combined with its biocompatible nature, make it an excellent choice as a coating agent (Farag and Leopold, 2011). Coating various fruits and vegetables with water-resistant material has been found to be effective in extending their postharvest shelf life. Because of the increasing awareness about natural products and evidence of health issues connected with synthetic waxes used for coating, shellac or shellac-based coating has been gaining popularity and has been reported in several studies (Du *et al.*, 2019). Chitravathi and co-workers did work on postharvest shelf-life extension of green chillies in 2014. It was found that if they are coated with shellac, starch, ethylenediaminetetraacetic acid (EDTA) and sodium alginate, the storage life of fresh green chillies could be increased to 12 days. It was also suggested that shellac mixed with sodium alginate and applied over chillies, kept at 26 ± 2°C, RH 68 ± 4% ambient temperature, gave the best results (Chitravathi *et al.*, 2014). Shellac was used as a coating material by Musa *et al.* (2018) in a study on the internal quality of chicken eggs. Different amounts of shellac solution were kept on different storage days at 40°C. No significant change occurred in the internal quality of the eggs. Coating mangosteens and limes with shellac-based formulation improved the gloss, reduced weight loss and preserved the fruits 14 and 41 days longer, respectively, when compared to the control (Accaseavorn *et al.*, 2006).

A shellac coating was applied to grapefruit for the biocontrol of *Penicillium digitatum*. The coating of shellac supports the population of the yeast *Candida oleophila*. On using shellac, the yeast population remained between 104 to 105 (CFU)/ml for 4 months, due to which the shelf life of grapefruit improved. Shellac has been used to coat apples for a long time, but according to a study, apple pieces coated with shellac can be kept for 30 days at 6 ± 1°C (Chauhan *et al.*, 2011). The surface coating results in reduced respiration and ethylene synthesis rates as well as reduced electrolyte leakage. Shellac can also be coated on cut slices as it reduces the enzymatic activity of polyphenol oxidase and peroxidase enzymes. Jo *et al.* (2014) used carnauba-shellac wax (CSW) containing lemongrass oil for the quality and microbial safety of fuji apples. Their study indicated no change in the hardness of coated apples that were stored for 5 months. Weight loss was 5.25%, total aerobic bacteria decreased and yeast and moulds were also not found, so a CSW coating is suitable for maintaining quality and extending shelf life. Shellac and aloe vera gel-based surface coating formulations for tomatoes were also developed and evaluated (Chauhan *et al.*, 2015). The coating was found to be effective in checking respiration, thereby reducing ethylene synthesis and delaying senescence. These coatings could extend the shelf life of tomatoes by 8–12 days when stored under ambient storage conditions. The effect of shellac, gelatin and Persian gum on the quality attributes of orange fruit was compared (Khorram *et al.*, 2017). Coatings successfully improved the quality of Valencia oranges as they decreased weight loss, retaining the firmness and flavour of the fruits. The shellac was the most promising coating agent out of different treatments due to its non-sticky, odourless, fast-drying and gloss-imparting properties.

Food packaging films have been widely used to protect food from environmentally induced deterioration. Shellac is a degradable film-forming agent with good barrier properties to achieve food protection. However, the poor mechanical strength of shellac films greatly limits their application. Fortunately, shellac modification or the preparation of shellac-based composite films addresses these shortcomings (Soradech *et al.*, 2013). Alternatively, shellac is applied as a layer on other biodegradable films such as agar and cassava starch films, thereby enhancing the performance of these films (Phan *et al.*, 2008). Generally, composite films maximize the advantages of various components, which can be easily adjusted according to specific applications. In the preparation of shellac composite films, the use of a polyethylene glycol 400 (PEG 400) plasticizer at a 20% w/w concentration effectively protects the active groups of shellac. This protection is provided via hydrogen bonds between the active components of shellac and the hydroxyl groups of PEG, thereby enhancing the properties of shellac films (Luangtana-anan *et al.*, 2010). Soradech *et al.* (2013) found that increasing the gelatin content in shellac–gelatin composite films improved the strength, flexibility and coating efficiency of the films on a hydrophilic substrate, and provided good storage stability. Despite the improved storage stability of the composite films, the mechanical properties of the films deteriorated during storage, as revealed in later studies. Fortunately, the use of 5% and 10% w/w PEG 400 prevents the polymerization of shellac and improves the mechanical properties of composite films. Similarly, PEG 400 and diethyl phthalate plasticizers were used to prepare shellac and pectin composites to overcome the shortcomings of pectin films, which include poor water retention, moisture resistance and mechanical properties (Luangtana-anan *et al.*, 2017). Du *et al.* (2019) fabricated konjac glucomannan–shellac composite films with improved hydrophilicity, good mechanical properties and thermal stability, suggesting possible applications as biodegradable packaging. Correspondingly, composite films prepared by blending shellac with soybean protein isolate showed significantly improved water resistance compared to pure soybean protein isolate films (Zhang *et al.*, 2020a). A new natural composite film was also recently prepared using carboxymethyl cellulose, gelatin and shellac. The composite film showed good uniformity, tensile strength, antibacterial properties, elongation, air permeability and thermal stability, low porosity and an inclination towards particle aggregation (Mohamed *et al.*, 2019). In the presence of two emulsifiers, stearic acid and lauric acid, shellac was successfully doped into hydroxypropyl methylcellulose (HPMC) to form a composite film with enhanced properties (Byun *et al.*, 2012). Similarly, shellac, stearic acid and Tween-20 were also used to improve the properties of pea starch–guar gum composite films (Saberi *et al.*, 2017). In another study, wrapping deacetylated konjac glucomannan film in a shellac/stearic acid emulsion efficiently produced water-resistant konjac glucomannan film with improved water/moisture resistance and mechanical properties. Such films have great potential for use in food preservation (Wei *et al.*, 2015). Recently, Ariyanto *et al.* (2019) developed a novel functional packaging by coating paper with a 1-methylcyclopropene controlled-release system that used shellac to wrap a 1-methylcyclopropene-loaded α-cyclodextrin. This functional packaging could reduce the ethylene production rate of apples and delay softening during storage. Studies have also shown that shellac modification during the preparation of films usually results in better performance. In the process of maintaining or improving its biodegradability, Ghoshal *et al.* (2010) observed that grafting various acrylic monomers on shellac using γ-rays yielded better quality shellac films with longer shelf lives. In recent years, the emergence of active and intelligent packaging systems that can monitor the quality of packaged food have gained wide attention (Bhargava *et al.*, 2020). The incorporation of naturally derived food pigments and food waste-based pigments, such as anthocyanins, curcumin, betalains, carotenoids, chlorophyll, brazilin and quercetin, into shellac-based packaging films will provide additional benefits such as pH sensitivity, time-temperature sensitivity, CO_2 sensitivity and ammonia sensitivity. Notably, this could provide a reference for the development of shellac-based intelligent biodegradable packaging films in the future.

Studies on food foaming agents have reported that micro-rod shellac-based particles have stronger foam-stabilizing capabilities than traditional foaming agents (Campbell *et al.*, 2008, 2009). To expand the application of shellac in the food industry, Campbell *et al.*, 2008 fabricated novel micro-rods containing a food-grade shellac wax and ethyl cellulose composite using shear emulsification in combination with solvent abrasion technology. The edible micro-rods revealed good adsorption and fusion capabilities at gas–liquid interface, which enhanced the stability of gas bubbles in foam. Changes in the shear rate and polymer concentration of the dispersed phase could control the diameter and aspect ratio of the micro-rods. Moreover, the morphology of the ethyl cellulose micro-rods can be transformed, from flat ribbons to long strings through the micro-rods, by changing the polarity of the solvent (Campbell *et al.*, 2008). To further enhance the foam-stabilizing abilities of these micro-rods and explore their potential applications, Campbell *et al.* (2009) incorporated solid particles, including sporopollenin, silica (with a porosity of approximately 95%), hydrated baker's yeast cells and sodium bicarbonate micro-crystals into the micro-rods. The presence of lumpy inclusions in the micro-rods reduced the foam drainage rate, which led to higher foam stability. The attained stability is attributed to the entanglement of the lumpy micro-rods, as compared to smooth rods at the foam film surfaces. The lumpy micro-rods adsorbed at the foam film surface, as well as those accumulated in the Gibbs Plateau borders, obstructed the flow, retarding liquid drainage. Notably, the incorporation of flavourants and functional substances in the shellac micro-rods may expand the applications of shellac-based food foaming agents to include enriching the taste, adjusting the appearance, increasing creativity and extending the storage period of food.

The preparation of oleogels by solidifying liquid edible oils has gradually attracted the interest of food researchers in recent years (Patel and Dewettinck, 2016). Patel *et al.* (2013) found that shellac can potentially be applied in the fabrication of oleogels with different textures and thermal properties by altering their concentrations. In a later study, it was reported that the oleogels resulted from the physical entrapment of liquid oil in crystal networks of shellac. This was achieved by cooling the hot oil-shellac dispersions to room temperature. In addition, the oleogels can be used in the fabrication of emulsifier-free emulsions with temperature-dependent melting behaviour and excellent storage stability. Interestingly, by regulating the water content, the oleogels can be used in the preparation of edible emulsifier-free emulsions, including structured water-in-oil (W/O) emulsions (spreads), oil-binder replacements in chocolate paste formulations and a shortening alternative for cake preparation (Patel *et al.*, 2014). The properties of the shellac oleogels were significantly different from those of HPMC, which are known as high-internal-phase emulsion oleogels. In short, the shellac oleogels exhibited pseudolipid structural properties owing to the reversible melting and crystallization of their wax crystals. The recovery of the shellac oleogels was significantly better than that of HPMC oleogels. Moreover, shellac oleogels showed good resistance to the incorporation of water and could therefore form stable emulsifier-free W/O emulsions, even at a water inclusion level of 60% by weight (Patel and Dewettinck, 2016). These emulsions were formed via the adsorption of shellac crystals at the oil–water interface to stabilize the droplets. Oil-in-water emulsions with pH-responsive switchable behaviour were also prepared using a complex formed by the strong interactions between xanthan gum and shellac, which showed Pickering properties and had a strong stabilizing effect on the emulsion (Patel *et al.*, 2013). Novel biocompatible amphiphilic alginate/shellac particles using the microfluidic method were fabricated to form colloidal surfactants that were able to stabilize both water-in-oil and oil-in-water emulsions (Chen *et al.*, 2017). Similarly, biocompatible amphiphilic dimer particles consisting of a hydrophobic PLA bulb and a hydrophilic shellac-PEG bulb were fabricated through controlled co-precipitation and phase separation. The easily adjustable size and ratio of the dimer particles made it possible to stabilize different types of Pickering emulsions, while the pH-responsive properties of shellac enhanced the characteristics of the entire emulsions (Sun *et al.*, 2020). Oil-gelling agents and emulsifiers have been the attractive applications of shellac in the food industry in recent years. However, the hydrophilicity of shellac needs to be regulated in future studies on shellac-based emulsions.

It is well known that delivery systems can comprehensively regulate the distribution of bioactive components in the body at specific locations, times and doses. They effectively improve the stability and bioavailability of bioactive components, and are thus widely used in the fields of medicine and food (Zhou *et al.*, 2014). Currently, the delivery of bioactive compounds is still the subject of intensive research, and many delivery systems have been proposed, including microcapsules, coating materials, nanofibres and nanoparticles. Oral application remains the simplest and most direct way to ingest active ingredients, attracting attention to food delivery systems. Generally, the materials used in food delivery systems must be safe and edible, limiting the use of irritant and toxic synthetic polymers and drawing increasing attention to natural products (Patel *et al.*, 2013). Shellac is a non-toxic, biodegradable, film-forming, pH-responsive, water-resistant and acid-resistant natural polymer (Limmatvapirat *et al.*, 2007; Tang *et al.*, 2019), which therefore makes it an ideal raw material for food delivery systems.

The application of shellac in microcapsules has shown great potential in the fields of food and drug delivery. Shellac-based microcapsules have been developed using various techniques, including solvent evaporation (Kong *et al.*, 2017), extrusion (Patel *et al.*, 2013; Ben Messaoud *et al.*, 2016), hot-melt extrusion (Gately and Kennedy, 2017; Khor *et al.*, 2017), fluidized bed drying (Silva *et al.*, 2016) and spray drying (Berg *et al.*, 2012; Oidtmann *et al.*, 2012). Many natural food-derived bioactive ingredients in the food industry, such as probiotics, polyphenols, natural pigments and enzymes, are sensitive to processing. They are therefore encapsulated in microcapsules to enhance their stability (Silva *et al.*, 2016; Ahn *et al.*, 2019), mask their unpleasant taste (Khor *et al.*, 2017) and control their release behaviour (Henning *et al.*, 2012). Shellac has advantages for use in preparing food-grade microcapsules. Shellac is a weak acid with a pKa<7, which can be dissolved in alkaline solutions to form ionized shellac. Ionized shellac is protonated and cross-linked by calcium ions, causing it to solidify and promoting the formation of microcapsules in acid and calcium chloride solution (Xue and Zhang, 2009). However, the carboxyl groups of shellac could lead to polymerization, resulting in poor stability and low permeability. Therefore, some researchers have attempted to combine shellac with other natural polymers (such as proteins and polysaccharides) to prevent polymerization and control the final physical and chemical properties (Berg *et al.*, 2012; Yang *et al.*, 2018). Patel *et al.* (2013) prepared gelatin–shellac microcapsules under acidic pH conditions using the acid resistance of shellac and the interaction between shellac and gelatin. They successfully encapsulated and released bioactive substances, flavour substances, colourants, flavourings and immobilized enzymes. A composite capsule wall composed of shellac, alginate and pectin was applied as an outer membrane to form coated capsules. In the composite capsule, the diffusion effects of calcium and hydrogen ions in the mixed solution of pectin and shellac affected the mechanical properties and thickness of the capsule membrane (Leick *et al.*, 2011). The co-extrusion of sodium alginate–shellac and coconut fat was found to reduce the porosity of microcapsules, effectively protecting *Lactobacillus paracasei* BGP-1, which reached a 95% survival rate under simulated gastrointestinal conditions (Silva *et al.*, 2016). Microcapsules with sodium alginate as the inner layer and shellac as the outer layer, especially after the cross-linking of shellac and calcium, showed the pH-dependent release of vitamins. As the heating temperature of the coated capsule was higher than the glass transition temperature of shellac, the vitamin retention efficiency also increased. This is because of the thermoplastic nature of shellac, which can form a viscous polymer precipitate at such temperatures, creating a dense network with a porosity lower than that of an unheated capsule (Ben Messaoud *et al.*, 2016). Apart from the traditional extrusion and fluidized bed technology, spray drying is a promising technology expected to be widely used in shellac-based microcapsule production because of its low cost and suitability for large-scale production. In addition, emerging microencapsulation technologies, such as layer-by-layer self-assembly, electrospray, supercritical fluid and microfluid technologies with the advantages of green health, energy-saving, environmental protection and customizable preparation (Bah *et al.*, 2020) can also be used in the further development of shellac-based microcapsules.

However, the low solubility of shellac in intestinal fluid (pH ~6.8) limits its use as an enteric coating polymer. Plasticizers and water-soluble polymers can improve the enteric coating properties of shellac. Shellac enteric coatings containing 5% w/w glycerol, 5% w/w sodium alginate or up to 20% w/w PVP, protected probiotics from gastric juice and provided the best release characteristics in simulated intestinal juices (Stummer *et al.*, 2010). Oehme *et al.* (2011) prepared acylated pectin beads coated with shellac and HPMC for colon-targeted anthocyanin delivery. The resistance of the beads in simulated gastric fluid was improved and the release rate reduced below 5% after the addition of the water-soluble polysaccharide HPMC to the shellac coating. Therefore, coating composition is essential to achieving drug stability in the gastrointestinal tract and the targeted release of anthocyanin. Shellac can also be used as a potential enteric coating for lactase microcapsules, which help to reduce the effects of lactose intolerance when incorporated in milk formulations (Ahn *et al.*, 2019). Notably, the inter- and intra-individual variability in gastrointestinal pH, gastrointestinal environment, gastric emptying and gastrointestinal transit time will inevitably affect the activity of bioactive substances coated with shellac-based enteric coatings. Thus, designing shellac-based enteric coatings targeting specific sites (such as the mucosa or receptor proteins) and combining them with the gastrointestinal environment is of great significance to improve the biological activity of active substances.

Electrospinning is a simple, effective and top-down processing technology for the production of nanofibre films (Thammachat *et al.*, 2013; Yu *et al.*, 2017). The nanofibre films prepared by this technology have the advantages of a large surface area to volume ratio, high porosity and good fibre uniformity. These properties have important application value in the fields of medicine and food (Thammachat *et al.*, 2010). Recently, some researchers fabricated shellac nanofibre films through electrospinning and applied them as food-grade delivery systems, including targeted delivery (Wang *et al.*, 2015) and edible packaging materials (Li *et al.*, 2020).

Shellac and its derivatives can easily form nanofibre films by electrospinning, owing to the high concentrations of shellac polymer chains generating entanglement which allows the maintenance of a continuous solution jet (Thammachat *et al.*, 2010, 2013). The concentration of the shellac solution was identified as the key parameter that determines spinnability. In addition, electrospinning can be used to generate various complex nanostructures, using the macrostructure spinneret as a template. This has great potential for the development of advanced materials. In coaxial electrospinning, the concentration of shellac solution used as the core fluid is important. For example, Wang *et al.*, 2015 prepared ferulic acid-loaded shellac nanofibre films with a release rate of <10% in a pH 2.0 solution. However, upon transfer to a neutral phosphate buffer solution, the shellac molecules absorbed water, causing the fibres to swell and gradually expanding the dense structure. The drugs were then continuously released into the solution. In addition, Li *et al.* used coaxial electrospinning to prepare a PVP/shellac fibre membrane loaded with eugenol. Because the PVP/shellac fibre membrane was furnished with good thermal stability and moisture resistance, it inhibited strawberry rotting and prolonged their shelf life (Li *et al.*, 2020). Because of the low production efficiency of electrospinning in the fabrication of shellac-based nanofibres, the technology requires advancement to expand shellac-based nanofibre applications in food and medicine.

The application of shellac nanoparticles for the encapsulation of hydrophobic bioactive chemicals is not only effective in controlling their release but also improves their stability. The loading of curcumin (a natural yellow colourant) in shellac nanoparticles showed various advantages, such as biocompatibility and enhanced stability, which enriched the applications of this natural colourant in drinks, with appealing colour and health benefits (Kong *et al.*, 2019a). However, shellac nanoparticles tend to aggregate at acidic pHs, therefore stabilizers are required to overcome this shortcoming (Patel *et al.*, 2011). The authors presented an anti-solvent precipitation method that uses an ethanolic shellac solution with an aqueous natural polymer (xanthan gum) solution to prepare shellac nanoparticles. The average particle size of the nanoparticles was controlled (approximately 150–300 nm).

In a study conducted, a shellac ethanol solution was added to an almond gum solution using the anti-solvent precipitation method to obtain shellac–almond gum nanoparticles. The nanoparticles showed good stability at pH 1.2, with diameters and encapsulation efficiencies of approximately 197 nm and 78%, respectively (Sedaghat Doost *et al.*, 2018, 2019). Structurally, the almond gum was adsorbed onto the surface of the shellac nanoparticles, and quercetin was converted from a crystalline to an amorphous structure in the nanoparticles. Interestingly, they also found that the addition of polysorbate 80 (Tween-80) or Quillaja surfactants effectively reduced the particle size and increased the encapsulation efficiency. Uniform shellac–almond gum nanoparticles using Tween-80 as a stabilizer were obtained with an encapsulation rate >97%. Furthermore, the quercetin encapsulated in the nanoparticles showed stronger antioxidant capacity, less degradation and higher absorption rates than free quercetin under gastrointestinal tract pH conditions (Sedaghat Doost *et al.*, 2018). Moreover, cinnamon extracts encapsulated using shellac–xanthan gum nanoparticles can effectively improve the antioxidant activity and physical stability of chocolate beverages (Muhammad *et al.*, 2019, 2020).

The anti-solvent co-precipitation technique has also been used to encapsulate chlorophyll in shellac nanoparticles. The chlorophyll-loaded nanoparticles were incorporated into calcite crystals grown from an agarose gel media. They enhanced the stability and promoted the targeted intestinal release of chlorophyll (Kong *et al.*, 2019b). Interestingly, Kong *et al.* (2019b) developed a versatile microfluidic strategy to incorporate natural colourants such as curcumin in shellac nanoparticles with controlled physico-chemical properties. The ability of shellac to solubilize in water at various pHs suggests the possibility of shellac nanoparticle preparation via other techniques; for example, shellac nanoparticles can be obtained by adjusting the pH of an aqueous solution. Shellac and rice protein biopolymers were dissolved at a pH of 12, followed by a pH adjustment to 7 to fabricate the nanoparticles. The obtained composite nanoparticles ranged from 50 to 500 nm in size and were capable of being induced by calcium ions to undergo a structural transition to ribbons or networks. These structures have shown positive effects in encapsulating and delivering apigenin with controllable cellular uptake in a HepG-2 cell model (Wang *et al.*, 2020). Shellac nanoparticles with sizes ranging from 100 to 300nm were also prepared by ionic cross-linking between the cationic molecules of chitosan and the anionic molecules of shellac. The obtained nanoparticles showed high encapsulation efficiency for bovine serum albumin (Kraisit *et al.*, 2013). Studies have also shown that shellac submicron particles (approximately 720 nm) can be produced by electrospraying (Yu *et al.*, 2013). Interestingly, oat protein–shellac nanoparticles could be prepared via water-in-water emulsion, followed by cold-gelation. The resultant particles have the capability to protect resveratrol during transportation through gastric fluid and control its release in the colon. Furthermore, these optimized nanoparticles could improve the cellular uptake, transport and bioavailability of resveratrol and control its effects, preventing liver toxicity induced by tetrachloromethane (Yang *et al.*, 2019).

Larger-sized or micro shellac particles (beads/spheres) are prepared by controlling the preparation conditions. Recently, shellac microparticles have shown great potential in the food industry, not only owing to their potential for the encapsulation and delivery of active substances, but also for their ability to form multiscale structures, consequently enhancing their encapsulation and delivery effects. Shellac microparticles loaded with β-carotene using the microfluidic method were fabricated. Shellac microparticles were produced that not only retain the health-promoting effects of the natural colouring agent but also fulfil its function as a food additive, i.e. control of the colour of food. For this, β-carotene was uniformly dispersed in densely packed shellac microparticles, reducing its exposure to oxygen and thereby preventing its oxidation (Chen *et al.*, 2017). Zein–shellac composite colloidal particles, prepared by the anti-solvent co-precipitation method, showed significantly higher encapsulation efficiencies for curcumin than pure zein particles. At a 1:1 mass ratio of zein to shellac, the curcumin in the complex particles exhibited significantly improved photochemical and thermal stability (Sun *et al.*, 2017). Mei *et al.* (2018) fabricated chitin–shellac composite microspheres based on the self-assembly of the polymers caused by thermally induced phase separation. The porous structure of these composite microspheres differed from that of the chitin microspheres, which made it possible to immobilize the yeast alcohol dehydrogenase with a loading capacity of 79.0 mg/g (enzyme/carrier). Moreover, the immobilized enzymes exhibited catalytic activity comparable to their free counterparts. Obviously, particles with multiscale structures are more attractive in food applications. As such, sodium shellac was used to solidify a complex vitamin E (VE) emulsion to generate multiscale particles using the protons, calcium ions and spray-drying methods. However, because of the high solid content of the nanoscale emulsion particles, they were difficult to wrap using the spray-drying method. The particles solidified by protons showed higher VE loading and encapsulation efficiency than those of the other methods. These multiscale particles can be widely used in the medical, food and cosmetic industries (Luo *et al.*, 2016). Similarly, an emulsion-encapsulated alginate–shellac composite formed microbeads with high oil encapsulation efficiency under calcium ion induction. The microbeads, prepared under optimized conditions, effectively encapsulated sunflower oil and controlled its release in the intestine. This system has important implications for the design of oil-encapsulated and oil-soluble compound delivery systems (Morales *et al.*, 2017). Most of the above-mentioned micro-scale shellac particles prepared from pure shellac or shellac complexes with polymers are only used to encapsulate a single active compound. However, multiple active compounds incorporated in the micro-scale shellac particles would exhibit more interesting characteristics to act as carriers for multiple active compounds.

Gums

As already mentioned, natural gums have various applications in the food industry due to their thickening, stabilizing, emulsifying and gelling properties. Here are some of the key applications. Natural gums are favoured in the food industry for being plant-derived and generally recognized as safe (GRAS), making them appealing to consumers looking for clean-label ingredients. Therefore, this section of the chapter covers various applications of important natural gums and the research carried out in recent years.

Guar gum

Guar gum is the cheapest hydrocolloid for use in food industries, with hardly any health hazards. In form of a food additive, guar gum binds water, prevents ice crystals in frozen products, emulsifies, moisturizes, thickens, stabilizes and suspends many liquid–solid systems (Takahashi *et al.*, 1994). Guar gum is used in beverages to control viscosity as well as to reduce calories in them. Guar gum is an important non-calorie source of dietary fibre and is used in capsules as such. The gum is safe for animals and widely used in pet foods too (Carlson *et al.*, 1962; Carlson and Ziegenfu, 1965). Guar gum is also used in salad dressings, sauces, relishes, ketchups and canned soup to improve stability and appearance. It controls gel creation, viscosity, bloom, glazing and moisture retention to produce the best grade confections (Carlson and Ziegenfu, 1965). The binding property of guar gum is used in frozen food and meat products. In ice cream, the gum improves the texture, chewiness and resistance to heat shock. Partially hydrolyzed guar gum (at a concentration level of 2–6%) decreases syneresis and hence improves the rheological and textural properties of low-fat yoghurt (Brennan and Tudorica, 2008). Furthermore, it reduces crystal formation and acts as a stabilizer to extend the shelf life of ice creams (Sutton and Wilcox, 1998). Guar gum is used for increasing the dough yield in baked goods. In pastry fillings, guar gum prevents weeping or syneresis of water in the filling, which keeps the pastry crust crisp (Miyazawa, 2006). It is used as a thickening agent in milk, yoghurt, soft cheese and other dairy products (Klis, 1966) and helps to maintain the homogeneity and texture of ice cream and sorbet (Brennan and Tudorica, 2008).

As a food additive, use of guar gum was tested in chapati at 0.75% to impart softness (Ghodke, 2009), in bread at 0.5% for softness and loaf volume (Ribotta *et al.*, 2001; Keskin *et al.*, 2007), in fried products at 0.5–1.0% for reducing oil uptake (Sakhale *et al.*, 2011), in yoghurt at 2.0% as a texture improver (Brennan and Tudorica, 2008), in cake at 0.15% as a fat replacer and to give firmness (Zambrano *et al.*, 2004), in sausage at 0.13–0.32% to impart softness (Andres *et al.*, 2006), in pasta at 1.5% as a texture improver (Raina *et al.*, 2005), in ice cream at 0.5% to get smaller ice crystals (Sutton and Wilcox, 1998) and in baked goods up to 1.0% as a dough improver (Kohajdová and Karovičová, 2009). In tomato ketchup at 0.5–1.0%, guar gum was superior as a consistency improver and serum loss reducer to other popular hydrocolloids like gum acacia, pectin, carboxymethyl cellulose and sodium alginate (Gujral *et al.*, 2002; Koocheki *et al.*, 2009). Similarly, guar gum also retards staling in chapati by controlling retrogradation of starch at room temperature as well as refrigerated temperature (Shaikh *et al.*, 2008). Guar gum, along with xanthan gum, checks staling in gluten-free rice cakes as it decreases the retrogradation enthalpy and weight loss (Sumnu *et al.*, 2010). The addition of guar gum to potato protein-based emulsions enhances stability, as demonstrated by the combination of different techniques like rheology, optical microscopy and multiple light scattering (Santos *et al.*, 2015). The effect of incorporating xanthan or guar gum in oil-in-water emulsions stabilized by soy soluble polysaccharide was studied. Addition of these polysaccharides improved emulsion creaming stability, droplet sizes and droplet polydispersion. Consequential continuous phase viscosity in xanthan or guar gum-containing emulsions improved homogenization efficiency and inhibited movement of emulsion droplets, resulting in greater stability (Chivero *et al.*, 2015). It is often used in ice cream, sauces, cake mixes, cheese spreads, fruit drinks and dressings at a concentration of <1% of food (Whistler and Hymowitz, 1979; Parija *et al.*, 2001). As excess concentration of guar gum in food may create digestive obstruction and other related problems due to highly viscous intestinal tract contents, its use for food applications is restricted to 2%. Even when guar gum is fed to animals in excess concentration (about 10–15% on weight basis), harmful effects have been observed. It has been observed that excess consumption can decrease feed intake by impairing digestion and thereby reduce the growth of the animal.

Gum arabic

GA is mainly used in the confectionery industry, where it is incorporated in a wide range of products. It has a long tradition of use in wine gums, where it produces a clarity that is higher than can be obtained with other hydrocolloids (Williams and Phillips, 2001). Furthermore, it prevents sucrose crystallization, provides a controlled flavour release and slows down melting in the mouth, making the wine gum long-lasting (Imeson, 1992). It also provides the appropriate texture to these sweets, which are easily deformed in the mouth but do not adhere to the teeth. In lower-calorie sweets, GA is used to compensate for the loss of texture, mouthfeel and body resulting from the replacement of sugars by artificial sweeteners (Brucker *et al.*, 1974). It is also used in chewing gum as a coating agent and as a pigment stabilizer (Cherukuri *et al.*, 1983; Huzinec and Graff, 1987). In aerated confectionery products, such as marshmallows, nougats and meringues, GA acts as a whipping and stabilizing agent. It is also used in toffees and caramels as an emulsifier, to maintain a uniform distribution of the fat across the product. In jelly products, it is used to provide a fibrous, fruit-like texture (Shigeo *et al.*, 1989). GA glazes are used as coatings for nuts, dragees and others. GA is widely used as an emulsifier in the manufacture of soft drinks. Due to its stability in acid conditions and its high solubility, GA is well suited for use in citrus and cola flavour oil emulsions (Growers, 1958). High levels of gum are used to ensure a complete coverage of the interface and to prevent flocculation and coalescence of oil droplets. Normally, a weighting agent is added to increase the oil-phase density, inhibiting destabilization due to creaming. GA can also form a stable cloud in the drink, imitating the effect of added fruit pulps and juices (Eng and Mackenzie, 1984). In beer, it is used as a foaming agent and to assist lacing (Tiss *et al.*, 2001). GA is used increasingly as a source of soluble fibre in low-calorie and dietetic beverages (Phillips, 1998). In powdered beverage mixes, GA is added to produce the same opacity, appearance, mouthfeel and palatability as natural fruit juices (Common and Manor, 1962). In microencapsulation, liquid, solid or gaseous substances are coated with a protective layer to prevent chemical deterioration and the loss of volatile compounds. It is a useful technique to convert liquid food flavours to flowable powders that can be used in dry food product (McNamee *et al.*, 1998). GA is an effective encapsulation agent because of its high water solubility, low viscosity and emulsification properties and is therefore extensively used in soups and desserts. GA is also used to prevent gelation in canned gravy-based pet foods, as it inhibits the extraction of proteins from the meat into the gravy (Glicksman and Farkas, 1975).

GA is easily soluble in water, forms a solution of very low viscosity even at higher concentrations (Stephen, 1990; Stephen and Churms, 1995) and maintains Newtonian behaviour at 50% concentration. At 50% concentration, GA solution has similar viscosity to a 1.5% solution of xanthan gum. In the confectionery industry, GA finds applications in different concentrations: in hard sweets, GA concentration varies between 40 and 50%, whereas, in soft sweets, such as chocolate bars, nougat and chocolates, it is between 30 and 35%. Safety regulation in the US permits GA application up to 85% in soft sweets (Imeson, 1992; Whistler, 1993; Jales, 2018). Generally, GA prevents sugar crystallization in confectionery products with higher sugar content, such as jujubes and pastilles. The gum also works as an emulsifier, ensuring even distribution of fat in products, and prevents the formation of oxidizable film. Other applications of GA in confectionary include as a whipping agent in products like marshmallows, meringues and nougat, a film-maker in liquorice preparation and a binding agent in chewing gum. Sometimes starch or pectin are used in place of GA in confectionary coating, but with a deterioration of product quality (Flowerman, 1985). GA's use in the wine industry is prevalent in UK and Ireland to maintain clarity and controlled flavour release (Williams and Phillips, 2001).

In beverages, GA prevents droplet flocculation and coalescence in concentrated solutions and emulsions. The concentrated emulsion of beverages may contain 5–20% GA, but in finished products it may be up to 1%. In the US, the use of GA in beverages is permitted up to 2% by weight (Jales, 2018). GA is applied in wine making to stabilize young red wines against precipitation of colour pigment and protein, iron and copper hazes (Teissedre, 2012).

Furthermore, GA is used as a flavour-encapsulating agent in dry foods like spices, desserts and soups (Krishnan *et al.*, 2005). The gum coating on flavour particles prevents the oxidation and evaporation of volatiles. The gum is also used as a stabilizer in frozen products such as ice cream to prevent ice crystal formation and delay melting, thus maintaining the fine texture (Sulieman, 2018). Eltahir *et al.* (2013) utilized GA as a stabilizer in Robe, an important fermented milk product in Sudan. Bashir (2013) investigated the effect of GA on the enzymatic coagulation of camel milk.

The application of GA as an emulsifier is found in pretzels, dressings and chops (Rinsky and Rinsky, 2009; Achouri *et al.*, 2012); as a thickener and stabilizer in toppings, creamers, coffee and non-dairy milk (Mariod, 2018); and as thickener, emulsifier and stabilizer in pudding, fillings and gelatines (Imeson, 2010).

Since it was first declared as a low-calorie fibre in foods and food ingredients several years ago, there has been a significant increase in the use of GA as a functional food (Mariod, 2018). Soluble fibres such as GA generally help to lower total cholesterol and low-density cholesterol. It has been shown that a lowering of total cholesterol as a result of the intake of GA helps to reduce the risk of coronary heart disease (McLean-Ross *et al.*, 1983).

Gum karaya

The use of gum karaya in foods has to be in accordance with the US Food and Drug Administration (FDA) Code of Federal Regulations (Title 21, Section 184.1349; Table 7). Due to its acid stability, high viscosity and suspension properties, gum karaya is well suited for stabilizing low-pH emulsions, such as sauces and dressings (Partyka, 1963). Gum levels of 0.6–1.0% are used to obtain the desired texture, colour and suspension, but during processing care must be taken regarding the heat- and shear-sensitivity of the gum. In French dressings, gum karaya functions as a stabilizer, increasing the viscosity of the aqueous phase of the oil-in-water emulsion (Dziezak, 1991). In cheese spreads, gum karaya is used as a binder to provide texture and spreadability and to prevent water separation. In salads, the gum acts as a stabilizer and prevents the weeping of the water from the oil-in-water emulsion (Beach and Salisbury, 1969). It is also incorporated in aerated dairy desserts and whipped cream, where it acts as a foam stabilizer. Gum karaya significantly reduces the bioavailability of calcium in milk-based foods, as demonstrated by Kelly and Potter (1990). In frozen desserts, such as sorbet, sherbet and ice lollies, gum karaya is used in concentrations typically varying between 0.2% and 0.4% (Weiping and Branwell, 2000). Here, it controls the formation of ice crystals, preventing them from growing too large, prevents the migration of free water and reduces the suck-out of colour and flavour during consumption. It is also found in ice cream, together with locust bean gum.

In sausages, gum karaya is used in concentrations ranging from 0.75% to 3%, performing several functions (Allen and McCaleb, 1938). It acts as an adhesive between meat particles and as a water-binder during preparation and storage. During smoking and cooking, the gum seals the pores of the casings, reducing the loss of volatile flavouring components. It also provides the appropriate sensorial characteristics to the product, such as a smooth texture, mouthfeel, body and appearance. In bakery products, gum karaya is mainly used to reduce the effect of variations in water addition and mixing time. Due to its water-binding capacity, it also slows down aging, extending the shelf life of baked goods.

Gum karaya is widely used in food, dairy and bakery products (Gautami and Bhat, 1992), as it possesses unique properties like high water retention capacity, intrinsic nature and high viscosity and is abundantly available (Le Cerf *et al.*, 1990). It's hydrocolloidal properties and unique water retention in large quantities makes it suitable for extending shelf life and decelerating aging of bakery products (Davidson, 1980; Verbeken *et al.*, 2003). A study on gum karaya effects in humans was carried out and found that a 10.5g daily intake, which was more than the typically small amounts of gum karaya present in foodstuffs, did not cause any toxic effects as observed in the wide range of the study carried out; furthermore, gum karaya had no metabolic effect of any consequence (Eastwood *et al.*, 1983). Gum karaya is widely used as a stabilizer, binder and adhesive, to control ice crystal formation and to improve water-binding capacity (Le Cerf *et al.*, 1990).

Gum ghatti

Gum ghatti is widely used in bakery, dairy and other food products, as it has medium water-retaining capacity, viscosity and is readily available. The hydrocolloidal as well as the water-retaining property of gum ghatti helps to enhance shelf life and retard ageing in products of the bakery industry. It works as an excellent emulsifier, stabilizer, thickener and adhesive, besides controlling ice crystal formation, reducing staling and binding water in products of the dairy, bakery and other food industries (Castellani *et al.*, 2010). It possesses excellent emulsification properties due to its comparatively high protein content. As it binds to oil, it works as a better emulsifier than GA (Ido *et al.*, 2008). Further, this gum has application as a stabilizer as well as an emulsifier in carbonated drinks, beverages, chewing gum, fruit jelly, gum drops, sweets, bakery products, butter, wine and flavour fixatives (Amar *et al.*, 2006). Gum ghatti is rich in dietary fibre and, unlike gum karaya, it doesn't risk blocking the intestine. It is often fed to camels and other cattle in desert areas to conserve water in their body and to save them from heatstroke, although this needs to be scientifically established. Some of the most common food applications of gum ghatti are discussed below.

The beverage industry uses gum ghatti for controlling the viscosity, thickness and optical clarity of their products. Optical clarity is one of the most important visual properties of beverages and is critical for attracting consumers and end-users. During enrichment or supplementation of a beverage by fat-soluble colourants such as carotenoids, turmeric pigments or other means, increased turbidity causes the formation of a separate layer at the top of the beverage, which needs to be diminished. At a concentration of 30–65 wt%, gum ghatti is very effective against fat-soluble active ingredients for the enrichment (vitamin A, D, E, K and derivatives), fortification and/or colouration of food and beverages, attenuating the separation phenomena and visible turbidity problems (Hitzfeld *et al.*, 2014). Gum ghatti may be added to prevent the adhesion of carotenoid pigments to manufacturing equipment or containers such as polyethylene terephthalate (PET) bottles during the production of liquid foods. Studies indicate that for beverage production different amounts of gum ghatti can be employed to inhibit adhesion. Other natural products, such as GA and modified starch, may not achieve such adhesion inhibition for carotenoid pigments (Miuchi and Nishino, 2013). Accumulation and sedimentation of turmeric pigment used to colour and enrich hydrous solutions is effectively prevented by the addition of gum ghatti in the concentration range of 1–10% (Miuchi *et al.*, 2012, 2016). In the field of brewed cappuccinos, espressos and flavoured beverages as well as cafés and coffee houses, purified or modified gum ghatti is utilized to enhance the flavour, creaminess and sweetness characteristics and acts as a thickening agent as well as an edible carrier for emulsified fat in non-foaming creamer (Sargent and Villagran, 2002). In the case of table syrup, gum ghatti may be used to achieve stable emulsions, improving resistance to separation of the oil and the aqueous medium. It also modifies the refractive index of the syrup until it becomes clear. Such stable emulsions are widely used in the food industry for salad dressings, sauces, sweet cream and syrup (Srivastava and Ray, 2015).

In the dairy industry, gum ghatti is used for the heat treatment of milk and stabilization of ice cream and frozen products. During the high-temperature short-time (HTST) process and retort sterilization of milk at temperatures exceeding 120°C, gum ghatti may be employed to retain the activity of the basic protein fraction in the milk (Urazono *et al.*, 2016). Gum ghatti is used for the stabilization of frozen products, especially for ice cream stabilization, because of its high water-binding properties. Since the stabilizer concentration affects the flavour of the products, gum ghatti is particularly suitable as only up to a 1% concentration of gum ghatti is required to serve as an efficient stabilizer for ice creams (Syed *et al.*, 2018). A plain ice cream mix with 10% fat, 11% non-fat solids, 15% sugar and different levels of gum ghatti (0.25, 0.50, 0.75 and 1.0%) as a stabilizer and emulsifier was prepared using buffalo milk, cream, skimmed milk powder and sugar. The concentration of the stabilizer was found to affect the viscosity of the ice cream mix, which in turn influenced the mouth feel (i.e. texture) of the finished product. Optimal usage of gum as a stabilizer based on these characteristics has been proposed for gum guar (0.96%), gum acacia (0.91%) and gum

Table 6.3. Food applications for gum ghatti and their characteristics in food and beverages industries. Author's own table.

Food application	Characteristics	Reference
Emulsifier for carbonated drinks	Stabilizes water and oil emulsions	(Amar *et al.*, 2006; Al-Assaf *et al.*, 2009)
Beverages (powder)	Encapsulation of flavour	(Shahidi and Han, 1993)
Wine	Colloidal stabilization and tannin suspension	(Kora *et al.*, 2012)
Flavours (powdered)	Encapsulation of oil entrapments	(Whistler, 1993)
Fruit jellies	Independent gelling of sugar	(Amar *et al.*, 2006)
Sweets	Source of fibre	(Parvathi *et al.*, 2009)
Coated sweets	Coating, film-forming and glazing	(Zhang *et al.*, 2016)
Gum drops	Prevents sugar crystallization	(Barak *et al.*, 2020)
Caramels	Improves chewability	(Amar *et al.*, 2006)
Bakery products	Enhances shelf life	(Kala, 2016)
Powdered medicines	Encapsulation and oil entrapment	(Kora *et al.*, 2012)

ghatti (0.95%) (Minhas *et al.*, 2002). Aerated frozen confectionery products comprising 0.2–3% by weight of gum ghatti stabilizer have also been manufactured (Bartels-ArntzAart *et al.*, 2016). Food applications for gum ghatti are presented in Table 6.3.

Bakery products. Emulsions of 5% gum ghatti show similar viscosity and emulsification properties as those using 15% GA due to presence of a higher proteinaceous component (1.4%) in gum ghatti (Al-Assaf *et al.*, 2008; Ido *et al.*, 2008). Further, the addition of 0.1–0.2% gum ghatti may enrich loaf volume, softness, texture properties and water absorption, as shown by a farinograph (Sharadanant and Khan, 2003).

Edible films and coating. The feasibility of creating an edible film using gum ghatti was investigated by Zhang *et al.* (2016), who found that it can be utilized as a novel film-producing material with increased usefulness and adaptability. Edible films prepared with gum ghatti may have 0.75–1.0% w/v gum and may employ either of the two plasticizers, glycerol or sorbital. Glycerol-based gum ghatti edible films were shown to have a more swollen network, whereas sorbital-based gum ghatti had a lower moisture content and water vapour permeability. The colour of gum ghatti films appears to be green and yellow, and a decrease in the whiteness index was observed with increasing concentration of the plasticizer.

In glycerol-based gum ghatti films an increase in the water activity of the film increased its moisture content and an increase in the temperature of the film decreased its moisture content, indicating that they may be useful as a novel fresh-food preservation technique (Zhang *et al.*, 2020b). The edible films based on gum ghatti may act as a barrier for moisture, ultraviolet light, deteriorating gases and temperature from the environment to enhance the shelf life of food products. Therefore, they can be integrated with antimicrobials, probiotics, antioxidants and aromatic compounds to improve their diverse application in nutrient-rich food for the food industry.

Gum tragacanth

GT is used in the preparation of low-viscosity, pourable dressings and sauces. Because of its high acid stability, the gum is well suited for the use in these low-pH products. It lowers the interfacial tension due to its surface-active properties and acts as a thickener in the continuous water phase, thus stabilizing the emulsion and providing long shelf life. Typical usage levels are at 0.4–0.8%, depending on the individual oil content (Imeson, 1992). In low-calorie, oil-free dressings, GT is used to imitate the creamy mouthfeel and body of dispersed oil droplets. In fish and citrus oil emulsions with a long shelf life, GT is used in combination with GA. The interaction between the two gums reduces viscosity, allowing the production of thin, pourable smooth emulsions. GT is used in ice cream to provide texture to the product (Weiping and Branwell, 2000). It also prevents texture changes during storage due to the formation of ice crystals. In water ices, sorbets and ice pops, the gum prevents the migration of flavour and colour components during storage and consumption. It is also used in milkshakes as a viscosity control agent, providing texture and stabilizing incorporated air. GT is used in bakery fruit-based toppings and fillings to give a shiny, clear appearance and a creamy texture (Dziezak, 1991). In chewy sweets, such as lozenges, it acts as a thickener and provides texture. The gum is used in fruit tablets, gum drops and pastilles as a binding agent during compression. It also provides body and mouthfeel and ensures good flavour-release during consumption. In icings, GT is used as a water binder, maintaining pliability and preventing cracks and breaks. It also provides consistency, smooth texture and creamy taste to the product.

GT has been extensively used in food processing as a polysaccharide food coating (Jafari *et al.*, 2018). The use as food coating results in decreased respiration, dehydration and enzymatic browning of individual plant products. This is achieved by the functioning of the gum as a semi-permeable barrier against water steam and gases, which results in a higher preservation period for foods such as freshly cut apple slices (Soto *et al.*, 2017). Due to the early perishability of fruits such as bananas, chewing gum was also used as an edible coating material (Mohebbi *et al.*, 2012). For example, dried banana slices coated with the gum showed less shrinkage, better texture, less water loss and better rehydration compared to uncoated samples (Farahmandfar *et al.*, 2017). In addition, gum tragacanth is widely used as a thickening agent to increase viscosity and improve emulsions in food. Due to its excellent stability against heat and acid and its effectiveness as an emulsifier, tragacanth stabilizes the viscosity of food systems (Phillips and Williams, 2009; Mostafavi *et al.*, 2016; Tonyali *et al.*, 2018).

GT is also widely used as a food additive, for example as a fat substitute in fat-free yogurt, white cheese and low-fat sausage, as a stabilizing agent in fermented milk drinks and ice cream mixtures and as a thickening agent in restructured mutton chops. It is also used as an active agent for altering and optimizing the rheological, physical and functional properties of Cheddar cheese, milk protein, flavoured milk drinks, kashk (dried yogurt paste) and tomato ketchup. Furthermore, GT works well as a thermal stability improvement agent in fish gelatin gel, as a carrier for the safeguarded and sustained release of micronutrients, as a natural coating for enhancement of the postharvest quality and shelf life of button mushrooms and as a hydrocolloid coating in production of low-fat shrimps (Abbasi *et al.*, 2015; Jain *et al.*, 2016). In addition, it exhibits potential prebiotic activity by promoting the growth of human intestinal flora. Thus, the gum may be used for the production of value-added prebiotic compounds in the manufacturing of functional foods (Gavlighi *et al.*, 2013).

Being natural and biodegradable, GT packaging of food products is an alternative to synthetic petroleum-based films that have disadvantages such as migration into food and difficulty of recovery along with environment pollution (Fazel *et al.*, 2012). Edible and biodegradable gum bioplastics have been used for food packaging purposes to extend food shelf life (Phillips and Williams, 2009; Mostafavi *et al.*, 2016; Tonyali *et al.*, 2018). The permeability and physical properties of GT-based films are poor and, hence, other biopolymers and nanomaterials have been added to overcome these issues. For example, GT/chitosan/graphene oxide bionanocomposite film was prepared via solution casting. These bionanocomposite films possess higher mechanical properties compared to neat GT/chitosan. In addition, vapour and CO_2 permeability declined by increasing graphene oxide nanoparticles (Talaei and Kiani, 2015). Biocomposite films based on GT in combination with other biopolymers, such as soy protein/nanocellulose, potato starch and chitosan, have been manufactured to improve mechanical properties and reduce film permeability (Fazel *et al.*, 2012; Separdar *et al.*, 2016).

Finally, by 1961 GT had been reported as 'generally recognized as safe' by the FDA (Vojdani, 2015). In accordance with the FDA Code of Federal Regulations, GT may be used as a food additive at the concentration of 0.2–1.3 wt%. The Scientific Committee for Food of the European Community approves GT with the number E413 on its list of additives (Ghayempour *et al.*, 2015).

References

Abbasi, M.K., Musa, N. and Manzoor, M. (2015) Mineralization of soluble P fertilizers and insoluble rock phosphate in response to phosphate-solubilizing bacteria and poultry manure and their effect on the growth and P utilization efficiency of chilli (*Capsicum annuum* L.). *Biogeosciences* 12(15), 4607–4619. Available at: https://doi.org/10.5194/bg-12-4607-2015

Abdel-Maksoud, G. and El-Amin, A.-R. (2011) A review on the materials used during the mummification processes in ancient Egypt. *Mediterranean Archaeology and Archaeometry* 11(2), 129–150.

Abdul-Ghani, R.A., Loutfy, N. and Hassan, A. (2009) Myrrh and trematodoses in Egypt: An overview of safety, efficacy and effectiveness profiles. *Parasitology International* 58(3), 210–214.

Abu-Zaiton, A.S. (2010) Anti-diabetic activity of *Ferula asafoetida* extract in normal and alloxan-induced diabetic rats. *Pakistan Journal of Biological Sciences* 13(2), 97–100.

Accaseavorn, P., Ampornratana, T., Boonsiri, A., Siwarungsun, N., Prichanont, S. *et al.* (2006) Development of shellac coating for extending shelf-life of mangosteen and lime (CV. Pan). *Agricultural Science Journal* 37(5), 42–45.

Achouri, A., Zamani, Y. and Boye, J.I. (2012) Stability and physical properties of emulsions prepared with and without soy proteins. *Journal of Food Research* 1(1). Available at: https://doi.org/10.5539/jfr.v1n1p254

Ahmadi, E., Sadrjavadi, K., Mohammadi, G. and Fattahi, A. (2018) De-esterified tragacanth microspheres loaded into eudragit s-100 coated capsules for colon-targeted delivery. *Iranian Journal of Pharmaceutical Research: IJPR* 17(2), 470–479.

Ahn, S.-I., Kim, A.Y. and Kwak, H.-S. (2019) Optimisation of the microencapsulation conditions and physicochemical properties of enteric-coated lactase microcapsules. *International Journal of Dairy Technology* 72, 23–35.

Alam, N.H., Meir, R., Schneider, H., Sarker, S.A., Bardhan, P.K. *et al.* (2000) Partially hydrolyzed guar gum–supplemented oral rehydration solution in the treatment of acute diarrhea in children. *Journal of Pediatric Gastroenterology and Nutrition* 31(5), 503–507.

Alam, N.H., Ashraf, H., Sarker, S.A., Olesen, M., Troup, J. *et al.* (2008) Efficacy of partially hydrolyzed guar gum-added oral rehydration solution in the treatment of severe cholera in adults. *Digestion* 78(1), 24–29.

Al-Assaf, S., Amar, V. and Philips, G. (2008) Characterization of gum ghatti and comparison with gum arabic. In: Williams, P. (ed.) *Gums and Stablizers for the Food Industry*. Royal Society of Chemistry, Cambridge, UK.

Al-Assaf, S., Phillips, G. and Amar, V. (2009) Gum ghatti. In: Phillips, G. and Williams, P. (eds) *Handbook of Hydrocolloids*. CRC Press, Boca Raton, Florida, pp. 477–494.

Al-Harrasi, A. and Al-Saidi, S. (2008) Phytochemical analysis of the essential oil from botanically certified oleogum resin of *Boswellia sacra* (Omani Luban). *Molecules* 13(9), 2181–2189.

Al-Howiriny, T., Al-Sohaibani, M., Al-Said, M., Al-Yahya, M., El-Tahir, K. *et al.* (2005) Effect of *Commiphora opobalsamum* (L.). *Journal of Ethnopharmacology* 98(3), 287–294.

Al-Jafari, A.H., Vila, R., Freixa, B., Costa, J. and Canigueral, S. (2012) Antifungal compounds from the rhizome and roots of *F. hermonis*. *Phytotherapy Research* 27(6), 911–915. Available at: https://doi.org/10.1002/ptr.4806

Allen, H.E. and McCaleb, A.G. (1938) A method or process of manufacturing sausages and sausage-filling composition. GB Patent Application 483,925.

Alonso-Sande, M., Teijeiro, D., Remuñán-López, C. and Alonso, M.J. (2009) Glucomannan, a promising polysaccharide for biopharmaceutical purposes. *European Journal of Pharmaceutics and Biopharmaceutics* 72(2), 453–462.

Al-Saidan, S.M., Krishnaiah, Y.S.R., Patro, S.S. and Satyanaryana, V. (2005) *In vitro* and *in vivo* evaluation of guar gum matrix tablets for oral controlled release of water-soluble diltiazem hydrochloride. *AAPS PharmSciTech* 6, 14–21.

Altaf, S.A., Yu, K., Parasrampuria, J. and Friend, D.R. (1998) Guar gum-based sustained release diltiazem. *Pharmaceutical Research* 15, 1196–1201.

Alves, J., Mantovani, A., Martins, M., Abrao, F., Lucarini, R. *et al.* (2015) Antimycobacterial activity of some commercially available plant-derived essential oils. *Chemistry of Natural Compounds* 51(2), 353–355.

Amar, V., Al-Assaf, S. and Phillips, G.O. (2006) An introduction to gum ghatti: Another proteinaceous gum. *Foods and Food Ingredients Journal of Japan* 211(3), 275–279.

Ammon, H.P. (2006) Boswellic acids in chronic inflammatory diseases. *Planta Medica* 72(12), 1100–1016.

Andres, S., Zaritzky, N. and Califano, A. (2006) The effect of whey protein concentrates and hydrocolloids on the texture and colour characteristics of chicken sausages. *International Journal of Food Science & Technology* 41(8), 954–961.

Angelini, P., Pagiotti, R., Venanzoni, R. and Granetti, B. (2009) Antifungal and allelopathic effects of asafoetida against *Trichoderma harzianum* and *Pleurotus* spp. *Allelopathy Journal* 23(2), 357–368.

Ariyanto, H.D., Chiba, M., Oguma, K., Tatsuki, M. and Yoshii, H. (2019) Release behavior of 1-methylcylopropene coated paper-based shellac solution in response to stepwise humidity changes to develop novel functional packaging for fruit. *Packaging Technology and Science* 32, 523–533.

Babu, G.V.M.M., Kumar, N.R., Sankar, K.H., Ram, B.J., Kumar, N.K. *et al.* (2002a) *In vivo* evaluation of modified gum karaya as a carrier for improving the oral bioavailability of a poorly water-soluble drug, nimodipine. *AAPS PharmSciTech* 3, 55–63.

Babu, G.V.M.M., Prasad, C.D. and Murthy, K.R. (2002b) Evaluation of modified gum karaya as carrier for the dissolution enhancement of poorly water-soluble drug nimodipine. *International Journal of Pharmaceutics* 234(1–2), 1–17.

Bagheri, S.M. and Dashti-R, M.H. (2015) Influence of asafoetida on prevention and treatment of memory impairment induced by d-galactose and $NaNO_2$ in mice. *American Journal of Alzheimer's Disease and Other Dementias* 30(6), 607–612.

Bagheri, S.M., Hejazian, S.H. and Dashti-R, M.H. (2014) The relaxant effect of seed's essential oil and oleo-gum-resin of *Ferula asafoetida* on isolated rat's ileum. *Annals of Medical and Health Sciences Research* 4(2), 238–241.

Bah, M.G., Bilal, H.M. and Wang, J. (2020) Fabrication and application of complex microcapsules: A review. *Soft Matter* 16, 570–590.

Banker, G.S. and Anderson, N.R. (1987) Tablets. In: Lachman, L., Lieberman, H.A. and Kanig, J.L. (eds) *The Theory and Practice of Industrial Pharmacy*. Varghese Publishing House, Bombay, India.

Bar, H. and Bianco-Peled, H. (2020) Modification of shellac coating using jeffamine® for enhanced mechanical properties and stability. *Progress in Organic Coatings* 141, 105559. Available at: https://doi.org/10.1016/J.PORGCOAT.2020.105559

Bartels-ArntzAart, M., Alting, A.C. and Westerbeek, J.M.M. (2016) A frozen aerated confectionery product and a method for preparing such a product. Patent WO 2016167660 A1.

Barak, S., Mudgil, D. and Taneja, S. (2020) Exudate gums: Chemistry, properties and food applications – A review. *Journal of the Science of Food and Agriculture* 100(7), 2828–2835. Available at: https://doi.org/10.1002/jsfa.10302

Bashir, A.Y.M. (2013) *Effect of Gum Arabic on Enzymatic Coagulation of Camel Milk*. Sudan University of Science & Technology, Khartoum, Sudan.

Beach, P.L. and Salisbury, M. (1969) Preparation of a protenaceous food salad. US Patent No. 3454405. Available at: https://doi.org/10.1145/178951.178972

Belo, G.M., Diniz, A.S. and Pereira, A.P. (2008) Effect of partially hidrolized guar-gum in the treatment of functional constipation among hospitalized patients. *Gastroenterology Archives* 45(1), 93–95.

Ben Messaoud, G., Sanchez-Gonzalez, L., Probst, L., Jeandel, C., Arab-Tehrany, E. *et al.* (2016) Physicochemical properties of alginate/shellac aqueous-core capsules: Influence of membrane architecture on riboflavin release. *Carbohydrate Polymers* 144, 428–437.

Bera, H., Boddupalli, S. and Nayak, A.K. (2015) Mucoadhesive-floating zinc-pectinate-sterculia gum interpenetrating polymer network beads encapsulating ziprasidone HCl. *Carbohydrate Polymers* 131, 108–118. Available at: https://doi.org/10.1016/j.carbpol.2015.05.042

Bera, H., Ippagunta, S.R., Kumar, S. and Vangala, P. (2017) Core-shell alginate-ghatti gum modified montmorillonite composite matrices for stomach-specific flurbiprofen delivery. *Materials Science and Engineering C* 76, 715–726. Available at: https://doi.org/10.1016/j.msec.2017.03.074

Berg, S., Bretz, M., Hubbermann, E.M. and Schwarz, K. (2012) Influence of different pectins on powder characteristics of microencapsulated anthocyanins and their impact on drug retention of shellac coated granulate. *Journal of Food Engineering* 108, 158–165.

Bhardwaj, T.R., Kanwar, M. and Gupta, A. (2000) Natural gums and modified natural gums as sustained-release carriers. *Drug Development and Industrial Pharmacy* 26, 1025–1038.

Bhargava, N., Sharanagat, V.S., Mor, R.S. and Kumar, K. (2020) Active and intelligent biodegradable packaging films using food and food waste-derived bioactive compounds: A review. *Trends in Food Science & Technology* 105, 385–401.

Bhatnager, R., Rani, R. and Dang, A.S. (2015) Antibacterial activity of *Ferula asafoetida*: A comparison of red and white type. *Journal of Applied Biology & Biotechnology* 3(2), 18–21. Available at: https://doi.org/10.7324/JABB.2015.3204

Bhosale, R.R., Osmani, R.A.M., Abu Lila, A.S., Khafagy, E.S., Arab, H.H. *et al.* (2021) Ghatti gum-base graft copolymer: A plausible platform for ph-controlled delivery of antidiabetic drugs. *RSC Advances* 11(24), 14871–14882. Available at: https://doi.org/10.1039/d1ra01536b

Boiteau, C. and Blondeel, G. (1981) Tampon having a surface covering. GB Patent Application 2,075,864 75–846.

Brennan, C.S. and Tudorica, C.M. (2008) Carbohydrate-based fat replacers in the modification of the rheological, textural and sensory quality of yoghurt: Comparative study of the utilisation of barley beta-glucan, guar gum and inulin. *International Journal of Food Science & Technology* 43(5), 824–833.

Brucker, C.E., Uhlarik, K.S., Lampe, J.W. and Bush, J.W. (1974) Lower calorie candy. US Patent Application 3,800,045.

Butt, M.S., Shahzadi, N., Sharif, M.K. and Nasir, M. (2007) Guar gum: A miracle therapy for hypercholesterolemia, hyperglycemia and obesity. *Critical Reviews in Food Science and Nutrition* 47, 389–396.

Byun, Y., Ward, A. and Whiteside, S. (2012) Formation and characterization of shellachydroxypropyl methylcellulose composite films. *Food Hydrocolloids* 27, 364–370.

Campbell, A.L., Holt, B.L., Stoyanov, S.D. and Paunov, V.N. (2008) Scalable fabrication of anisotropic micro-rods from food-grade materials using an in shear flow dispersion–solvent attrition technique. *Journal of Materials Chemistry* 18, 4074–4078.

Campbell, A.L., Stoyanov, S.D. and Paunov, V.N. (2009) Fabrication of functional anisotropic food-grade micro-rods with micro-particle inclusions with potential application for enhanced stability of food foams. *Soft Matter* 5, 1019–1023.

Carlson, W.A. and Ziegenfu, E.M. (1965) Effect of sugar on guar gum as a thickening agent. *Food Technology* 19, 64–68.

Carlson, W.A., Overton, J.D. and Ziegenfuss, E. (1962) Compatibility and manipulation of guar gum. *Food Technology* 16, 50–54.

Castellani, O., Al-Assaf, S., Axelos, M., Phillips, G.O. and Anton, M. (2010) Hydrocolloids with emulsifying capacity. Part 2 – adsorption properties at the n-hexadecane–water interface. *Food Hydrocolloids* 24(2–3), 121–130. Available at: https://doi.org/10.1016/J.FOODHYD.2009.07.006

Chamarthy, S.P. and Pinal, R. (2008) Plasticizer concentration and the performance of a diffusion-controlled polymeric drug delivery system. *Colloids and Surfaces A: Physicochemical and Engineering Aspects* 331, 25–30.

Chauhan, O.P., Raju, P.S., Singh, A. and Bawa, A.S. (2011) Shellac and aloe-gel-based surface coatings for maintaining keeping quality of apple slices. *Food Chemistry* 126, 961–966.

Chauhan, O.P., Nanjappa, C., Ashok, N., Ravi, N., Roopa, N. *et al.* (2015) Shellac and *Aloe vera* gel based surface coating for shelf life extension of tomatoes. *Journal of Food Science and Technology* 52(2), 1200–1205. Available at: https://doi.org/10.1007/S13197-013-1035-6

Chaurasia, M., Chourasia, M.K., Jain, N.K., Jain, A., Soni, V. *et al.* (2006) Cross-linked guar gum microspheres: A viable approach for improved delivery of anticancer drugs for the treatment of colorectal cancer. *AAPS PharmSciTech* 7, 1–9.

Chavan, S.S., Pavlov, V.A. and Tracey, K.J. (2017) Mechanisms and therapeutic relevance of neuro-immune communication. *Immunity* 46(6), 927–942. Available at: https://doi.org/10.1016/j.immuni.2017.06.008

Chen, D., Zhao, C.-X., Lagoin, C., Hai, M., Arriaga, L.R. *et al.* (2017) Dispersing hydrophobic natural colourant β-carotene in shellac particles for enhanced stability and tunable colour. *Royal Society Open Science* 4(12), 170919.

Cherbut, C., Michel, C., Raison, V., Kravtchenko, T. and Severine, M. (2009) Acacia gum is a bifidogenic dietary fibre with high digestive tolerance in healthy humans. *Microbial Ecology in Health and Disease* 15(1), 43–50. Available at: https://doi.org/10.1080/08910600310014377

Cherukuri, S.R., Friello, D.R., Parker, E., Hopkins, W. and Mackay, D.A.M. (1983) Stable liquid red beet color and chewing gum containing same. US Patent No. 4,371,549.

Chevrier, M.R., Ryan, A.E., Lee, D.Y., Zhongze, M., Wu Yan, Z. *et al.* (2005) *Boswellia carterii* extract inhibits TH1 cytokines and promotes TH2 cytokines *in vitro*. *Clinical and Vaccine Immunology* 12(5), 575–580.

Chinatangkul, N., Limmatvapirat, C., Nunthanid, J., Luangtana-anan, M., Sriamornsak, P. *et al.* (2018) Design and characterization of monolaurin loaded electrospun shellac nanofibers with antimicrobial activity. *Asian Journal of Pharmaceutical Sciences* 13(5), 459–471. Available at: https://doi.org/10.1016/j.ajps.2017.12.006

Chitravathi, K., Chauhan, O.P. and Raju, P.S. (2014) Postharvest shelf-life extension of green chillies (Capsicum annuum L.) using shellac-based edible surface coatings. *Postharvest Biology and Technology* 92, 146–148.

Chivero, P., Gohtani, S., Yoshii, H. and Nakamura, A. (2015) Effect of xanthan and guar gums on the formation and stability of soy soluble polysaccharide oil-in-water emulsions. *Food Research International* 70, 7–14. Available at: https://doi.org/10.1016/j.foodres.2015.01.025

Chudzikowski, R.J.C. (1971) Guar gum and its application. *Journal of the Society of Cosmetic Chemists* 22, 43–60.

Chuenbarn, T., Tuntarawongsa, S., Janmahasatian, S. and Phaechamud, T. (2021) Bleached shellac *in situ* forming micro-particle fabricated with different oils as antibacterial delivery system for periodontitis treatment. *Materials Today: Proceedings* 47, 3546–3553. Available at: https://doi.org/10.1016/J.MATPR.2021.03.549

Common, J.L. and Manor, B. (1962) Powdered instant beverage mix. US Patent no.3,023,106.

Contreras-Oliva, A., Perez-Gago, M.B. and Rojas-Argudo, C. (2012) Effects of chitosan coatings on physicochemical and nutritional quality of clementine mandarins cv. *'Oronules'*. *Food Science and Technology International* 18, 303–315.

Coon, J.T. and Ernst, E. (2003) Herbs for serum cholesterol reduction: A systematic review. *Database of Abstracts of Reviews of Effects*. Available at: https://www.ncbi.nlm.nih.gov/books/NBK70118/ (accessed 9 May 2025).

Cui, J., Huang, L.I., Zhao, A., Lew, J.L., Yu, J. *et al.* (2003) Guggulsterone is a farnesoid X receptor antagonist in coactivator association assays but acts to enhance transcription of bile salt export pump. *Journal of Biological Chemistry* 278(12), 10214–10220.

Cushman, D.R. and Schick, J.W. (1975) New Jersey Patent No.3,891,620.

Dall'Alba, V., Silva, F.M., Antonio, J.P., Steemburgo, T., Royer, C.P. *et al.* (2013) Improvement of the metabolic syndrome profile by soluble fibre – guar gum – in patients with type 2 diabetes: A randomised clinical trial. *British Journal of Nutrition* 110(9), 1601–1610.

Dauqan, E. and Abdullah, A. (2013) Utilization of gum arabic for industries and human health. *American Journal of Applied Sciences* 10(10), 1270–1279. Available at: https://doi.org/10.3844/ajassp.2013.1270.1279

Davidson, R.L. (1980) *Handbook of Water-Soluble Gums and Resins*. McGraw-Hill, New York.

De Rapper, S., Van Vuuren, S., Kamatou, G., Viljoen, A. and Dagne, E. (2012) The additive and synergistic antimicrobial effects of select frankincense and myrrh oils – a combination from the pharaonic pharmacopoeia. *Letters in Applied Microbiology* 54(4), 352–358.

de Souza, A.G., Cesco, C.T., de Lima, G.F., Artifon, S.E.S., Rosa, D. dos S. *et al.* (2019) Arabic gum-based composite hydrogels reinforced with eucalyptus and pinus residues for controlled phosphorus release. *International Journal of Biological Macromolecules* 140, 33–42. Available at: https://doi.org/10.1016/j.ijbiomac.2019.08.106

Deb, A. and Vimala, R. (2018) Natural and synthetic polymer for graphene oxide mediated anticancer drug delivery – a comparative study. *International Journal of Biological Macromolecules* 107, 2320–2333.

Dehpour, A.A., Ebrahimzadeh, M.A., Fazel, N.S. and Mohammad, N.S. (2009) Antioxidant activity of the methanol extract of *Ferula asafoetida* and its essential oil composition. *Grasas Aceites* 60(4), 405–412.

Deore, S.L. and Khadabadi, S.S. (2008) Standardisation and pharmaceutical evaluation of *Chlorophytum borivilianum* mucilage. *Rasayan Journal of Chemistry* 1, 887–892.

Deshmukh, A.S., Setty, C.M., Badiger, A.M. and Muralikrishna, K.S. (2012) Gum ghatti: A promising polysaccharide for pharmaceutical applications. *Carbohydrate Polymers* 87(2), 980–986. Available at: https://doi.org/10.1016/j.carbpol.2011.08.099

Dey, P., Bal, T. and Gupta, R.N. (2020) Fabrication and invitro evaluation of electrospun gum ghatti-polyvinyl alcohol polymeric blend green nanofibre mat (GG-PVA NFM) as a novel material for polymeric scaffolds in wound healing. *Polymer Testing* 91, 106826. Available at: https://doi.org/10.1016/J.POLYMERTESTING.2020.106826

Divya, K., Ramalakshmi, K., Murthy, P.S. and Jagan Mohan Rao, L. (2014) Volatile oils from *Ferula asafoetida* varieties and their antimicrobial activity. *LWT - Food Science and Technology* 59(2P1), 774–779. Available at: https://doi.org/10.1016/j.lwt.2014.07.013

Dolara, P., Corte, B., Ghelardini, C., Pugliese, A.M., Cerbai, E. *et al.* (2000) Local anaesthetic, antibacterial and antifungal properties of sesquiterpenes from myrrh. *Planta Medica* 66(4), 356–358.

Drake, S.R. and Nelson, J.W. (1990) Storage quality of waxed and nonwaxed "Delicious" and "Golden Delicious." *Journal of Food Quality* 13, 331–341.

Du, Y., Wang, L., Mu, R., Wang, Y., Li, Y. *et al.* (2019) Fabrication of novel konjac glucomannan/shellac film with advanced functions for food packaging. *International Journal of Biological Macromolecules* 131, 36–42. Available at: https://doi.org/10.1016/J.IJBIOMAC.2019.02.142

Dubey, R., Shaikh, S., Dhande, S., Joshi, Y.M. and Kadam, V.J. (2012) *Anogeissus latifolia* - an overview. *Research Journal of Pharmacognosy and Phytochemistry* 4(6), 287–290.

Dziezak, J.D. (1991) A focus on gums. *Food Technology (Chicago)* 45(3), 116–132.

Eastwood, M.A., Brydon, W.G. and Anderson, D.M.W. (1983) The effects of dietary gum karaya (*Sterculia*) in man. *Toxicology Letters* 17(1–2), 159–166. Available at: https://doi.org/10.1016/0378-4274(83)90052-8

El Ashry, E.S.H., Rashed, N., Salama, O.M. and Saleh, A. (2003) Components, therapeutic value and uses of myrrh. *Die Pharmazie-An International Journal of Pharmaceutical Sciences* 58(3), 163–168.

Eldeen, A.M.G., Amer, H. and Helmy, W.A. (2006) Cancer chemopreventive and anti-inflammatory activities of chemically modified guar gum. *Chemico-Biological Interactions* 161(3), 229–240.

Eltahir, M.E.S., Elsayed, M.E.O. and Hamad, M.A.A. (2013) Assessment of local knowledge and traditional uses of *Acacia senegal* in rural areas of North Kordofan, Sudan. In: *Proceedings of the Conference on International Research on Food Security, Natural Resource Management and Rural Development: Agricultural Development within the Rural-Urban Continuum, September 17-19, 2013, Stuttgart-Hohenheim, Germany*.

Eng, J.L. and Mackenzie, K.M. (1984) Glyceride fat based clouds for ready-to-drink beverages. US Patent Application 4, 479, 971.

Farag, Y. (2010) Characterization of different shellac types and development of shellac coated dosage forms. PhD dissertation, Staats-und Universitätsbibliothek Hamburg Carl von Ossietzky, Germany.

Farag, Y. and Leopold, C.S. (2011) Development of shellac-coated sustained release pellet formulations. *European Journal of Pharmaceutical Sciences* 42, 400–405.

Farahmandfar, R., Mohseni, M. and Asnaashari, M. (2017) Effects of quince seed, almond, and tragacanth gum coating on the banana slices properties during the process of hot air drying. *Food Science & Nutrition* 5(6), 1057–1064.

Fatehi, M., Farifteh, F. and Fatehi-Hassanabad, Z. (2004) Antispasmodic and hypotensive effects of *Ferula asafoetida* gum extract. *Journal of Ethnopharmacology* 91(2–3), 321–324. Available at: https://doi.org/10.1016/j.jep.2004.01.002

Fazel, M., Azizi, M.H., Abbasi, S. and Barzegar, M. (2012) Effect of tragacanth, glycerol and sunflower oil on potato starch-based edible films. *Iranian Journal of Food Science and Technology* 9(24), 97–106.

Flowerman, P.M. (1985) *Marketing Sudanese Gum Arabic in the USA - Facts and Options.* The U.S. Agency for International Development Khartoum, Sudan.

Foglio Bonda, A., Regis, L., Giovannelli, L. and Segale, L. (2020) Alginate/maltodextrin and alginate/shellac gum core-shell capsules for the encapsulation of peppermint essential oil. *International Journal of Biological Macromolecules* 162, 1293–1302. Available at: https://doi.org/10.1016/J.IJBIOMAC.2020.06.194

Francis, J.A., Raja, S.N. and Nair, M.G. (2004) Bioactive terpenoids and guggulusteroids from *Commiphora mukul* gum resin of potential anti-inflammatory interest. *Chemistry & Biodiversity* 1(11), 1842–1853.

Freitas, A.A.R., Ribeiro, A.J., Santos, A.C., Veiga, F., Nunes, L.C.C. *et al.* (2020) *Sterculia striata* gum as a potential oral delivery system for protein drugs. *International Journal of Biological Macromolecules* 164, 1683–1692. Available at: https://doi.org/10.1016/j.ijbiomac.2020.07.276

Friend, D.R., Altaf, S.A., Yu, K.L., Gebert, M.S., and Proc. (1997) Development of a zero-order nifedipine dosage form using COSRx technology. *International Symposium on Controlled Release of Bioactive Materials* 24, 311–312.

Fu, G., Tian, X., Xu, H., Guo, Y., Liu, M. *et al.* (2013) Application of silica aerogel-shellac wax composite in mango preservation. In: Sung, W.P., Kao, J.C.M. and R, C. (eds) *Frontiers of Manufacturing Science and Measuring Technology III*. Trans Tech Publications, Switzerland, pp. 2096–2100.

Gately, N. and Kennedy, J. (2017) The development of a melt-extruded shellac carrier for the targeted delivery of probiotics to the colon. *Pharmaceutics* 9(4), 38.

Gatenby, S.J., Ellis, P.R., Morgan, L.M. and Judd, P.A. (1996) Effect of partially depolymerized guar gum on acute metabolic variables in patients with non-insulin-dependent diabetes. *Diabetic Medicine* 13(4), 358–364.

Gautami, S. and Bhat, R.V. (1992) *A Monograph on Gum Karaya*. Silver Prints, Uppal, Hyderabad, India.

Gavlighi, H.A., Meyer, A.S. and Mikkelsen, J.D. (2013) Tragacanth gum: Functionality and prebiotic potential. *Agro Food Industry Hi Tech* 24(2), 46–48.

George, M. and Abraham, T.E. (2007) pH sensitive alginate-guar gum hydrogel for the controlled delivery of protein drugs. *International Journal of Pharmaceutics* 335(1–2), 123–129. https://doi.org/10.1016/j.ijpharm.2006.11.009

Ghayempour, S., Montazer, M. and Mahmoudi, M. (2015) Tragacanth gum as a natural polymeric wall for producing antimicrobial nanocapsules loaded with plant extract. *International Journal of Biological Macromolecules* 81, 514–520.

Ghodke, S.K. (2009) Effect of guar gum on dough stickiness and staling in chapatti - An Indian unleavened flat bread. *International Journal of Food Engineering* 5(3).

Ghoshal, S., Khan, M.A., Khan, R.A., Gul-E-Noor, F. and Chowdhury, A.M.S. (2010) Study on the thermo-mechanical and biodegradable properties of shellac films grafted with acrylic monomers by gamma radiation. *Journal of Polymers and the Environment* 18(3), 216–223. https://doi.org/10.1007/S10924-010-0182-3

Glicksman, M. and Farkas, E.H. (1975) Method of preventing gelation in canned gravy-based pet foods. US Patent No. 3,881,031.

Gliko-Kabir, I., Yagen, B., Baluom, M. and Rubinstein, A. (2000a) Phosphated crosslinked guar for colon-specific drug delivery: II. *In vitro* and *in vivo* evaluation in the rat. *Journal of Controlled Release* 63, 129–134.

Gliko-Kabir, I., Yagen, B., Penhasi, A. and Rubinstein, A. (2000b) Phosphated crosslinked guar for colon-specific drug delivery: I. Preparation and Physicochemical Characterization. *Journal of Controlled Release* 63, 121–127.

Goswami, S. and Naik, S. (2014) Natural gums and its pharmaceutical application. *Journal of Scientific and Innovative Research* 3(1), 112–121.

Goyal, S., Sharma, P., Ramchandani, U., Shivastanva, S.K. and Dubey, P.K. (2011) Novel anti inflammatory topical herbal gels containing *Withania somnifera* and *Boswellia serrata*. *International Journal of Pharmaceutical and Biological Archive* 2(4), 1087–1094.

Grossmith, F. (1956) Process for the production of jellies or viscous solutions. GB Patent No. 750, 126.

Growers, S. (1958) Flavour preservation of essential oils. US Patent No. 802,538.

Gujral, H.S., Sharma, A. and Singh, N. (2002) Effect of hydrocolloids, storage temperature, and duration on the consistency of tomato ketchup. *International Journal of Food Properties* 5(1), 179–191.

Guo, J., Skinner, G.W., Harcum, W.W. and Barnum, P.E. (1998) Pharmaceutical applications of naturally occurring water-soluble polymers. *Pharmaceutical Science & Technology Today* 1(6), 254–261.

Gupta, A.P. and Verma, D.K. (2014) Guar gum and their derivatives: A research profile. *Int. J. Adv. Res* 2(1), 680–690.

Gupta, S., Saurabh, C.K., Variyar, P.S. and Sharma, A. (2015) Comparative analysis of dietary fiber activities of enzymatic and gamma depolymerized guar gum. *Food Hydrocolloids* 48, 149–154. Available at: https://doi.org/10.1016/j.foodhyd.2015.02.013

Hadavi, M., Hasannia, S., Faghihi, S., Mashayekhi, F., Zadeh, H.H. *et al.* (2017) Novel calcified gum arabic porous nano-composite scaffold for bone tissue regeneration. *Biochemical and Biophysical*

Research Communications 488(4), 671–678. Available at: https://doi.org/10.1016/j.bbrc.2017.03.046

Haeri, S.M.J., Sadeghi, Y., Salehi, M., Farahani, R.M. and Mohsen, N. (2016) Osteogenic differentiation of human adipose-derived mesenchymal stem cells on gum tragacanth hydrogel. *Biologicals* 44(3), 123–128.

Heini, A.F., Lara-Castro, C., Schneider, H., Kirk, K.A., Considine, R.V. *et al*. (1998) Effect of hydrolyzed guar fiber on fasting and postprandial satiety and satiety hormones: A double-blind, placebo-controlled trial during controlled weight loss. *International Journal of Obesity* 22(9), 906–909.

Henning, S., Leick, S., Kott, M., Rehage, H. and Suter, D. (2012) Sealing liquid-filled pectinate capsules with a shellac coating. *Journal of Microencapsulation* 29, 147–155.

Hitzfeld, A., Leuenberger, B.H. and Vidoni, O. (2014) Compositions of fat-soluble active ingredients containing gum ghatti. US Patent No. 8680161 B2.

Huang, Y., Yu, H. and Xiao, C. (2007) PH-sensitive cationic guar gum/poly (acrylic acid) polyelectrolyte hydrogels: Swelling and *in vitro* drug release. *Carbohydrate Polymers* 69(4), 774–783. Available at: https://doi.org/10.1016/j.carbpol.2007.02.016

Hutmacher, D.W. (2000) Scaffolds in tissue engineering bone and cartilage. *Biomaterials* 21(24), 2529–2543. Available at: https://doi.org/10.1016/S0142-9612(00)00121-6

Huzinec, R.J. and Graff, A.H. (1987) Coatings for chewing gums containing gum arabic and a soluble calcium salt. US Patent No. 4,681,766.

Ido, T., Ogasawara, T., Katayama, T., Sasaki, Y., Al-Assaf, S. *et al*. (2008) Emulsification property of gatifolia (gum ghatti) used for emulsions in food products. *Foods and Food Ingredients Journal of Japan* 213, 365–372.

Imeson, A. (1992) *Thickening and Gelling Agents for Food*. Chapman and Hall, London.

Imeson, A. (2010) *Food Stabilisers,Thickeners and Gelling Agents*, Vol. 34. Wiley-Blackwell, Chichester, UK.

Iranshahi, M., Amin, G., Salehi Sourmaghi, M., Shafiee, A. and Hadjiakhoondi, A. (2006) Sulphur-containing compounds in the essential oil of the root of *Ferula persica* Willd var. *persica*. *Flavour Fragrance Journal* 21(2), 260–261.

Iranshahy, M. and Iranshahi, M. (2011) Traditional uses, phytochemistry and pharmacology of asafoetida (*Ferula assa-foetida* oleo-gum-resin) – A review. *Journal of Ethnopharmacology* 134(1), 1–10. Available at: https://doi.org/10.1016/j.jep.2010.11.067

Jabri, T., Imran, M., Rao, K., Ali, I., Arfan, M. *et al*. (2018) Fabrication of lecithin-gum tragacanth mucoadhesive hybrid nano-carrier system for *in-vivo* performance of Amphotericin B. *Carbohydrate Polymers* 194, 89–96.

Jacoby, A., Morrison, K.A., Hooper, R.C., Asanbe, O., Joyce, J. *et al*. (2017) Fabrication of capillary-like structures with pluronic F127® and *Kerria lacca* resin (shellac) in biocompatible tissue-engineered constructs. *Journal of Tissue Engineering and Regenerative Medicine* 11(8), 2388–2397. Available at: https://doi.org/10.1002/TERM.2138

Jafari, S., Hojjati, M. and Noshad, M. (2018) Influence of soluble soybean polysaccharide and tragacanth gum based edible coating to improve the quality of fresh cut apple slices. *Journal of Food Processing and Preservation* 42(6), e13638.

Jagtap, S.D., Deokule, S.S., Pawar, P.K. and Harsulkar, A.M. (2009) Traditional ethnomedicinal knowledge confined to the Pawra tribe of Satpura Hills, Maharashtra, India. *Ethnobotanical Leaflets* 13, 98–115.

Jain, N.K. and Dixit, V.K. (1988) Studies on gums and their derivatives as binding agent. *Indian Journal of Pharmaceutical Sciences* 50, 113–114.

Jain, A., Thakur, D., Ghoshal, G., Katare, O.P. and Shivhare, U.S. (2016) Characterization of microcapsulated β-carotene formed by complex coacervation using casein and gum tragacanth. *International Journal of Biological Macromolecules* 87, 101–113.

Jales, M. (2018) *Commodities at a Glance: Special Issue on Gum Arabic*. UNCTAD, Geneva, Switzerland.

Jayaramudu, J., Reddy, G.S.M., Varaprasad, K., Sadiku, E.R., Sinha Ray, S. *et al*. (2013) Preparation and properties of biodegradable films from *Sterculia urens* short fiber/cellulose green composites. *Carbohydrate Polymers* 93(2), 622–627. Available at: https://doi.org/10.1016/j.carbpol.2013.01.032

Jenkins, D.J.A., Derek, T., Hockaday, R., Howarth, R., Apling, E.C. *et al*. (1977) Treatment of diabetes with guar gum: Reduction of urinary glucose loss in diabetics. *The Lancet* 310(8042), 779–780.

Jo, W.-S., Song, H.-Y., Song, N.-B., Lee, J.-H., Min, S.C. *et al*. (2014) Quality and microbial safety of "Fuji" apples coated with carnauba-shellac wax containing lemongrass oil. *LWT - Food Science and Technology* 55, 490–497.

Joshi, M.G., Setty, C.M., Deshmukh, A.S. and Bhatt, Y.A. (2010) Gum ghatti: A new release modifier for zero order release in 3-layered tablets of diltiazem hydrochloride. *Indian Journal of Pharmaceutical Education and Research* 44(1), 78–85.

Kaith, B.S., Sharma, K., Kumar, V., Kalia, S. and Swart, H.C. (2014) Fabrication and characterization of gum ghatti-polymethacrylic acid based electrically conductive hydrogels. *Synthetic Metals* 187(1), 61–67. Available at: https://doi.org/10.1016/j.synthmet.2013.10.021

Kala, C.P. (2016) Important gum yielding species *Anogeissus latifolia* (Roxb.) bedd., *Boswellia serrata* roxb. and *Sterculia urens* roxb.: Ethnobotany, population density and management. *Applied Ecology and Environmental Science* 4(3), 61–65. Available at: https://doi.org/10.12691/aees-4-3-2

Kamble, V.A. and Patil, S.D. (2008) Spice-derived essential oils: Effective antifungal and possible therapeutic agents. *Journal of Herbs, Spices & Medicinal Plants* 14(3–4), 129–134.

Kavoosi, G. and Rowshan, V. (2013) Chemical composition, antioxidant and antimicrobial activities of essential oil obtained from *Ferula asafoetida* oleo-gum-resin: Effect of collection time. *Food Chemistry* 138(4), 2180–2187.

Kays, S.E., Morris, J.B. and Kim, Y. (2006) Total and soluble dietary fiber variation in *Cyamopsis tetragonoloba* (L.) Taub. (guar) genotypes. *Journal of Food Quality* 29, 383–391.

Kelly, B.J. and Potter, N.N. (1990) Dialyzable calcium from milk processed with soluble fiber-containing gums, thickeners, and cocoa. *Journal of Food Science* 55(4), 1004–1007.

Kenyon, C.J., Nardi, R.V., Wong, D., Hooper, G., Wilding, I.R. *et al.* (1997) Colonic delivery of dexamethasone: A pharmacoscintigraphic evaluation aliment. *Pharmacology & Therapeutics* 11, 205–213.

Keskin, S.O., Sumnu, G. and Sahin, S. (2007) A study on the effects of different gums on dielectric properties and quality of breads baked in infrared-microwave combination oven. *European Food Research and Technology* 224, 329–334.

Khazdair, M.R. and Boskabady, M.H. (2015) The relaxant effect of *Ferula asafoetida* on smooth muscles and the possible mechanisms. *Journal of HerbMed Pharmacology* 4(2), 40–44.

Khor, C.M., Ng, W.K., Kanaujia, P., Chan, K.P. and Dong, Y. (2017) Hot-melt extrusion microencapsulation of quercetin for taste-masking. *Journal of Microencapsulation* 34, 29–37.

Khorram, F., Ramezanian, A. and Hosseini, S. (2017) Shellac, gelatin and Persian gum as alternative coating for orange fruit. *Scientia Horticulturae* 225, 22–28.

Klis, J.B. (1966) Woody's Chunk O'Gold cold-pack cheese food weeps no more. *Food Processing Marketing* 27, 58–59.

Kohajdová, Z. and Karovičová, J. (2009) Application of hydrocolloids as baking improvers. *Chemical Papers* 63(1), 26–38. Available at: https://doi.org/10.2478/s11696-008-0085-0

Kong, L., Amstad, E., Hai, M., Ke, X., Chen, D. *et al.* (2017) Biocompatible microcapsules with a water core templated from single emulsions. *Chinese Chemical Letters* 28, 1897–1900.

Kong, L., Chen, R., Wang, X., Zhao, C.X., Chen, Q. *et al.* (2019a) Controlled co-precipitation of biocompatible colorant-loaded nanoparticles by microfluidics for natural color drinks. *Lab on a Chip* 19(12), Article 12. Available at: https://doi.org/10.1039/C9LC00240E

Kong, L., Jin, X., Hu, D., Feng, L., Chen, D. *et al.* (2019b) Functional delivery vehicle of organic nanoparticles in inorganic crystals. *Chinese Chemical Letters* 30, 2351–2354.

Koocheki, A., Ghandi, A., Razavi, S.M., Mortazavi, S.A. and Vasiljevic, T. (2009) The rheological properties of ketchup as a function of different hydrocolloids and temperature. *International Journal of Food Science & Technology* 44(3), 596–602.

Kora, A.J., Beedu, S.R. and Jayaraman, A. (2012) Size-controlled green synthesis of silver nanoparticles mediated by gum ghatti (*Anogeissus latifolia*) and its biological activity. *Organic and Medicinal Chemistry Letters* 2(1), 1–10. Available at: https://doi.org/10.1186/2191-2858-2-17

Kouassi, K.H.S., Bajji, M. and Jijakli, H. (2012) The control of postharvest blue and green molds of citrus in relation with essential oil-wax formulations, adherence and viscosity. *Postharvest Biology and Technology* 73, 122–128.

Kovacs, E.M.R., Westerterp-Plantenga, M.S., Saris, W.H.M., Goossens, I., Geurten, P. *et al.* (2001) The effect of addition of modified guar gum to a low-energy semisolid meal on appetite and body weight loss. *International Journal of Obesity* 25(3), 307–315.

Kraisit, P., Limmatvapirat, S., Nunthanid, J., Sriamornsak, P. and Luangtana-anan, M. (2013) Nanoparticle formation by using shellac and chitosan for a protein delivery system. *Pharmaceutical Development and Technology* 18(3), 686–693. Available at: https://doi.org/10.3109/10837450.2012.685657

Krishnaiah, Y.S.R., Satyanarayana, S., Rama Prasad, Y.V. and Narasimha Rao, S. (1998a) Evaluation of guar gum as a compression coat for drug targeting to colon. *International Journal of Pharmaceutics* 171(2), 137–146. Available at: https://doi.org/10.1016/S0378-5173(98)00172-0

Krishnaiah, Y.S.R., Satyanarayana, S., Rama Prasad, Y.V. and Narasimha Rao, S. (1998b) Gamma scintigraphic studies on guar gum matrix tablets for colonic drug delivery in healthy human volunteers. *Journal of Controlled Release* 55(2–3), 245–252. Available at: https://doi.org/10.1016/S0168-3659(98)00057-1

Krishnaiah, Y.S.R., Satyanaryana, S. and Prasad, Y.V.R. (1999) Studies of guar gum compression-coated 5-aminosalicylic acid tablets for colon-specific drug delivery. *Drug Drug Development and Industrial Pharmacy* 25, 651–657.

Krishnaiah, Y.S.R., Devi, A.S., Rao, L.N., Reddy, P.R.B., Karthikeyan, R.S. *et al.* (2001a) Guar gum as a carrier for colon specific delivery; influence of metronidazole and tinidazole on *in vitro* release of albendazole from guar gum matrix tablets. *Journal of Pharmacy & Pharmaceutical Sciences* 4(3), 235–243.

Krishnaiah, Y.S.R., Veer Raju, P., Dinesh Kumar, B., Bhaskar, P. and Satyanarayana, V. (2001b) Development of colon targeted drug delivery systems for mebendazole. *Journal of Controlled Release* 77(1–2), 87–95. Available at: https://doi.org/10.1016/S0168-3659(01)00461-8

Krishnaiah, Y.S.R., Reddy, P.R.B., Satyanarayana, V. and Karthikeyan, R.S. (2002a) Studies on the development of oral colon targeted drug delivery systems for metronidazole in the treatment of amoebiasis. *International Journal of Pharmaceutics* 236, 43–55.

Krishnaiah, Y.S.R., Satyanarayana, V., Dinesh Kumar, B. and Karthikeyan, R.S. (2002b) *In vitro* drug release studies on guar gum-based colon targeted oral drug delivery systems of 5-fluorouracil. *European Journal of Pharmaceutical Sciences* 16(3), 185–192. Available at: https://doi.org/10.1016/S0928-0987(02)00081-7

Krishnan, S., Kshirsagar, A.C. and Singhal, R.S. (2005) The use of gum arabic and modified starch in the microencapsulation of a food flavoring agent. *Carbohydrate Polymers* 62(4), 309–315. Available at: https://doi.org/10.1016/J.CARBPOL.2005.03.020

Kulanthaivel, S., Sharan Rantha, V.S., Agarwal, T., Pradhan, S., Pal, K. *et al.* (2017) Gum tragacanth-alginate beads as proangiogenic-osteogenic cell encapsulation systems for bone tissue engineering. *Journal of Materials Chemistry B* 5(22), 4177–4189.

Kumar, P. and Singh, D.K. (2006) Molluscicidal activity of *Ferula asafoetida*, *Syzygium aromaticum* and *Carum carvi* and their active components against the snail *Lymnaea acuminata*. *Chemosphere* 63(9), 1568–1574. Available at: https://doi.org/10.1016/j.chemosphere.2005.08.071

Kumar, A., Balakrishna, T. and Rajiv, J. (2011) Formulation and evaluation of mucoadhesive microcapsules of metformin HCL with gum karaya. *International Journal of Pharmacy and Pharmaceutical Sciences* 3(3), 150–155.

Kuo, D.-C., Hsu, S.-P. and Chien, C.-T. (2009) Partially hydrolyzed guar gum supplement reduces high-fat diet increased blood lipids and oxidative stress and ameliorates FeC13-induced acute arterial injury in hamsters. *Journal of Biomedical Science* 16(1). Available at: https://doi.org/10.1186/1423-0127-16-15

Lampe, J.W., Effertz, M.E., Larson, J.L. and Slavin, J.L. (1992) Gastrointestinal effects of modified guar gum and soy polysaccharide as part of an enteral formula diet. *Journal of Parenteral and Enteral Nutrition* 16(6), 538–544.

Lamsen, M.R.L., Wang, T., D'Souza, D., Dia, V., Chen, G. *et al.* (2020) Encapsulation of vitamin D3 in gum arabic to enhance bioavailability and stability for beverage applications. *Journal of Food Science* 85(8), 2368–2379. Available at: https://doi.org/10.1111/1750-3841.15340

Le Cerf, D., Irinei, F. and Muller, G. (1990) Solution properties of gum exudates from *Sterculia urens* (Karaya gum). *Carbohydrate Polymers* 13(4), 375–386. Available at: https://doi.org/10.1016/0144-8617(90)90037-S

Lee, C.L., Chiang, C.L., Cheng, L.H. and Liaw, C.C. (2009) Influenza a (H1N1) antiviral and cytotoxic agents from *Ferula asafoetida*. *Journal of Natural Products* 72(9), 1568–1572.

Leick, S., Kott, M., Degen, P., Henning, S., Paesler, T. *et al.* (2011) Mechanical properties of liquid-filled shellac composite capsules. *Physical Chemistry Chemical Physics* 13, 2765–2773.

Lemenith, M. and Teketay, D. (2005) Frankincense and myrrh resources of Ethiopia: II. Medicinal and industrial uses. *Ethiopian Journal of Science* 26(2), 161–172.

Lett, J.A., Sagadevan, S., Shahnavaz, Z., Latha, M.B., Alagarswamy, K. *et al.* (2020) Exploration of gum ghatti-modified porous scaffolds for bone tissue engineering applications. *New Journal of Chemistry* 44(6), 2389–2401. Available at: https://doi.org/10.1039/c9nj05575d

Li, X., Wu, W., Wang, J. and Duan, Y. (2006) The swelling behavior and network parameters of guar gum/poly (acrylic acid) semi-interpenetrating polymer network hydrogels. *Carbohydrate Polymers* 66, 473–479.

Li, X., Wu, W. and Liu, W. (2008) Synthesis and properties of thermo-responsive guar gum/poly (N-isopropylacrylamide) interpenetrating polymer network hydrogels. *Carbohydrate Polymers* 71, 394–402.

Li, Y., Dong, Q., Chen, J. and Li, L. (2020) Effects of coaxial electrospun eugenol loaded core-sheath PVP/shellac fibrous films on postharvest quality and shelf life of strawberries. *Postharvest Biology and Technology* 159, 111028.

Leupold, C.W., Kellner, W. and Hellmuth, J. (1962) Material with deodorising action. GB Patent no.901,554.

Limmatvapirat, S., Limmatvapirat, C., Luangtana-anan, M., Nunthanid, J., Oguchi, T. *et al.* (2004) Modification of physicochemical and mechanical properties of shellac by partial hydrolysis. *International Journal of Pharmaceutics* 278, 41–49.

Limmatvapirat, S., Limmatvapirat, C., Puttipipatkhachorn, S., Nuntanid, J. and Luangtana-anan, M. (2007) Enhanced enteric properties and stability of shellac films through composite salts formation. *European Journal of Pharmaceutics and Biopharmaceutics* 67(3), 690–698. Available at: https://doi.org/10.1016/j.ejpb.2007.04.008

Limmatvapirat, S., Limmatvapirat, C., Puttipipatkhachorn, S., Nunthanid, J., Luangtana-anan, M. *et al.* (2008a) Modulation of drug release kinetics of shellac-based matrix tablets by *in-situ* polymerization through annealing process. *European Journal of Pharmaceutics and Biopharmaceutics* 69(3), 1004–1013. Available at: https://doi.org/10.1016/j.ejpb.2008.01.027

Limmatvapirat, S., Nunthanid, J., Puttipipatkhachorn, S. and Luangtana-anan, M. (2008b) Effect of alkali treatment on properties of native shellac and stability of hydrolyzed shellac. *Pharmaceutical Development and Technolgy* 10(1), 41–46. Available at: https://doi.org/10.1081/PDT-35897

Limmatvapirat, S., Thammachat, T., Sriamornsak, P., Luangtana-Anan, M., Nunthanid, J. *et al.* (2011) Preparation and characterization of shellac fiber as a novel material for controlled drug release. *Advanced Materials Research* 152–153, 1232–1235. Available at: https://doi.org/10.4028/WWW.SCIENTIFIC.NET/AMR.152-153.1232

Lin, Y., Wu, S., Li, X., Wu, X., Yang, C. *et al.* (2018) Microstructure and performance of Z-scheme photocatalyst of silver phosphate modified by MWCNTs and Cr-doped $SrTiO_3$ for malachite green degradation. *Applied Catalysis B: Environment and Energy* 227, 557–570.

Lloyd, A.O. and Read, A.E. (1964) Improvements in or relating to glaze mixes. GB Patent Application 977,838.

Luangtana-anan, M., Limmatvapirat, S., Nunthanid, J., Wanawongthai, C., Chalongsuk, R. *et al.* (2007) Effect of salts and plasticizers on stability of shellac film. *Journal of Agricultural and Food Chemistry* 55(3), 687–692.

Luangtana-anan, M., Nunthanid, J. and Limmatvapirat, S. (2010) Effect of molecular weight and concentration of polyethylene glycol on physicochemical properties and stability of shellac film. *Journal of Agricultural and Food Chemistry* 58, 12934–12940.

Luangtana-anan, M., Soradech, S., Saengsod, S., Nunthanid, J. and Limmatvapirat, S. (2017) Enhancement of moisture protective properties and stability of pectin through formation of a composite film: Effects of shellac and plasticizer. *Journal of Food Science* 82, 2915–2925.

Luján-Medina, G.A., Ventura, J., Ascacio-Valdés, J.A., Cerqueira, M.A., Boone Villa, D. *et al.* (2015) Microencapsulation of ellagic acid from pomegranate husk and karaya gum by spray drying. *International Journal of Pharmacy and Pharmaceutical Sciences* 7(1), 212–216.

Luo, Q., Li, K., Xu, J., Li, K., Zheng, H. *et al.* (2016) Novel biobased sodium shellac for wrapping disperse multiscale emulsion particles. *Journal of Agricultural and Food Chemistry* 64, 9374–9380.

Magalhães, G.A., Moura Neto, E., Sombra, V.G., Richter, A.R., Abreu, C.M.W.S. *et al.* (2016) Chitosan/*Sterculia striata* polysaccharides nanocomplex as a potential chloroquine drug release device. *International Journal of Biological Macromolecules* 88, 244–253. Available at: https://doi.org/10.1016/j.ijbiomac.2016.03.070

Mahendra, P. and Bisht, S. (2012) *Ferula asafoetida*: Traditional uses and pharmacological activity. *Pharmacognosy Reviews* 6(12), 141–146.

Maia, F.R., Musson, D.S., Naot, D., Silva, L.P., Bastos, A.R. *et al.* (2018) Differentiation of osteoclast precursors on gellan gum-based spongy-like hydrogels for bone tissue engineering. *Biomedical Materials* 13(3), 35012.

Mallikarjuna, G.U., Dhanalakshmi, S., Raisuddin, S. and Rao, A.R. (2003) Chemomodulatory influence of *Ferula asafoetida* on mammary epithelial differentiation, hepatic drug metabolizing enzymes, antioxidant profiles and N-methyl-N-nitrosourea-induced mammary carcinogenesis in rats. *Breast Cancer Research and Treatment* 81(1), 1–10. Available at: https://doi.org/10.1023/A:1025448620558

Manna, P.J., Mitra, T., Pramanik, N., Kavitha, V., Gnanamani, A. *et al.* (2015) Potential use of curcumin loaded carboxymethylated guar gum grafted gelatin film for biomedical applications. *International Journal of Biological Macromolecules* 75, 437–446. Available at: https://doi.org/10.1016/j.ijbiomac.2015.01.047

Mariod, A.A. (2018) *Gum Arabic: Structure, Properties, Application and Economics*. Academic Press, New York.

Marsan, A.E. (1967) Sealing pad for a post-surgical drainage pouch. US Patent no.3,302,647.

Mathe, C., Connan, J., Archier, P., Mouton, M. and Vieillescazes, C. (2007) Analysis of frankincense in archaeological samples by gas chromatography mass spectrometry. *Annali Di Chimica* 97(7), 433–445.

McLean-Ross, H.A., Eastwood, M.A. and Brydon, W.G. (1983) A study of the effects of dietary gum Arabic in humans. *American Journal of Clinical Nutrition* 37(3), 368–375.

McNamee, B.F., O'Riordan, E.D. and O'Sullivan, M. (eds) (1998) Emulsification and microencapsulation properties of gum arabic. *Journal of Agricultural and Food Chemistry* 46, 4551–4555.

Mei, S., Han, P., Wu, H., Shi, J., Tang, L. *et al.* (2018) One-pot fabrication of chitin-shellac composite microspheres for efficient enzyme immobilization. *Journal of Biotechnology* 266, 1–8.

Minhas, K.S., Sidhu, J.S., Mudahar, G.S. and Singh, A.K. (2002) Flow behavior characteristics of ice cream mix made with buffalo milk and various stabilizers. *Plant Foods for Human Nutrition* 57(1), 25–40.

Miuchi, T. and Nishino, M. (2013) Method for preventing carotenoid dye from adhering to container. US Patent No. 20130115345 A1.

Miuchi, T., Nishino, M., Sasaki, Y., Morimoto, T. and Tanaka, Y. (2012) Turmeric pigment composition and method for preparing same. US Patent No. 20120195949 A1.

Miuchi, T., Nishino, M., Sasaki, Y., Morimoto, T. and Tanaka, Y. (2016) Turmeric pigment composition and method for preparing same. Patent CA 2775665 C.

Miyazawa, T. (2006) Hydrocolloid structures, which allow more water interactions through hydrogen bonding. *Carbohydrate Research* 341, 870–877.

Moghaddam, M. and Farhadi, N. (2015) Influence of environmental and genetic factors on resin yield, essential oil content and chemical composition of *Ferula assa-foetida* L. populations. *Journal of Applied Research on Medicinal and Aromatic Plants* 2(3), 69–76. Available at: https://doi.org/10.1016/j.jarmap.2015.04.001

Moghbel, A., Hemmati, A.-A., Agheli, H., Rashidi, I. and Amraee, K. (2005) The effect of tragacanth mucilage on the healing of full-thickness wound in rabbit. *Archives of Iranian Medicine* 8(4), 257–262.

Mohamed, S.A.A., El-Sakhawy, M., Nashy, E.L.S.H.A. and Othman, A.M. (2019) Novel natural composite films as packaging materials with enhanced properties. *International Journal of Biological Macromolecules* 136, 774–784.

Mohebbi, M., Ansarifar, E., Hasanpour, N. and Amiryousefi, M.R. (2012) Suitability of aloe vera and gum tragacanth as edible coatings for extending the shelf life of button mushroom. *Food and Bioprocess Technology* 5, 3193–3202.

Moradi, S., Sadrjavadi, K., Frahdian, N., Hossainzadeh, L. and Shahlaei, M. (2018a) Easy synthesis, characterization and cell cytotoxicity of green nano carbon dots using hydrothermal carbonization of Gum Tragacanth and chitosan bio-polymers for bioimaging. *Journal of Molecular Liquids* 259, 284–290. DOI: 10.1016/j.molliq.2018.03.054.

Moradi, S., Taran, M. and Shahlaei, M. (2018b) Investigation on human serum albumin and Gum Tragacanth interactions using experimental and computational methods. *International Journal of Biological Macromolecules* 107, 2525–2533.

Morales, E., Rubilar, M., Burgos-Díaz, C., Acevedo, F., Penning, M. *et al.* (2017) Alginate/shellac beads developed by external gelation as a highly efficient model system for oil encapsulation with intestinal delivery. *Food Hydrocolloids* 70, 321–328.

Morkhade, D.M., Fulzele, S.V., Satturwar, P.M. and Joshi, S.B. (2006) Gum copal and gum damar: Novel matrix forming materials for sustained drug delivery. *Indian Journal of Pharmaceutical Sciences* 68(1), 53–58.

Mostafa, Z., Soheil, P., Mahdi, J. and Mahmoodi, S. (2013) Antifungal effects of asafoetida seed essential oil on *in vitro* growth of five species of plant pathogenic fungi. *International Research Journal of Applied and Basic Sciences* 4(5), 1159–1162.

Mostafavi, F.S., Kadkhodaee, R., Emadzadeh, B. and Koocheki, A. (2016) Preparation and characterization of tragacanth–locust bean gum edible blend films. *Carbohydrate Polymers* 139, 20–27.

Muhammad, D.R.A., Gonzalez, C.G., Sedaghat Doost, A., Walle, D., Meeren, P. *et al.* (2019) Improvement of antioxidant activity and physical stability of chocolate beverage using colloidal cinnamon nanoparticles. *Food and Bioprocess Technology* 12, 976–989.

Muhammad, D.R.A., Sedaghat Doost, A., Gupta, V., Sintang, M.D., Walle, D. *et al.* (2020) Stability and functionality of xanthan gum–shellac nanoparticles for the encapsulation of cinnamon bark extract. *Food Hydrocolloids* 100, 105377.

Munday, D.L. and Cox, P.J. (2000) Compressed xanthan and karaya gum matrices: Hydration, erosion and drug release mechanisms. *International Journal of Pharmaceutics* 203(1–2), 179–192. Available at: https://doi.org/10.1016/S0378-5173(00)00444-0

Murthy, S.N., Hiremath, S.R.R. and Paranjothy, K.L.K. (2004) Evaluation of carboxymethyl guar films for the formulation of transdermal therapeutic systems. *International Journal of Pharmaceutics* 272, 11–18.

Musa, H.H., Ahmed, A.A. and Musa, T.H. (2018) Chemistry, biological, and pharmacological properties of gum arabic. In: Mérillon, J.-M. and Ramawat, K.G. (eds) *Bioactive Molecules in Food*. Springer, Cham, Switzerland, pp. 1–18. Available at: https://doi.org/10.1007/978-3-319-54528-8_11-1

Nagarajan, M., Waszkuc, T.W. and Sun, J. (2001) Simultaneous determination of E- and Z-guggulsterones in dietary supplements containing *Commiphora mukul* extract (guggulipid) by liquid chromatography. *Journal of AOAC International* 84(1), 24–28.

Nawinda, C., Sirikarn, P., Suchada, P., Chutima, L. and Sontaya, L. (2020) Development of electrospun shellac and hydroxypropyl cellulose blended nanofibers for drug carrier application. *Key Engineering Materials* 859, 239–243. Available at: https://doi.org/10.4028/www.scientific.net/KEM.859.239

Nebergall, W.H. (1956) Dentrifice preparations. GB Patent Application 746,550.

Nur, M., Ramchandran, L. and Vasiljevic, T. (2016) Tragacanth as an oral peptide and protein delivery carrier: Characterization and mucoadhesion. *Carbohydrate Polymers* 143, 223–230.

Oehme, A., Valotis, A., Krammer, G., Zimmermann, I. and Schreier, P. (2011) Preparation and characterization of shellac-coated anthocyanin pectin beads as dietary colonic delivery system. *Molecular Nutrition & Food Research* 55, 75–85.

Oidtmann, J., Schantz, M., Maeder, K., Baum, M., Berg, S. *et al.* (2012) Preparation and comparative release characteristics of three anthocyanin encapsulation systems. *Journal of Agricultural and Food Chemistry* 60, 844–851.

Otadi, M. and Mobayen, S. (2011) The survey of kinetic behavior of immobilized glucose oxidase on gum tragacanth carrier. *World Applied Sciences Journal* 14, 15–19.

Otady, M. and Vaziri, A. (2005) Gum tragacanth gels as a new supporting matrix for immobilization of whole-cell. *Iranian Journal of Chemistry and Chemical Engineering* 24(4), 1–7.

Padhy, S., Rai, S., Lamba, N.N.H.K. and Upadhyay, M. (2014) Spices as potent antibacterial agents against *Staphylococcus aureus*. *Journal of Science and Technology* 4, 46–51.

Padil, V.V.T. and Černík, M. (2013) Green synthesis of copper oxide nanoparticles using gum karaya as a biotemplate and their antibacterial application. *International Journal of Nanomedicine* 8, 889–898. Available at: https://doi.org/10.2147/IJN.S40599

Padmanabhan, V.P., Prakash, N., Sankara Narayanan, T.S.N., Kulandaivelu, R., Mohammad, F. *et al.* (2021) Drug delivery and *in vitro* biological effects of gum ghatti-modified hydroxyapatite nanoporous composites. *Materials Chemistry and Physics* 263, 124385. Available at: https://doi.org/10.1016/J.MATCHEMPHYS.2021.124385

Pal, P., Singh, S.K., Mishra, S., Pandey, J.P. and Sen, G. (2019) Gum ghatti based hydrogel: Microwave synthesis, characterization, 5-Fluorouracil encapsulation and '*in vitro*' drug release evaluation. *Carbohydrate Polymers* 222, 114979. Available at: https://doi.org/10.1016/j.carbpol.2019.114979

Pandey, R. and Khuller, G.K. (2004) Polymer based drug delivery systems for mycobacterial infections. *Current Drug Delivery* 1(3), 195–201.

Parija, S., Misra, M. and Mohanty, A.K. (2001) Studies of natural gum adhesive extracts: An overview. *Journal of Macromolecular Science, Part C: Polymer Reviews* 41(3), 175–197.

Parisi, G., Bottona, E., Carrara, M., Cardin, F., Faedo, A. *et al.* (2005) Treatment effects of partially hydrolyzed guar gum on symptoms and quality of life of patients with irritable bowel syndrome. A multicenter randomized open trial. *Digestive Diseases and Sciences* 50(6), 1107–1112.

Partyka, A. (1963) Salad dressing. GB Patent no.936531.

Parvathi, K.M.M., Ramesh, C.K., Krishna, V., Paramesha, M. and Kuppast, I.J. (2009) Hypolipidemic activity of gum ghatti of *Anogeissus latifolia*. *Pharmacognosy Magazine* 5(19), 11–14.

Patel, A.R. and Dewettinck, K. (2016) Edible oil structuring: An overview and recent updates. *Food & Function* 7, 20–29.

Patel, A., Heussen, P., Hazekamp, J. and Velikov, K.P. (2011) Stabilisation and controlled release of silibinin from pH responsive shellac colloidal particles. *Soft Matter* 7, 8549–8555.

Patel, A.R., Schatteman, D., Vos, W.H.D. and Dewettinck, K. (2013) Shellac as a natural material to structure a liquid oil-based thermo reversible soft matter system. *RSC Advances* 3, 5324–5327.

Patel, D.M., Prajapati, D.G. and Patel, N.M. (2007) Seed mucilage from *Ocimum americanum* linn. as disintegrant in tablets: Separation and evaluation. *Indian Journal of Pharmaceutical Sciences* 69, 431–435.

Patel, J.J., Karve, M. and Patel, N.K. (2014) Guar gum: A versatile material for pharmaceutical industries. *International Journal of Pharmacy and Pharmaceutical Sciences* 6(8), 13–19.

Patil, S.D., Shinde, S., Kandpile, P. and Jain, A.S. (2015) Evaluation of antimicrobial activity of asafoetida. *International Journal of Pharmaceutical Sciences and Research* 6(2), 722–727.

Patil, V.D., Nayak, U.R. and Dev, S. (1972) Chemistry of ayurvedic crude drugs – I: Guggulu (resin from *Commiphora mukul*) – 1: Steroidal constituents. *Tetrahedron* 28(8), 2341–2352.

Pearnchob, N., Siepmann, J. and Bodmeier, R. (2003) Pharmaceutical applications of shellac: Moisture-protective and taste-masking coatings and extended-release matrix tablets. *Drug Development and Industrial Pharmacy* 29, 925–938.

Pearnchob, N., Dashevsky, A. and Bodmeier, R. (2004) Improvement in the disintegration of shellac-coated soft gelatin capsules in simulated intestinal fluid. *Journal of Controlled Release* 94, 313–321.

Pedram Rad, Z., Mokhtari, J. and Abbasi, M. (2018) Fabrication and characterization of PCL/zein/gum arabic electrospun nanocomposite scaffold for skin tissue engineering. *Materials Science and Engineering C* 93, 356–366. Available at: https://doi.org/10.1016/j.msec.2018.08.010

Peng, L., Xiang, R.C., Jia, W.W., Dong, X.X. and Wang, G.E. (2016) Preparation and evaluation of porous chitosan/collagen scaffolds for periodontal tissue engineering. *Journal of Bioactive and Compatible Polymers* 21(3), 207–220. Available at: https://doi.org/10.1177/0883911506065100

Phaechamud, T. and Choncheewa, C. ek. (2016) Double-layered matrix of shellac wax-lutrol in controlled dual drug release. *AAPS PharmSciTech* 17(6), 1326–1335. Available at: https://doi.org/10.1208/s12249-015-0468-9

Phaechamud, T. and Setthajindalert, O. (2018) Antimicrobial *in-situ* forming gels based on bleached shellac and different solvents. *Journal of Drug Delivery Science and Technology* 46, 285–293. Available at: https://doi.org/10.1016/j.jddst.2018.05.035

Phaechamud, T., Mahadlek, J. and Chuenbarn, T. (2016) *In situ* forming gel comprising bleached shellac loaded with antimicrobial drugs for periodontitis treatment. *Materials and Design* 89, 294–303. Available at: https://doi.org/10.1016/j.matdes.2015.09.138

Phaechamud, T., Lertsuphotvanit, N. and Praphanwittaya, P. (2018) Viscoelastic and thermal properties of doxycycline hyclate-loaded bleached shellac *in situ*-forming gel and–microparticle. *Journal of Drug Delivery Science and Technology* 44, 448–456. Available at: https://doi.org/10.1016/j.jddst.2018.01.021

Phaechamud, T., Senarat, S., Puyathorn, N. and Praphanwittaya, P. (2019) Solvent exchange and drug release characteristics of doxycycline hyclate-loaded bleached shellac *in situ*-forming gel and microparticle. *International Journal of Biological Macromolecules* 135, 1261–1272. Available at: https://doi.org/10.1016/j.ijbiomac.2018.11.098

Phan The, D., Debeaufort, F., Luu, D. and Voilley, A. (2008) Moisture barrier, wetting and mechanical properties of shellac/agar or shellac/cassava starch bilayer biomembrane for food applications. *Journal of Membrane Science* 325, 277–283.

Phillips, G.O. (1998) Acacia gum (gum arabic): A nutritional fibre; metabolism and calorific value. *Food Additives & Contaminants* 15(3), 251–264.

Phillips, A.O. and Phillips, G.O. (2011) Biofunctional behaviour and health benefits of a specific gum arabic. *Food Hydrocolloids* 25(2), 165–169. Available at: https://doi.org/10.1016/J.FOODHYD.2010.03.012

Phillips, G.O. and Williams, P.A. (2009) *Handbook of Hydrocolloids*, 2nd edn. CRC Press, Boca Raton, Florida.

Phillips, G.O., Ogasawara, T. and Ushida, K. (2008) The regulatory and scientific approach to defining gum arabic (*Acacia senegal* and *Acacia seyal*) as a dietary fibre. *Food Hydrocolloids* 22(1), 24–35. Available at: https://doi.org/10.1016/J.FOODHYD.2006.12.016

Postulkova, H., Chamradova, I., Pavlinak, D., Humpa, O., Jancar, J. *et al.* (2017) Study of effects and conditions on the solubility of natural polysaccharide gum karaya. *Food Hydrocolloids* 67, 148–156. Available at: https://doi.org/10.1016/j.foodhyd.2017.01.011

Prajapati, V.D., Jani, G.K., Moradiya, N.G. and Randeria, N.P. (2013) Pharmaceutical applications of various natural gums, mucilages and their modified forms. *Carbohydrate Polymers* 92(2), 1685–1699. Available at: https://doi.org/10.1016/j.carbpol.2012.11.021

Puri, V., Sharma, A., Kumar, P., Singh, I. and Huanbutta, K. (2021) Synthesis and characterization of thiolated gum ghatti as a novel excipient: Development of compression-coated mucoadhesive tablets of domperidone. *ACS Omega* 6(24), 15844–15854. Available at: https://doi.org/10.1021/acsomega.1c01328

Qurishi, Y., Hamid, A., Zargar, M.A., Singh, S.K. and Saxena, A.K. (2010) Potential role of natural molecules in health and disease: Importance of boswellic acid. *Journal of Medicinal Plants Research* 4(25), 2778–2785.

Qussi, B. and Suess, W.G. (2006) The influence of different plasticizers and polymers on the mechanical and thermal properties, porosity and drug permeability of free shellac films. *Drug Development and Industrial Pharmacy* 32(4), 403–412. Available at: https://doi.org/10.1080/03639040600559099

Raina, C.S., Singh, S., Bawa, A.S. and Saxena, D.C. (2005) Textural characteristics of pasta made from rice flour supplemented with proteins and hydrocolloids. *Journal of Texture Studies* 36(4), 402–420.

Raja, A.F., Ali, F., Khan, I.A., Shawl, A.S. and Arora, D.S. (2011) Acetyl-11-keto-β-boswellic acid (AKBA): Targeting oral cavity pathogens. *BMC Research Notes* 4, 406.

RamaPrasad, Y.V., Krishnaiah, Y.S. and Satyanarayana, S. (1998) *In vitro* evaluation of guar gum as a carrier for colon-specific drug delivery. *Journal of Controlled Release* 51, 281–287.

Ramawat, K.G., Marthur, M., Dass, S. and Suthar, S. (2008) Guggulsterone: A potent natural hypolipidemic agent from *Commiphora wightii* – problems, perseverance, and prospects. In: Ramawat, K.G. and Merillon, J.M. (eds) *Bioactive Molecules and Medicinal Plants*. Springer, Cham, Switzerland, pp. 101–121.

Ramirez, E., Sanchez, A.H., Romero, C. and Brenes, M. (2015) Combined use of nitrogen and coatings to improve the quality of mechanically harvested Manzanilla olives. *Food Chemistry* 171, 50–55.

Rani, A., Jain, S. and Dureja, P. (2009) Synergistic fungicidal efficacy of formulations of neem oil, nicotinic acid and *Ferula asafoetida* with α, β-unsaturated carbonyl compounds against *Sclerotium rolfsii* ITCC 5226 & *Macrophomina phaseolina* ITCC 0482. *Journal of Pest Science* 34(4), 253–258.

Ranjbar-mohammadi, M. and Bahrami, S.H. (2016) Electrospun curcumin loaded poly (ε-caprolactone)/gum tragacanth nanofibers for biomedical application. *International Journal of Biological Macromolecules* 84, 448–456.

Ravi, V., Kumar, T.P. and Shivakumar, H. (2013) Investigation of Kondagogu gum and ghatti gum as binders in formulating metoprolol tartrate tablets. *Research Journal of Pharmaceutical, Biological and Chemical Sciences* 4(2), 1110–1121.

Ray, S., Roy, G., Maiti, S., Bhattacharyya, U.K., Sil, A. *et al.* (2017) Development of smart hydrogels of etherified gum ghatti for sustained oral delivery of ropinirole hydrochloride. *International Journal of Biological Macromolecules* 103, 347–354. Available at: https://doi.org/10.1016/j.ijbiomac.2017.04.059

Raymond, C.R., Paul, J.S. and Siân, C.O. (eds) (2006) *Handbook of Pharmaceutical Excipients*, 5th edn. McGraw-Hill, New York.

Reddy, O.S., Subha, M.C.S., Jithendra, T., Madhavi, C. and Chowdoji Rao, K. (2019) Fabrication of gelatin/karaya gum blend microspheres for the controlled release of distigmine bromide. *Journal of Drug Delivery & Therapeutics* 9(3-s), 1–11.

Ribotta, P.D., León, A.E. and Añón, M.C. (2001) Effect of freezing and frozen storage of doughs on bread quality. *Journal of Agricultural and Food Chemistry* 49(2), 913–918.

Rideout, T.C., Harding, S.V., Jones, P.J.H. and Fan, M.Z. (2008) Guar gum and similar soluble fibers in the regulation of cholesterol metabolism: Current understandings and future research priorities. *Health Risk Management* 4, 1023–1033.

Rijkers, T., Ogbazghi, W., Wessel, M. and Bongers, F. (2006) The effect of tapping for frankincense on sexual reproduction in *Boswellia papyrifera*. *Journal of Applied Ecology* 43(6), 1188–1195.

Rinsky, L.H. and Rinsky, G. (2009) *The pastry chef's companion: A Comprehensive Resource Guide for the Baking and Pastry Professional*. Wiley, Chichester, UK.

Rubinstein, A., Gliko-Kabir, I., Penhasi, A. and Yagen, B. (1997) Enzyme dependent release of budesonide from crosslinked guar hydrogels. In: *Proceedings of the International Symposium on Controlled Release of Bioactive Materials*, pp. 839–840 (Vol. 24).

Saberi, B., Chockchaisawasdee, S., Golding, J.B., Scarlett, C.J. and Stathopoulos, C.E. (2017) Development of biocomposite films incorporated with different amounts of shellac, emulsifier, and surfactant. *Food Hydrocolloids* 72, 174–184.

Saeed, S., Al-Reza, H.M., Fatemeh, A.N. and Saeideh, D. (2012) Antihyperglycemic and antihyperlipidemic effects of guar gum on streptozotocin-induced diabetes in male rats. *Pharmacognosy Magazine* 8(29), 65–72.

Saha, T., Sakhawat Hossain, M., Kumar Mondal, S., Uddin Masum, Z., Abu Jobaer, M. *et al.* (2018) Application of natural polymers as pharmaceutical excipients. *Global Journal of Life Sciences and Biological Research* 4, 1. Available at: https://doi.org/10.24105/gjlsbr.2018.4.1

Sakata, Y. and Shimbo, S. (2006) How much does partially hydrolyzed guar gum affect the weight, moisture and hardness of feces? *Japanese Journal of Public Health* 53(4), 257–264.

Sakhale, B.K., Badgujar, J.B., Pawar, V.D. and Sananse, S.L. (2011) Effect of hydrocolloids incorporation in casing of samosa on reduction of oil uptake. *Journal of Food Science and Technology* 48, 769–772.

Saleem, M., Alam, A. and Sultana, S. (2001) Asafoetida inhibits early events of carcinogenesis: A chemopreventive study. *Life Sciences* 68(16), 1913–1921. Available at: https://doi.org/10.1016/S0024-3205(01)00977-8

Samavati, V., Emam-Djomeh, Z. and Mehdinia, A. (2014) Thermodynamic and kinetic study of volatile compounds in biopolymer based dispersions. *Carbohydrate Polymers* 99, 556–562.

Sanderson, G.R. (1996) Gums and their use in food systems. *Food Technology (Chicago)* 50(3), 81–84.

Santos, J., Calero, N., Guerrero, A. and Muñoz, J. (2015) Relationship of rheological and microstructural properties with physical stability of potato protein based emulsions stabilized by guar gum. *Food Hydrocolloids* 44, 109–114. Available at: https://doi.org/10.1016/j.foodhyd.2014.09.025

Sarathchandiran, I. and Suresh Kumar, P. (2013) Characterization and standardization of gum karaya. *International Journal of Biopharmaceutics* 4(3), 180–189.

Sargent, J.A. and Villagran, F.V. (2002) Beverage brewing devices for preparing creamy beverages. Patent WO 2002074143 A2.

Sarika, P.R., Cinthya, K., Jayakrishnan, A., Anilkumar, P.R. and James, N.R. (2014) Modified gum arabic cross-linked gelatin scaffold for biomedical applications. *Materials Science and Engineering* C 43, 272–279. Available at: https://doi.org/10.1016/J.MSEC.2014.06.042

Satyavati, G.V., Dwarakanath, C. and Tripathi, S.N. (2013) Experimental studies on the hypocholesterolemic effect of *Commiphora mukul*. Engl. (*Guggul*). *Indian Journal of Medical Research* 137(2), 403.

Sedaghat Doost, A., Muhammad, D.R.A., Stevens, C.V., Dewettinck, K. and Meeren, P. (2018) Fabrication and characterization of quercetin loaded almond gum-shellac nanoparticles prepared by antisolvent precipitation. *Food Hydrocolloids* 83, 190–201.

Sedaghat Doost, A., Kassozi, V., Grootaert, C., Claeys, M., Dewettinck, K. *et al.* (2019) Self-assembly, functionality, and *in-vitro* properties of quercetin loaded nanoparticles based on shellac-almond gum biological macromolecules. *International Journal of Biological Macromolecules* 129, 1024–1033. Available at: https://doi.org/10.1016/J.IJBIOMAC.2019.02.071

Seeli, D.S. and Prabaharan, M. (2016) Guar gum succinate as a carrier for colon-specific drug delivery. *International Journal of Biological Macromolecules* 84, 10–15. Available at: https://doi.org/10.1016/j.ijbiomac.2015.12.002

Sen, G., Mishra, S., Jha, U. and Pal, S. (2010) Microwave initiated synthesis of polyacrylamide grafted guar gum (GG-g-PAM) – characterizations and application as matrix for controlled release of 5-amino salicylic. *International Journal of Biological Macromolecules* 47, 164–170.

Senarat, S., Wai Lwin, W., Mahadlek, J. and Phaechamud, T. (2021) Doxycycline hyclate-loaded *in situ* forming gels composed from bleached shellac, Ethocel, and Eudragit RS for periodontal pocket delivery. *Saudi Pharmaceutical Journal* 29(3), 252–263. Available at: https://doi.org/10.1016/j.jsps.2021.01.009

Separdar, H., Rahimi, E. and Shahabi Ghahfarrokhi, I. (2016) An investigation of mechanical and physical properties of soy protein isolate-tragacanth-nanocellulose based nanocomposite film. *Iranian Journal of Biosystems Engineering* 47(3), 431–439.

Shahidi, F. and Han, X.Q. (1993) Encapsulation of food ingredients. *Critical Reviews in Food Science and Nutrition* 33(6), 501–547.

Shahiwala, A. and Zarar, A. (2018) Pharmaceutical product lead optimization for better *in vivo* bioequivalence performance: A case study of diclofenac sodium extended release matrix tablets. *Current Drug Delivery* 15(5), 705–715.

Shaikh, I.M., Ghodke, S.K. and Ananthanarayan, L. (2008) Inhibition of staling in chapati (Indian unleavened flat bread). *Journal of Food Processing and Preservation* 32(3), 378–403.

Sharadanant, R. and Khan, K. (2003) Effect of hydrophilic gums on the quality of frozen dough: II. Bread characteristics. *Cereal Chemistry* 80(6), 773–780.

Sharma, K., Kumar, V., Kaith, B.S., Kumar, V., Som, S. *et al.* (2015) Evaluation of a conducting interpenetrating network based on gum ghatti-g-poly(acrylic acid-aniline) as a colon-specific delivery system for amoxicillin trihydrate and paracetamol. *New Journal of Chemistry* 39(4), 3021–3034. Available at: https://doi.org/10.1039/c4nj01982b

Shelly, Ahuja, M. and Kumar, A. (2013) Gum ghatti-chitosan polyelectrolyte nanoparticles: Preparation and characterization. *International Journal of Biological Macromolecules* 61, 411–415. Available at: https://doi.org/10.1016/j.ijbiomac.2013.07.022

Shigeo, O., Hiroya, Y., Sumio, T. and Keiyi, M. (1989) A method of manufacturing jelly products having fibrous texture. *European Patent* 0 344 466 A1.

Shishodia, S. and Aggarwal, B.B. (2004) Guggulsterone inhibits NF-κB and IκBα kinase activation, suppresses expression of anti-apoptotic gene products, and enhances apoptosis. *Journal of Biological Chemistry* 279(45), 47148–47158.

Shrivastava, V., Bhardwaj, U., Sharma, V., Mahajan, N., Sharma, V. *et al.* (2012) Antimicrobial activities of asafoetida resin extracts (a potential Indian spice). *Journal of Pharmacy Research* 5(5), 5022–5024.

Siddiqul, M.Z. (2011) Boswellia serrata, a potential antiinflammatory agent: An overview. *Indian Journal of Pharmaceutical Sciences* 73(3), 255–261.

Silva, M.P., Tulini, F.L., Ribas, M.M., Penning, M., Favaro-Trindade, C.S. *et al.* (2016) Microcapsules loaded with the probiotic *Lactobacillus paracasei* BGP-1 produced by co-extrusion technology using alginate/shellac as wall material: Characterization and evaluation of drying processes. *Food Research International* 89, 582–590.

Singh, B. and Pal, L. (2008) Development of sterculia gum based wound dressings for use in drug delivery. *European Polymer Journal* 44(10), 3222–3230. Available at: https://doi.org/10.1016/j.eurpolymj.2008.07.013

Singh, B. and Sharma, N. (2008) Development of novel hydrogels by functionalization of sterculia gum for use in anti-ulcer drug delivery. *Carbohydrate Polymers* 74(3), 489–497. Available at: https://doi.org/10.1016/j.carbpol.2008.04.003

Singh, B. and Vashishtha, M. (2008) Development of novel hydrogels by modification of sterculia gum through radiation cross-linking polymerization for use in drug delivery. *Nuclear Instruments and Methods in Physics Research, Section B: Beam Interactions with Materials and Atoms* 266(9), 2009–2020. Available at: https://doi.org/10.1016/j.nimb.2008.03.086

Singh, B. and Sharma, N. (2009) Mechanistic implication for cross-linking in sterculia-based hydrogels and their use in GIT drug delivery. *Biomacromolecules* 10(9), 2515–2532. Available at: https://doi.org/10.1021/bm9004645

Singh, B. and Sharma, N. (2011) Design of sterculia gum based double potential antidiarrheal drug delivery system. *Colloids and Surfaces B: Biointerfaces* 82(2), 325–332. Available at: https://doi.org/10.1016/j.colsurfb.2010.09.004

Singh, B. and Sharma, V. (2014) Influence of polymer network parameters of tragacanth gum-based pH responsive hydrogels on drug delivery. *Carbohydrate Polymers* 101, 928–940.

Singh, B. and Singh, B. (2018) Modification of sterculia gum polysaccharide via network formation by radiation induced crosslinking polymerization for biomedical applications. *International Journal of Biological Macromolecules* 116, 91–99. Available at: https://doi.org/10.1016/j.ijbiomac.2018.05.032

Singh, B., Sharma, V. and Chauhan, D. (2010) Gastroretentive floating sterculia-alginate beads for use in antiulcer drug delivery. *Chemical Engineering Research and Design* 88(8), 997–1012. Available at: https://doi.org/10.1016/j.cherd.2010.01.017

Singh, B., Sharma, V. and Pal, L. (2011) Formation of sterculia polysaccharide networks by gamma rays induced graft copolymerization for biomedical applications. *Carbohydrate Polymers* 86(3), 1371–1380. Available at: https://doi.org/10.1016/j.carbpol.2011.06.041

Singh, R.B., Niaz, M.A. and Ghosh, S. (1994) Hypolipidemic and antioxidant effects of *Commiphora mukul* as an adjunct to dietary therapy in patients with hypercholesterolemia. *Cardiovascular Drugs and Therapy* 8, 659–664.

Sinha, V.R., Mittal, B.R., Bhutani, K.K. and Kumria, R. (2004) Colonic drug delivery of 5-fluorouracil: An *in vitro* evaluation. *International Journal of Pharmaceutics* 269, 101–108.

Sirisha, V.N.L., Eswariah, M.C. and Rao, A.S. (2017) Formulation and evaluation of karaya gum microspheres of cromolyn sodium: Treatment of ulcerative colitis. *International Research Journal of Pharmacy* 8(10), 182–188. Available at: https://doi.org/10.7897/2230-8407.0810205

Soppimath, K.S., Kulkarni, A.R. and Aminabhavi, T.M. (2001) Chemically modified polyacrylamide-g-guar gum-based crosslinked anionic microgels as pH-sensitive drug delivery systems: Preparation and characterization. *Journal of Controlled Release* 75(3), 331–345. Available at: https://doi.org/10.1016/S0168-3659(01)00404-7

Soppimath, K.S. and Aminabhavi, T.M. (2002) Water transport and drug release study from cross-linked polyacrylamide grafted guar gum hydrogel microspheres for the controlled release application. *European Journal of Pharmaceutics and Biopharmaceutics* 53(1), 87–98. Available at: https://doi.org/10.1016/S0939-6411(01)00205-3

Soradech, S., Nunthanid, J., Limmatvapirat, S. and Luangtana-anan, M. (2012) An approach for the enhancement of the mechanical properties and film coating efficiency of shellac by the formation of composite films based on shellac and gelatin. *Journal of Food Engineering* 108(1), 94–102.

Soradech, S., Limatvapirat, S. and Luangtana-anan, M. (2013) Stability enhancement of shellac by formation of composite film: Effect of gelatin and plasticizers. *Journal of Food Engineering* 116(2), 572–580. Available at: https://doi.org/10.1016/J.JFOODENG.2012.12.035

Soto, A.M., Schaeberle, C., Maier, M.S., Sonnenschein, C. and Maffini, M.V. (2017) Evidence of absence: Estrogenicity assessment of a new food-contact coating and the bisphenol used in its synthesis. *Environmental Science & Technology* 51(3), 1718–1726.

Sparberg, M., Prohaska, J. and Kirsner, J.B. (1966) Solid state karaya gum ring for use in disposable and permanent ileostomy appliances. *American Journal of Surgery* 111(4), 610–611. Available at: https://doi.org/10.1016/0002-9610(66)90300-X

Srichamroen, A. (2007) Influence of temperature and salt on viscosity property of guar gum. *Naresuan University Journal: Science and Technology (NUJST)* 15(2), 55–62.

Srivastava, S. and Ray, D.P. (2015) Natural gum resources in India and their commercial importance. *International Journal of Bioresource Science* 2(2), 151–155.

Steinhardt, A. and Goldwater, F.A. (1962) Gelatin adhesive pharmaceutical preparations. US Patent no.3.029.

Stephen, A.M. (1990) Structure and properties of exudate gums. In: Phillips, G.O., Williams, P.A. and Wedlock, D.J. (eds) *Gums and Stabilisers for the Food Industry*. Oxford University Press, Oxford, UK, pp. 3–16.

Stephen, A.M. and Churms, S.C. (1995) Gums and mucilages. In: Stephen, A.M. (ed.) *Food Polysaccharides and Their Applications*. Marcel Dekker, New York, pp. 377–440.

Stummer, S., Salar-Behzadi, S., Unger, F.M., Oelzant, S., Penning, M. *et al.* (2010) Application of shellac for the development of probiotic formulations. *Food Research International* 43, 1312–1320.

Su, S., Wang, T., Duan, J.-A., Zhou, W., Hua, Y.-Q. *et al.* (2011) Anti-inflammatory and analgesic activity of different extracts of *Commiphora myrrha*. *Journal of Ethnopharmacology* 134(2), 251–258.

Su, S., Hua, Y., Wang, Y., Gu, W., Zhou, W. *et al.* (2012) Evaluation of the anti-inflammatory and analgesic properties of individual and combined extracts from *Commiphora myrrha*, and *Boswellia carterii*. *Journal of Ethnopharmacology* 139(2), 649–656.

Sulieman, A.M.E.H. (2018) Gum arabic as thickener and stabilizing agents in dairy products. In: Mariod, A.A. (ed.) *Gum Arabic: Structure, Properties, Application and Economics*. Academic Press, New York, pp. 151–165. Available at: https://doi.org/10.1016/B978-0-12-812002-6.00013-0

Sumnu, G., Koksel, F., Sahin, S., Basman, A. and Meda, V. (2010) The effects of xanthan and guar gums on staling of gluten-free rice cakes baked in different ovens. *International Journal of Food Science & Technology* 45(1), 87–93.

Sun, C., Xu, C., Mao, L., Wang, D., Yang, J. *et al.* (2017) Preparation, characterization and stability of curcumin-loaded zein-shellac composite colloidal particles. *Food Chemistry* 228, 656–667.

Sun, Z., Yang, C., Wang, F., Wu, B., Shao, B. *et al.* (2020) Biocompatible and pH responsive colloidal surfactants with tunable shape for controlled interfacial curvature. *Angewandte Chemie International Edition* 59(24), 9365–9369.

Sutton, R.L. and Wilcox, J. (1998) Recrystallization in ice cream as affected by stabilizers. *Journal of Food Science* 63(1), 104–107.

Suzuki, T. and Hara, H. (2004) Ingestion of guar gum hydrolysate, a soluble and fermentable nondigestible saccharide, improves glucose intolerance and prevents hypertriglyceridemia in rats fed fructose. *Journal of Nutrition* 134(8), 1942–1947.

Syed, Q.A., Anwar, S., Shukat, R. and Zahoor, T. (2018) Effects of different ingredients on texture of ice cream. *Journal of Nutritional Health & Food Engineering* 8(6), 422–435.

Szapary, P.O., Wolfe, M.L., Bloedon, L.T., Cucchiara, A.J., DerMarderosian, A.H. *et al.* (2003) Guggulipid for the treatment of hypercholesterolemia: A randomized controlled trial. *JAMA* 290(6), 765–772.

Takahashi, H., Yang, S.I., Fujiki, M., Kim, M., Yamamoto, T. *et al.* (1994) Toxicity studies of partially hydrolyzed guar gum. *Journal of the American College of Toxicology* 13(4), 273–278.

Takahashi, T., Yokawa, T., Ishihara, N., Okubo, T., Chu, D.C. *et al.* (2009) Hydrolyzed guar gum decreases postprandial blood glucose and glucose absorption in the rat small intestine. *Nutrition Research* 29(6), 419–425.

Takeoka, G. (2001) Volatile constituents of Asafoetida. *ACS Symposium Series* 794, 33–44. Available at: https://doi.org/10.1021/bk-2001-0794.ch004

Talaei, S. and Kiani, A. (2015) Study on permeability of bionanocomposite film based on tragacanth gum-chitosan-graphene oxide. *Indian Journal of Fundamental and Applied Life Sciences* 5(S3), 25–31.

Tang, L., Hong, B., Li, T. and Huang, B. (2019) Development of bilayer films based on shellac and esterified cellulose nanocrystals for buccal drug delivery. *Cellulose* 26, 1157–1167.

Tayeboon, G.S., Tavakoli, F., Hassani, S., Khanavi, M., Sabzevari, O. *et al.* (2013) Effects of *Cymbopogon citratus* and *Ferula asafoetida* extracts on glutamate-induced neurotoxicity. *In Vitro Cellular & Developmental Biology* 49(9), 706–715.

Teissedre, P. (2012) Gum arabic in oenology. In: Kennedy, J.F., Phillips, G.O. and Williams, P.A. (eds) *Gum Arabic: Proceedings of the World Conference on New Developments in Acacia Gums Research and Products*, Royal Society of Chemistry, Cambridge, UK, pp. 311–314.

Thakur, S., Chauhan, G.S. and Ahn, J.H. (2009) Synthesis of acryloyl guar gum and its hydrogel materials for use in the slow release of l-DOPA and l-tyrosine. *Carbohydrate Polymers* 76, 513–520.

Thammachat, T., Sriamornsak, P., Luangtana-anan, M., Nunthanid, J., Limmatvapirat, C. *et al.* (2010) Preparation and characterization of shellac fiber as a novel material for controlled drug release. *Advanced Materials Research* 152–153, 1232–1235.

Thammachat, T., Tabcharoen, S. and Soradech, S. (2013) Preparation and characterization of shellac/polyvinylpyrrolidone nanofiber as a material for medical application. *Advanced Materials Research* 652–654, 278–281.

Tiss, A., Carrière, F. and Verger, R. (2001) Effects of gum arabic on lipase interfacial binding and activity. *Analytical Biochemistry* 294(1), 36–43. Available at: https://doi.org/10.1006/abio.2001.5095

Tiwari, A. and Prabaharan, M. (2010) An amphiphilic nanocarrier based on guar gum-graft-poly (ε-caprolactone) for potential drug-delivery applications. *Journal of Biomaterials Science* 21, 937–949.

Tonyali, B., Cikrikci, S. and Oztop, M.H. (2018) Physicochemical and microstructural characterization of gum tragacanth added whey protein based films. *Food Research International* 105, 1–9.

Toti, U.S. and Aminabhavi, T.M. (2004) Modified guar gum matrix tablet for controlled release of diltiazem hydrochloride. *Journal of Controlled Release* 95(3), 567–577. Available at: https://doi.org/10.1016/j.jconrel.2003.12.019

Tripathi, Y.B., Malhotra, O.P. and Tripathi, S.N. (1984) Thyroid stimulating action of Z-guggulsterone obtained from *Commiphora mukul*. *Planta Medica* 50(1), 78–80.

Tripathi, Y.B., Tripathi, P., Malhotra, O.P. and Tripathi, S.N. (1988) Thyroid stimulatory action of (Z)-guggulsterone: Mechanism of action. *Planta Medica* 54(4), 271–277.

Tripathy, S. and Das, M.K. (2013) Guar gum: Present status and applications. *Journal of Pharmaceutical and Scientific Innovatation* 2(4), 24–28.

Tsai, R.Y., Chen, P.W., Kuo, T.Y., Lin, C.M., Wang, D.M. *et al.* (2014) Chitosan/pectin/gum arabic polyelectrolyte complex: Process-dependent appearance, microstructure analysis and its application. *Carbohydrate Polymers* 101(1), 752–759. Available at: https://doi.org/10.1016/J.CARBPOL.2013.10.008

Ulbricht, C., Basch, E., Szapary, P., Hammerness, P., Axentsev, S. *et al.* (2005) Guggul for hyperlipidemia: A review by the natural standard research collaboration. *Complementary Therapies in Medicine* 13(4), 279–290.

Upadhyay, R.K. (2017) Nutritional, therapeutic, and pharmaceutical potential of plant gums: A review. *International Journal of Green Pharmacy* 11(1), S30–S48.

Urazono, H., Moria, Y., Ueda, N., Ueno, H., Katoh, K. *et al.* (2016) Protein composition and production process therefor. US Patent No. 20160044949 A1.

Varshosaz, J., Tavakoli, N. and Eram, S.A. (2006) Use of natural gums and cellulose derivatives in production of sustained release Metoprolol tablets. *Drug Delivery* 13, 113–119.

Verbeken, D., Dierckx, S. and Dewettinck, K. (2003) Exudate gums: Occurrence, production, and applications. *Applied Microbiology and Biotechnology* 63(1), 10–21. Available at: https://doi.org/10.1007/s00253-003-1354-z

Verma, A., Kumar, P., Rastogi, V. and Mittal, P. (2021) Preparation and evaluation of polymeric beads composed of Chitosan-Gellan Gum-Gum Ghatti/-Gum Karaya polyelectrolyte complexes as polymeric carrier for enteric sustained delivery of Diclofenac sodium. *Future Journal of Pharmaceutical Sciences* 7, 196. Available at: https://doi.org/10.1186/s43094-021-00343-y

Viinanen, A., Salokannel, M., Lammintausta, K. and Beezhold, D.H. (2011) Case report: Gum arabic as a cause of occupational allergy. *Journal of Allergy* 2011. Available at: https://doi.org/10.1155/2011/841508

Vojdani, C. (2015) Immune reactivities against gums. *Alternative Therapies in Health and Medicine* 21, 1–95.

Wang, T., Yang, Y., Feng, W., Wang, R. and Chen, Z. (2020) Co-folding of hydrophobic rice proteins and shellac in hydrophilic binary microstructures for cellular uptake of apigenin. *Food Chemistry* 309, 125695.

Wang, X., Yu, D.-G., Li, X.-Y., Bligh, S.W.A. and Williams, G.R. (2015) Electrospun medicated shellac nanofibers for colon targeted drug delivery. *International Journal of Pharmaceutics* 490, 384–390.

Waters, D., Freedman, D., Lesperance, J. and Theroux, P. (1987) Design features of a controlled clinical trial to assess the effect of a calcium entry blocker upon the progression of coronary artery disease. *Controlled Clinical Trials* 8(3), 216–242.

Wei, X., Pang, J., Zhang, C., Yu, C., Chen, H. *et al.* (2015) Structure and properties of moisture-resistant konjac glucomannan films coated with shellac/stearic acid coating. *Carbohydrate Polymers* 118, 119–125.

Weiping, W. and Branwell, A. (2000) Tragacanth and karaya. In: Philips, G.O. and Williams, P.A. (eds) *Handbook of Hydrocolloids*. CRC Press, Boca Raton, Florida, pp. 231–246.

Whistler, R.L. and Hymowitz, T. (1979) *Guar: Agronomy, Production, Industrial Use, and Nutrition*. Purdue University Press, West Lafayette, Indiana.

Whistler, R.L. (1993) Exudate gums. In: Whistler, R.L. and BeMiller, J.N. (eds) *Industrial Gums: Polysaccharides and Their Derivatives*, 3rd edn. Academic Press, New York, pp. 309–339.

Williams, P. and Phillips, G. (2001) Gum arabic: Production, safety and physiological effects, physiochemical characterization, functional properties and food applications. In: Cho, S.S. and Dreher, M.L. (eds) *Handbook of Dietary Fiber*. Marcel Dekker, New York, pp. 675–694.

Wong, D., Larrabeo, S., Clifford, K., Tremblay, J. and Friend, D.R. (1997) USP dissolution apparatus III (reciprocating cylinder) for screening of guar-based colonic delivery formulations. *Journal of Controlled Release* 47, 173–179.

Xue, J. and Zhang, Z. (2009) Physical, structural, and mechanical characterization of calcium-shellac microspheres as a carrier of carbamide peroxide. *Journal of Applied Polymer Science* 113, 1619–1625.

Yang, C., Wang, Y., Lu, L., Unsworth, L., Guan, L.L. *et al.* (2018) Oat protein shellac beads: Superior protection and delivery carriers for sensitive bioactive compounds. *Food Hydrocolloids* 77, 754–763.

Yang, C., Wang, Y., Xie, Y., Liu, G., Lu, Y. *et al.* (2019) Oat protein-shellac nanoparticles as a delivery vehicle for resveratrol to improve bioavailability *in vitro* and *in vivo*. *Nanomedicine* 14, 2853–2871.

Yu, D., Qian, W., Wang, X., Li, Y., Lu, W. *et al.* (2013) Ferulic acid-loaded shellac microparticles prepared using electrohydrodynamic atomization. In: Zhang, H., Jin, D. and Zhao, X.J. (eds) *Advanced Research on Industry, Information System and Material Engineering*, Vol. 675. Trans Tech Publications, Zurich, Switzerland, pp. 326–330.

Yu, D.-G., Li, J.-J., Zhang, M. and Williams, G.R. (2017) High-quality Janus nanofibers prepared using three-fluid electrospinning. *Chemical Communications* 53, 4542–4545.

Zambrano, F., Despinoy, P., Ormenese, R.C.S.C. and Faria, E.V. (2004) The use of guar and xanthan gums in the production of 'light' low fat cakes. *International Journal of Food Science & Technology* 39(9), 959–966.

Zhang, P., Zhao, Y. and Shi, Q. (2016) Characterization of a novel edible film based on gum ghatti: Effect of plasticizer type and concentration. *Carbohydrate Polymers* 153, 345–355. Available at: https://doi.org/10.1016/j.carbpol.2016.07.082

Zhang, P., Zhao, Y., Zhang, X., Zhu, L., Fang, Z. *et al.* (2020b) Thermodynamic properties and state diagram of gum ghatti-based edible films: Effects of glycerol and nisin. *Polymers* 12(2), 449. Available at: https://doi.org/10.3390/polym12020449

Zhang, Y., Li, T., Zhang, H., Zhang, H., Chi, Y. *et al.* (2020a) Blending with shellac to improve water resistance of soybean protein isolate film. *Journal of Food Process Engineering* 43(11), e13515.

Zhong, S.P., Zhang, Y.Z. and Lim, C.T. (2010) Tissue scaffolds for skin wound healing and dermal reconstruction. *Wiley Interdisciplinary Reviews: Nanomedicine and Nanobiotechnology* 2(5), 510–525. Available at: https://doi.org/10.1002/WNAN.100

Zhou, L., Huang, J., He, B., Zhang, F. and Li, H. (2014) Peach gum for efficient removal of methylene blue and methyl violet dyes from aqueous solution. *Carbohydrate Polymers* 101(1), 574–581. Available at: https://doi.org/10.1016/j.carbpol.2013.09.093

Zhu, Y., Fang, B., Huang, L., Guan, C. and Yang, G. (2008) Study on the selective removal of plasma low-density lipoprotein and fibrinogen by degraded guar sulfate. *Journal of Biomedical Engineering* 25(5), 1135–1140.

7 Quality Aspects and Analysis

Michael Keusgen[1]*, Victor S. Jayakumar[2,3] and Hartwig Schulz[4]
[1]Philipps University Marburg, Institute for Pharmaceutical Chemistry, Marbacher Weg 8, 35032 Marburg, Germany; [2]Deepthi, Amaravila P O, Thiruvananthapuram 695122, Kerala, India; [3]St Thomas College of Engineering & Technology, Kozhuvalloor, PO 689521, Chengannur, Kerala, India; [4]Consulting and Project Management for Medicinal and Aromatic Plants, Waltraudstraße 4, Stahnsdorf, Germany

Abstract

Resins are typically complex mixtures of lipophilic and hydrophilic substances, including polymeric structures. They consist of water-soluble polysaccharides as well as proteoglycans and lipophilic compounds like volatile essential oil substances as well as non-volatile phenolic, di- and triterpenic substances. Resins with a high level of essential oils and soft consistency are named oleoresins. If chromatographic separation techniques should be applied, the crude resin has to be fractionated in an appropriate manner. If the main interest of analysis is in the volatile fraction consisting of essential oil compounds, hydro-distillation is most often used, but also headspace techniques are rather common. Gas chromatography (GC) and thin-layer chromatography (TLC) can be applied as separating techniques. For GC, a mass spectrometer (MS) as detector is most suitable, and identification of individual compounds can be performed by dedicated databases. However, the result is not always convincing, and interpretation of MS data needs some experience. For the separation of non-volatile, lipophilic compounds, a fractionation of the raw resin is strongly recommended prior to analysis by high-performance liquid chromatography (HPLC), GC with pre-column derivatization of the sample and TLC. Many of the di- and triterpenoids bear a carboxy substituent leading to an acidic pH, by which these compounds can be specifically separated. Detection after HPLC separation can be performed either by UV detector or by a mass sensitive device (MSD). In contrast, vibrational spectroscopy like infrared (IR) spectroscopy, near-infrared (NIR) spectroscopy and Raman spectroscopy, often in combination with Fourier transformation (FT), require little or no sample preparation. These techniques allow a differentiation between plant species and their origins, as well as differentiation between gums, resins and volatile compounds that come from them. Also, fossil resin and resins from archaeological artefacts can be analysed in order to determine the plant origin. Some of theses methods can also combined with chromatographic techniques. Proper data evaluation is only possible if a comprehensive collection of reference spectra exists.

Introduction

Resins and gums are highly complex mixtures of natural products. Numerous resins with significant medicinal importance consist of polysaccharides, proteoglycans and more or less lipophilic, low molecular weight compounds like phenols and terpenoids (mono-, sesqui-, di- and triterpenoids). Mono- and sesquiterpenes usually form an essential oil, which gives a soft consistency as well as a characteristic smell to the resin. In this case, the resin is more or less clear and named an oleoresin. If the lipophilic part of a resin is extracted, the mostly water-soluble residue is named gum.

*Corresponding author: keusgen@staff.uni-marburg.de

 Natural Gums and Resins: Botany and Sustainable Uses in Medicine, Nutrition, Perfumery and Cosmetics (eds Hartwig Schulz *et al.*)
DOI: 10.1079/9781800626690.0007

Di- and triterpenoids show a very high diversity of chemical structures and many of them expose some bioactivity. Low molecular phenolic compounds are usually responsible for the colour of the resin, exposing some bioactivities. A good example for this is the intense red-coloured resin ('Dragon's blood') of the Dragon tree *Dracaena draco* (Camarda *et al.*, 2011).

On the other hand, resins which are more an emulsion or suspension with a relatively low content of essential oil are named gum. Some gums contain considerable amounts of polyisoprenes, which form natural rubber. Under the influence of light and oxygen, resin components, especially diterpenoids, can polymerize, finally leading to resin fossilization called amber (Langenheim, 1990).

Because resins are typically complex mixtures of numerous classes of substances, several methods have to be applied to get full chemical information. Spectroscopic methods as described later in this chapter only require a simple sample preparation. Also, mass spectrometry (MS) can be performed using a rather crude sample. However, if a full spectrum of compounds is required, the resin has to be fractionated by different techniques, followed by chromatographic analysis.

If essential oil compounds are of interest, steam distillation is the method of choice. After that, more or less lipophilic compounds like non-volatile terpenoids (di- and triterpenoids) can be extracted by ethanol and/or dichloromethane. If phenolic compounds are of interest, they can be extracted by ethanol–water mixtures under mild conditions (moderate heat, exclusion of oxygen if possible). Highly lipophilic compounds like essential oil substances can be removed by liquid–liquid extraction with cyclohexane.

The extraction of water-soluble compounds requires more care. The resin has to be defatted under mild conditions. Extensive heating has to be avoided because proteins can denaturize. Appropriate solvents are dichloromethane followed by ethanol 96% (v/v). The remaining residue is than subsequently extracted by cold water, 70°C hot water and aqueous sodium hydroxide (0.1 mol/l). The obtained polysaccharides and proteoglycans have to be dialyzed and further separated by anion-exchange chromatography (AEC) followed by gel permeation chromatography (GPC). Further on, glycoproteins can be separated by gel electrophoresis. The resulting purified polysaccharides and proteoglycans can be sequenced according to standard techniques (Herrmann *et al.*, 2012; Hosain *et al.*, 2019).

The following paragraphs are focused on the analysis of low-molecular-weight compounds. Volatiles can be typically obtained by water steam distillation of the crude resin, where as non-volatiles must be extracted by an appropriate solvent and further purified by liquid–liquid extraction, as described later in this chapter.

Analysis of Volatiles

The most prominent constituents of resins are essential oils in various amounts (Langenheim, 1990; Camarda *et al.*, 2011; Murthy, 2021; Sura and Cheng, 2024). Theses essential oils are the main reason why resins have been highly estimated by humans since ancient times as fumigants, incense, food ingredients and medicine. However, the content of essential oils in resins is highly variable. Besides factors like age and constitution of the bearing plant, time between the excretion and harvest of the resin, temperature, humidity and rain as well as sun exposure are of tremendous influence. This means that a constant quality of a distinct resin can only be achieved if the factors mentioned above are also stable. Of particular importance is the time between resin excretion and harvest. Later on, significant amounts of the essential oil can get lost during transportation and storage. In any case, a harvested resin should be stored at cold temperature and with light protection.

Essential oil constituents can be analysed directly from the raw material or after a dedicated sample preparation. Analytical methods of choice are gas chromatography (GC) and also thin-layer chromatography (TLC). The essential oil can be separated from the raw resin either by solvent extraction with a highly lipophilic solvent (preferably hexane), by steam distillation or thermally by another appropriate method such as headspace sample preparation in combination with GC analysis.

A brief overview of these methods is given by Rubiolo *et al.* (2010). Most common is the dedicated analytical equipment based on circulatory distillation as introduced by Clevenger. In theory, the recoveries of volatiles are quantitative for an infinite distillation time. But in practice, distillation times are 2 or 4 hour, and oxygen-bearing substances have a more or less good solubility in water, meaning that plain hydrocarbons are more present in the obtained distillate than oxygenated substances. Further on, thermal composition depends on the type of heater and its heating power as well as on distillation time and on the material the distillery is made of (glass, metal). The 'classical' apparatus can be also miniaturized so that distillation time and heat impact can be reduced. However, it must be considered that small, pure hydrocarbonic compounds like monoterpenes without oxygen functionality are distilled first, followed by others with higher molecular weight and more oxygen functionalities. This means that the essential oil obtained will change over distillation time. Therefore, the exact properties of the apparatus used as well as distillation time have to be given, otherwise the results are not reproducible.

In order to overcome most of these constrains, the headspace (HS) technique has been developed. In contrast to conventional techniques based a two-step approach (first step sample preparation, second step analysis), HS is suitable for fully automatic approaches in which sample preparation and analysis are integrated, better known under the acronym TAS (total analysis system).

Traditionally, HS sampling can be operated either in the static (S-HS) or in the dynamic mode (D-HS), but both techniques can be replaced by so-called 'high-concentration-capacity headspace' techniques (HCC-HS). This approach is based on either the static or the dynamic accumulation of volatile(s) on polymers operating in sorption and/or adsorption modes or, less frequently, on solvents. It has to be pointed out that the recoveries with HCC-HS differ from one essential oil component to another, since these techniques are based on partition between more than two phases (matrix, mostly the resin sample, headspace and polymer) and, in consequence, the sampled volatile fraction does not represent either the composition of the headspace or that of the essential oil. The HCC-HS techniques were immediately successful, mainly because they are as simple, fast, easy to automate and reliable as S-HS, while at the same time showing analyte concentration factors that are very often comparable to those of D-HS. Again, it must be stressed that the 'true' essential oil composition is not always reflected by analytical results and, therefore, precise experimental conditions have to be followed to allow comparison of data.

Several techniques based on this HS approach are now available, including: HS solid-phase microextraction (HS-SPME), in-tube sorptive extraction (INCAT, HS–SPDE), headspace sorptive extraction (HSSE), solid-phase aroma concentrate extraction (SPACE), headspace liquid-phase microextraction (HS–LPME) and large surface area HCC-HS sampling (MESI, MME, HS-STE). Further approaches for the isolation of essential oils prior to analysis are given by Hassan *et al.* (2016).

A further, rather innovative technique is 'cryogenic trapping'. This is an effective dynamic approach that was developed to isolate the volatile fraction of a vegetable matrix. Obtained results are close to those obtained by HS. Among others, two techniques based on cryogenic condensation of the stripped volatiles have successfully been applied to the flavour and fragrance field: cryogenic vacuum trapping, based on the stripping of the volatile fraction under vacuum, and Aquaspace® techniques, based on its stripping by humidified air. However, cryogenic trapping is not the method of first choice when it comes to the analysis of resins.

As a general remark about GC sample preparation, it must be guaranteed that the prepared sample only contains substances that are volatile under the chosen conditions, typically at the temperature of the injection block, e.g. 250°C. Additionally, the sample must be free of particles which would block the GC column. Filtration is mandatory (if not using HS or cryogenic trapping), but before that it must tested that the filter does not alter the analytical result, either by retaining certain essential oil compounds or by leaching additives of the filter membrane into the sample. Miniaturized sample preparation techniques as well as HS allow high sample throughput, which is required in cultivation field testing in order to find the appropriate point of harvest. High sample throughput is also required in routine analysis.

For the analysis of defined essential oil compounds, GC is the method of choice (Bartle and Myers, 2002). The first GC experimental set-up was published by Martin and James in 1951, but others also performed similar experiments at that time. Similar to liquid chromatography (LS), the core part of the chromatograph is a column. The column is named 'stationary phase' and the used gas (typically helium, nitrogen or hydrogen) is called 'mobile phase'. First, columns were packed with sorbents coated with the stationary liquid phase, like modified silicone polymers or polyethylene glycol. The length of the packed columns is typically 1–5m and the diameter 2 mm. The advantage of them is their robustness against contaminants and high loading capacity also allowing application of less sensitive detectors (see below). The column is placed in an oven with adjustable temperature, allowing temperature gradients.

However, the packed GC column has some drawbacks, mainly a restricted resolution and a back pressure, depending on the length of the column. These obstacles were recognized rather soon, and consequently, Martin and others came up with the idea of a capillary column in 1956. In our day, modern capillary columns have a length typically of 10–30 m, an inner diameter of 0.1–0.8 mm and a film thickness (stationary phase) of 0.1–1 µm, whereas the stationary phase can be solid or liquid. For essential oil analysis, oven temperatures between 40 and 250°C were chosen and mostly a temperature gradient was applied.

For essential oil analysis, the most common columns with apolar, but sometimes also polar, stationary phases are (Rubiolo *et al.*, 2010): methyl polysiloxanes (SE30, OV-1, OV 101, DB-1, HP-1, PS 347.5, etc.), methylphenylpolysiloxanes (SE-52, SE-54, DB-5, HP-5, PS-086, etc.) and polyethyleneglycol (PEG-20M, CW-20M, DB-Wax, etc.) as a polar stationary phase. Identification is generally carried out either by chromatographic data (Kováts indices, linear retention indices, relative retention time, retention time locking), by spectrometric (MS) or by spectroscopic (infrared, IR) detectors. MS and IR detectors give structural information, which is very helpful for the elucidation of unknown compounds. In order to allow comparison of MS spectra, the standard settings for the detector should be 70 eV.

If identification of peaks is done by reference substances, less sensitive and less expensive detectors can also be used (Bartle and Myers, 2002). The most common flame ionization detector (FID) was introduced in 1958, allowing detection in the sub-ng range. The detector is robust and inexpensive, so it rapidly became the most popular detector. A constrain is that the detector needs hydrogen gas for operation, but only in small amounts which can be produced by a rather simple hydrogen generator.

Another, highly sensitive alternative to the FID is the electron-capture detector (ECD). Based on the ability of a molecule to capture free electrons from a small β-radiation source, the ECD has since found wide acceptance as an extremely sensitive detector (pg to fg range) for halogen-containing compounds, with wide applications to trace pesticide residues (e.g. analysis of resin contaminants). Among a number of non-radioactive ionization detectors, the photoionization detector (PID) and the alkali flame or thermionic detector (TID) have also found long-standing applications. Depending on the discharge gas, which determines photon energy, and the materials of the optical window, the PID allows selective detection for, for example, aromatics over aliphatic compounds, which can be helpful for some special questions around essential oils from resins.

Since the 1980s, the GC-MS combination has been preeminent. The major challenge was the GC-MS interface, because the stream of carrier gas did not allow sensitive MS detection. All MS detectors have an ionization source and the formed charged molecules are separated by their mass to charge (m/z) ratio. For electron impact (EI) ionization, the standard electron voltage is 70 eV, allowing easy comparison of mass spectra. Besides a single MS detector, two-dimensional MS detectors (MS/MS) are also available, which have some advantages in the differentiation between isomers of essential oil compounds or still unknown substances. In the meantime, numerous further MS detectors have become available, among which the time-of-flight (TOF) detector is rather prominent, some of them allowing high-resolution (HR) measurements, giving the exact mass of the analysed compound. Further MS detectors suitable for GC analysis are discussed by Beale *et al.* (2018).

A particular problem of essential oil analysis (and therefore also the precise analysis of resins) is that many substances exist as stereoisomers, which cannot be separated by the column materials listed above. Therefore, so-called 'chiral recognition' was one of the most important milestones reached in essential oil analysis in recent decades (Rubiolo *et al.*, 2010). Enantiomer separation and determination is important not only because optical isomers can have different odours (e.g. limonene, carvone, etc.) but also because these isomers enable us to define the biosynthetic and geographical origins of the matrix investigated, and to investigate technological treatments undergone by and/or the authenticity of most essential oils.

Enantiomer separation in routine enantioselective GC (ES-GC) analysis has been made possible by the introduction of cyclodextrin derivatives (CDs) as chiral selectors. CDs were first introduced for packed columns, but are also available for capillary columns in numerous variations. A universal CD derivative suitable to provide the separation of most significant racemates in this field has not yet been found, mainly because of the intrinsic mechanism of chiral recognition in GC, which is based on a host–guest interaction of each enantiomer of a racemate with the CD selector, the separation depending on the rather small difference in the energy of interaction of each enantiomer with the CD chiral selector. Unfortunately, a laboratory specialized in these questions must therefore have available at least two columns coated with different CD derivatives to enable separation of at least 80% of the most common racemates in the flavour and fragrance field.

The most effective CD derivatives are nowadays those belonging to the so-called 'second generation', consisting of β-cyclodextrins substituted in position 6 (i.e. the 'narrow' side of the CD) with a bulky group (tert-butyldimethylsilyl- or tert-hexyldimethylsilyl-) and with alkylated and acylated groups in positions 2 and 3 (mainly methyl, ethyl and acetyl) of its 'wide' side. The effectiveness of CD derivatives as chiral selectors for ES-GC has to be demonstrated by is chiral reference substances.

A rather common problem in the analysis of resin-born essential oils is that the one-dimensional (1D) GC chromatograms are rather complex and peaks are heavily overlapping. Reasons for this can be, for example, transformation of primary compounds into secondary by the influence of light and heat over time, which is rather normal for natural resins. The problem can be partially solved by optimizing the temperature programme and perhaps changing the column, but a full solution can often be gained by two-dimensional (2D) GC (Rubiolo *et al.*, 2010; Eiserbeck *et al.*, 2012). First attempts started in a manner with each peak sliced automatically every 4–8 s, trapped by cryogenic focusing and newly injected. Modern applications work in a way whereby 'slices' of a couple of seconds are automatically transferred onto a second, much smaller column, typically a very short (< 1.5 m) narrow-bore column. The result will be a 2D contour plot in a kind of heat map, where the x-axis represents the first dimension ('long column') and the y-axis represents the second dimension ('short column').

Another important issue of essential oil GC analysis is the time of analysis. For less complex mixtures, GC time of analysis can be as short as 15 min. However, essential oil fractions of resins are always very complex, giving much longer times for analysis. How can this be accelerated? Some ideas are offered by Cramers *et al.* (1999). The duration of a chromatographic run is primarily influenced by the length of the column. If the column is shortened, retention times will also be shorter, but resolution might be affected in a negative way. This can be compensated by a narrow-bore column. But for this, a higher gas pressure is required, which might cause some other problems. This issue can be solved by a gas, which allows high velocity. Under this aspect, hydrogen is the gas of choice; the second best is helium. However, for the use of hydrogen, special precautions have to be taken to avoid explosions. The chromatographic speed can also be improved by applying a vacuum at the column outlet. In conclusion, standard GC instrumentation should be used if there is no special need for rapid analysis.

Another crucial point is the evaluation of obtained GC-MS data. If the full spectrum should be evaluated, it is not efficient to do this manually. Some ideas for computer-assisted evaluation are given by Beale *et al.* (2018). In order to allow for further processing of the obtained MS data, they have to be converted to an open standard format such as mzML. Also, open-source programs like msConvert can be used for this purpose.

Peaks of GC-MS chromatograms with standardized EI ionization (70 eV) can be identified by using databases like the National Institute of Standards and Technology (NIST) Mass Spectral Database (Borland *et al.*, 2010) and the Wiley Registry™ (Hoboken, USA) of Mass Spectral Data (now part of the Wiley Spectra Laboratory) or databases especially dedicated to essential oils, like the König database from the University of Hamburg (Joulain and König, 1998). Also, the establishment of an in-house database for special questions is highly recommended. The highly reproducible retention indices (RIs) from GC can also be used for orthogonal confirmation of isobaric compounds that often produce similar mass spectra but are distinctly separate in the chromatographic domain.

Currently, the published literature that describes essential oils is populated by works that fail to present the data with calculated retention indices (syn. arithmetic index; Sadgrove *et al.*, 2022). This is the first indicator that the data has not been interpreted and corrected, but rather, reported without modification from the library search report. Authors need to be reminded that NIST library matches are merely suggestions that need to be confirmed or nullified by comparing calculated retention indices against published values. Authors should also avoid using the enantiomeric descriptors (R, S, +, −, D, L) that are suggested in NIST library searches. These values need to be removed from the compound identities unless further work is done to determine their respective enantiomer identity. As already mentioned above, the NIST database has only a limited set of MS data sets related to essential oils, easily leading to misinterpretation.

An example for this is given in Fig. 7.1. Peaks of a GC-MS chromatogram of raw resin from *Boswellia sacra* are assigned by the NIST14 database without any further proof. α-Pinene can be considered as the main essential oil compound (Sura and Cheng, 2024), but some of the other assignments are questionable (isomeric substances).

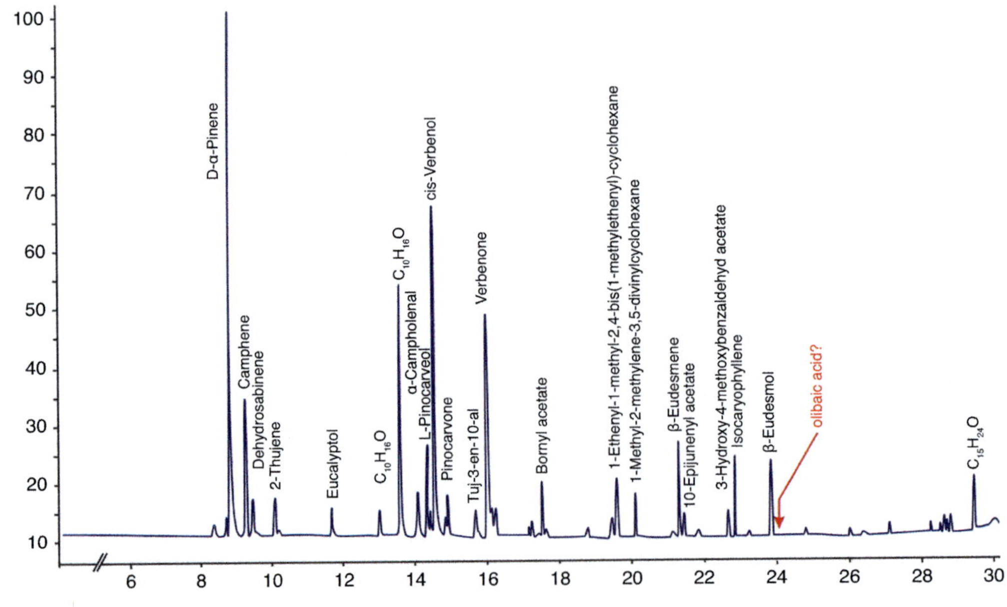

Fig. 7.1. GC-MS chromatogram of a raw sample of *Boswellia sacra* showing the essential oils, mostly mono- and sesquiterpens. The assignment has been made by using NIST14. It must be mentioned that this NIST database only contains a selection of essential oil compounds and therefore the assignment has to be proved either by authentic reference materials or by databases more dedicated to essential oil compounds. The arrow at 24 min points out a possible key odorant. Author's own image.

In the published literature there is also some misunderstanding over the relative quantities of components in essential oil mixtures (Sadgrove *et al.*, 2022). The values determined by integration of a chromatogram, generated from mass spectral detection, are not absolute; they are merely qualitative and are used as a general guide on the quality of an essential oil. The quantity is strongly dependent on the age of the sample, including exposure to sun and heat, the biological origin and the applied sample preparation. The values are also affected by the integration parameters used, which can be adjusted to select peaks above a specified quantitative value. Minor components that are not recognized by the integrator will not be included in the quantitative analysis. Thus, authors should avoid reporting their quantitative values to several decimal places. A single decimal place is used for convention, but in reality, the quantitative value is only useful as a whole integer.

Significant natural variation of essential oil component quantities is expected. Even if the same essential oil is analysed two times, the values of component quantities might change by several integers. More significant differences will be evident between two specimens of the same species. Even when plant material is harvested from the same plant specimen to make essential oil for a second time in succession, the respective essential oil will be different from the first. Thus, authentic essential oils do not need to adhere to exact quantities specified in the published literature.

Conversely, authentication scientists should be aware that confirming a single known component in an essential oil (out of a resin) does not automatically qualify it as authentic. For example, it is impossible to count the number of essential oils that contain in nature highly abundant compounds like 1,8-cineole, limonene, α-pinene and so forth. However, detection of uncommon or rare components that belong to the essential oil increases the likelihood of authenticity. In order to prove authenticity of resins, the non-volatile compounds like those listed below are more suitable.

Since literature repeatedly reported variations in the chemical composition of frankincense essential oils, Van Vuuren *et al.* (2010) analysed 20 commercially available essential oils hydro-distilled from the gum-resin of different *Boswellia* species (*Boswellia carteri*, *B. neglecta*, *B. sacra*, *B. thurifera*, *B. frereana*) using GC-MS. The essential oils examined proved to be qualitatively similar in terms of their main components. However, there were large quantitative fluctuations in certain essential oil substances. The components identified by the authors and their proportions in the oils include α-pinene (2.0–64.7%), α-thujene (0.3–52.4%), β-pinene (0.3–13.1%), myrcene (1.1–22.4%), sabinene (0.5–7.0%), limonene (1.3–20.4%), *p*-cymene (2.7–16.9%) and β-caryophyllene (0.1–10.5%). However, there were large quantitative fluctuations in certain essential oil substances, on the basis of which it seems possible to distinguish, at least to some extent, between the different *Boswellia* species and origins mentioned.

As mentioned above, resins are used by mankind because of their intensive and characteristic smell, which is mainly caused by the essential oil. The essential oil can be extracted by several procedures and analysed by GC-MS or comparable methods. The obtained pattern of substances, combined with information about quantities of certain compounds, does not tell anything about the olfactory properties of the resin. However, some compounds can be correlated to a distinct smell, e.g. the rather common L-limonene to the smell of turpentine, similar to the highly abundant α-pinene.

Therefore, some olfactometry in combination with GC is also required (Van Ruth, 2001; Baldovini and Chaintreau, 2020). For this approach, a carefully trained group of assessors is required. They must be able to describe a smell precisely. In parallel to the physico-chemical detection, the GC must also contain a so-called 'sniff port'. It is rather important to combine the hot GC effluent with humidified air to reduce nasal dehydration. Only a small number of assessors is required.

Valuable information can be gained by four different approaches, but also alternatives are applied (Van Ruth, 2001): (i) In dilution analysis, an extract is diluted, usually as a series of 1:2 or 1:3 dilutions, and each dilution is sniffed until no significant odour is detected. Several injections are required to reach a dilution of the aroma extract in which odorous regions are no longer detected. The test is performed with pure substances. (ii) The detection frequency methods overcome the limitations of a small number of assessors and the use of detection thresholds, because a group of assessors is required. The number of assessors detecting an odour-active compound at the sniff port simultaneously (the frequency of detection, see above) is used as a measure for the intensity of a compound. A sniffing chromatogram can be composed which cumulates the number of detections of the compound (the more assessors, the higher the significance of the results). (iii) The posterior intensity method involves the recording of the odour intensity on a scale after a peak has eluted from the column. This allows better quantification of the odour, but requires specific training of the assessors. (iv) Time-intensity methods are based on magnitude estimation of the odour intensity. Trained assessors directly recorded the intensity and duration of each odour-active compound detected at the sniff port and described the odour perceived. They used a knob (or comparable) with a pointer moving along a 16-point category scale. A simultaneous computerized graphical feedback of the settled position of the cursor helps the assessors to adjust this position to the perceived intensity.

It is not always given that the essential oil compound with the greatest proportion also determines the characteristic smell of a resin. Again, a good example is α-pinene. The smell can be characterized as woody, needle-tree-like, like turpentine. But our nose can easily differentiate between pine, spruce and frankincense, all of them containing α-pinene. This means that minor components or even trace components, so-called 'key odorants', are responsible for the characteristic smell of a resin. To give an example, some years ago, (+)-cis- and (+)-trans olibanic acids from *Boswellia sacra* were identified as key odorants responsible for the typical old church-like base note of frankincense (Fig. 7.2; Baldovini and Chaintreau, 2020). However, these acids can hardly pass the GC and are not unambiguously visible on the GC-MS chromatogram, as shown in Fig. 7.2. A very weak signal with *m/z* 198 can be found at 24.08 min, which can eventually be attributed to olibanic acid. To enrich this compound, the essential oil fraction solved in hexane or dichloromethane has to be extracted with aqueous alkaline solution. Then, the alkaline solution is acidified and extracted with dichloromethane or diethyl ether. Finally, the volatile acids are in the organic layer and can be analysed. For physico-chemical analysis, sialylation or methylation is favourable. For olfactometry, these type of compounds can't be derivatized and rather high oven temperatures are required. Nevertheless, carbonylated hydrocarbons are prone the thermal decarboxylation.

In conclusion, the GC analysis of essential oil compounds obtained from natural resins looks rather easy, but there are pitfalls in all steps:

- **Sample materials**: Are the samples authentic? Is the place of collection and the trading route known? Because the macroscopic identification of a specific resin is difficult, falsifications and adulterations can be frequently observed. Is there any information about exposure to heat and sun available? How old is the sample?
- **Sample preparation:** There are several methods available in order to gain the pure essential oil. The pharmacopoeias of the

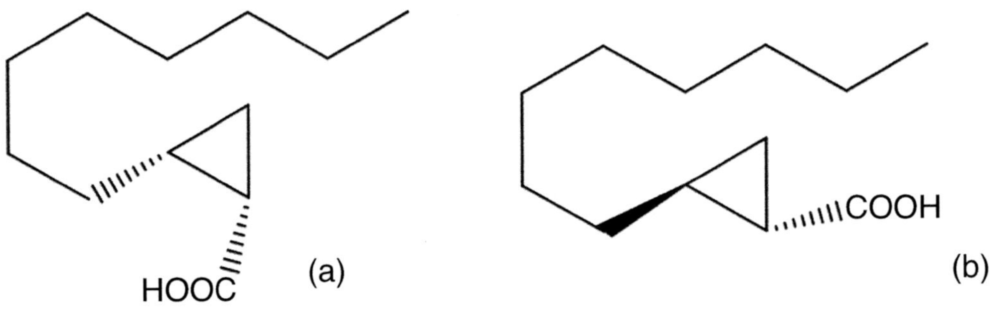

Fig. 7.2. Chemical structures of the two key odorants: a. (+)-cis-; b. (+)-trans olibanic acids ($C_{12}H_{22}O_2$, MW 198,3) from *Boswellia sacra*. Author's own image.

world prefer a steam distillation, but this is not always the best method. Native essential oil fractions can be gained by very short distillation times or by direct extraction of the sample without further evaporation of the solvent. In any case, a representative number of analytical samples has to be taken from the raw resin.

- **Chromatography:** Proper selection of the column. Possible are DB-1, DB-5, WAX or CD columns. If stereoisomers have to be properly identified, at least two CD columns are required. For complex mixtures, 2D GC-MS can be required. MS–MS and high-resolution time-of-flight (HRTOF) might deliver additional structural information.
- **Evaluation of the chromatogram and mass spectra:** Besides the mass spectra, the RI is also of importance. The standard databases do not deliver all essential oil compounds. It should be considered that an obtained mass spectrum can be a mixture of two or more compounds. It should also be considered that quantification of peaks is of limited value and strongly depends on the sample itself and the applied analytical method. Identification of a distinct essential oil by the relative amounts of individual compounds is not always goal-leading.

In summary, highly automated analytical equipment and obviously comprehensive databases seem to enable easy analysis of essential oils, but this is not the case. Actually, dedicated training over years is required to understand the world of resins and the essential oils derived from them.

Analysis of Non-Volatiles

As mentioned in the introduction, the crude resin has to be partitioned prior to further chromatographic analysis of non-volatile compounds. Main fractions are lipophilic and hydrophilic compounds as well as insoluble material, depending on the polymerization grade and age of the resin (polymerization due to oxygen and light). Hydrophilic, organic compounds are mainly sugars and proteins, whereas more lipophilic compounds are phenols, diterpenes and triterpenes.

An example for the extraction and fractionation of lipophilic compounds is given in Fig. 7.3 (Keusgen, 2021). Crushed, crude resin obtained from *Boswellia* sp. (frankincense) is extracted by an appropriate organic solvent like ethanol or diethyl ether at moderate temperature. Then the solvent is evaporated under reduced pressure, together with the majority of the essential oil compounds. In the next step, the obtained extract is resolved in dichloromethane (DCM) or diethyl ether and extracted with an alkaline, aqueous solution. Di- and triterpenic acids as well as some phenolic compounds will move into the aqueous layer, whereas the neutral compounds will remain in the organic layer. In the final step, the aqueous layer will be acidified and again extracted with an appropriate organic solvent. The di- and triterpenic acids will move into the organic layer and can be analysed by reversed-phase high-performance liquid chromatography (RP-HPLC) after removing the organic solvent under reduced pressure. Because of the amphiphilic nature of many of the triterpenic acids, partitioning between the organic and aqueous layer has to be done with special care because of foaming.

Because of their distinct biological activities, di- and triterpenic acids are of special analytical interest. Diterpenic acids like derivatives of abietic acid are characteristic for the plant families of *Pinaceae* and *Cupressaceae* (needle trees) as well as for amber (Langenheim, 1990). Triterpenic acids like boswellic acids are the active principle of *Boswellia* sp. It must be mentioned that *B. serrata* is the only *Boswellia* species listed in several pharmacopoeias around the globe. Also, other commonly used *Boswellia* species like *B. sacra* and *B. papyrifera* contain a comparable pattern of substances. These compounds can be nicely separated by RP-HPLC as depicted in Fig. 7.4 (Keusgen, 2021). In this kind of investigation, β-glycyrrhetinic acid (β-Glyc) can be used as internal standard.

The concentrations found for different species do vary in a rather huge range. For instance, keto-β-boswellic acid (β-KBA) was found in highest concentrations with 1.4% in *B. serrata*, whereas 3-O-acetyl-9,11-dehydro-β-boswellic acid (Ad-β-BA) was highest for *B. sacra* with 4.8%. Tirucallic acids (TAs) were most abundant in *B. serrata* and *B. papyrifera*, totalling more than 20% in *B. serrata* and 9%

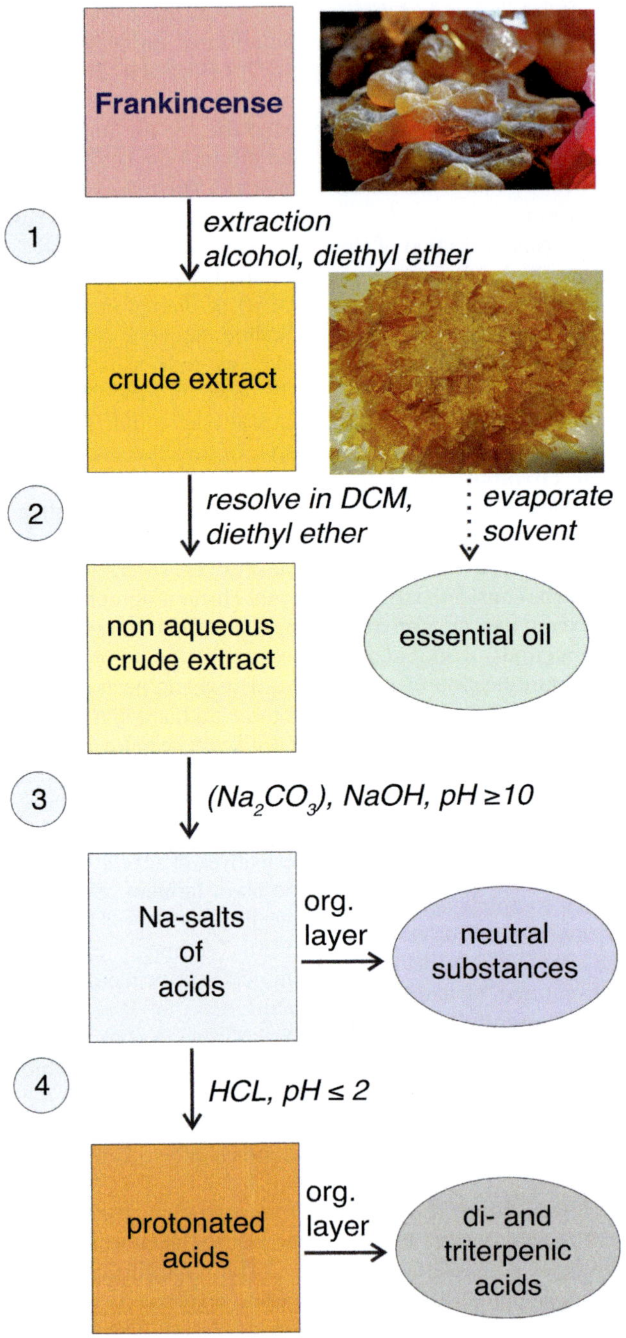

Fig. 7.3. Route of extraction and fractionation of crude *Boswellia* (frankincense) resin, which can also be applied to further resins, which typically contain neutral and acidic terpenoids besides essential oil compounds. (By M. Keusgen.) Used with permission.

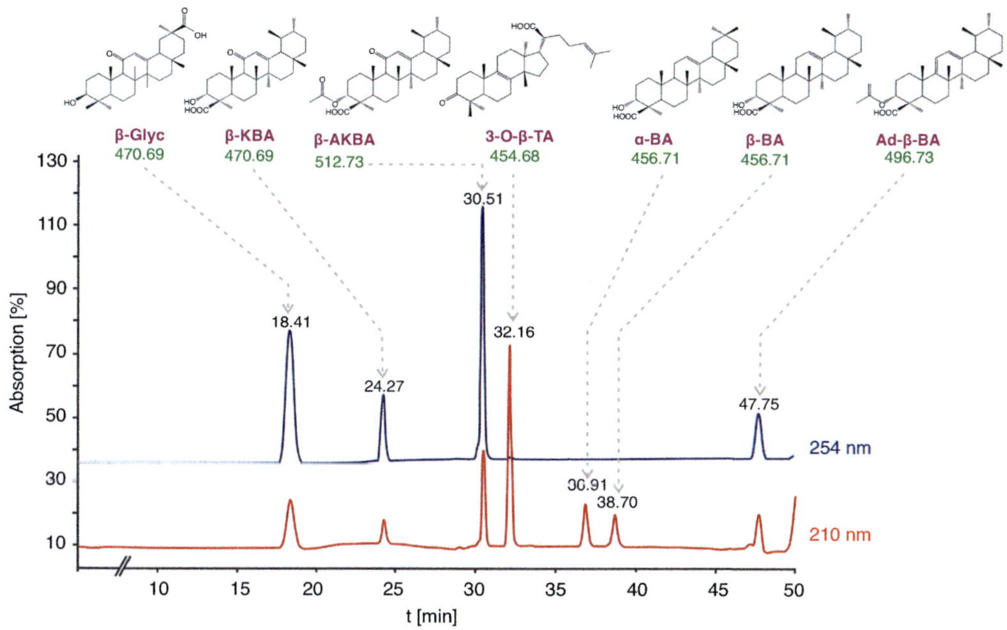

Fig. 7.4. HPLC chromatograms of triterpenic acid standards related to the genus *Boswellia*. Molecular weights are given below the molecular structures. RP-18 column, 254 × 4 mm, water-acetonitrile gradient with 0.1% trifluoroacetic acid (TFA), flow 1 ml. β-Glyc: β-glycyrrhetinic acid (internal standard), β-KBA: keto-β-boswellic acid, β-AKBA: 3-acetyl-11-keto-β-boswellic acid, 3-O-β-TA: 3-oxo-β-tirucallic acid, α-BA: α-boswellic acid, β-BA: β-boswellic acid, Ad-β-BA: 3-O-acetyl-9,11-dehydro-β-boswellic acid. (From Keusgen, 2021.) Used with permission.

in *B. papyrifera*. These findings are of some importance, because until now keto-boswellic acids (KBAs) have been considered the bioactive principle of frankincense. This should be reconsidered because the concentration of TAs and other triterpenic acids is much higher than that of boswellic acids.

A comprehensive overview of terpenoids in *Boswellia* is also given by Sura and Cheng (2024). In total, 15 triterpenoids of the lupane type, 14 triterpenoids of the dammarane type, 35 compounds of the ursane type including boswellic acids, 7 compounds of the oleanane type and 31 triterpenes of the tirucallane type have been identified until now. This confirms that the majority of compounds belongs to boswellic acids (ursane type) and tirucallic acids (tirucallane type). In addition, two phenylpropanoid derivatives, one flavonoid and one lignan have been found in *B. serrata*. These phenol-derived compounds can also be separated by RP-HPLC.

Detection of triterpenoids (and also diterpenoids) in combination with HPLC separation can be performed either by UV absorption at 210 nm and 254 nm in parallel or by a mass sensitive device (MSD). Because many of these substances contain an acid functionality, a low amount of TFA has to be added to all solvent components in exactly the same concentration. This is important because UV detection at 210 nm is also sensitive to TFA.

Diterpenoids and triterpenoids including their acids can be also analysed by GC-MS. For sample preparation, the resin has to be milled and extracted in an appropriate solvent as displayed in Fig. 7.3, step 1 (left-hand side). Suitable are diethyl ether, ethanol, dichloromethane or a mixture of chloroform and methanol 2:1 (v/v) under ultrasonication (Baeten *et al.*, 2014). The insoluble residue has to be removed by centrifugation and the supernatant has to be filtered through a suitable 0.45 μm pore size filter. However, plasticizer might also be eluted out of filter materials and, therefore, alternative routes using glass wool, etc. are recommended. The clear extract is concentrated by a gentle stream of nitrogen and then derivatized by N,O-bis(trimethylsilyl)trifluoracetamid (BSTFA) with 1% of trimethylsilyl chloride (TMSC) for 60 min at 60°C. The usage of this mixture for pre-column silylation is important, because the di- and triterpenic acids are often sterically hindered at the silylation site (especially boswellic acids). If this procedure is used for quantitative determination, the whole process has to be validated carefully with pure reference materials because silylation might be incomplete. In any case, air humidity has to be excluded and only water-free solvents can be used. If the result is not satisfying, water-free pyridine can be used in order to support the reaction. Finally, the obtained trimethyl (TMS) derivatives as well as the underivatized substances (mostly terpenoids without alcohol or acid functionality) can be solved in an appropriate solvent (toluene) and are ready for GC analysis.

A standard GC, ideally with an MS detector, can be used for this type of analysis. A capillary column of DB-5-type (30 m × 0.25 mm × 0.25 μm) or comparable can be used; high-temperature columns should preferably be chosen. For analysis of di- and triterpens, the temperature gradient might start at 80°C and goes up to 320°C or more with high-temperature columns. It also possible to use an 'old', partially deactivated DB-5-column. Then, the maximum temperature could be 280°C. It is also possible to analyse mono- and sesquiterpens with the same GC run, but then the temperature gradient should start at 40–60°C (compare the section 'Analysis of Volatiles'). Using a gradient between 60 and 280°C also allows the GC analysis of phenolic lipids, which are typical for *Pistacia* resins as well as further Anarcadiaceae.

Another good example for phenolic-rich resins is from the *Dracaea* species: the resin is called 'Dragon's blood' (Camarda *et al.*, 2011; Hu *et al.*, 2015; Sun *et al.*, 2019; Akbulut *et al.*, 2024; Sura and Cheng, 2024). According to literature, flavans, homoisoflavans, homoisoflavones, chalcones and dihydrochalcones have been isolated. From the pulverized material, phenols can be extracted by alcohol–water mixtures and gentle heating. The solvent will be evaporated under reduced pressure and the residue is resolved in methanol. A standard RP-18 column can be used for HPLC separation, and the detection wavelength is at 278 nm. A gradient of slightly acidified water with methanol or acetonitrile is most suitable for separation.

Another resin showing numerous phenolic compounds besides diverse terpenoids is myrrh (*Commiphora myrrha*) and other *Commiphora* species (Hanuš *et al.*, 2005; Batiha *et al.*, 2023). Like *Boswellia*, the species belongs to the family of Burseraceae. Similar to frankincense, the raw oleoresin can be divided into a very complex lipophilic fraction as well as water-soluble gum, which is mainly made up of polysaccharides and proteoglycans. Therefore, the fractionating scheme given in Fig. 7.3 can also be applied. Mono- and sesquiterpenes might be analysed by GC, other less volatile compounds such as TMS derivatives and higher GC oven temperatures up to 320°C as already described above.

In contrast to frankincense, *Commiphora* resins are more or less rich in phenolic compounds belonging to the family of lignans. Most of them are methylated and can be analysed by HPLC as described above with UV detection at 254 nm or 278 nm. Because of the methylation, they are more lipophilic than unsubstituted phenolic compounds. Some lignans belong to the polygamatin type, and a podophyllotoxin skeleton has also been observed. Also, lactone rings are rather common structural elements of *Commiphora* lignans.

Another rather special feature of *Commiphora* resins are sesquiterpenoid dimers (Zhu *et al.*, 2019; Sura and Cheng, 2024). Sesquiterpenes with 15 carbon atoms can be easily analysed by GC, but their dimers are less volatile and additionally highly oxygenated. GC separation on a DB-5 column seems to be still possible, but the oven temperature has to be around 300°C. In total, 27 of theses dimers have been reported until now. Many of them are cardinane-cardinane, guaiane-cardinane and germacrene-germacrane, which can be connected in various ways, sometimes forming a spiro-centre. HPLC-analysis can be performed on RP-18 columns, using water–methanol or water–acetonitrile gradients without acidic modifier. Under isocratic conditions, the substances can be eluted with aqueous methanol 80% or aqueous acetonitrile 70%. Typical UV absorption for detection is between 228 nm and 254 nm. Also, an MSD can be used for detection.

Another valuable source for complex resins is the genus *Pistacia*, having its main distribution from the Mediterranean to Central Asia. The genus *Pistacia* belongs to the family of Anacardiaceae, and 11 different species can be distinguished. Most important species for resin production are *P. lentiscus* ('mastic tree'), *P. terebinthus*, *P. atlantica* ('Mt. Atlas mastic tree') and *P. eurycarpa*. Besides an essential oil, the resins contain a huge variety of tetra- and pentacyclic triterpenic compounds like oleanonic acid, masticdienoic acid, pistagremic acid and pistacigerrimones A–D (Rauf *et al.*, 2017). These compounds are partially acidic and can be fractionated as well separated, as described above for *Boswellia*.

Typical for the family of Anacardiaceae are alkylphenols, and several have been described for the genus *Pistacia* (Saitta *et al.*, 2009). Typical substances of this class are: anacardic acids, cardols and cardanols. Because of the amphiphilic character of these compounds, separation by HPLC seems to be difficult. The neutral alkylphenols can be separated by normal phase TLC using n-hexane: diethylether (60:40 v/v; R_f-value *c.*0.5) as mobile phase. The underivatized sample can be further analysed by GC on a DB-5 column (or comparable), with a temperature gradient rising from 60°C up to 275°C. The alkylphenols can also be converted into the corresponding trimethylsilyl derivatives, also allowing analysis of the acidic compounds.

Needle trees are mainly represented by the families of Pinaceae, Cupressaceae and Araucariaceae, with a nearly global distribution on both hemispheres. The most important genera are *Pinus*, *Abies* and *Cedrus*. The chemical composition of resins from needle trees is rather unique: The lipophilic, non-volatile fraction is not dominated by triterpenic compounds but instead by diterpenic substances (Azémard *et al.*, 2016; Sahin and Yalcin, 2017; El Omari *et al.*, 2021; Chauiyakh *et al.*, 2022). The diterpenic substances can be either neutral or acidic and, therefore, the separation scheme given for triterpenic compounds can also be applied to this chemical group. Prominent examples for diterpenic acids are abietic acid, neoabietic acid, palustric acid, pimaric acid, agathic acid and pinifolic acid, belonging to the class of abietanes, pimaranes and labdanes.

Most comfortable is the analysis of the diterpenic compounds by GC after derivatization of samples. The compounds can be transferred into the corresponding trimethylsilyl derivatives or methyl derivatives. For silylation, the following procedure is suitable (Azémard *et al.*, 2016): a few milligrams of a completely dry sample were mixed with 200 µL of N,O-bis(trimethylsilyl)-trifluoroacetamide (BSTFA) +1% trimethylchlorosilane (TMCS). Such solutions were left for 30 min at 70°C. Afterwards, the trimethylsilylated extracts were dried under a stream of nitrogen, dissolved in 1 ml of hexane (or comparable) and filtered. The methylation procedure was similar to the trimethylsilylation one (30 min, 70°C) with trimethylsulfonium hydroxide (TMSH) as reactant.

Suitable GC settings are as follows: A DB-5 column or comparable (5% phenyl, 95% dimethylpolysiloxane), film thickness 0.25µm, is most suitable. The GC temperature programme was held to 50°C (2 min), ramped at 7 °C min^{-1} until 250°C and finally increased at 2.5 °C min^{-1} until 330°C. Because 330°C is the temperature limit of most GC columns of this type, a high-temperature modification is recommended. The injector and the transfer line temperatures were set, respectively, at 250°C and 300°C. It is important that the temperature of the transfer line is sufficiently high. Basically, these analytical conditions are comparable with those for boswellic acids, as described above.

Raw and fractionated resins can also be analysed by TLC (high performance: HPTLC). However, fractionation is recommended because otherwise the zones are not clear. The use of reference substances (standards) is recommended in order to optimize chromatographic conditions.

An example for *Boswellia* is given in Fig. 7.5 (Keusgen, 2021). There are several points which must be considered for the development of a suitable TLC system. First of all, complex mixtures like resins rather often require a sample preparation. Therefore, a fractionation like that shown in Fig. 7.3 is recommended. Even with a fractionation, some amounts of the other substances are still present. For instance, if the triterpenoids are of interest, essential oil compounds can be separated simply by removing the

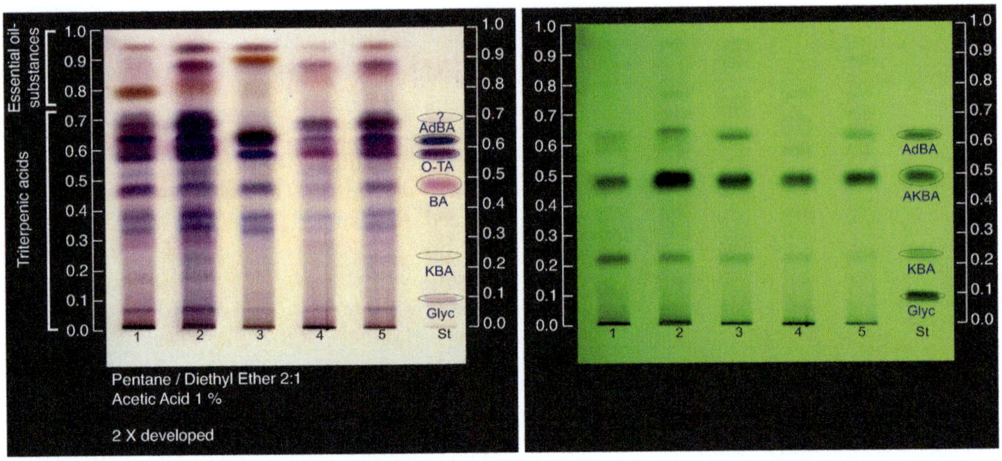

Fig. 7.5. HPTC plates showing five different samples of *Boswellia* resins, enriched by alkaline extraction. Plates were developed twice with pentane/diethyl ether (2:1 v/v) and 1% acetic acid as modifier. Detection has been done under UV light 254 nm (right) or spraying with anis aldehyde (left). St: standard mixture, consisting of Glyc: β-glycyrrhetinic acid (internal standard), KBA: keto-β-boswellic acid, AKBA: 3-acetyl-11-keto-β-boswellic acid, O-TA: 3-oxo-β-tirucallic acid, BA: boswellic acids, AdBA: 3-O-acetyl-9,11-dehydro-β-boswellic acid. (From Keusgen, 2021.) Used with permission.

extraction solvent. However, the evaporation of essential oil compounds is incomplete, as shown in Fig. 7.5. The major ones are still present.

Secondly, the chemical nature of the compounds of interest must be respected adequately. In resins, many of the non-volatile but lipophilic compounds are acidic, while others are more or less neutral. In this case, an acidic modifier like acetic acid has to be added.

As a serious problem, many of these acids are amphiphilic, which means that a sharp zone can hardly be obtained. Therefore, extensive experiments with different solvent mixtures serving as mobile phase are required.

The last point of concern is a suitable detection of the zones. Comfortable detection at 254 nm is only possible if the separated substances have a sufficient number of double bonds, which for triterpenoids is not always the case. Also, the detection of fluorescence at 366 nm is mostly not goal-leading, because most substances do not show a fluorescence. The last option is the use of spraying reagents like anis aldehyde or vanillin, or the use of an iodine chamber. These spraying reagents are rather universal, but as can be seen in Fig. 7.5, rather often two different detection methods are required (for instance, UV 254 nm and spraying with anis aldehyde).

In conclusion, HPLC or GC (after silylation or methylation of the sample) can be used for the analysis of the lipophilic, non-volatile fraction. The analysis of the sugar fraction requires special methods, mainly a combination of different chromatographic techniques. GC-MS seems to be most convenient for the lipophilic fraction, but it has to be considered that even after derivatization, rather high temperatures are required in order to facilitate a suitable separation. A temperature gradient up to 330°C is rather often necessary. Therefore, the usage of a high-temperature GC column is recommended. In addition, TLC methods are rather convenient for the identification of resin materials, but development of a suitable method is mostly rather time-consuming.

Vibrational Spectroscopy

For more than 20 years, various IR and Raman spectroscopic methods have been successfully used to analyse valuable plant constituents or to characterize various quality parameters in horticultural and agricultural crops. In most cases, vibrational spectroscopic measurements can be carried out directly on plant tissues or fractions that have been previously isolated from the plant material by hydro-distillation or solvent extraction. In general, spectra can be obtained with both spectroscopy techniques that show some characteristic key bands of individual components. These bands provide information about the chemical composition of the analysed samples, including the primary and secondary metabolites. Furthermore, spectroscopic analyses based on such markers related to individual plant compounds allow, in principle, the discrimination of different species and even chemotypes within the same species. In this context, the combination of vibrational spectroscopy and hierarchical cluster analysis offers a fast, simple and reliable method for chemotaxonomic characterization. The ability to quickly determine different plant components also enables efficient selection of high-quality individual plants from wild populations and from the offspring of crossing experiments in the context of plant breeding. In addition, the methods of vibrational spectroscopy can also be used by the processing industry to carry out rapid quality control of incoming raw materials and continuous monitoring in production (Schulz and Baranska, 2007; Schulz, 2020).

In addition, Raman and IR macrospectroscopy and microspectroscopy in combination with mapping or imaging techniques offer the particular advantage of measuring entire plant organs such as seeds, fruits, leaves, flowers or stems non-destructively and thus also enable the analysis of plant cell architecture and biochemical changes at the cellular level. In particular, the introduction of Fourier transform (FT)-Raman spectroscopy with excitation in the near-infrared (NIR) range has enabled the development of numerous applications on fresh plant material, as the problem of fluorescence and photochemical degradation, which normally occurs when using ultraviolet-visible (UV)-Vis lasers, can be minimized. In addition to characterizing plant quality, vibrational microspectroscopy methods are also suitable for investigating interactions between plant tissue and various pathogens such as fungi, bacteria or viruses (Schulz *et al.*, 2014).

Infrared and Raman spectroscopic analysis of resins and gums

Fourier transform infrared (FTIR) spectroscopy was used to measure the IR absorption spectra of the exudates obtained from different types of gymnosperms (Tappert *et al.*, 2011). Based on the spectra, it was possible to distinguish three compositionally different groups of exudates: resins, essential oils (oleoresins) and gums. These natural materials consist mainly of terpenoids, while polysaccharides make up the majority of gums. Resins and volatile resins are mainly found in conifers (Pinophyta). Gums, on the other hand, are produced by some conifer families, including Araucariaceae and Podocarpaceae, by non-conifer gymnosperms, including *Ginkgo biloba* (Ginkgophyta), cycads (Cycadophyta) and *Welwitschia mirabilis* (Gnetophyta), and by angiosperms. Based on the ratio of the spectral bands, Tappert *et al.* (2011) quantitatively divided the resins of the conifers into two different resin types, which reflect differences in the composition of their terpenoid components and largely correspond to the different conifer families. The first resin type (pine resin) is produced by the pine family (Pinaceae) and consists mainly of diterpenes based on abietan/pimaran skeletal structures. The second resin type (cupressaceous resin) is produced by members of the Cupressaceae, Sciadopityaceae, Araucariaceae and Podocarpaceae and consists mainly of diterpenes based on labdanoid structures. It was found that the variability of the resin spectrum correlates with the number of free hydroxyl groups and can therefore be used to assess the degree of polymerization of the resin. Resins with a high content of volatile components (essential oil) were found exclusively in the Pinaceae, which the authors associate with a generally higher content of volatile mono- and sesquiterpenoids in resins of this family.

First attempts to use IR spectroscopy in combination with HPLC technique normally used to distinguish between the resins of different *Boswellia* species were successfully carried out by Archier and Vieillescazes (2000). They found remarkable differences due to the different geographical origin of the resin samples from India, Somalia, Sudan and Yemen. Furthermore, they identified that *Boswellia sacra* from India differs from the other frankincense samples in the 450–900 cm^{-1} range, in particular by the intense bands at 779 and 692 cm^{1}, respectively, and by the absorption band at 459 cm^{-1}. In addition, characteristic differences in the C = O vibrational band region around 1740 and 1720 cm^{-1} were registered in the resin samples from Sudan and Somalia for the different *Boswellia* species and origins.

Edwards and Falk (1997) describe the use of FT-Raman spectroscopy for the non-destructive analysis of frankincense and myrrh. Since both resins are multi-component systems, the authors compared the spectra of the two resins with those of the individual components present in each resin. In this way, frankincense and myrrh can be identified using FT-Raman spectroscopy, each of which has different concentrations of the individual components. No sample preparation is required for the spectroscopic analysis described.

Daferera *et al.* (2002) succeeded in quantifying the two most valuable compounds, α-pinene and β-myrcene, in the essential oil distilled from mastic gum using FT-Raman spectroscopy. The two monoterpenes are usually determined by GC-MS, whereby the β-myrcene content is a decisive factor for the marketability of mastic gum. For the quantification of α-pinene and β-myrcene, the bands at 1658 and 1633 cm^{-1} were used and corresponding linear calibration curves were calculated (Fig. 7.6). The correlation coefficient for α-pinene was 0.992 and for β-myrcene 0.997 (for the concentration range of 30–80% and 3–45%, respectively). The efficiency of the Raman rapid method was demonstrated using various mastic oils from the island of Chios (Greece), with α-pinene and β-myrcene contents of 38–70% and 5–58%, respectively.

Various studies have already been carried out on fresh resins (Langenheim and Beck, 1965; Salome-Abarca *et al.*, 2018; Wagner-Wysiecka, 2018). Furthermore, a few papers deal with the investigation of the degradation process by FTIR analysis (Azémard *et al.*, 2014), so a comparison of aged resins can be performed to precisely evaluate the dependence of the identification criteria on the degradation processes. This issue is of crucial importance when it comes to the identification of archaeological artefacts.

Georgiou *et al.* (2022) successfully demonstrated the complementary use of X-ray Raman spectroscopy (XRS), FTIR and GC-MS to analyse important genus-specific chemical

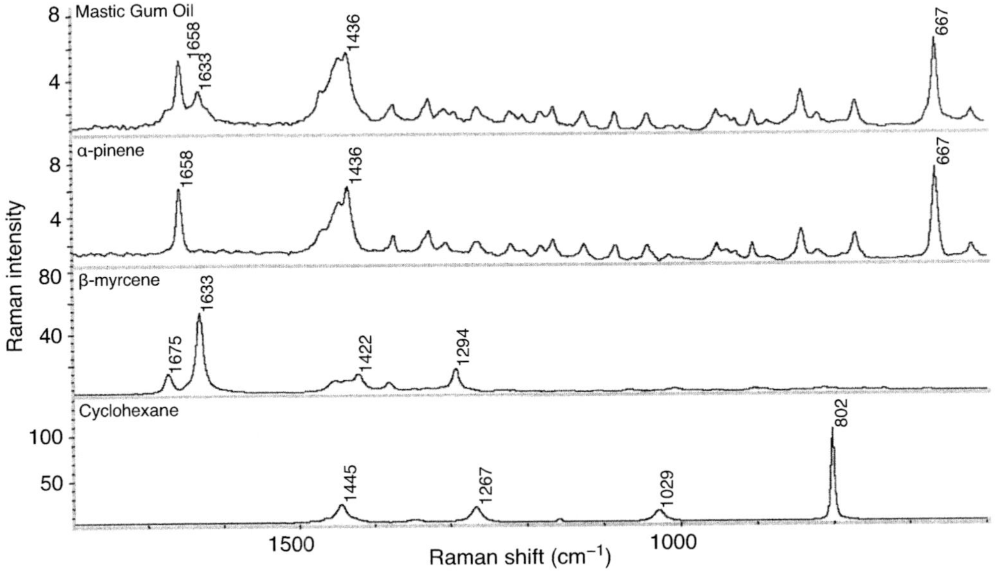

Fig. 7.6. FT-Raman spectra (1800–600 cm^{-1}) of mastic gum oil, α-pinene, β-myrcene and cyclohexane. (From Daferera *et al.*, 2002.). Used with permission.

characteristics for selected native Australian plant exudates obtained from *Acacia*, *Callitris*, *Eucalyptus* and *Xanthorrhoea* species. In the multi-analytical study, the analysed plant exudates were successfully divided into different chemical classes: phenolic and aromatic compounds (genera *Eucalyptus* and *Xanthorrhoea*), polysaccharides (genus *Acacia*) and terpenoids (genus *Callitris*). XRS was used to elucidate basic binding information, while FTIR provided important information about various functional groups and GC-MS supplied information about the molecular structure by characterizing the molecular pattern. The combination of these three techniques enabled a meaningful characterization and chemical classification of different plant exudates and their properties, with further potential for the identification of works of art, cultural artefacts and museum collections.

By using near-infrared spectroscopy (NIRS) in combination with multivariate statistical analysis, Rehman *et al.* (2020) were able to rapidly quantify incensol acetate in the resins of five different *Boswellia* species (*B. papyrifera*, *B. serrata*, *B. sacra*, *B. elongata* and *B. socotrana*). The authors could show that the acetylated fraction of the resin of *Boswellia papyrifera* yielded the largest amount of incensol acetate (32.87 %), while no incensol acetate was found in the resins of the four other *Boswellia* species. The results obtained by the non-destructive NIRS method are in good agreement with the results of the HPLC reference method, so that it can be assumed that a rapid quantitative determination of incensol acetate in *B. papyrifera* can be reliably performed.

A systematic study on the natural ageing of some *Pinus* resins and related compounds was carried out using FTIR (Sanchidrian, 2016). The oxidative processes and molecular changes that take place during ageing were identified. The results confirm the initial transformation of abietic acid into dehydroabietic acid and then into more oxidized species. In advanced stages of conversion, reactions and interactions between these molecules occur. On the one hand, the hydroxyl group of 15-hydroxy-7-oxo-dehydroabietic acid forms intermolecular bonds. On the other hand, the reaction between two carboxylic acid groups from different molecules (abietanes or pimaranes) can lead to the formation of anhydride acids. This explains the decrease in the amount of pimaranes in the late stages of ageing more

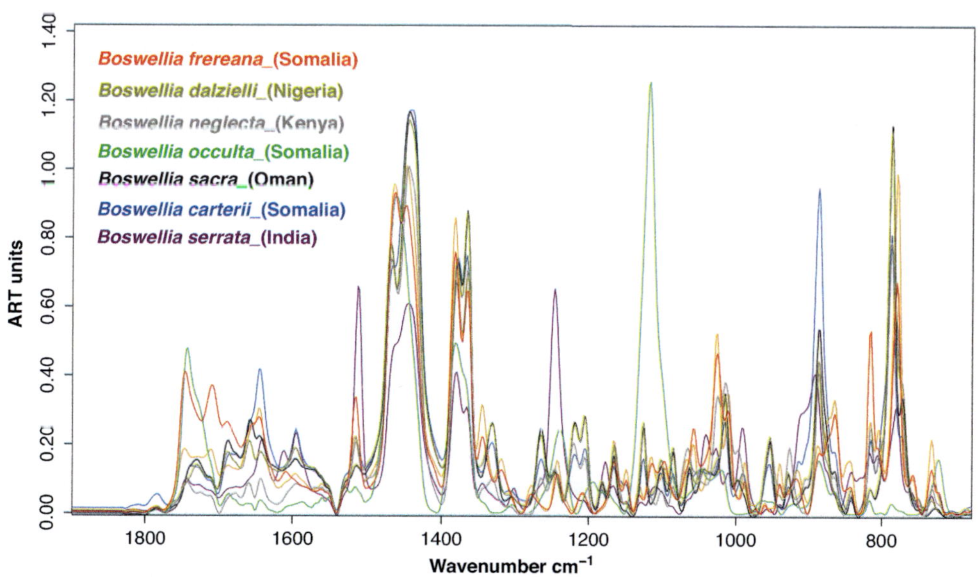

Fig. 7.7. ATR-IR spectra of the essential oils extracted from the resins of different *Boswellia* species show characteristic signals of the different terpenoid composition, which can be used to classify different *Boswellia* species and origins. (From Krähmer *et al.*, 2024.) Used with permission.

reliably than the presumed polymerization between them. Consequently, the resins of the different species evolve to a similar molecular composition (oxidized abietanes and high-molecular-weight compounds).

Vandenabeele *et al.* (2007) used Raman spectroscopy to analyse some natural artist materials that had been purchased at a local market in Mexico. The natural plant materials analysed were chia oil, linseed oil, Campeche wax, beeswax, white copal, dammar resin, rosin, mastic resin, *chapopote*, *chucum*, gum gutta, peach gum and gum arabic. The Raman spectrum of *chapopote*, the local name for bitumen, contained significant features of carbonaceous and terpenoid substances. Unfortunately, strong fluorescence was observed in the Raman spectra of *chapopote* and *chucum*, so the resulting Raman spectra of these materials were very noisy.

Infrared and Raman spectroscopic analysis of the volatile fraction of resins and gums

Since the described chromatographic methods are comparatively time-consuming and expensive, there were some attempts to provide rapid spectroscopic methods that enable a simple characterization of the various resins and related products obtained therefrom. In this context the chemical profiles of various *Boswellia* and *Commiphora* essential oils were analysed using attenuated total reflection-IR (ATR-IR) and NIR spectroscopy based on GC reference data (Fig. 7.7).

Furthermore, NIRS was proved for rapid and non-destructive analyses to distinguish between the resins of these two genera. Despite the sometimes very heterogeneous colour and structural composition of the individual resin samples, different species can be clearly distinguished in the spectral range of 5000–5800 cm^{-1} (Fig. 7.8). On the basis of spectroscopic analysis and chemometric data evaluation, it is generally possible to identify endangered species such as *Boswellia papyrifera* and thus carry out a regional classification of different *Boswellia* origins. In addition, the spectroscopic methods can be used for on-site measurements, for quality assurance at producers and traders and at customs when exporting the resins (Krähmer *et al.*, 2024).

Most natural resins cannot be used directly in end products, but must first be chemically modified to achieve the properties required for the respective application (e.g. degree of cross-linking, softening point, viscosity, chemical stability). In industry, these resin modification reactions are usually discontinuously monitored applying various physical sum parameters such as viscosity, turbidity, refractive index and colour or chemical parameters (acid number, hydroxyl number, resin and fatty acids). These methods are generally time-consuming, consume

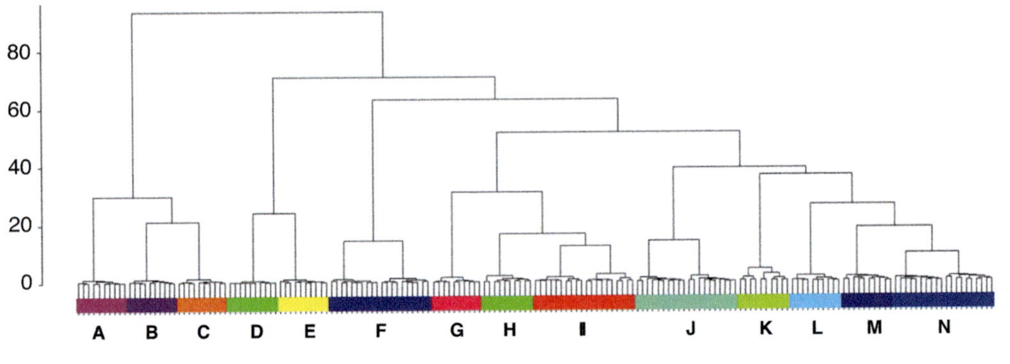

Fig. 7.8. Hierarchical cluster analysis (HCA) of the NIRS spectra of untreated resins of various *Boswellia* and *Commiphora* species as well as gum arabic obtained from *Acacia senegal* as a reference. A: *B. frereana*, B: *B. rivae*, C: *C. kua*, D: *G. arabicum*, E: *C. myrrha*, F. *B. neglecta*, G: *C. confusa*, H: *B. dalzielli*, I: *B. sacra*, J: *B. papyrifera*, K: *C. kataf*, L: *B. sacra*, M: *B. neglecta*, N: *B. carteri*. (From Krähmer *et al.*, 2024.) Used with permission.

comparatively large quantities of chemicals and cause generate waste. Therefore, a new powerful NIR spectroscopic method was developed to monitor the progress of the resin modification process using samples taken at different process times (Blanco and Villaescusa, 2007).

The authors found that NIR spectroscopy can be successfully used as an effective alternative to conventional wet analytical methods in the control of industrial resin modification processes, as it allows the simultaneous determination of the parameters normally used to evaluate the individual reaction state. In addition, analysis times can be significantly reduced and, in most cases, production cycles can be shortened.

Infrared and Raman spectroscopic analysis of fossil resins

FT-Raman spectra of 10 fossil resin samples were obtained, which differ in terms of localities, exact palaeoecological situation and geological age (Jehlička *et al.*, 2004). The analysed series includes amber originating from freshwater deposits in Moravia and Bohemia of Cenomanian age. As comparative material, unaltered amber from Yantarnaya (Baltic Russia), highly matured amber, rumanite and simetite as well as přílepite from the Westphalian age (Carboniferous) were analysed by the authors. The FT-Raman spectra clearly show the similarity between Bohemian and Moravian fossil resins and Baltic and Lebanese ambers from the Cretaceous period. Overall, considerable differences in the FT-Raman spectra obtained could be observed. However, similarities in the spectra of fossil resins of different ages and differences between similar materials collected from nearby locations suggest that post-depositional conditions of the tree resins and the associated transformations are mainly responsible for the character of the resulting fossil resins (Badea *et al.*, 2015; Casadio *et al.*, 2016; Stach *et al.*, 2019; Stach *et al.*, 2020; Drag *et al.*, 2022).

FT-Raman spectra of some natural waxes and resins found in different preparations of archaeological material were identified non-destructively for the first time by Edwards *et al.* (1996). In this context, suggestions were also made for the assignment of the individual vibration modes, which may be helpful for the identification of the materials used. In one case, a paper sample impregnated with a wax component, the presence of beeswax could be clearly demonstrated. Edwards *et al.* (1997) obtained FT-Raman spectra of two unknown archaeological resin samples dated to about 250 BC and compared the spectral features with those of contemporary resins from Vietnam. The authors point out that the bands in the 1750–1500 cm^{-1} range are particularly useful for characterizing the resins. Samples from a comparable situation, i.e. sealing compounds on burial vessels, from North America were also used for comparison.

Edwards *et al.* (2001) recorded the FT-Raman spectra of the resins of *Dracaena draco* and related Dragon's blood samples (Fig. 7.9).

These spectra mainly show bands in the wavenumber ranges 2600–3400 and 200–1800 cm^{-1}, which can be assigned to the C-H stretching and skeletal stretching and deformation modes, respectively. 'Dragon's blood' is a deep red resin that has been used by many cultures for several centuries. The resin has always been highly valued for its rarity, depth of colour and alchemical associations. The original source of Dragon's blood resin is said to be *Dracaena cinnabari* from Socotra in Africa, but since the Middle Ages there have been several alternatives from various geographical areas, from the Canary Islands to the East Indies. Edwards *et al.* (2001) compared the Raman spectra of Dragon's blood resins obtained from the Liliaceae tree *Dracaena draco* growing in various locations in the Mediterranean and Middle East with the resins of alternative botanical sources such as *Daemonorops draco*, *Dracaena cinnabari* and *Eucalyptus terminalis*, all of which are grouped under the name 'Dragon's blood'.

Edwards and Ali (2011) used selected examples to show how Raman spectroscopy can provide valuable support as a screening technique for the non-destructive analysis of ancient resins in artefacts and samples from archaeological excavations. They also show that the analytical capabilities of Raman spectroscopy depend largely on the different excitation wavelengths from the visible to the NIR range. Fig. 7.10 shows a stack plot of standard diterpenoid (frankincense) and triterpenoid (*Pistacia terebinthus*) resins and the Amarna resin for

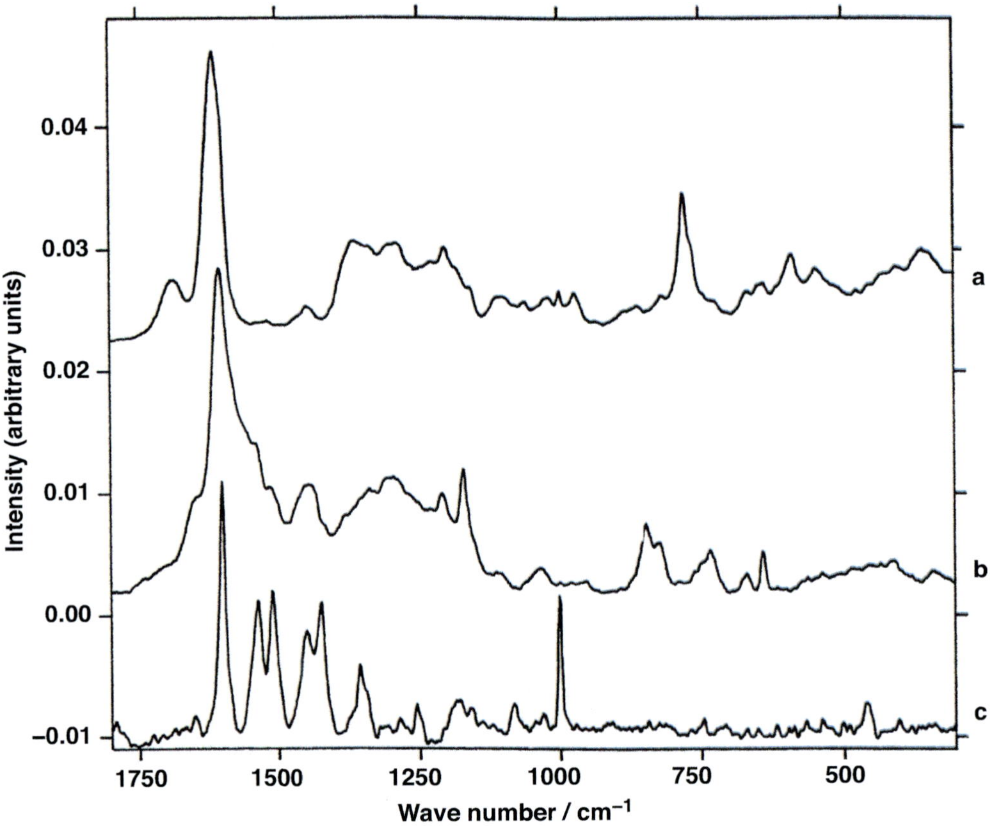

Fig. 7.9. Stack-plotted Raman spectra of three Dragon's blood resins from different botanical specimens: a. *Eucalyptus terminalis* (Australia); b. *Dracaena cinnabari* (Socotra, East Africa); c. *Daemonorops draco* (East Indies). Excitation wavelength: 1064 nm; spectral resolution: 4 cm^{-1}; 4000 scans accumulated; wavenumber range: 200–1800 cm^{-1}. Author's own image.

comparison over a wavenumber range critical to differentiating between terpenoids; the characteristic bands for diterpenoid resins at 1303, 970, 740, 370 and 305 cm^{-1} are absent from the Amarna resin spectrum, indicating that this specimen probably originates from a triterpenoid source.

The analysed resin samples from artworks and artefacts can sometimes span a period of up to 7000 years and are exposed to different influences during this time, such as environmental degradation and various chemical reactions in the soil, which are usually also reflected in the observed Raman and IR spectra (Daher *et al.*, 2010; Teodor *et al.*, 2010; Truică *et al.*, 2012; Peris-Díaz *et al.*, 2018; Guo *et al.*, 2021). The most important molecular Raman spectral features, which are crucial for the differentiation between contemporary resins, are considered in the context of these factors, which also illustrates the difficulties in creating a Raman spectral database for ancient resins.

Edwards *et al.* (1998) reported on the FT-Raman spectra of four technically important gums, namely locust bean gum, gum karaya, gum tragacanth and gum ghatti (Fig. 7.11). Characteristic vibrational bands were identified in the Raman spectra for each gum and, as far as possible, assigned to the respective molecular vibration modes, which can then be used for their identification. The authors demonstrate that powerful databases for the rapid and non-destructive analysis of gums and resins can be

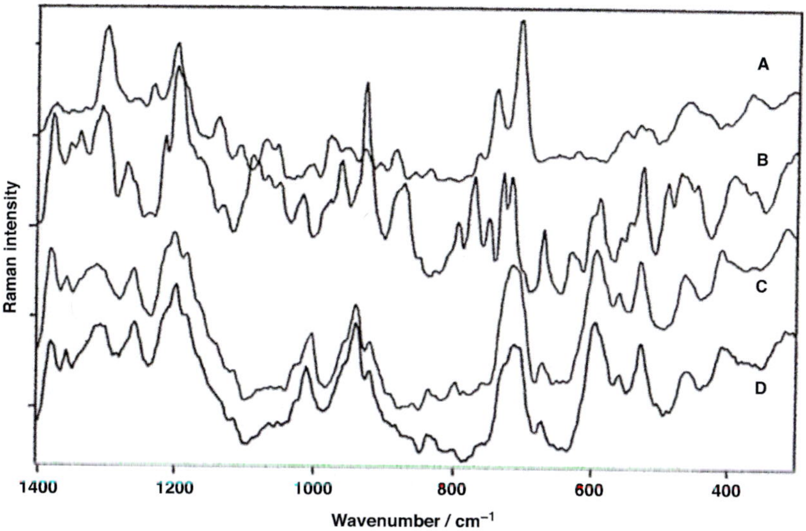

Fig. 7.10. FT-Raman spectral stackplot of (A) *Pinus halepensis* resin, (B) frankincense resin, (C) *Pistacia terebinthus* resin, (D) Gurob resin, amber coloured; wavenumber region: 300–1400 cm^{-1}. (From Edwards and Ali, 2011.) Used with permission.

constructed based on these individual Raman spectra.

Raman spectroscopy was also used with great success in a study of various ethnographic resins used by different indigenous peoples in Brazil. The aim of this study was to determine the composition of the resins and understand their use by comparing them with data collected on artefacts in museum collections representative of Brazilian indigenous peoples. In this context, de Faria *et al.* (2021) analysed eight ethnographic resins from different sites in the Amazon region and central Brazil using FT-Raman spectroscopy to determine their chemical composition non-invasively. Based on the results, the resins could be categorized into five different groups, each of which had a number of common spectroscopic characteristics. Diterpenoid resins, which are similar in chemical composition to Brazilian copal, a triterpenoid type (similar to dammar resin) containing some carotenoid compounds and two gums containing mainly polysaccharides could be clearly distinguished by their individual spectral signature. In general, the results show that Raman spectroscopy enables a much better understanding of the culture of indigenous peoples, e.g. with regard to the origin, use and processing of different resins. The authors also show that unknown resin samples can be reliably identified on the basis of spectral databases.

The polystyrene character of the fossilized tree resin 'Siegburgite' was proven on the basis of Raman spectroscopic data at various locations in Germany (Winkler *et al.*, 2003). In this context, the presence of cinnamic acid as the main component of the fossil resin was clearly demonstrated on the basis of reference spectra. The rapid method described by the authors appears to be generally suitable for the non-destructive differentiation of various fossil resins, especially when different resin types occur side by side in the same deposit.

For archaeological samples with complex and mixed composition, generally Raman and mid-infrared spectroscopy can provide important information on natural resins (Daher *et al.*, 2013; Azémard *et al.*, 2014; Invernizzi *et al.*, 2018) but usually require databases or standard samples as references.

There have been some studies on the characterization of resins and gums to obtain more precise information about their exact nature and origin. Daher *et al.* (2013), for example, used FT-Raman and IR spectroscopy to identify and differentiate these natural products. Although

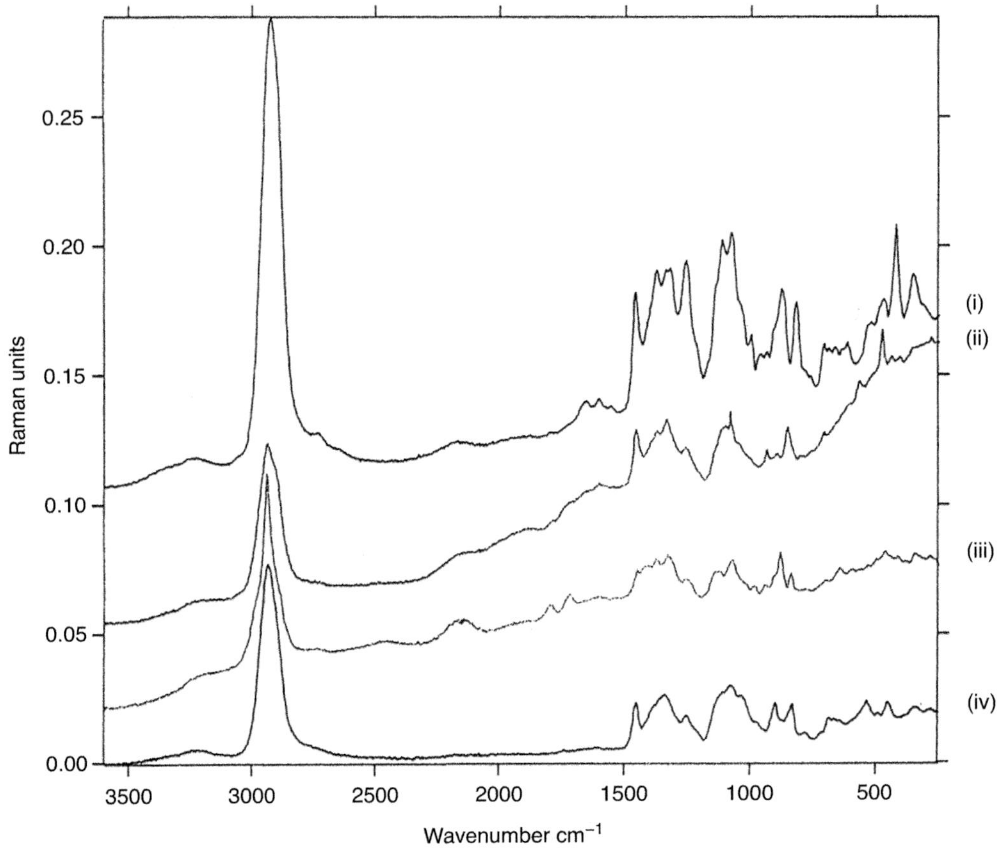

Fig. 7.11. FT-Raman spectra of representative samples of (i) locust bean gum, (ii) gum tragacanth (amber), (iii) gum karaya (amber), (iv) gum ghatti (yellow); excitation: 1064 nm, spectral resolution: 4 cm^{-1}, laser power: 70 mW, 2500 spectral scans accumulated. Wavenumber range: 200–3600 cm^{-1}. (From Edwards *et al.*, 1998.) Used with permission.

the resulting spectra show very similar vibrational signatures due to their related functional groups, the authors were able to overcome this limitation using sophisticated statistical analyses and obtained a reliable objective discrimination method for identification.

Fossil resins from Miocene coal deposits (Sarolangun mine, Jambi province, Sumatra, Indonesia) were measured using Raman spectroscopy, FTIR, ^{13}C nuclear magnetic resonance and fluorescence spectroscopy to improve their diagnostic features (Drzewicz *et al.*, 2020). Similarly, glessite, a fossil resin from the upper Oligocene Bitterfeld deposit (Saxony-Anhalt, Germany) derived from similar botanical sources (i.e. angiosperms), was analysed using the same analytical methods to determine similarities and differences between the two resin types. Overall, the authors found that the IR and Raman spectra of the resins were very similar, but the Raman spectra showed a slight difference in the intensity ratio of the 1650/1450 cm^{-1} bands (0.56 and 0.68 for the Sumatran and German resins, respectively), which is explained by a different maturation process (the Sumatran resins appear to be more mature than the German glessite).

Guiliano *et al.* (2007) developed an ATR-IR spectroscopic method to distinguish amber of different origins. They were also able to discriminate amber from immature resins such as copal and recognize most forgeries in the marketing of amber.

As an alternative to other methods for the characterization of natural resins, Prati *et al.* (2011) proposed a new far-infrared spectroscopy method. For this the authors analysed standards of natural resins belonging to four different categories (sesquiterpene resins such as elemi and shellac; diterpene resins such as rosin and Venice turpentine; diterpene resins with polymerized components such as copal and sandarac; triterpene resins such as mastic and dammar) using far-infrared spectroscopy in ATR mode. The same samples were also analysed in the mid-IR range, but the spectral differences between the four categories were less clear in these cases. Furthermore, the new method was tested on historical samples of the painting 'La Battaglia di Cialdiran' (16th century) and a gilded leather from the 17th century. In the first case, far-infrared spectroscopy was able to confirm the results obtained by analytical pyrolysis. In the latter case, the new spectroscopy method was able to identify the resin layer remaining on the leather, which could not be detected by chromatographic analyses.

References

Akbulut, H.F., Almaghrebi, E., Obali, I., Vatansev, H., Vatansev, H. *et al.* (2024) Evaluation the organic acid, tocopherol and phenolic profiles of *Dracaena cinnabari* resin extracts obtained by different solvent extraction. *Latin American Applied Research-An International Journal* 54(2), 195–200. DOI: 10.52292/j.laar.2024.2865.

Archier, P. and Vieillescazes, C. (2000) Characterisation of various geographical origin incense based on chemical criteria. *Analysis* 28, 233–237. DOI: 10.1051/analusis:2000112.

Azémard, C., Vieillescazes, C. and Ménager, M. (2014) Effect of photodegradation on the identification of natural varnishes by FT-IR spectroscopy. *Microchemical Journal* 112, 137–149. DOI: 10.1016/j.microc.2013.09.020.

Azémard, C., Menager, M. and Vieillescazes, C. (2016) Analysis of diterpenic compounds by GC-MS/MS: Contribution to the identification of main conifer resins. *Analytical and Bioanalytical Chemistry* 408, 6599–6612. DOI: 10.1007/s00216-016-9772-9.

Badea, G.I., Caggiani, M.C., Colomban, P., Mangone, A., Teodor, E.D. *et al.* (2015) Fourier transform Raman and statistical analysis of thermally altered samples of amber. *Applied Spectroscopy* 69(12), 1457–1463. DOI: 10.1366/15-07866.

Baeten, J., Deforce, K., Challe, S., De Vos, D. and Degryse, P. (2014) Holy smoke in medieval funerary rites: Chemical fingerprints of frankincense in southern Belgian incense burners. *PLoS One* 9(11), e113142. DOI: 10.1371/journal.pone.0113142.

Baldovini, N. and Chaintreau, A. (2020) Identification of key odorants in complex mixtures occurring in nature. *Natural Product Reports* 37(12), 1589–1626. DOI: 10.1039/d0np00020e.

Bartle, K.D. and Myers, P. (2002) History of gas chromatography. *TrAC Trends in Analytical Chemistry* 21(9–10), 547–557. DOI: 10.1016/S0165-9936(02)00806-3.

Batiha, G.E.S., Wasef, L., Teibo, J.O., Shaheen, H.M., Zakariya, A.M. *et al.* (2023) Commiphora myrrh: A phytochemical and pharmacological update. *Naunyn-Schmiedeberg's Archives of Pharmacology* 396(3), 405–420. DOI: 10.1007/s00210-022-02325-0.

Beale, D.J., Pinu, F.R., Kouremenos, K.A., Poojary, M.M., Narayana, V.K. *et al.* (2018) Review of recent developments in GC–MS approaches to metabolomics-based research. *Metabolomics* 14, 1–31. DOI: 10.1007/s11306-018-1449-2.

Blanco, M. and Villaescusa, V. (2007) Use of NIR spectroscopy in the production of modified industrial resins. *Talanta* 71, 1333–1338. DOI: 10.1016/j.talanta.2006.07.028.

Borland, L., Brickhouse, M., Thomas, T. and Fountain III, A.W. (2010) Review of chemical signature databases. *Analytical and Bioanalytical Chemistry* 397(3), 1019–1028. DOI: 10.1007/s00216-010-3606-y.

Camarda, L., Di Stefano, V. and Pitonzo, R. (2011) Natural resins: Chemical constituents and medicinal uses. In: Song, D.B. (ed.) *Resin Composites: Properties, Production and Applications*. Nova Science Publishers, New York, pp. 353–374. Available at: https://de.scribd.com/document/820255790/Ebook-Resin-Composites-Properties-Production-and-Applications-Properties-Production-and-Applications-by-Deborah-B-Song-ISBN-9781622570195-162 (accessed 8 May 2025).

Casadio, F., Daher, C. and Bellot-Gurlet, L. (2016) Raman spectroscopy of cultural heritage materials: Overview of applications and new frontiers in instrumentation, sampling modalities, and data processing. *Topics in Current Chemistry* 374, 62. DOI: 10.1007/s41061-016-0061-z.

Chauiyakh, O., Et-Tahir, A., Kettani, K., Cherrat, A., Benayad, A. *et al.* (2022) Review on health status, chemical composition and antimicrobial properties of the four species of the genus *Cedrus*. *International Wood Products Journal* 13(4), 272–285. DOI: 10.1080/20426445.2022.2118652.

Cramers, C.A., Janssen, H.G., Deursen, M.M. and Leclercq, P.A. (1999) High-speed gas chromatography: An overview of various concepts. *Journal of Chromatography A* 856(1–2), 315–329. DOI: 10.1016/S0021-9673(99)00227-7.

Daferera, D., Pappas, C., Tarantilis, P.A. and Polissiou, M. (2002) Quantitative analysis of α-pinene and β-myrcene in mastic gum oil using FT-Raman spectroscopy. *Food Chemistry* 77, 511–515.

Daher, C., Paris, C., Le Hô, A.-S., Bellot-Gurlet, L. and Echard, J.-P. (2010) A joint use of Raman and infrared spectroscopies for the identification of natural organic media used in ancient varnishes. *Journal of Raman Spectroscopy* 41(11), 12041209. DOI: 10.1002/jrs.2693.

Daher, C., Bellot-Gurlet, L., Le Hô, A.-S., Paris, C. and Regert, M. (2013) Advanced discriminating criteria for natural organic substances of cultural heritage interest: Spectral decomposition and multivariate analyses of FT-Raman and FT-IR signatures. *Talanta* 115, 540–547. DOI: 10.1016/j.talanta.2013.06.014.

de Faria, D.L.A., Maier, M.S., Parera, S.D., Afonso, M.C., Lima, S.C. *et al.* (2021) Non-invasive and non-destructive Raman spectroscopic characterization of some Brazilian ethnographic resins. *Journal of Raman Spectroscopy* 52, 2262–2271. DOI: 10.1002/jrs.6270.

Drag, K., Mroczkowska-Szerszeń, M., Dumańska-Słowik, M. and Żukowska, G. (2022) Identification of treated Baltic amber by FTIR and FT-Raman - a feasibility study. *Spectrochimica Acta Part A* 279, 121404. DOI: 10.1016/j.saa.2022.121404.

Drzewicz, P., Naglik, B., Natkaniec-Nowak, L., Dumańska-Słowik, M. and Stach, P. (2020) Chemical and spectroscopic signatures of resins from Sumatra (Sarolangun mine, Jambi Province) and Germany (Bitterfeld, Saxony-Anhalt). *Scientific Reports* 10, 18283. DOI: 10.1038/s41598-020-74671-z.

Edwards, H.G.M. and Falk, M.J. (1997) Fourier-transform Raman spectroscopic study of frankincense and myrrh. *Spectrochimica Acta A* 53, 2393–2401. DOI: 10.1016/S1386-1425(97)00179-0.

Edwards, H.G.M. and Ali, E.M.A. (2011) Raman spectroscopy of archaeological and ancient resins: Problems with database construction for applications in conservation and historical provenancing. *Spectrochimica Acta A* 80(1), 49–54. DOI: 10.1016/j.saa.2011.01.030.

Edwards, H.G.M., Farwell, D.W. and Daffner, L. (1996) Fourier-transform Raman spectroscopic study of natural waxes and resins. I. *Spectrochimica Acta A* 52(12), 1639–1648. DOI: 10.1016/0584-8539(96)01730-8.

Edwards, H.G.M., Sibley, M.G. and Heron, C. (1997) FT-Raman spectroscopic study of organic residues from 2300-year-old Vietnamese burial jars. *Spectrochimica Acta A* 53(13), 2373–2382. DOI: 10.1016/S1386-1425(97)00177-7.

Edwards, H.G.M., Falk, M.J., Sibley, M.G., Alvarez-Benedi, J. and Rull, F. (1998) FT-Raman spectroscopy of gums of technological significance. *Spectrochimica Acta A* 54, 903–920. DOI: 10.1016/S1386-1425(98)00018-3.

Edwards, H.G.M., de Oliveira, L.F.C. and Anita Quye, A. (2001) Raman spectroscopy of coloured resins used in antiquity: Dragon's blood and related substances. *Spectrochimica Acta A* 57, 2831–2842.

Eiserbeck, C., Nelson, R.K., Grice, K., Curiale, J. and Reddy, C.M. (2012) Comparison of GC–MS, GC–MRM-MS, and GC × GC to characterise higher plant biomarkers in Tertiary oils and rock extracts. *Geochimica et Cosmochimica Acta* 87, 299–322. DOI: 10.1016/j.gca.2012.03.033.

El Omari, N., Guaouguaou, F.E., El Menyiy, N., Benali, T., Aanniz, T. *et al.* (2021) Phytochemical and biological activities of *Pinus halepensis* Mill., and their ethnomedicinal use. *Journal of Ethnopharmacology* 268, 113661. DOI: 10.1016/j.jep.2020.113661.

Georgiou, R., Popelka-Filcoff, R.S., Sokaras, D., Beltran, V., Bonaduce, I. *et al.* (2022) Disentangling the chemistry of Australian plant exudates from a unique historical collection. *Proceedings of the National Academy of Sciences* 119(22), e2116021119. DOI: 10.1073/pnas.2116021119.

Guiliano, M., Asia, L., Onoratini, G. and Mille, G. (2007) Applications of diamond crystal ATR FTIR spectroscopy to the characterization of ambers. *Spectrochimica Acta A* 67(5), 1407–1411. DOI: 10.1016/j.saa.2006.10.033.

Guo, L., Zhang, Y., Wang, L., Zhao, X., Yang, F. *et al.* (2021) A multi-analytical approach for the identification of the natural resin from the Ming tomb in Shaanxi, China. *Heritage Science* 9, 109. DOI: 10.1186/s40494-021-00586-1.

Hanuš, L.O., Řezanka, T., Dembitsky, V.M. and Moussaieff, A. (2005) Myrrh-commiphora chemistry. *Biomedical Papers* 149(1), 3–28. DOI: 10.5507/bp.2005.001.

Hassan, W., Rehman, S., Noreen, H., Gul, S., Neelofar, A.N. *et al.* (2016) Gas chromatography mass spectrometric (GCMS) analysis of essential oils of medicinal plants. *Advances in Animal and Veterinary Sciences* 4(8), 420–437. DOI: 10.14737/journal.aavs/2016/4.8.420.437.

Herrmann, A., König, S., Lechtenberg, M., Sehlbach, M., Vakhrushev, S.Y. *et al.* (2012) Proteoglycans from *Boswellia serrata* Roxb. and *B. carteri* Birdw. and identification of a proteolytic plant basic secretory protein. *Glycobiology* 22(11), 1424–1439. DOI: 10.1093/glycob/cws107.

Hosain, N.A., Ghosh, R., Bryant, D.L., Arivett, B.A., Farone, A.L. *et al.* (2019) Isolation, structure elucidation, and immunostimulatory activity of polysaccharide fractions from *Boswellia carterii* frankincense resin. *International Journal of Biological Macromolecules* 133, 76–85. DOI: 10.1016/j.ijbiomac.2019.04.059.

Hu, L., Wang, F.F., Wang, X.H., Yang, Q.S., Xiong, Y. *et al.* (2015) Phytoconstituents from the leaves of *Dracaena cochinchinensis* (Lour.) SC Chen. *Biochemical Systematics and Ecology* 63, 1–5. DOI: 10.1016/j.bse.2015.09.012.

Invernizzi, C., Fichera, G.V., Licchelli, M. and Malagodi, M. (2018) A non-invasive stratigraphic study by reflection FT-IR spectroscopy and UV-induced fluorescence technique: The case of historical violins. *Microchemical Journal* 138, 273–281. DOI: 10.1016/j.microc.2018.01.021.

Jehlička, J., Villar, S.E.J. and Edwards, H.G.M. (2004) Fourier transform Raman spectra of Czech and Moravian fossil resins from freshwater sediments. *Journal of Raman Spectroscopy* 35(8–9), 761–767. DOI: 10.1002/jrs.1191.

Joulain, D. and König, W.A. (1998) *The Atlas of Spectral Data of Sesquiterpene Hydrocarbons*. E.B. Verlag, Hamburg, Germany.

Keusgen, M. (2021) Methodenentwicklung zur qualitativen und quantitativen Analyse der Triterpensäuren von Boswellia sacra FLÜCK.(Burseraceae) sowie verwandter Arten. PhD dissertation, Philipps-Universitaet Marburg, Germany.

Krähmer, A., Obersat, G.I., Beleites, C. and Schulz, H. (2024) Characterization of various frankincense and myrrh species applying ATR-FTIR and FT-NIR spectroscopy methods. In: Junghanns, W. and Marthe, F. (eds) *34th Bernburg Winter Seminar on Medicinal and Spice Plants, February 20–21, 2024*. DOI: 10.5073/20240131-085854-0.

Langenheim, J.H. (1990) Plant resins. *American Scientist* 78, 16–24. Available at: https://www.jstor.org/stable/29773859

Langenheim, J.H. and Beck, C.W. (1965) Infrared spectra as a means of determining botanical sources of amber. *Science* 149(3679), 52–54. DOI: 10.1126/science.149.3679.52.

Murthy, H.N. (2021) Chemical constituents and applications of gums, resins, and latexes of plant origin. In: Murthy, H.N. (ed.) *Gums, Resins and Latexes of Plant Origin*. Reference Series in Phytochemistry, Springer, Cham, Switzerland, pp. 1–21. DOI: 10.1007/978-3-030-76523-1_1-1.

Peris-Díaz, M.D., Łydżba-Kopczyńska, B. and Sentandreu, E. (2018) Raman spectroscopy coupled to chemometrics to discriminate provenance and geological age of amber. *Journal of Raman Spectroscopy* 49, 842–851. DOI: 10.1002/jrs.5357.

Prati, S., Sciutto, G., Mazzeo, R., Torri, C. and Fabbri, D. (2011) Application of ATR-far-infrared spectroscopy to the analysis of natural resins. *Analytical and Bioanalytical Chemistry* 399(9), 3081–3091. DOI: 10.1007/s00216-010-4388-y.

Rauf, A., Patel, S., Uddin, G., Siddiqui, B.S., Ahmad, B. *et al.* (2017) Phytochemical, ethnomedicinal uses and pharmacological profile of genus Pistacia. *Biomedicine & Pharmacotherapy* 86, 393–404. DOI: 10.1016/j.biopha.2016.12.017.

Rehman, N.U., Al-Shidhani, S., Al-Harrasi, A., Al-Rahawi, A., Mabood, F. *et al.* (2020) Analysis of incensole acetate in *Boswellia* species by near infrared spectroscopy coupled with partial least squares regression and cross-validation by high-performance liquid chromatography. *Journal of Near Infrared Spectroscopy* 28, 18–24. DOI: 10.1177/0967033519895689.

Rubiolo, P., Sgorbini, B., Liberto, E., Cordero, C. and Bicchi, C. (2010) Essential oils and volatiles: Sample preparation and analysis. A Review. *Flavour and Fragrance Journal* 25(5), 282–290. DOI: 10.1002/ffj.1984.

Sadgrove, N.J., Padilla-González, G.F. and Phumthum, M. (2022) Fundamental chemistry of essential oils and volatile organic compounds, methods of analysis and authentication. *Plants* 11(6), 789. DOI: 10.3390/plants11060789.

Sahin, H. and Yalcin, O. (2017) Chemical composition and utilization of conifer needles - a review. *Journal of Applied Life Sciences International* 14(3), 1–11. DOI: 10.9734/JALSI/2017/37076.

Saitta, M., Giuffrida, D., La Torre, G.L., Potortì, A.G. and Dugo, G. (2009) Characterisation of alkylphenols in pistachio (*Pistacia vera* L.) kernels. *Food Chemistry* 117(3), 451–455. DOI: 10.1016/j.foodchem.2009.04.043.

Salome-Abarca, L.F., van der Pas, J., Kim, H.K., van Uffelen, G.A., Klinkhamer, P.G.L. *et al.* (2018) Metabolic discrimination of pine resins using multiple analytical platforms. *Phytochemistry* 155, 37–44. DOI: 10.1007/s00216-010-4388-y.

Sanchidrian, V.B. (2016) Vibrational spectroscopic study of Pinus resin in materials from cultural heritage objects. Doctoral thesis, Universitat Politècnica de Catalunya, Spain.

Schulz, H. and Baranska, M. (2007) Identification and quantification of valuable plant substances by IR and Raman spectroscopy. *Vibrational Spectroscopy* 43, 13–25. DOI: 10.1016/j.vibspec.2006.06.001.

Schulz, H., Krähmer, A., Naumann, A. and Gennadi Gudi, G. (2014) Infrared and Raman spectroscopic mapping and imaging of plant materials. In: Salzer, R. and Siesler, H.W. (eds) *Infrared and Raman Spectroscopic Imaging*, 2nd edn. pp. 225–294.

Schulz, H. (2020) Analysis of secondary metabolites in breeding research and plant breeding. In: Novak, J. and Blüthner, W.D. (eds) *Medicinal, Aromatic and Stimulant Plants. Handbook of Plant Breeding*, Vol. 12. Springer, Cham, Switzerland.

Stach, P., Martinkutė, G., Šinkūnas, P., Natkaniec-Nowak, L., Drzewicz, P. *et al.* (2019) An attempt to correlate the physical properties of fossil and subfossil resins with their age and geographic location. *Journal of Polymer Engineering* 39(8), 716–728. DOI: 10.1515/polyeng-2019-0159.

Stach, P., Natkaniec-Nowak, L., Wagner, M., Dumańska-Słowik, M., Mroczkowska-Szerszeń, M. *et al.* (2020) A study on the formation environment of the La Cumbre amber deposit, from Santiago Province, the northwestern part of the Dominican Republic. *Minerals* 10(9), 736. DOI: 10.3390/min10090736.

Sun, J., Liu, J.N., Fan, B., Chen, X.N., Pang, D.R. *et al.* (2019) Phenolic constituents, pharmacological activities, quality control, and metabolism of Dracaena species: A review. *Journal of Ethnopharmacology* 244, 112138. DOI: 10.1016/j.jep.2019.112138.

Sura, M.B. and Cheng, Y.X. (2024) Medicinal plant resin natural products: Structural diversity and biological activities. *Natural Product Reports* 10, 1471–1542. DOI: 10.1039/D4NP00007B.

Tappert, P., Wolfe, A.P., McKellar, R.C., Tappert, M.C. and Muehlenbachs, K. (2011) Characterizing modern and fossil gymnosperm exudates using micro-Fourier transform infrared spectroscopy. *International Journal of Plant Sciences* 172, 120–138. DOI: 10.1086/657277.

Teodor, E.S., Teodor, E.D., Virgolici, M., Manea, M.M., Truică, G. *et al.* (2010) Non-destructive analysis of amber artefacts from the prehistoric Cioclovina hoard (Romania). *Journal of Archaeological Science* 37, 2386–2396. DOI: 10.1016/j.jas.2010.04.011.

Truică, G.I., Teodor, E.D., Teodor, E.S., Liţescu, S.C. and Radu, G.L. (2012) FTIR and statistical studies on amber artefacts from three Romanian archaeological sites. *Journal of Archaeological Science* 39(12), 3524–3533. DOI: 10.1016/j.jas.2012.05.031.

Van Ruth, S.M. (2001) Methods for gas chromatography-olfactometry: A review. *Biomolecular Engineering* 17(4–5), 121–128. DOI: 10.1016/S1389-0344(01)00070-3.

Van Vuuren, S.F., Kamatou, G.P.P. and Viljoen, A.M. (2010) Volatile composition and antimicrobial activity of twenty commercial frankincense essential oil samples. *South African Journal of Botany* 76, 686–691. DOI: 10.1016/j.sajb.2010.06.001.

Vandenabeele, P., Ortega-Avilès, M., Tenorio Castilleros, D. and Moens, L. (2007) Raman spectroscopic analysis of Mexican natural artists' materials. *Spectrochimica Acta Part A: Molecular and Biomolecular Spectroscopy* 68(4), 1085–1088. DOI: 10.1016/j.saa.2007.01.031.

Wagner-Wysiecka, E. (2018) Mid-infrared spectroscopy for characterization of Baltic amber (succinite). *Spectrochimica Acta A* 196, 418–431. DOI: 10.1016/j.saa.2018.02.053.

Winkler, W., Musso, M. and Kirchner, E.C. (2003) Fourier transform Raman spectroscopic data on the fossil resin siegburgite. *Journal of Raman Spectroscopy* 34(2), 157–163. DOI: 10.1002/jrs.971.

Zhu, C.Z., Hu, B.Y., Liu, J.W., Cai, Y., Chen, X.C. *et al.* (2019) Anti-mycobacterium tuberculosis terpenoids from *Resina Commiphora*. *Molecules* 24(8), 1475. DOI: 10.3390/molecules24081475.

8 'Going Green': How Can We Move to Ecologically Sustainable Use of Gums and Resins in Commercial Supply Chains?

Anthony B. Cunningham[1,2*]
[1]School of Life Sciences, University of KwaZulu-Natal, King Edward Avenue, Pietermaritzburg, 3209, South Africa; [2]School of Veterinary and Life Sciences, Murdoch University, Murdoch, WA 6150, Australia

Abstract

Awareness of the need for ethical sourcing and the 'triple bottom line' goals (social responsibility, environmental sustainability and economic viability) is growing. Yet while demand increases, wild populations of many gum and resin sources are in decline. Compared to growing fast-growing crops, economically viable production of tree species in agroforestry systems or plantations can be challenging, particularly in the dry woodlands and shrublands that are sources of frankincense and myrrh. In higher rainfall areas, cultivation has been successful in the case of *Aquilaria* and *Shorea javanica*. In arid and semi-arid areas, large-scale cultivation can be a mirage, distracting attention from what is needed most: 'growing the wild' through wise management, better monitoring, fair prices and local community involvement. In this chapter, I examine some of the major challenges preventing sustainable use of gum and resin resources, and discuss how these challenges can be confronted.

Growth in Demand: But What About Supplies From Wild Stocks?

At the most basic level, ecologically sustainable use is a question about whether growing demand can be met by supplies. The problem is that most companies focus primarily on growing demand for their products in order to increase sales and profit margins. There are fewer incentives, however, to manage wild stocks of many of the slow-growing species to meet this growing demand in an ecologically sustainable way. In Asia, particularly China, India, Pakistan and Vietnam, for example, government support for development and modernization of traditional medical systems has been an important factor in growing demand for medicinal gums and resins. In India, as Viale and Vicol (2023) point out in the case of 'brand Ayurveda', what was a holistic healing and knowledge system is increasingly regarded as a profit-making opportunity. Twenty years ago, there were 7500 factories producing thousands of Ayurvedic and Unani formulae (Bode, 2006). The projected value of the domestic Indian Ayurveda market was expected to be US$8.0 billion by 2022, with an annual export value of over US$780 million (Viale and Vicol, 2023). In Europe and North America, the significance of the flavour and fragrance market is illustrated by the fact that the top eight large companies had a combined revenue in excess of US$25.6 billion per year (Cunningham and DeCarlo, 2022). In

*Corresponding author: tonybcunningham@gmail.com

 Natural Gums and Resins: Botany and Sustainable Uses in Medicine, Nutrition, Perfumery and Cosmetics (eds Hartwig Schulz *et al.*)
DOI: 10.1079/9781800626690.0008

Europe, some of those companies were established over a century ago. In France, for example, Mane SA was founded in 1871. In Germany, Haarmann & Reimer (H&R), now incorporated into Symrise GmbH, was founded in 1874. In Switzerland, both Givaudan and Firmenich were founded in 1895. In contrast, big companies in the USA are much more recent. International Flavors and Fragrances Inc. was established in 1958, Young Living in 1993, while dōTERRA was only established in 2008. Through multi-level marketing (MLM), these two most recent companies have grown demand spectacularly, adding significantly to the global demand for products such as frankincense (Cunningham and DeCarlo, 2022).

Yet, as Sir David Attenborough reportedly said: 'anyone who thinks you can have infinite growth in a finite environment is either a madman or an economist'. So, what about the supply side? Wild populations of many gums and resin species are not only finite, some are in steep decline. Based on the most comprehensive data set available for any gum- or resin-producing species anywhere, for example, Bongers *et al.* (2019) provided compelling evidence that *Boswellia papyrifera* was in decline in Ethiopia, Eritrea and Sudan, suggesting population collapse of *B. papyrifera* in these countries, due to little natural regeneration for decades and a halving of frankincense production in 20 years' time.

Why Should Companies Bother About 'Going Green' and Ecologically Sustainable Use?

Why should commercial companies bother about the supply side? Well, firstly, growing numbers of consumers realize that their buying choices affect both the environment and society. Driven by this consumer awareness, many commercial companies and their natural product suppliers are trying to stand out in the marketplace by 'going green'. They hope to achieve this through ethical sourcing so as to meet corporate social responsibility (CSR) goals and the three 'triple bottom line' requirements: social responsibility, environmental sustainability and economic viability.

Secondly, there is growing awareness of multiple threats to trees at a global scale (Rivers *et al.*, 2023; Boonman *et al.*, 2024), including many important gum- and resin-producing species.

Thirdly, the economically viable production of many gum and resin species in plantations or agroforestry systems at a large scale can be challenging. There are exceptions in the moist tropics, as we discuss later in this chapter. *Aquilaria* is one of these, due to the combination of high prices, relatively fast growth and chemical stimulants to promote resin yields (Persoon and van Beek, 2008; Abdulah *et al.*, 2022). However, economically viable cultivation of long-lived resin- or gum-producing trees in arid and semi-arid areas is a major challenge. Consequently, wise management of the wild populations that are the primary sources of gums and resins in arid and semi-arid areas is essential, rather than exploitation of populations in long-term decline.

Fourthly, in arid and semi-arid drylands, the economic importance of gums and resins to local people's livelihoods, while often a supplemental rather than primary income, is still significant and should not be discounted as a poverty trap. Loss of this 'green social security', for the complex reasons we discuss below, can be serious. Therefore, we should all bear in mind what Hoskins (1990) wrote over 30 years ago in her paper on forestry and food security: 'All research and management by outsiders must remember that their activities come and go, but food security – land and resources surety – is a long-term, life and death issue for rural peoples'.

Demand Exceeding Supply in Commercial Trade is Not New: But Can We Learn from the Past?

Plant resource over-harvesting or, in extreme cases, species extinctions due to over-exploitation for commercial trade, is not a new issue. The case of silphium is a good example. This unidentified *Ferula* species (Umbelliferae) was the commercial source of an asafoetida-like product trade from Cyrenaica (in today's Libya) for over 2000 years, with trade being so important that silphium plants featured on Cyrenian coins (Fig. 8.1).

Silphium harvests were destructive, however, and by Roman times, quantities of Cyrenaic silphium in trade were shrinking (Kiehn, 2007). This resulted in a shift to an alternative source species with a wider distribution: *Ferula asafoetida*

Fig. 8.1. Which species will be the next silphium through 'knowing but not doing'? A. Silver coin showing Cyrenaic silphion, which went extinct 2000 years ago when demand exceeded supply of this destructively harvested, habitat-specific species. B. Will overgrazing, branch lopping and climate change make the already Critically Endangered *Boswellia* species (*B. hesperia* and *B. nana*) on Socotra (Yemen) the next silphium? C. Or will it be an *Aquilaria* species that is Critically Endangered in the wild due to trade and habitat loss (such as A. *malaccensis*, *A. khasiana* or *A. sinensis*)? (Photos: ArchaiOptix, CC BY-SA 4.0, https://creativecommons.org/licenses/by-sa/4.0 (A), S. Johnson (B), A.B. Cunningham (C).

(sourced from Iran). Finally, the end came, with Pliny the Elder pointing out almost 2000 years ago that Cyrenaic silphium 'has not been found in this country for many years ... as far as I remember only one plant was found and sent to Emperor Nero' (Pliny, *The Natural History* XIX, 39, cited in Kiehn, 2007, 4). Demand had exceeded supply, and destructive harvest of a habitat-specific species (possibly also affected by desertification and sterile seed; Johnson, 2022) drove Cyrenaic silphium to extinction. At this time, awareness of geography and plant conservation was poor. So, driving a plant species to extinction and shifting to an alternative 'resource-rich' frontier was understandable.

Today, there is no excuse for ignorance of basic ecological information, or for 'knowing but not doing'. Published, peer-reviewed research on sustainable harvest of wild plant populations is literally at our fingertips through the internet. The recent assessment by Maděra *et al.* (2024) of the complex factors affecting *Boswellia* populations on Socotra (Yemen) despite sustainable harvests (due to hand-picking of naturally exuded oleoresins) is just one example of this scientific evidence. In the 21st century, we have access to databases that cover everything from correct species names (International Plant Names Index, IPNI) to species distributions (Global Biodiversity Information Facility, GBIF) to the conservation status (such as International Union for the Conservation of Nature (IUCN) Red listings) of species threatened by trade. So, for commercial trade to contribute to species extinctions today is inexcusable.

Whether companies are small or large, what is required across the board is to deal with ten challenges to the ecologically sustainable use of gums and resins.

Which Ten Challenges Need to Be Overcome for Truly 'Green' Supply Chains?

Challenge 1: Confusion about which species (or even genera) are involved in trade

Knowing what species are involved is the most basic step in environmental management and monitoring plans (EMMPs) for a harvested resource. So, this is not a trivial matter. Within a single trade name category, gum and resin sources can vary in their commonness or rarity, and in the factors that influence their abundance and the impact of harvest on the plant species. Three reasons for lack of clarity on species in commercial trade are that firstly, a single species such as *Commiphora myrrha* can have many trade names: myrrh, African myrrh, Somali myrrh, Common myrrh, Hirabol myrrh and true myrrh, while its pharmacopoeial names are Myrrha, Gummi Myrrha, Gummi Resina, Myrrha and Resina Commiphorae Myrrhae. Secondly, a single trade name category (such as benzoin, copal, dammar, elemi, frankincense, gaharu, gum arabic, gum olibanum, kino, myrrh and opopanax) can refer to more than one species (and sometimes different genera or plant families) (Mabberley, 2008). What is opopanax, for example? It could be oleoresin from *Commiphora guidottii* (also known as bisabol or bissabol) in the Burseraceae. It can also refer to oleoresin from *Commiphora kataf*, or gum opopanax from *Opopanax chironium* (Apiaceae), or gum from *Vachellia farnesiana* (Fabaceae) (Mabberley, 2008). Similarly, the widely used trade name 'rosewood' (from the Portuguese name 'pau-rosa' for *Aniba rosodora* (Lauraceae), the source of 'rosewood oil') is another example of the confusion caused by trade names, as the trade name 'rosewood' generally refers to trees in the genus *Dalbergia* (250 species). This confusion is not a trivial matter. *Aniba rosodora*, for example, is an endangered species on the IUCN Red List and on CITES (Convention on International Trade in Endangered Species of Wild Fauna and Flora) Appendix II (Barstow, 2021; Chapter 10). In 2017, for example, the multi-level marketing company Young Living broke both the Lacey Act and the Endangered Species Act and was fined US$760,000 by the US Supreme Court in a court case related to importing *Aniba rosodora* wood and exporting ('spikenard') oil (Office of Public Affairs, 2017). The good news is that Young Living now have a Lacey Act Compliance Program, which is the first in the essential oils industry to be accepted by the US government (see https://www.youngliving.com). The bottom line is that clarity on scientific species names is essential, not only because 'knowing your species' by its 'world name' (the scientific name) is a key to peer-reviewed research on everything from active ingredients to health and safety factors and conservation issues, but for legal and reputational reasons as well.

Yet for two main reasons, there are impediments to clarity on scientific names. Firstly, changes in plant taxonomy are frequent, yet trade names and names used on pharmacopeia often stay the same. Gum arabic, for example, was officially required to be sourced from *Acacia senegal* and *Acacia seyal* (Williams *et al.*, 2024), yet the controversial taxonomic changes across the genus *Acacia* (see Moore *et al.*, 2010) mean that the same species sources are now in two genera (*Senegalia senegal* and *Vachellia seyal*). Furthermore, supply chains from 'forest to factory' can be long and complex. Yet knowing which species are being used is crucial in terms of what is sustainable or not. For common, widely distributed wild species, the costs and complexity of resource management and monitoring can be low. Slow-growing, destructively harvested species with small populations, on the other hand, are very vulnerable to over-harvesting. For slow-growing species with small populations, there is a very fine line between over-exploitation and sustainable harvest.

Secondly, because in some habitats there is a great diversity of co-occurring gum- or resin-producing species, some of these are very similar to each other, particularly in dry deciduous woodlands, where trees are leafless for much of the year. This increases the chances of either inadvertent or deliberate mixing. In some landscapes, the knowledge of local harvesters combined with less diverse vegetation makes things simpler (Fig. 8.2). Diverse forests, woodlands and shrublands are more challenging. *Commiphora*, the largest genus in the Burseraceae (with *c.*190 species and 323 synonyms for species) and considered the 'impossible genus' (Weeks and Simpson, 2007), illustrates how challenging the start of the supply chain can be if the goal is to avoid mixing of oleoresins from different species. *Commiphora* oleoresins from the semi-arid *Senegalia-Vachellia-Commiphora* woodlands and shrublands typical of north-eastern Africa are a good example of this challenge. In the Somali region of Ethiopia, for example, in addition to *Commiphora myrrha* (one of the sources of myrrh), *Commiphora gileadensis* (the source of Arabian balsam, also known as Balm of Gilead or Mecca myrrh) and *Commiphora guidotti* (a source of bisabol or opopanax), the dry woodlands can have a profusion of other *Commiphora* species such as *Commiphora africana*, *C. fluviflor*, *C. habessinica*, *C. paolii*, *C. crenulata*, *C. boranensis*, *C. erythraea*, *C. schimperi*, *C. ogadensis*, *C. rostrata*, *C. serrulata*, *C. gileadensis*, *C. hildebrandtii*, *C. erosa*, *C. cyclophylla* and *C. corrugate* as well as several *Boswellia* species (*Boswellia microphylla*, *B. ogadensis*, *B. neglecta* and *B. rivae*).

Challenge 2: A widespread lack of ecological literacy: A complex interplay of factors influences ecologically sustainable use

Orr (1991) sums up this challenge very well, pointing out that everything depends on whether the public recognizes the links between human well-being and the health of the natural systems. Sadly, a lack of understanding – of ecological literacy – is widespread amongst many suppliers, commercial companies and users of products from gums and resins. And that's aside from the 'knowing but not doing' gap (Knight *et al.*, 2008), where there is knowledge of unsustainable use, yet this is neither monitored nor managed. The fact is that without effective environmental awareness campaigns, many consumers and businesses will continue to be more worried about economic issues (such as affordability or profitability) than plant conservation or sustainable wild collection. One outcome of this lack of ecological literacy amongst consumers is that many commercial companies get away with 'greenwashing' (see Challenge 5). Even the words 'sustainable' and 'sustainably' have become clichés that have lost almost all meaning, which is why we deliberately use the term 'ecologically sustainable use' in the title of this chapter. Forty years ago, the words 'sustainable' and 'sustainability' were used primarily by ecologists. Today, as Kiss (2011, p. 1) points out, 'You can hardly read a text or an interview by a leading economist or politician where sustainability is not used several times. By now its original meaning has faded away and been forgotten. It simply means 'good', a synonym for everything that is positive. One can read and hear about a sustainable state budget, exchange rate, interest rate, exports, financing, sustainable society, social health, and pension policies. The worst of everything is 'sustainable economic growth', which is the oxymoron of economics'. As Fuller (2010, p. 8) wrote, 'It is ironic that the words 'sustainability' and 'sustainable', so vital for the continuation of life as we know it, should now have fallen into the category of the lifeless'. What is needed to deal with this challenge when it comes to ecologically sustainable use of wild harvested gums and resins? We suggest the following important steps.

1. Customers and companies need to understand that gums and resins are not plant 'by-products' for humans to simply tap and trade

Instead, resins can play multiple roles in plant survival. For example, resins covering plant leaves can reduce moisture loss from the plant through lower cuticular evapotranspiration (for example in *Cannabis* and *Humulus* (Langenheim, 2003) and spinifex (*Triodia*) species (Wickens *et al.*, 1985)). Resins covering leaves can also increase leaf reflectance and decrease the penetration of solar radiation, thus lowering internal temperatures in the plant. In addition, resins contribute to plant defences against herbivores or pathogens, whether on leaves or in bark, roots

Fig. 8.2. Clarity on which species are the source of gums and resins at the start of supply chains is crucial. Yet this can be challenging, particularly where there is a high diversity of co-occurring gum- or resin-producing species, some of them very similar to each other. A. In north-western Namibia, Himba women, who are local conservancy members, sustainably harvest naturally exuded oleoresin from *Commiphora wildii*. In this less diverse landscape, good local folk taxonomy of *omumbiri*, the local name for *C. wildii*, ensures that mixing with oleoresins from other species does not occur. B. *Boswellia papyrifera* in the Amhara region, Ethiopia. Many *Boswellia* species occur in mono-dominant stands, making things simpler, unless deliberate adulteration takes place. C. Samburu women next to *Boswellia neglecta* in dry deciduous woodland that also includes several *Commiphora* species. D. Dry woodlands of the Somali region of Ethiopia have a high diversity of co-occurring gum- and resin-producing species. E. Tropical rainforests in Laos are very species diverse but have relatively few gum- and resin-producing species, although some, such as *Aquilaria* species, are highly valued and over-exploited. F. Rainforest resources are important to the livelihoods of local people. (Photos: A.B. Cunningham (A, E, F), B-E van Wyk (B), S. Canney Davison (C) and S. Johnson (D).) Used with permission.

or stems (Fig. 8.3). Herms and Mattson (1992, p. 284) summed this up well: the dilemma faced by all plants is that 'they must grow fast enough to compete, yet maintain the defences necessary to survive in the presence of pathogens and herbivores'. Sixty years ago, Ehrlich and Raven (1964) pointed out that, over millions of years, plants have evolved different lines of defence against herbivores. An extensive body of research now contributes to our understanding of the evolution and ecology of plant defences (Janzen, 1975; Agrawal, 2011; Johnson, 2011; Mithöfer and Boland, 2012). Plant exudates therefore play important roles in plant 'budgets' and plant survival. And those 'budgets' are drained by high-intensity, high-frequency tapping (Rijkers *et al.*, 2006; Bongers *et al.*, 2019).

2. Customers and companies need to understand that different resin and gum extraction methods have different impacts

Langenheim (2003) gives an excellent summary of different types of exogenous and endogenous resin structures. So, for the sake of brevity, this section does not deal with the extraction of

Fig. 8.3. The evolutionary 'arms race': resins have important functions, including plant defence against predation. A. When wounded, many Burseraceae, including this frankincense (*B. sacra*), exudate monoterpene-rich oleoresins to prevent insect and fungal attack. B. In some cases, there is a truce: where oleoresin exudation is stimulated by beetle larvae (Buprestidae and Cerambycidae) that do not appear to have a long-term detrimental effect on the trees (Sommerlatte and van Wyk, 2022). C. With incorrect, high-intensity tapping, the cambium is damaged and trunks are exposed to beetle attack. D. Wood-boring beetle and larvae. E. Despite their unpleasant odour, *Commiphora krauseliana* are eaten by elephants in Namibia (as well as *C. multijuga, C. wildii* and *C. saxicola*) (Viljoen, 1989). F. Elephant damage to *Boswellia serrata*, showing holes where woodborers have entered the trunk. G. Destructive resin tapping to *Canarium strictum* (South India). (Photos: A.B. Cunningham (A, C, E, F), Sommerlatte (B), K. Soumya (D) and A. Varghese (G).) Used with permission.

exogenous resins (for example from *Cannabis*) or steam extraction of endogenous oleoresins from fruits or leaves, as these can be more easily managed. The focus here is on wounding bark for gum or resin production, as the type of wounding as well as the intensity and frequency of wounding influences tree health and whether yields are sustainable or not. In his excellent ecological primer on sustainable use of non-timber plant resources in tropical moist forests, Peters (1994, p. 28) pointed out that 'When properly conducted, the tapping of latexes, resins, and gums does not disturb the forest canopy, kill the exploited tree, or remove its seeds from the site. In theory, this activity probably comes the closest to conforming to the ideal of sustainable non-timber forest product extraction. In actual practice, the exploitation of plant exudates can be very destructive'. On top of this challenge, conflicting uses, such as tree felling for timber, lopping for fodder or grazing of seedlings and saplings, also affect tapped tree populations. It is essential to take these uses into account in deciding whether populations of tree species that are commercial sources of gums or resins are viable or not, and whether conflicting uses can be managed or not. Few consumers at the end of supply chains are aware of the range of resin collection techniques and their varying impacts. Historically, digging up accumulated 'fossil' resin (for example in the Congo Basin (*Copaifera* and *Guibourtia*), East Africa (*Hymenaea*) and New Zealand (*Agathis*)) was widespread (Langenheim, 2003), with impact at the habitat level rather than at the individual plant level.

Hand-picking resin exuded naturally (for example on Socotra (Yemen) (Maděra *et al.*, 2024) as well as in Namibia (from *Commiphora wildii*) and Kenya (from *Boswellia neglecta*)) has no impact on the plants (Fig. 8.4). Neither does handpicking resin exuded due to insect damage (for example due to weevil (Curculionidae) larvae in Amazonia on *Protium* and Buprestid and Cerambycid beetle larvae in Kenya (*Commiphora, Boswellia neglecta*) (Sommerlatte and van Wyk, 2022).

Tapping by regular cutting has an impact, but can be sustained at lower intensities and frequencies. However, high-intensity, high-frequency tapping is of concern. This includes cases where traditional oleoresin collection techniques are undermined, for instance when an Italian-owned commercial business in Kenya supplied cutting tools to Samburu women that could increase the risk of destructive harvest (Cunningham and DeCarlo, 2022), or when men (in this case, former soldiers) were encouraged to tap *Boswellia papyrifera* without training on the right methods (Fig. 8.5). These cases underscore the need for regular monitoring of impacts (see step 7, below).

Tapping by cutting combined with chemicals that stimulate oleoresin flow is also of concern at high intensities and frequencies, as is tapping by boxing and burning, as was practised in the USA for *Pinus elliottii* and *P. palustris*, resulting in over-exploitation and tappers having to steadily shift to resource-rich areas that had not been tapped (Langenheim, 2003). In India, lopping trees to extract gum guggul has had a devastating impact on *Commiphora wightii* populations since the 1960s (Cunningham *et al.*, 2018). Atal *et al.* (1975, p. 211), for example, recorded that a 'large number of plants of *C. mukul* (= *C. wightii*) were present some 10–15 years ago in Rajasthan, but the number has come down considerably, largely due to the faulty techniques of tapping employed by the local inhabitants', with the annual production of gum guggul dropping from 500 t in 1960 to just 5 t by 2010 (Thomas *et al.*, 2010). By 2015, *C. wightii* was Critically Endangered (Ved *et al.*, 2015) and the cultivation has faced constraints, including termites eating the cuttings (Cunningham *et al.*, 2018). Consequently, by late 2023, most of the gum guggul (from *C. wightii*) sold in Khari Bouli market in Delhi was imported from Pakistan due to severe decline of the Indian *C. wightii* population (Fig. 8.6).

3. *Habitat loss due to agricultural expansion*

From elemi (*Canarium* spp.) and *Aquilaria* species in South-east Asia to *Boswellia* woodland habitats in Ethiopia (*B. papyrifera*) and India (*B. serrata*), this is a major factor contributing to the demise of resin- and gum-producing species. Through much of South-east Asia, high-diversity rainforests have been converted to low-diversity farmland (see Fig. 8.2E) or rubber plantations. Several studies in Ethiopia record the extensive conversion of *Boswellia* woodlands to other land uses. In the Metema area of Ethiopia, for example, even if the conservative official land allocation of 2 ha per household was taken into account, then at least 37,172 ha

Fig. 8.4. Hand-picking of naturally exuded gum is ecologically sustainable and in Africa's arid and semi-arid areas is predominantly done by women. A. Picking gum from *Commiphora wildii* in north-western Namibia. B. Samburu women with *Boswellia neglecta* gum in north-eastern Kenya. C. Samburu women climbing a *B. neglecta* tree to collect naturally exuded gum. (Photos: A.B. Cunningham (A), S. Canney-Davison (B,C).) Used with permission.

of woodland in the district was lost to farmland in just 4 years (Lemenih and Kassa, 2011b). In the same area, over a 30–35 year period, almost 303,180 ha of woodland was cleared for farmland and in Tigray Regional State of Ethiopia, and over 177,000 ha of *B. papyrifera* woodland was converted to farmland over a 20 year period (Gebrehiwot, 2003). The main drivers of this land conversion are firstly, that it is allowed (and often encouraged by government) and secondly, that farmers wrongly anticipated higher financial returns from crop production than from frankincense and gum-resins tapped in intact woodlands. However, Dejene *et al.* (2013) compared financial returns for two crop production options (sesame and cotton) and woodland use in an area of northern Ethiopia (Metema district) well known for frankincense production. While net income was highest for sesame, it was lowest for cotton production. Income from frankincense was intermediate, but more consistent. Expectations of consistently higher returns from crop production were incorrect.

4. The combined effects of uncontrolled grazing, browsing and lopping for fodder have serious impacts on the long-term survival of populations of many resin-producing species

Grazing and browsing by unmanaged livestock can be a major threat to accessible *Boswellia*, *Dracaena cinnabari* and *Commiphora myrrha* populations. On Socotra (Yemen), for example,

Fig. 8.5. Poorly thought-out projects that may have intended to improve income for people in the short term but can have serious consequences, in this case, giving tapping rights to former soldiers without training on how to tap properly. A. *Boswellia papyrifera* trees devastated by a combination of lopping for fodder and intense tapping in Ethiopia. B–C. Individual trees showing intensively tapped trunks and reduced crown shape after lopping for fodder. D. Close-up of a *B. papyrifera* trunk showing signs of high-intensity, high-frequency tapping. (Photos: S. Johnson.) Used with permission.

populations of *Dracaena cinnabari*, the source of 'Dragon's blood', are in decline, mainly due to grazing by livestock (Fig. 8.9), with Adolt and Pavliš (2004, p. 52) concluding that most of the world's largest *D. cinnabari* woodland at Firmihin would 'be in the stage of intensive disintegration within 30 to 77 years'. As local people graze their livestock in the moist uplands of Socotra when drier areas have no grazing, this situation may be compounded by climate change. Grazing animals, particularly goats and camels, consume *Boswellia* leaves and bark and can inflict considerable damage on adult trees and kill seedlings and immature trees. *Boswellia* leaves and bark are both palatable. While *Boswellia* species can resprout vigorously after browsing, this cannot be sustained indefinitely. This major threat to recruitment of all *Boswellia* species due to unmanaged browsing and grazing livestock (goats, camels and, in some cases, cattle) becomes the primary threat in arid and semi-arid landscapes when limited fodder is available. Grazing is also one of the primary causes of blocked regeneration, as seen in many *B. papyrifera* populations (Bongers *et al.*, 2019) and in other species of *Boswellia* (e.g. *B. serrata*, *B. dalzielii* and *B. sacra*). The same applies to the restricted range *Boswellia* species on Socotra, the global centre for *Boswellia* diversity (Miller and Morris, 2004), as well as to *Commiphora* species on the archipelago (La Montagna *et al.*, 2023). In Oman, grazing and bark browsing by camels is also an issue of concern in some parts of Dhofar (Farah, 2008; Fig. 8.8). And in India, although *Boswellia serrata* has the largest natural

Fig. 8.6. A 'lose-lose' situation due to unmanaged and ecologically unsustainable exploitation. A. *Commiphora wightii* is now a Critically Endangered species on the IUCN Red List, with tiny populations inside conservation areas. B. Trying to grow *C. wightii* from cuttings in Mayda Pradesh faced problems due to termites. C. One of the destructive techniques that resulted in over-exploitation of wild populations was to cut the branches off *C. wightii* for steam distillation. D. Intense wounding to get gum guggul is another problem. E. Gum guggul from *C. wightii* imported from Pakistan and for sale in Khari Bouli market, Delhi. F. Family-owned businesses in Khari Baouli market have sold herbal ingredients, including gum guggul, for generations. G. Offloading bulk-packed cardamom in Khari Bouli market, also the largest herbal medicine market in South Asia. (Photos: A.B. Cunningham (A–C, E–G) and R. Kuloli (D)). Used with permission.

distribution of any *Boswellia* species, grazing is also a major factor outside of forest reserves (Fig. 8.7). This is because India has the largest concentration of livestock in the world, having one-third of the world's cattle on approximately 3% of the world's land area, with both cattle and water buffalo populations increasing (Fox, 1999; Soju and Meena, 2017).

This poses what managers and policy makers term a 'wicked problem': problems so socially complex that there may be no solutions (Lönngren and Van Poeck, 2021). For example, how can grazing by India's cattle and water buffalo be managed in India when cattle are sacred? What can be done when camels are so culturally important in Oman that owners spend thousands of dollars buying fodder to maintain camel stocks in over-grazed landscapes (Fig. 8.8)? These are situations that are particularly complex when the cattle or camel herds are owned by wealthy people with vastly more power and influence that resin tappers.

In Mayda Pradesh in India for example, the *adivasi* ('tribal') people who commercially tap certainly have customary rights over *Boswellia serrata* trees, but have little control over the major long-term impact: the livestock that eat the seedlings and saplings. This undermines

Fig. 8.7. Across arid and semi-arid areas, overgrazing and intensive browsing are causing population collapse in many landscapes. This is huge challenge. A. Goats breed fast, are culturally important and provide a good economic return. B. Water buffalo are highly prized, yet in high numbers graze all non-toxic seedlings and saplings. C. Shrublands in Kitui, Kenya, degraded by goats. D. A 'standing dead' Boswellia serrata in an overgrazed landscape with no young trees. E. Sacred cattle in India. F. A browsing camel in the Arabian Peninsula. (Photos: A.B. Cunningham (A, B, D–F) and S. Canney-Davison (C).) Used with permission.

claims of long-term sustainability of *B. serrata* supplies (Schultz, 2022), particularly when the third-party audit was done by questionnaire without the auditor visiting India to ground-truth the situation (Flaster, T., personal communication to ABC, 2022) (see Challenge 4).

5. Increase awareness that wild populations of gum- and resin-producing tree species can be affected by multiple 'non-trade' factors (such as climate change, habitat loss, grazing by livestock and fire) as well as high-intensity, high-frequency tapping

In his classic book A Sand County Almanac, Aldo Leopold (1949, p. 21) wrote that 'One of the penalties of an ecological education is that one lives alone in a world of wounds. Much of the damage inflicted on land is quite invisible to laymen. An ecologist must either harden his shell and make believe that the consequences of science are none of his business, or he must be the doctor who sees the marks of death in a community that believes itself well and does not want to be told otherwise'. The future of ecologically sustainable resin and gum supplies requires that *all* of the multiple factors that negatively affect vulnerable species populations are taken into account. These range from action at a local scale, such as monitoring and managing fire, grazing and unsustainable tapping practices, where local community involvement is essential for

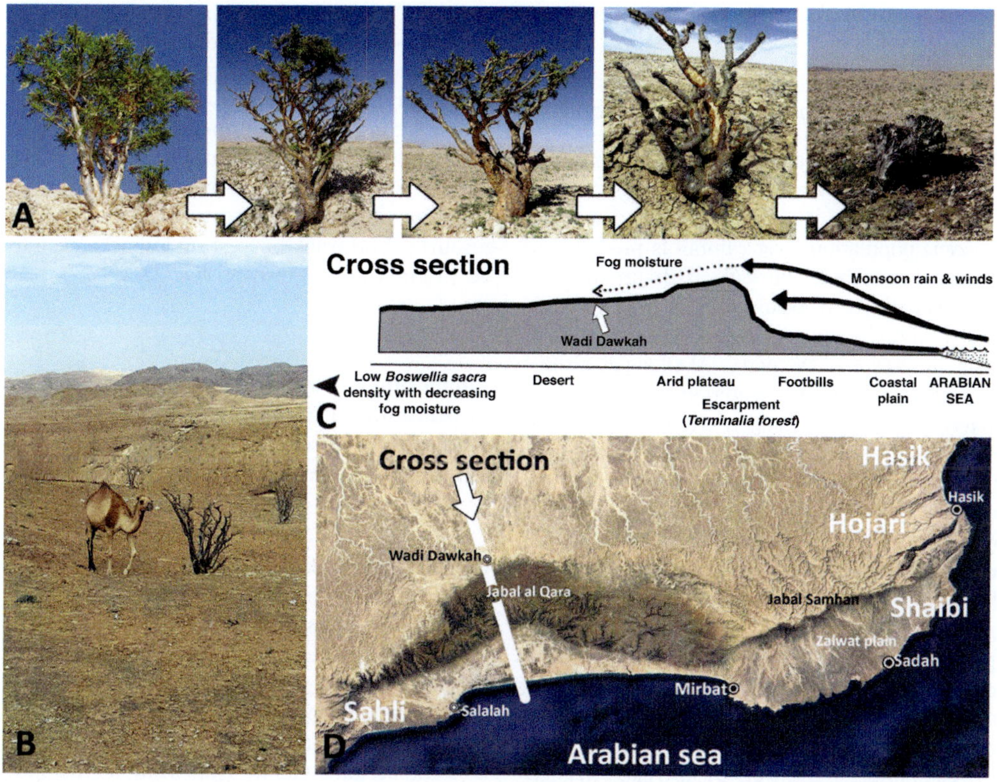

Fig. 8.8. The biggest concern for many gum- and resin-producing species is poor livestock management, not tapping. This applies to the rarest and most threatened *Boswellia* species (such as *B. bullata*, *B. dioscoridis*, *B. elongata*, *B. nana*, *B. popoviana* and *B. socotrana*) on Socotra (Yemen), which are all threatened by unmanaged grazing by goats, or much larger populations in Oman, which are affected by grazing and bark browsing by camels. In Oman, *Boswellia sacra* populations of several landscape types are healthy, with good recruitment and no camel browsing or tapping. But in the landscape type north-west of the Jabal al Qara (C, D), at the margins of monsoon-driven fog moisture, excessive browsing by camels is a major issue (A, B). (From Cunningham and DeCarlo, 2022.) Used with permission.

implementation, to implementation on national and international policies (including climate change, which has to be tackled at a global scale). Over-simplification and mischaracterization of the challenges facing species or groups of species is widespread, misinforming policy processes to the detriment of species conservation and to the detriment of people along supply chains (Challender *et al.*, 2022). The multiple factors causing decline are best documented by Bongers *et al.* (2019) in the large data set from decades of study of *Boswellia papyrifera* in three countries that are the important exporters of frankincense (Ethiopia, Eritrea and Sudan). *B. papyrifera* woodlands are in decline due to clearing for farmland, fire, grazing by livestock and opportunistic resin harvest rather than careful tapping (Lemenih *et al.*, 2014), which in turn leads to infestation by wood-boring beetles (Lemenih and Kassa, 2011a). Multiple threat factors often apply to threatened trees in general (Rivers *et al.*, 2023). Boonman *et al.* (2024) recently assessed changes in six anthropogenic threats to 32,090 tree species over the last 20 years, and suggest that over half (54.2%) of the assessed tree species faced increased threats. Significantly, only 8.7% of these tree species were on the IUCN Red List as threatened. Wild populations of slow-growing, habitat-specific tree species tapped for the resin trade are no exception, particularly those in the subtropics and tropics.

6. *Climate change: droughts, cyclones*

The East African–Asian Monsoon (EAM) is one of the world's main weather systems and is critical to wild *Boswellia, Commiphora, Hymenaea* and *Dracaena* populations, yet rainfall is predicted to become increasingly variable with an increase in the severity of tropical cyclones (Kazora *et al.*, 2023). In his study of *Dracaena cinnabari* populations on Socotra (Yemen), Al-Okaishi (2020) not only found that the traditional management system of tapping 'Dragon's blood' tree resin (Fig. 8.9) is not sustainable, but also that due to their heavy crowns, *D. cinnabari* (Fig. 8.9 A,B) have an increased chances of being blown over by strong monsoon winds after their trunks are weakened by traditional tapping. Also, on Socotra, the centre of diversity for *Boswellia*, seven endangered and four critically endangered *Boswellia* species, due to the high frequency and intensity of cyclones, are also affecting those populations in addition to habitat fragmentation, grazing impacts and long drought periods (Maděra *et al.*, 2024). Changes in the EAM are also likely to affect *Boswellia* species along the maritime escarpments of northern Somalia and the 'fog oases' of Dhofar (Oman) and the Hawf region (Yemen). In South Asia, the effects of climate change have been modelled for *Boswellia serrata*, with predictions of decreases in the area of land suitable for this species (Rajpoot *et al.*, 2020).

7. *A low awareness in the commercial sector that management and monitoring plans are needed*

A clear understanding of what additional information is required for management and monitoring of ecologically sustainable use is essential. This is not easily achieved, however, in a world of low ecological literacy, 'greenwashing' and limited commitment to the triple bottom line. Scientific research on what steps (and data) are required for ecologically sustainable harvest were distilled into publicly available manuals decades ago (such as Peters, 1994; Cunningham, 2001; Stockdale, 2005). Despite this, a massive 'knowing but not doing gap' remains. For many species, ecological monitoring is woefully inadequate. This is despite the fact that user-friendly, open-source programs such as CyberTracker (https://cybertracker.org/partner/smart-conservation/) have been used for over 30 years, including for monitoring impacts of tree populations (Fig. 8.10; Cunningham, 2001). ArcGIS Collector and other data collection apps are also available to map environmental resources and track harvesting impacts, even in very remote areas (Johnson *et al.*, 2023).

Challenge 3: Plantations and agroforestry production at commercially viable scales requires very specific circumstances for success. Achieving this in arid and semi-arid areas is hugely challenging

Cultivation is widely seen as the solution to the challenge of demand exceeding supply. And for fast-growing species, it often is. Cultivation of fast-growing essential sources from flowers (*Cannabis*, roses, jasmine), leaves (lemongrass, mint, *Ocimum* species) and leaves and stems (geranium, patchouli) for essential oils often is commercially viable. But what about slow gum- and resin-producing growing trees, where it can take years before the farmers get financial returns and risks can be high?

With trees, it is worth taking a long view on trade and cultivation, learning from history as we look to how to sustain supplies to meet growing demand in the 21st century. After all, there are several large flavour and fragrance companies in Europe that have been around since the late 19th century and are presumably in for the long haul. This includes asking ourselves some pertinent questions. What happened to the successful, collectively large-scale agroforestry production of resins such as damar (*Shorea javanica*), when social and economic changes took place? What circumstances resulted in successful Aquilaria (*oud* or *gaharu*) production in plantations and agroforestry systems? And given the huge challenges of large-scale production of gums and resins in arid and semi-arid areas, have plantation experiments distracted from what is clearly necessary: 'growing the wild' through wise management and monitoring of wild populations, fair prices and local community involvement? Perhaps the most pertinent question of

Fig. 8.9. A lethal combination: goats, heavy crowns, weakened trunks and monsoon winds. A. The otherworldly Dragon trees (*D. cinnabari*), with their heavy crowns, are a remarkable feature of Socotra. B. Where are the seedlings and saplings? Eaten by goats. C. 'Dragon's blood' resin from Socotra. D. A heavily tapped trunk. Recent research (Al-Okaishi, 2020) suggests that traditional tapping weakens the trunks and makes trees susceptible to being blown over by strong monsoon winds. E. Close up of a tapping scar. (Photos: S. Johnson.) Used with permission.

all is where and under what circumstances does Brazilian economist Alfredo Homma's (1992) domestication-intensification pathway work or fail in terms of a transition from the wild to plantations (or to substituting forest products with synthetics)? For the sake of brevity, the wet tropics are considered first, then cultivation in arid and semi-arid areas.

Higher rainfall landscapes: agroforestry systems and plantations

In the wet tropics, many gum- and resin-producing species are cultivated in small-scale farming systems for the multiple products they provide, often in complex agroforestry systems. In West and Central Africa, for example, *Dacroydes edulis* is a common feature of agroforestry systems (Leakey *et al.*, 2002), as are many *Canarium* species in the Asia-Pacific region (Weeks, 2009). But what about expanding cultivation to a large enough scale to meet market demand? Although pine resin tapping has a bleak history (Langenheim, 2003; Zallen, 2019), there are successful cases of pine resin production both from plantations and from natural pine forests, where *Pinus* is monodominant. Yet even in China, low resin prices and relatively high labour costs are a challenge to the economic viability of resin production (Liu *et al.*, 2022). As Chapter 4 covers pine resin production in detail, we focus below on other examples of gum and resin production in agroforestry systems or plantations.

Fig. 8.10. Why are commercial companies so reluctant to use a tried and tested method that tappers can be trained to use? A. Designed for use by people with great local knowledge but low numeracy or literacy, this app was tested in African forests 30 years ago. B. Icons showing levels of bark damage make the app user friendly and modifiable for monitoring in the case of resin tapping (such as C–E). F. Tree crown health can also be a feature, as shown in G–I. (From Cunningham and DeCarlo, 2022.) Used with permission.

Damar (*Shorea javanica*) in complex agroforestry systems: lessons from the past 140 years. In Sumatra, Indonesia, for example, the *Shorea javanica* (damar) agroforests were started in the 1880s due to reduced quantities of wild trees and growing commercial demand for damar resin from commercial varnish and paint industries (Michon *et al.*, 2000). By 1998, with enrichment planting of coffee and pepper, these mature agroforests were 50,000 ha in extent (Michon *et al.*, 2000). By 1996, however, despite relatively stable demand and prices for *S. javanica* resin, a study of 223 households showed that most (63%) people preferred to earn an income in other ways (Wollenberg and Nawir, 2005), despite the fact that most household income came from the sale of damar resin (Kusters *et al.*, 2008). Low prices paid to farmers were one driving factor (see Challenge 6). Weak tenure and increased theft of damar (including by extended family members) was another (see Challenge 7), resulting in shorter tapping rotations and lower damar yields, with some farmers converting damar agroforests to grow coffee and pepper (Kusters *et al.*, 2008).

Growing gaharu (oud) for the world's most expensive wood and oil: more than a century of lessons. Although the first *Aquilaria* plantation trial was begun in Malaysia in 1927, the recent rise in plantation and agroforestry production of several *Aquilaria* for oud (or gaharu) has been nothing short of spectacular. The success with *Aquilaria* cultivation has been driven by four main factors. Firstly, that there is a huge demand for the oil and wood, whether for perfume and wood chips in the Middle East (Fig. 8.11) or for *koh doh* incense ceremonies in Japan. Agarwood is the most expensive wood in the world and the market is huge. Worldwide sales of agarwood chips were predicted to reach US$64 billion by 2029 (Ash, 2020).

Secondly, retail prices are extremely high for both the wood and the oil. Oud oil and wood chips are highly valued in the Middle East (Fig. 8.12; Persoon and van Beek, 2008; Thompson *et al.*, 2022). Despite predictions in 2006 that between 2010 and 2020 cultivated *Aquilaria* production would replace exploitation of wild *Aquilaria* populations and that a 50% drop in price would occur, this did not happen (Soeharto *et al.*, 2016). Oud oil prices vary with geographic origin and grade (for example oil graded as AAA, AA, A+, A, A1). There is also variation in the cost of *Aquilaria* wood chips. Those from India cost US$7121/kg, Indonesia US$5106/kg and Vietnam US$2334/kg, with an average of US$4421/kg. In 2023, the price of oud oil from Cambodia was US$510 per tola (one tola = 12 ml) (so US$42,500 per litre), from India US$391 per tola (US$32,500 per litre) and from Vietnam US$280 per tola (US$23,333 per litre), with an average price across all producing countries of US$439 per tola (US$36,583) (Oud Oil Trading, n.d.). This is 140 times more than cost of frankincense oil per litre.

Thirdly, *Aquilaria* trees are fast-growing and can be grown at high densities. They take about 8 years to reach 10 cm in diameter at breast height (DBH), the minimum trunk diameter for inoculation, after which it takes at least 15–24 months for the agarwood to develop (Thompson *et al.*, 2022). Although wild *Aquilaria* populations occur at very low densities (0.50 and 0.58 trees > 10 cm DBH/ha) even in remote parts of Kalimantan (Soehartono, 1999), densities in multi-species stands can be 2517–3500 trees/ha and 4140 trees/ha in plantations.

Fourthly, huge strides have been made in the science of producing plantation agarwood over the past 30 years (Thompson *et al.*, 2022). In order to scale up their pioneering efforts (Fig. 8.11), The Rainforest Project (TRP) designed agarwood inducement kits that are now widely available in South-east Asia.

Finally, there has been significant support, from both government and the private sector. The Assam Agarwood Promotion Policy (Government of Assam, 2021) is one example. The partnership between Jalali Agarwood and the Swiss company Firmenich to grow *Aquilaria malaccensis* is another.

What about large-scale tree cultivation in arid and semi-arid areas? In arid and semi-arid Africa the Burseraceae (*Boswellia* and *Commiphora* species) and gum arabic (from *Acacia senegal* and *A. seyal*) are leading non-timber products. They are also important in terms of carbon sequestration, with *Boswellia* species often a dominant (or even mono-dominant) component. Gum arabic production won't be dealt with here as this comes from farmed fallow rotation systems that even before the current conflict in Sudan were at risk due to low prices paid to collectors (Elmqvist *et al.*, 2005; Challenge 6). So what factors stand in the way of plantation production of *Commiphora* or *Boswellia* oleoresins at a large enough scale to meet demand from national and international markets?

The frankincense trade provides a useful reality check. Firstly, it is likely to take a massive increase in price before frankincense or myrrh cultivation would be viable at scale without massive subsidies for fencing, irrigation and other costs. The current price of frankincense oil on the wholesale international market averages US$260–275 per litre. That's more than 140 times less than the wholesale price of oud oil from *Aquilaria* (at an average of US$36,583 per litre). And in tropical forest gardens and plantations, neither irrigation nor fencing against camels are needed! Furthermore, in

Fig. 8.11. A. Perfumes are a wonderful aspect of Arabic culture. The exceptionally high demand and high prices for oud (both oil and wood chips) are factors that have supported the expansion of *Aquilaria* agroforestry and plantations. Prices for both oil and wood chips vary according to geographic origin and grade, such as wood chips classed as Super Deluxe, Super Double or Super. These 2025 retail prices (all negotiable) are in tola (one tola = 11.663 g). B. Cambodi fakher ('Cambodian pride') at 39.408 Omani Rials (OR) per tola (*c.*US$3400 per kg). C. Hindi malaki ('Hindi (or Indian) angel') at 41.958 OR per tola (*c.*US$3600 per kg). D. Calimantan super (referring to Super quality from Kalimantan (Indonesian Borneo) (at 43.205 OR per tola (*c.*US$3700 per kg). (Photos: A.B. Cunningham.) Used with permission.

landscapes such as Dhofar, wild *Boswellia sacra* trees are abundant. So why plant when there is major competition from wild trees and prices are low? The *Aquilaria* case described above is very different. Relatively few wild *Aquilaria* trees were left to compete with production from cultivation, and prices were astronomically high.

Secondly, customary law can also be a barrier. Across much of Africa, for example, tree planting is associated with laying claim to land. This is a highly sensitive issue that can lead to deliberate felling of *Boswellia* trees in plantations. Additionally, *Boswellia* plantation production at a scale large enough to take pressure off wild stocks will require permanent access to very large tracts of land, with secure tenure of the land and the trees on the land for decades. For example, the only way that an expatriate entrepreneur from Somaliland (currently living in the USA) was able to plant *B. carteri* on a modest scale was to plant cuttings in black plastic nursery bags so that from a cultural perspective that they were not planted out on the land (Cunningham and DeCarlo, 2022).

Thirdly, in Somaliland, customary rights even applied to cuttings from taken *Boswellia* trees, with clan leaders reluctant to allow the cuttings to be 'taken away' without an undertaking that they would be returned (Cunningham and DeCarlo, 2022). If replanted

Fig. 8.12. Cultivation of *Aquilaria* was pioneered in Vietnam by The Rainforest Project (TRP) and is a process ranging from seedling production (top left), plantations and chemical stimulation of resin impregnated wood through to production of the final product. (Photos: H. Heuveling van Beek.) Used with permission.

cuttings are protected by thorny brushwood 'anti-browse barriers' that are maintained by community members, then this could work in favour of enrichment planting and restoration of *Boswellia* woodlands affected by poor recruitment. However, it could be a point of conflict if cuttings are taken for plantations established in other locations.

Finally, there is the issue of water rights. *Boswellia* plantations, even on a modest scale, require access to water for years. Yet water rights (and conflicts over access to water) are particularly acute in arid landscapes where the priority is water for people then livestock, and only then trees. This could be the focus of deliberate damage if water-use conflicts are not resolved. Growing at a large scale requires lots of land and water.

How much land and how many *Boswellia* trees would be needed? Although exceptional individual trees may produce more, the average frankincense yield averages about 500 g per tree per year (Table 8.1). This equates to 2000 trees to harvest 1 t of frankincense. Ethiopia exports 3500–7900 t/

year of frankincense and Somalia 1400–3000t/year (Cunningham and DeCarlo, 2022). If even half of the frankincense exports from Ethiopia and Somalia were to come from planted trees, this would require an estimated 4.9 million to 10.9 million *Boswellia* trees. At a tree planting density of 400 trees/ha (which is higher than the 6 m × 6 m spacing used in Oman (S. Ajeeb, personal communication, 2022)), this would still represent a vast area of land (*c*.12,250–27,250ha), posing huge challenges in terms of water rights (particularly during the establishment phase) and grazing rights (with fencing to keep out livestock). These risks will apply whether there are many large plantations or thousands of individual investments in smaller-scale planting.

Plantations trials are certainly possible where government agencies have tenure over the land and where the economic costs of fencing (to keep out livestock) and irrigation water for the trees are subsidized. In the 1970s in Sudan, for example, a 10 ha *Boswellia papyrifera* plantation trial was implemented at a government research station (Nour, 2002). In 1982, under the centrally controlled government of Mohamed Siad Barre, the then-president of the Somali Democratic Republic, a plantation of *Boswellia carteri* was established by a frankincense cooperative in Somalia at Ambarsare (north-east of Gardo (or Qardho) in the Bari region of the autonomous state of Puntland. According to Thulin and Warfa (1987) there were 10,000 trees in the plantation. The fate of these government-funded plantations today is unknown. In 1998, a *B. sacra* plantation research trial was established in Oman by the Ministry of Fisheries and Agriculture (Fig. 8.13A). At the World Heritage Site at Wadi Dowkah in Oman, more *B. sacra* are cultivated in a fenced and watered area (Fig. 8.13B).

Essentially, Homma's (1992) from wild to cultivated transition model can apply in the tropics but

Table 8.1. Types of harvesting techniques across different species in different locations and countries with average yields of oleoresins per tree (g) and average annual quantities collected per family (kg). Note that most harvesting is during the dry season. Includes unpublished field records. Author's own table.

Species	Main geographical region	Harvesting practice	Average yield/ mature tree/ year	Average yearly kg/harvester	References
Boswellia sacra	Oman	Tapping	*c*.500 g	559 kg	Farah, 2008
Boswellia sacra	Somalia/ Somaliland	Tapping	*c*.500 g	Average 474 kg; 280–600 kg for both *B. sacra* and *B. frereana*	Stephen Johnson (field records); FAO (2016)
Boswellia frereana	Somalia/ Somaliland	Tapping		270 kg	Farah, 2008
Mixed *Acacia* / *Commiphora* gums	Adadle, Ethiopia	Mixed tapping and natural exudate		219.82 kg per family	Yusuf *et al.*, 2023
Three *Boswellia* species	Adadle, Ethiopia	Natural exudate		58.27 kg per family	Yusuf *et al.*, 2023
Boswellia papyrifera		Tapping	41–840 g/year (from 7 tapping rounds) or 185–1826 g/ year (from 14 rounds).		Wassie, 2011
Senegalia senegal and *Vachellia seyal*	Sudan	Natural exudate and tapping	100–250 g		Abtew *et al.*, 2014

is unlikely to apply in arid systems, where the priority should be to focus on managing and monitoring wild populations. *Boswellia* cultivation promoted by international companies may have more to do with public relations benefits than economically viable cultivation at scale.

Challenge 4: A lack of transparency in many supply chains

A lack of transparency is common in gum and resin supply chains, making it difficult to identify where the materials have come from, who harvested them and the impact of their collection on the plants and broader ecosystems. Often, there is a complete lack of sourcing information, aside from the country of origin. Where additional traceability information is available, it is often incomplete. Buyers may be able to identify a specific region or town where the material originated, but rarely are specific farms or wild harvesting sites identified. This lack of specificity makes it impossible to determine what harvesting and biodiversity management practices were used to produce a particular batch of product, and to track the collectors and how they were paid. As a result, in many supply chains it is near impossible to connect purchasing a specific batch with supporting ethical, social and environmental practices. There are various initiatives that audit traceability and the 'triple bottom line'. One is through an alliance of the Union for Ethical Biotrade-Rainforest Alliance (UEBT/RA) and Rainforest Alliance Certification Platform (RACP), which provides UEBT/RA members access to MultiTrace (Version 1.0, February 2024). Tracing crops like lavender, tumeric or vanilla is a lot simpler, however, than tracing dispersed, wild populations of gum- and resin-producing species.

Traceability is possible, even in complex wild-harvested plant systems. The FairWild standard is one example of this (FairWild Foundation, 2023). It simply means that more cost and effort is involved, but even some large companies have not been willing to make that investment yet. They are instead focused on cost reduction, treating gums and resins as commodity products where one batch is the same as the next, regardless of any social or environmental implications. Starting in 2015, for example, a USAID-funded programme in Somalia considered frankincense as a high-potential value chain product. In 2019, the frankincense sector was dropped, reportedly because key players were uncooperative (BIFAD, 2019).

Yet there are some Somali companies that have implemented voluntary sustainability standards, such as the FairWild Standard (FairWild Foundation, 2023) and Ecocert (Fig. 8.14).

These require mapping of collection areas, resource assessment and management, audits by accredited independent third-party inspection and certification control bodies. However, organic certification has been shown in Somalia to be ineffective at differentiating between species, as areas containing *Boswelia occulta* were certified as organic *Boswellia carteri* before the species was 'discovered' and described as clearly distinct (Thulin *et al.*, 2019), raising serious questions as to whether it can guarantee sustainable harvesting practices (Johnson *et al.*, 2019). And FairWild certification generally does not publicly release audit reports, raising the question of whether this standard will run into similar problems with on-ground verification of practices (Cunningham and DeCarlo, 2022). Claims of 'third-party' audits when there has been little or no ground-truthing also need to be avoided. A 'third-party' audit of *Boswellia serrata* by questionnaire from the USA, without the auditor 'ground-truthing' in the field, does not add credence to Schultz's (2022) claim of sustainable supplies. The same applies to the FairWild certification (by Ecocert) of 500 t of frankincense from Somalia, where seven tappers were interviewed: a tiny sample size of tappers (0.7%) considering that 500t of frankincense would be collected from *c.*1 million trees by around 1000 tappers (Cunningham and DeCarlo, 2022). There are much better ways of doing things, and the reputation of certifiers is at stake. A good way of getting around this is to use the expertise of tappers themselves, backed up by smartphone-based technology such as Cybertracker (see www.cybertracker.org; Fig. 8.10A). Yet despite promising discussions I had with a major USA-based multi-level marketing company during 2024, including booking a training course for some of the frankincense tappers in the use of Cybertracker, this

Fig. 8.13. Government and international aid can cover the costs of fencing and drip irrigation in dry areas. But expecting frankincense or myrrh plantations to expand to scale without significant subsidies can be a mirage. A. A 23-year-old *Boswellia sacra* plantation trial funded by Oman's Ministry of Fisheries and Agriculture in a fenced and irrigated area (in 1998 in Dhofar, Oman). B. *B. sacra* established at the UNESCO World Heritage Site at Wadi Dowkah in Oman in 1990, with major investment in drip irrigation and fencing. C. Collection of branches for cuttings in Somaliland for development aid funded project hoping to start *Commiphora myrrha* cultivation. D and E. Fruits on a female tree opening up prior to dispersal by birds attracted by the pseudaril. *C. myrrha* are dioecious (with separate female and male trees), so large-scale lopping for cuttings risks an Allee effect, with reduced pollination success. F. The propagation stage is easy. But surviving goats and limited water once the development aid dries up is another thing entirely. (Photos: A.B. Cunningham (A, B, E, F) and U. Feiter (C,D). Used with permission.

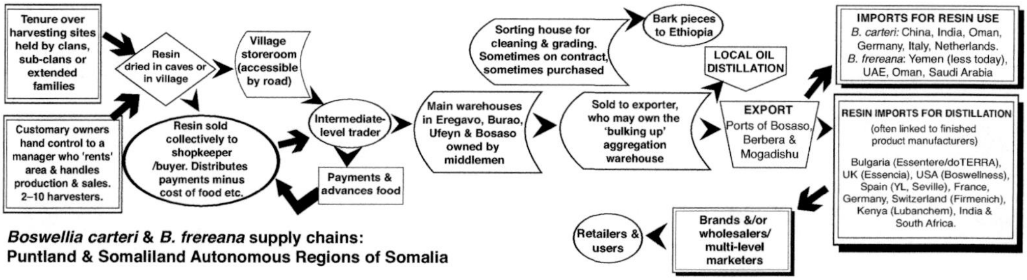

Fig. 8.14. Complexity can lead to opacity: a diagram of the frankincense supply chains from Somaliland and Somalia. (Reproduced from Cunningham and DeCarlo, 2022). Used with permission.

opportunity was lost. This occurred when the company decided not to proceed with this opportunity. Consequently, the course was cancelled, despite this being a crucially important test case of a recommendation made to the CITES Secretariat, and the implications that companies failing to do the right thing would have for people and the frankincense trade from Ethiopia and Somalia if a CITES Appendix II listing occurred (see Challenge 9).

Local traders and brokers, frequent actors in gum and resin supply chains, may also benefit from complex (Fig. 8.15) and often opaque supply chains, as this protects their sources and their businesses from competition. The relationship is not entirely one-sided. Many gum- and resin-bearing species are widespread throughout the harvesting regions and often found in very remote areas. Gum and resin production is typically labour intensive, necessarily involving a large number of people spread across many communities, and materials are collected across a large land area (see Challenge 6). Exporters, wholesalers and processors, therefore, often find it more expedient to simply work with a smaller pool of local traders or brokers to get the material they need.

Likewise, companies further downstream may also use a lack of transparency and layers of middle traders as an alternative to tackling thorny environmental and human rights issues. These issues are often complex, and companies are rarely well equipped to address them (see Challenge 2). If they do manage to identify specific harvesting sites and find a problem with unsustainable extraction or broader land management, what then? The threats and drivers influencing gum- and resin-bearing species can be complex, and influencing the outcome requires expertise, investment and a long-term commitment. The risk is that companies may see a lack of transparency as a matter of efficiency, gambling that if such problems are exposed, they can deny knowledge or involvement. This was the strategy taken by international cacao trading companies when news of child and forced labour in the supply chain came to light (Schrage and Ewing, 2005).

There are positive examples of commercial companies that are committed to transparent, ecologically sustainable and socially equitable supply chains where this works in practice. In both north-western Namibia and northern Kenya, sustainable, low-impact harvest of *Boswellia* and/or *Commiphora* resins occurs within conservancies. In Somaliland, a coalition of stakeholders including Lush, Pacha Soap, FairSource Botanicals and Save Frankincense are working with the Dayaxa Frankincense Export Company to partner with harvesters and incentivize sustainable harvest methods, make traceable digital payments to harvesters to confirm fair pricing and engage with responsible buyers and importing companies. In addition, this group have pioneered a combination of harvester-friendly monitoring with new technologies, where tapped trees are photographed using mobile phones to document tapping practices, marking the GPS locations of harvested trees. This allows the status of harvested *B. carteri* populations to be traced through the supply chain using blockchain technology, and payments are made to harvesters via a mobile banking platform for transparency.

Larger quantities are under organic certification by Ecocert, for example in Oman (Salalah Frankincense Oil) and Somalia (Neobotanika

Fig. 8.15. Gum olibanum organically certified by Ecocert in a warehouse in Somaliland. Ecocert is a well-known organic certifier in France, where they certify an estimated 70% of organic food. (Photo: U. Feiter.) Used with permission.

and Kobac General Trading). These efforts are still at a limited scale, with only a small percentage of total oleoresin exports sourced through these transparency and certification initiatives (Cunningham and DeCarlo, 2022). However, these examples demonstrate that such approaches are viable and could be scaled up.

Challenge 5. 'Mind the green gap': companies claiming to be 'sustainable' or meeting CSR or ESG goals without taking enough detail or monitoring to furnish sufficient proof

Independent monitoring and evaluation can be extremely challenging in gum and resin source countries such as Somalia or Sudan. This creates opportunities for actors along value chains with multiple staging points to paint a different picture to what the reality is on the ground. This has led to terms such as 'greenwashing'. Greenwashing in gum and resin supply chains can take many forms. For example:

- the retailer using non-definable or non-verifiable terms such as 'naturally sourced' or 'eco-friendly', or relying on in-house self-certification;
- highlighting the sustainability of a specific verifiable value chain and engaging an overall catch phrase in the hopes customers will assume it is the same for all products;
- producing well-meaning glossy brochures exemplifying best environmental and employment practices in places that customers are unlikely to ever go to check the reality on the ground;
- claiming to be supporting women in places were 30% or more of harvesters are already women (e.g. Soumya *et al.*, 2019), without looking deeper as to whether these women are able to decide on the use of the income;
- giving numbers of trees planted, with no follow-up of the extent to which these trees survived in the years after planting; and
- to guarantee their supply chain, companies and aggregators may contract whole

villages and areas and discourage any outside visitors from verifying sustainability claims.

A recent example of a gap between promoting responsible sourcing practices and local livelihoods and the sourcing reality has been in the large essential oil company dōTERRA, known for their co-impact sourcing initiative (dōTERRA, 2019). For over a decade, first indirectly through a US supplier, then directly, dōTERRA mostly relied on a sole supplier, Asli Maydi, for their *Boswellia carterii* supply out of Somaliland and Puntland (dōTERRA 2019). After some in-depth investigation into forced labour conditions, in November 2024, US Customs and Border Protection issued an order detaining any imports of Asli Maydi frankincense (US Customs and Border Protection, 2024). An NGO report went deeper into the allegations, suggesting that unsustainable and unethical practices may have been going on for some years (Corporate Accountability Lab, 2024). This raises the question: What prevents well-resourced large companies from finding out the actual realities on the ground and acting quickly to dissociate from or change any unsustainable, unethical or illegal practices or malpractice? A local export company is embedded in the local economy and can more easily frustrate monitoring and evaluative activities. This can make gathering of factual evidence of unsustainable practices or malpractice, necessary for the company and/or appropriate judicial systems to follow up, quite challenging

The size and intent of a company also matters in achieving optimal environmental and social governance. Large companies tend to have multiple supply chains, so can have greater positive impact with proactive good intent, monitoring and the due diligence of corporate leadership. There is also the argument that because of the bargaining power a large firm has over a supplier, suppliers of larger firms are more likely to adopt environmental goals, such as emissions reduction, than suppliers of small firms (Homroy and Rauf, 2024). Smaller companies, dealing with one or a few products, who want to gain market advantage by demonstrating very high verifiable environmental, social and governance (ESG) standards may have a greater incentive to create more direct value chains sourcing from specific forests, farms or cooperatives (Inês *et al.*, 2023).

A step in the right direction is that many large corporations set specific corporate social responsibility (CSR) goals to minimize harm and create societal benefit beyond employment and profit. This can also be seen as a way to create social and reputational advantage through large companies advertising their 'green' credentials. In Bulgaria, for example, Esseterre Bulgaria EOOD (Dobrich, Bulgaria) (Esseterre, n.d.), owned by the US company dōTERRA (Pleasant Grove, Utah), says on its website that 'all Esseterre's employees are familiar with and committed to following the company's policy for sustainable environmental development ....' and that 'With its eco-policies and voluntary eco-initiatives, Esseterre strives to set a leading example and inspire other followers in the region in a responsible way of protecting the environment'. In France, Mane SA (Mane, n.d.)'apply a responsible purchasing policy aimed at raising awareness among suppliers of the CSR policy and to assess their environmental and social performance'. While in Germany, Symrise AG is a member of Sedex (Sedex, n.d.; Symrise n.d.), which has a 'risk assessment tool' to investigate raw material supply chains and follow a Sedex Members Ethical Trade Audit (SMETA) system. According to their website, they also follow Sustainable Development Goals (SDG) 12 and 13 on sustainable sourcing (Symrise, n.d.), aiming to have transparent, traceable supply chains and stating that all of their suppliers have to complete a code of conduct. In Switzerland, Firmenich belongs to the Union for Ethical BioTrade (UEBT), producing regular ESG reports using a 'Path2Farm' digital traceability tool (dsm-firmenich, n.d.), while Givaudan has a 'Sourcing4Good' responsible sourcing programme (Givaudan, n.d.).

In the USA, where the Lacey Act and the Pelly Amendment apply to global supply chains (for example to species endangered by trade), there also are policy developments that will require large North American companies to develop transparent supply chains that improve on existing internal company 'green' or ESG approaches. Since 2012, for example, the California Transparency in Supply Chains Act has enabled consumers to take action against unfair labour practices in supply chains of both manufacturers and retailers. These are relevant to large frankincense importing companies such as dōTERRA, Young Living and International Flavors and Fragrances Inc.

In the Asia-Pacific region, it is uncertain how this may apply to branches of US companies in China (such as those of dōTERRA and Young Living). Supply chain transparency legislation does, however, apply in India and Australia. In India, effective from April 2021, the top 1000 listed companies (by market capitalization) needed to comply with the new Business Responsibility and Sustainability Reporting legislation, which incorporates requirements of the National Guidelines on Responsible Business Conduct (NGRBC) and of the previous legislation on Business Responsibility Reporting (BRR). In Australia, by 31 March 2021, all Australian businesses earning over AUD$100 million had to comply with the Modern Slavery Act 2018, which is focused on human rights issues along supply chains, enabling Australian Customs can seize products that do not comply.

Consumers, investors and governments need to apply greater scrutiny on ESG standards in organizational strategies and provide evidence of green supply chain management (GSCM). Increasing numbers of finance institutions have also voluntarily joined the UN Principles for Responsible Investment (PRI) set up in 2005, which works to include six ESG principles in fiduciary investment decision making. Others have signed up to the United Nations Global Compact covering human rights, labour, environment and corruption, with ESG data analytics organizations increasingly providing support.

Another positive development in some gum and resin supply chains is the role of the diaspora. A strong contribution to sustainable development can be achieved when young, educated professionals with business experience in importing countries establish commercial organizations sourcing from their countries of origin. In the frankincense trade, examples include Dayaxa Frankincense Export Company in Somaliland (see Challenge 4) and Nature-is-Unique in Germany. These two companies also bring in foreign technical support and financing and have a stronger motivation and understanding of how to work through challenges to create sustainable supply chains with positive environmental and social impact.

As described earlier, some national and international legislation in importing countries, such as the EU Corporate Sustainability Due Diligence Directive (CSDDD), is moving to enforce greater accountability of the end buyer for the entire and interconnected supply chain. Even so, the regulatory landscape around sustainability claims along the entire supply chain is often inadequate. Informed consumers, customers and media are the most likely drivers to question greenwashing. They can put pressure on corporate leadership to conduct in-depth due diligence in the choice of suppliers and to invest in training and capacity building along the supply chain and upstream to source.

Challenge 6. Harvesting, sorting and processing gums and resins is labour intensive, with low returns to labour

With the exception of more recent industrial extraction of resins (such as pine resins) from pulpwood, all gum and resin tapping is labour intensive (Figs 8.4 and 8.1), often with low prices paid and at the start of the supply chain. At the same time, this labour-intensive gum and resin collection provides an important source of income to remote rural households. In the East Golis Mountains of the Somaliland Autonomous Region of Somalia, for example, 225,750 people in the Bari and Sanaag areas earned 55–60% of their income from frankincense, goats and fish, with wealthier households earning 30% more from frankincense than poorer ones (FAO, 2016). In southern Ethiopia, frankincense harvesting provided the second most important livelihood activity (after livestock) to Borana households, and while the income range across households was wide (4–77%, average 35%), this income was particularly important to poorer households (50% of total cash income) (Berhanu *et al.*, 2021).

Even in mixed farming systems, while net income was highest for sesame, it was intermediate and more consistent for frankincense, and lowest for cotton production (Dejene *et al.*, 2013). It is likely that >5,00,000 people earn an income from the frankincense trade in the Horn of Africa (based on *c.*200,000 tappers in Ethiopia and >2,25,750 tappers in just two frankincense source areas, with unknown but significant numbers of commercial tappers in Eritrea and Sudan.

Income from gum arabic is also important both to pastoralists who collect the gum from *Senegalia senegal* and *Vachellia seyal* and to the women who sort and grade the product (Jales, 2018). In Burkina Faso, for example, gum cleaning and sorting was the most important income source for the women doing this task (Shackleton *et al.*, 2011). Although data on income levels is limited, the quantities are significant. According to Jales (2018), exports of crude and semi-processed gum arabic almost tripled in 25 years, from 35,000 t/year between 1992 and 1994 to 102,000 t/year between 2014 and 2016, and exports of gum arabic more than tripled, from 17,000 t to 53,000 t, over the same period.

Who are the harvesters?

Due to the remote areas where wild populations of gum or resin source trees occur, tappers living in the harvesting areas can be marginalized in terms of access to resources, education or the market. Nevertheless, with secure tenure arrangements, knowledge of correct tapping methods and adequate returns on labour, harvesters in communities closest to the trees are often motivated to manage the trees they tap. Harvesting areas are often reached on foot or using pack animals such as donkeys or camels, and harvesting forays can be either daily or involve walking for days at a time. Gums and resins are also harvested as pastoralists walk through dry woodlands or shrublands with livestock, searching for grazing.

Most commercial volume supply chains of tree gums and resins are based on cuts or incisions made in the outer bark to reach resin ducts in the cambium, resulting in the tree producing resin to heal the wound (Fig. 8.3). Over a harvesting season, this can mean visiting a tree several times to deepen the wound, depending on the tree species and harvesting technique. The time spent and distance that needs to be and can be covered by harvesters is also determined by the quantities harvested from each tree (Table 8.1), with frequency and intensity of harvest influenced by expertise of the tapper and tree tenure (Challenge 7). Traditional resin harvesting methods can also break down under the pressure of opportunistic commercialization (Soumya *et al.*, 2019) or due to changes in National Forest Department policies. This results in a loss of income and trees.

Understanding the cultural and socio-economic dynamics at the harvesting level in each location of a species is key to long-term, sustainable supply chains. In Ethiopia alone, seven different harvesting models were identified in the harvesting of *Boswellia papyrifera*. Some models benefitted the harvesters and trees, while others led to exclusion, loss of income and degradation (Habebo, 2011).

Many harvesters only have access to minimal educational, literacy and financial resources, with traditional harvesting knowledge passed down generationally. As literacy, urban migration, financial pressure, changes in prices and markets and exposure to other sources of income increase, knowledge about sustainable harvesting, motivation for regenerative forest management practices and careful harvesting can decrease (Kusters *et al.*, 2008; Rajpoot *et al.*, 2020).

Postharvest labour: Graders, sorters and bulk packing

The number of postharvest actors can be complex. In Ethiopia, for example, harvesters take the resin they collect to village traders who sort, weigh, buy and collect for intermediary agents who use daily labourers to further sort, pack, load and transport the product to retailers or exporters (Yusuf *et al.*, 2023). In supply chains where source trees are sparse and harvesting communities have little access to transport, supply chains rely on a series of ‘bulkers’. Sorting bulked gums and resins, either by species or by grades, is usually done by hand or a combination of sorting machines (ecgranules) and manual sorting, often by women. In some well-developed supply chains, the sorting and grading of clean resin is safe, while in others, basic conditions persist, such as having to sit on the floor for long hours and inhaling dust, increasing health risks for sorters.

Ultimately, the major shift will come when large cooperatives, state organizations and wholesale and retail private companies make the effort to ensure, follow up and monitor fair pay and working conditions and/or site-specific collaborative community-based development, as well as providing training in sustainable harvesting techniques and resources for tree regeneration in their key sourcing areas. As explored in the next section, the exact land and tree tenure rights legal framework in each harvesting location can make a significant difference to both environmental and social outcomes.

Fortunately, the need to do better is being driven by legislation in Europe, North America and parts of the Asia-Pacific region. This places responsibility on large companies to be accountable for correcting negative social and environmental factors along their supply chains, with serious financial sanctions and reputational damage if factors such as exploitative labour practices or environmental damage (including unsustainable harvest) are not corrected. Good examples are the Lacey Act in the USA and the CSDDD in the EU, which came into force on 25 July 2024 (Directive 2024/1760). In Switzerland, the Responsible Business Initiative (RBI) is making a comeback, after getting 50.7% of the popular vote, having been defeated in 2020 at the canton level (Business & Human Rights Resource Centre, 2020). A group of more than 90 human rights, environmental and aid organizations has called on the Swiss government for a new RBI. If they can collect 100,000 valid signatures by 7 July 2026, then people will vote on the new RBI nationally (Business & Human Rights Resource Centre, 2020). This would bring the RBI in step with the CSDDD, which requires large companies to identify and deal with the negative social and environmental impacts of their businesses. This directive affects about 6000 large EU limited liability companies and partnerships (that have >1000 employees and a > €450 million turnover (net) worldwide as well as about 900 large non-EU companies with > €450 million turnover (net) in the EU. The directive may be relevant to some large resin supply chains of flavour and fragrance companies. And while CSDDD legislation does not yet apply to smaller companies, there is no reason why they shouldn't do the right thing too, as this can make business sense in terms of giving an edge over less ethical competitors.

The French company Nexira (originally established by Charles Dondain in 1895, later becoming CNI (Colloides Naturels International) and then Nexira) is a good example. As a medium-sized, family-owned company, CSDDD legislation does not yet apply to Nexira. However, Nexira is a major buyer and processor of gum arabic with a turnover in 2020/21 of €224.3 million and over 200 employees. They chose to participate in the United Nations Global Compact and worked on socially and ecologically responsible programmes in the 1970s, including with SOS Sahel (Nexira, n.d.a). The objective of the third phase of their Acacia Program in Chad, for example, is to work across 200 villages with 50,000 producers to improve the quality of acacia gum and increase productivity from 2600 tons to 5000 t, (Nexira, n.d.b). Other companies can do the same. At a much smaller scale in both north-western Namibia and northern Kenya, sustainable, low-impact harvest of *Boswellia* and/or *Commiphora* resins is done by small companies working within local women's groups in conservancies (Fig. 8.2). In Namibia, Mane SA (France) is involved in a transformed *Commiphora wildii* supply chain, with harvesters getting paid 67% of the wholesale price for the resins they collect (Galloway *et al.*, 2016). Furthermore, value-adding *Commiphora wildii* oil extraction takes place at a local factory in Opuwo, north-western Namibia. In Kenya, Arbor Oils of Africa distils *Boswellia* and *Commiphora* essential oils, paying the women harvesters directly. This is a vast improvement on exporting graded, bulk-bagged gums and resins with minimal value adding.

Challenge 7. Sustainability depends on clear land and tree tenure coupled to harvester rights and responsibilities

In many ways, the term 'natural resource management' is misleading. Ecologically sustainable use is often more about 'people management' than plant populations. This applies across a range of tenure types, from private ownership to common property in communal areas. Customary (or 'traditional') conservation practices across the world are a good example. In those cases, there are good examples where useful plants have been managed for centuries on the basis of local rules and tenure systems. The worst situation in terms of 'people management' is an 'open access' situation, where anybody can access the land or resources on that land. With commercialization, this can stimulate a 'winner takes all' scramble to harvest resources for short-term gain before anyone else can, or when tenure over land and tenure over the trees on that land are separated, particularly when resin or gum tappers who have tenure over the trees have little power to control more powerful interests, such as livestock owners, whose livestock have a major impact on seedlings and saplings (Challenge 2; Figs 8.7 and 8.8)

The factors that can lead to successful common property resource management (Ostrom, 1990) were highlighted in 2009 when the Nobel Prize in economics was awarded to Elinor Ostrom. As village economies shift from subsistence to developing links with external commercial markets, the increasing involvement of opportunistic 'outsiders' and local elites can increase environmental risks. These risks are reduced when:

- each gum or resin harvester group is clearly defined and has clear user rights to their area;
- the local community values the resource and is dependent to some extent on the landscape, so that the majority of community and harvester group(s) are concerned enough to invest time and effort in resource management and monitoring;
- the harvester group, with community back-up, are able to develop and enforce rules related to resource use and management of the area;
- the community and resin resource harvester group(s) have well-developed conflict resolution mechanisms that can mediate cases where there is conflict over gum or resin resource use; and
- the gum or resin harvester group can draw on external technical and institutional support when necessary, pooling knowledge and experience for more effective outcomes in an integrated approach to land and resource management.

Land tenure can be seen as distinct and different from 'tree tenure', which can include ownership, inheritance, planting, management, clearing and produce harvesting rights as well as the right to exclude others (and the livestock that reduce tree regeneration). Simplistic approaches such as formal documentation and land registration may encourage community forest management, but long-term community investment needs stronger security of long-term ownership (Larson *et al.*, 2023).

There are positive examples, even under tough circumstances. In Somalia, for example, 'xeer' customary law can play a significant role and it includes the kinship and social contracts that determine land and tree tenure. Each frankincense (*Boswellia sacra* and *B. frereana*) stand belongs to a clan with a patrilineal line of inheritance. While government records may lack specific delineations, the owners have precise knowledge of tenure arrangements (DeCarlo *et al.*, 2020). Elders select a dedicated custodian committee and district authorities play a significant role in supporting them to uphold 'Xeer'. 'Xeer' includes unwritten conservation rules that have evolved over 200 years to reduce natural resource tenure conflict, exploitation and abuse, as well as manage the frankincense industry. Amongst many beliefs and taboos that effectively support the trees, it is also taboo to allow livestock to graze in these frankincense fields, supported by the trees often being in areas inaccessible to livestock. There is evidence, however, that under different pressures, such as high commercial demand, changing lifestyle patterns and lack of alternative livelihoods, 'xeer' can break down leaving the trees vulnerable to over exploitation and degradation (DeCarlo *et al.*, 2020). Similar changes have occurred in *Shorea javanica* agroforests in Indonesia (Kusters *et al.*, 2008). Proactively integrating clear socially and legally agreed land and tree tenure with local ecological knowledge is more likely to result in longer-term sustainable value chains (Haq *et al.*, 2023; Goshu and Gebre, 2024).

This applies as much in developing countries as elsewhere. In Spain, for example, *c.*1200 known resin collectors (*resineros*) currently have no formal allocation or ownership of the trees in the mostly municipally owned forests of north-west Spain. The national government focus is rather on creating sustainable employment that also results in sustainable forest management (Soliño *et al.*, 2018). Rebuilding the resin industry from collapse in the 1970s and 1980s is seen by the Spanish government not only as a way of repopulating towns that faced depopulation in the resin harvesting area (Perez *et al.*, 2013), but also as a way of sustaining the pine forests. To monitor the supply chain and create visibility for the harvesters, Spanish research and forestry institutions have created ResinApp to help the National Resin Collectors Association, as well as designing ergonomic harvesting tools and providing technical knowledge in order to optimize both resin and timber outcomes for the same trees. A territorial contract for regional administration has also been drafted to enable resin workers to engage in multifunctional forest management for secure annual income and bring the industry into mainstream EU policy and funding (CESEFOR, n.d.).

Changes in government tenure policies and implementation can also reduce supply. In Eritrea, *Boswellia papyrifera* tree populations and regeneration declined due to land clearance for agriculture, the intensity of tapping reducing viable seed production and unregulated grazing. Additionally, smallholders were banned from tapping or benefitting from *Boswellia papyrifera* trees as the central government handed out tapping rights to concessionaires. Furthermore, the traditional village or collective land ownership 'dessa' system that requires the redistribution of arable land amongst villagers every 5–7 years provides almost no incentive for farmers to carry out permanent improvement of the land. Lack of tree tenure was cited as the cause of villagers felling the trees on their allocated lands and allowing livestock to graze freely in the state-owned woodlands, leading to no regeneration except in areas inaccessible to livestock and tappers (Ogbazghi, 2001; Ogbazghi *et al.*, 2006).

Challenge 8. Armed conflicts and poor security in some gum and resin source areas

Security and conflict issues often play a major role in determining the social and environmental outcomes of gum and resin supply chains (Johnson and Bongers, 2024). While companies are often loath to admit it due to the negative perception of 'conflict materials' (such as 'conflict minerals'), security challenges fundamentally shape how they can and do source materials. In some cases, war may completely cut off trade routes, such as the traditional trade in frankincense and myrrh from Somalia to Yemen (DeCarlo *et al.*, 2020); in others, trade continues, but with a variety of impacts that make achieving sustainable extraction of botanical resources difficult or impossible.

Violent conflict often severely degrades traceability and transparency within a supply chain. While lack of transparency is a frequent feature of non-timber forest product (NTFP) supply chains (Challenge 7), conflict makes things worse. International personnel are often restricted from visiting the sourcing areas or countries. For instance, company staff may rarely visit Somalia or Sudan where frankincense, myrrh and gum arabic are produced, making it difficult to conduct supplier audits or work with local personnel to improve practices.

Collection of gums and resins may also be reduced if the armed conflict excludes collectors from harvest sites (Adam *et al.*, 2004; Burgess *et al.*, 2015). This restricts the amount of material on the market, forcing companies who use it to accept material wherever they can get it. Furthermore, gums and resins can become a source of revenue for armed parties, either by the direct takeover of trade or indirectly through 'taxes' on local populations. While the use of natural resources by armed actors typically focuses on very high-value materials such as ivory (Beyers *et al.*, 2011), rhino horn (Anagnostou and Doberstein, 2022) or cocaine (Clerici *et al.*, 2020), it is opportunistic and can also encompass broader forest revenues (Baral and Heinen, 2005) or resins such as frankincense (Johnson and Bongers, 2024). Accidentally funding (directly or indirectly) armed conflict is clearly a negative, and it is difficult to imagine armed groups taking a more sustainability- and human rights-oriented approach than the average harvester. The impacts of armed groups and conflict on conservation are very rarely positive (Gaynor *et al.*, 2016). In addition to ethical considerations, there are very real regulatory concerns for companies sourcing from conflict areas. For example, many importing countries have laws against illegal harvesting of forest materials (e.g. USA-based Lacey Act), corruption (UK

Bribery Act or US Foreign Corrupt Practices Act) or child and forced labour (US Section 1307, EU Due Diligence Act). The involvement of conflict and degraded traceability exposes sourcing companies to increased regulatory risk (Cunningham and DeCarlo, 2022) and potential negative publicity (Hassan, 2022), with limited options to effectively reduce these risks.

Beyond the immediate difficulties of sustainable sourcing, armed conflicts have multi-dimensional and often far-reaching impacts on ecosystems and wild harvesting systems. Gum- and resin-producing species and their ecosystems are often damaged either unintentionally as a by-product of bombing, artillery, etc. (Henig, 2012; Daiyoub *et al.*, 2023) or destroyed intentionally to deny resources to enemy combatants (Negash and Birhane, 2023; Johnson and Bongers, 2024). Fighting often results in massive displacement of people, potentially both excluding knowledgeable harvesters (Burgess *et al.*, 2015) and introducing refugees who will be dependent on forest resources for their basic needs (McNeely, 2003; Langer *et al.*, 2015). This process weakens land tenure, reduces incentives for long-term sustainable management and degrades both traditional and formal governance systems that would normally prevent destructive practices. Even when acute violent conflict ends, the impacts of this knowledge and governance loss can continue, particularly as a result of psychological trauma sustained by local people throughout the conflict (Anckermann *et al.*, 2005; Boege and Franks, 2011).

The impacts of armed conflict are clearly profound and wide-ranging. Resolving these conflicts is beyond the reach of commercial supply chains, and in many cases, conflicts continue for decades. Even when active fighting ceases, there are often lasting effects on the environment, culture and psychology of local people that can make creating or continuing sustainable systems difficult. Where conflict impacts are present, much greater due diligence and supply chain investment are necessary to avoid ecosystem degradation, exploitation or corruption, if engagement is possible at all.

Challenge 9. Policy failures, complex realities and the policy–implementation gap

As the preceding challenges outlined in this chapter describe, the real world of gum and resin wild harvests is a complex one, far more complex than the ideal conditions where policies were developed. It should come as no surprise therefore, that policies face unexpected challenges and may have unintended consequences in these complex systems (Fig. 8.16). Hudson *et al.* (2019) and Mueller (2020) describe the many reasons for policy failure or a lack of implementation. Many of these will be familiar to people involved in gum and resin supply chains: porous borders, corruption and political power plays, a lack of coordination between different government departments or a lack of capacity or even political will to implement policy. The bottom line is that policies most commonly fail when, as Mueller (2020, p. 2) puts it, 'complex problems are addressed using standard linear and reductionist approaches that presuppose more knowledge and control than is ever possible in such situations'.

Without a full understanding of the threats to gum and resin species populations, combined with action at local, national and international scales, the risk of failure is high. What is required is a nuanced understanding of complexities and local realities (Johnson *et al.*, 2023). These apply at local, national and international levels. There is simply not enough space here to synthesize the local, national and regional challenges that can lead to implementation failure across the many countries that are gum and resin sources. What can be done is to highlight a few examples of lessons that can be learned from case studies carried out at these different scales.

Local level: Participatory research and economic insights from ethiopia

Excellent applied research in Ethiopia provides examples of what needs to be done in other places. The best land-use options for sustainable use of *Boswellia papyrifera* were developed through Gidey *et al.* (2020) working with local stakeholders to assess the different land-use options for woodland management in Tigray (in

Fig. 8.16. Postharvest grading, sorting and bulk packing are all labour-intensive processes in Africa, Asia and the Arabian Peninsula. A. Sorting in Somaliland and Somalia is generally carried out by women. B. Breaking up large lumps of *Boswellia sacra* frankincense is done by hand (Salalah, Oman). C. A male migrant worker from Bangladesh in Oman busy sieving and grading *B. sacra* frankincense. D. High-quality, naturally exuded black dammar (*Canarium strictum*) resin, showing scale in centimetres (Nilgiri Hills, India). E. Large lumps of high-grade gum arabic from *Senegalia senegal* in contrast to small, less valuable gum from the same species (Burkina Faso). F. Packing into standard size bags for wholesale is common practice with many gums and resins. In this case, 25-kg bags of gum arabic (*Senegalia senegal*) at a warehouse in Burkina Faso. (Photos: U. Feiter (A), A.B. Cunningham (B-F)). Used with permission.

an area with a 'low' *Boswellia papyrifera* density of 266 trees/ha). The best option nominated to prevent poor management (due to overgrazing, agricultural expansion, over-tapping and the associated wood-boring beetle attack) was 'Area Exclosure with Medium Tapping Resting period' (AEMTR). The next best option was 'Area Exclosure with Long Rapping Resting period' (AELTR) (Gidey *et al.*, 2020). Using contingent valuation methods, Gelaye *et al.* (2011) and Gelaye (2012) also showed that rural households in Ethiopia were willing to pay for and contribute labour towards *B. papyrifera* woodlands. What is important is that the consultative processes and valuation methods pioneered for *B. papyrifera* woodlands by Gelaye *et al.* (2011), Gidey *et al.* (2020) and Dejene *et al.* (2013) should be repeated in other landscapes. In Oman, Somalia and Yemen, for example, similar processes could be followed on land-use options that deal with a combination of frankincense production, beekeeping and improved camel management (Cunningham and DeCarlo, 2022).

National policies, power and permits

This topic is a book in itself and a brief overview must suffice here. A useful lens through which to view policy weaknesses is Médard (1992) 'Big Man' analysis that deals with power, politics and African development, where he describes the 'political entrepreneur' who merges the roles of politician, government official and businessman. This lens was certainly very useful in understanding challenges in the case of a gum producing species better known for its medicinal bark: *Prunus africana* (Cunningham *et al.*, 2016). The power dynamics and corruption context that (Ingram, 2014, p. 221,) eloquently described for *P. africana* may resonate with harvesters and middle-level of gums or resins in other places too:

> Corruption feeds off and thwarts the statutory permitting procedures controlling access to the resource, monitoring and sanctioning. The high level of regulation of this high-value trade and the small number of harvest areas, channels and actors through which the trade flows provide a lucrative opportunity for corrupt officials. This then further exacerbates regulatory and customary failures. Corruption aggravates the challenges actors have to operate legally, promoting rather informality and illegality as a (slightly) easier and apparently profitable way of doing business. Corruption thus becomes an insidious, predictable but incalculable governance arrangement and may continue to be so under the revised management plan, as unprecedented power is channelled via government officials controlling inventories rather than participative inventories.

Unlike local- or regional-level bulking and trading agents, harvesters in remote areas face logistic difficulties in obtaining the licences from Forest Departments. In particular, journey-specific transport licence offices can be far from the harvesting site. With little bargaining power, harvesters need to then rely on a fair and inclusive local cooperative, agent or middleman. The reality is something different, whether the supply chains are charcoal (Kambewa *et al.*, 2007), bark (Cunningham *et al.*, 2016) or gums and resins. There is one important distinction: gum and resin harvesting, if done right, has a low impact (and none if gums and resins are hand-picked). In Kenya, the intention of a national-level licence was to enable those transporting bulk resins to cross county access points without further payments.

The reality was different. Consequently, discussions are taking place to encourage policy makers to rework the legislation towards a more inclusive, adaptive approach before the regulations affecting gum and resin harvest are gazetted nationally.

International policy: Complexities and CITES in the Boswellia case

While Chapter 10 in this book focuses on four species that are sources of aromatic oils and oleoresins, the case of frankincense illustrates the need to understand complex realities (Johnson *et al.*, 2024) and avoid 'unintended consequences'. As mentioned at the start of this chapter, there is compelling evidence that *Boswellia papyrifera* is in decline Ethiopia, Eritrea and Sudan (Bongers *et al.*, 2019). Under 'normal' circumstances, if *B. papyrifera* or the genus *Boswellia* were listed on CITES Appendix II, international trade would still continue but would be monitored through data in export permits or re-export certificates granted by national CITES authorities. But the governance situation in several countries where *Boswellia* occurs is far from normal, and instead of continued trade, a CITES Appendix II listing for *Boswellia* would result in CITES-imposed trade suspensions for both Djibouti (CITES Notifications No. 2011/010 and No. 2018/015) and Somalia (CITES Notification No. 2019/35), as neither country has fulfilled its obligations of regular reporting to the CITES Secratatriat for decades. A trade ban triggered by CITES would also effectively block Ethiopian *B. papyrifera* oleoresin exports through Djibouti, risking significant unintended consequences, including livelihood implications, in a region affected by both conflict and drought, and a massive decline in market-related incentives to manage trees for frankincense. This, in turn, would risk that that they would be converted to charcoal instead, as the value of trees would plummet following a CITES-triggered trade suspension. In this complex situation, we suggest that in fragile states, positive outcomes may be more likely from voluntary approaches by ethical companies, highlighting the need for a nuanced approach and, as Cooney *et al.* (2021) point out, to 'think before you act'.

Challenge 10. Lack of a systems approach that considers the whole supply chain and coordination in terms of local, national and international policies to achieve ecologically sustainable use

Ecologically sustainable harvest cannot be assumed, particularly with commercial harvest of high-value resources where land or resource tenure is weak and the risks of 'elite capture' are high. In theory, sustainable harvesting of plants from wild populations is possible, but as the 10 challenges describe, it is often more complex than commercial companies realize. As is often the case, the devil is in the details. Sustainable management of wild plant use depends as much upon an understanding of the biological component as it does on the social and economic aspects of wild plant use (Fig. 8.17). Without an understanding of ecological, political and socio-economic factors and coordination at different scales, plans for sustainable use are likely to fail, making things worse for local people and the wild plant resources they use. Inadequate monitoring does not help. What is needed is practical, informed planning based on resource assessments and managed harvests in response to market demand for clean, green natural products. The overarching principle, as in medical ethics, is 'first do no harm'.

Conclusions

On the surface, the commercial demand vs supply equation seems simple. In reality, maintaining supplies of wild harvested gums and resins through ecologically sustainable use is challenging. Dealing with the 10 challenges in this chapter not only avoids the risk

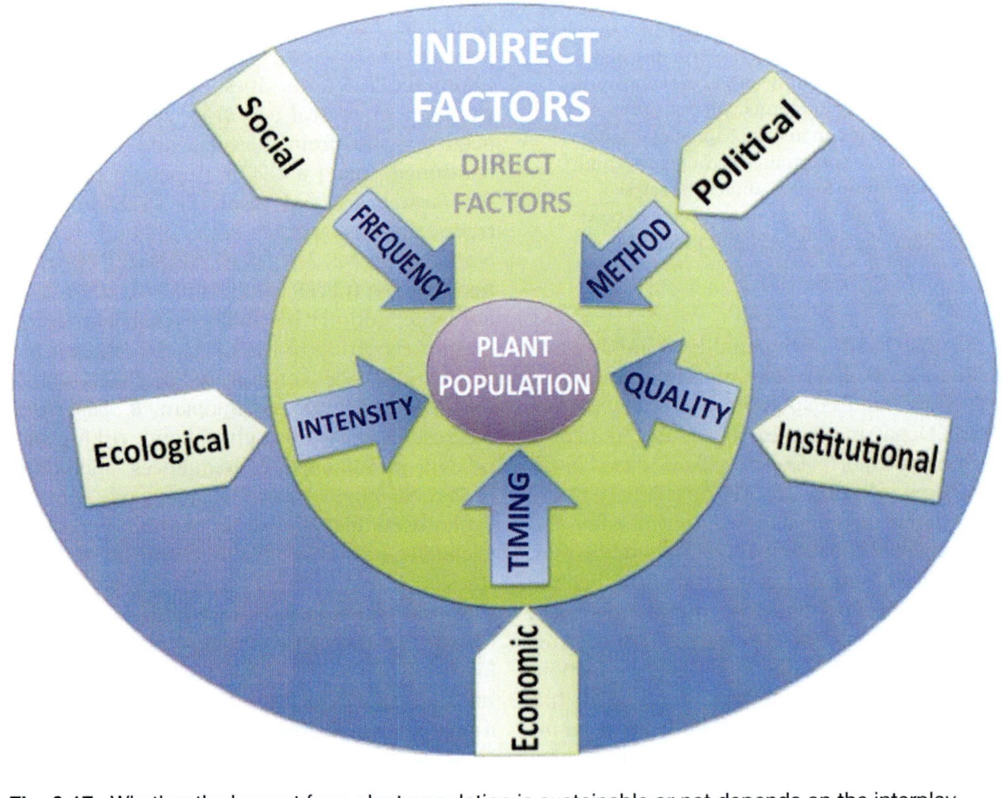

Fig. 8.17. Whether the harvest from plant population is sustainable or not depends on the interplay between a variety of indirect and direct factors that influence harvest. Modified and redrawn from Lawrence *et al*. (2008).

of 'resource mining' rather than resource management, it also helps avoid reputational risk in a world where consumers are increasingly aware of unethical practices, ecologically unsustainable supply chains and 'green washing'. Improved traceability and better monitoring now are preferable to having the 'benefit' of hindsight in the midst of a devastated resource. So let us take precautions to 'do no harm', learning from history to avoid the next Cyrenaic silphion. And to the consumers or organizations who read this chapter and have any concerns about sustainable use, remember Edward Abbey's saying: 'Sentiment without action is the ruin of the soul' (Abbey, 2015). Ask yourselves whether what you see is to the long-term benefit of people and wild plant populations? If not, become better informed and learn what is required. Then take action on each of these 10 challenges in the most strategic and best-informed ways you can. And don't lose hope.

Acknowledgements

Without the support of many people over many decades across Africa, Asia and the Middle East, this chapter would not have been possible. Too many to thank by name here, but know that the lessons we learned have been invaluable, particularly in the field, away from desktop computers. Barbara Seele is thanked for her edits, as are Sue Canney Davison and Stephen Johnson, who also contributed in other ways as well. The photographers who contributed are acknowledged after each Figure.

References

Abbey, E. (2015) *A voice crying in the wilderness: vox clamantis in deserto: Notes from A secret journal*. Rosetta Books, New York.

Abdulah, L., Susanti, R., Rahajoe, J.S., Atikah, T.D., Subarudi, S. *et al*. (2022) Feasibility of agarwood cultivation in indonesia: dynamic system modelling approach. *Forests* 13(11), 1869.

Abtew, A.A., Pretzsch, J., Secco, L. and Mohamod, T.E. (2014) Contribution of small-scale gum and resin commercialization to local livelihood and rural economic development in the drylands of eastern Africa. *Forests* 5(5), 952–977. DOI: 10.3390/f5050952.

Adam, A.A., Ahmed, A.E. and El Tayeb, A.M. (2004) Phenology of Boswellia papyrifera (Del.) Hochst (olibanum gum tree) in Jebel Marra, Darfur, Sudan. *Sudan Silva* 10(1), 72–80.

Adolt, R. and Pavliš, J. (2004) Age structure and growth of *Dracaena cinnabari* populations on Socotra. *Trees Structure and Function* 18(1), 43–53.

Agrawal, A.A. (2011) Current trends in the evolutionary ecology of plant defence. *Functional Ecology* 25, 421–433.

Al-Okaishi, A. (2020) Local management system of dragon's blood tree (*Dracaena cinnabari* Balf. F.) resin in Firmihin Forest, Socotra Island, Yemen. *Forests* 11(4), 389.

Anagnostou, M. and Doberstein, B. (2022) Illegal wildlife trade and other organised crime: A scoping review. *Ambio* 51, 1615–1631.

Anckermann, S., Dominguez, M., Soto, N., Kjaerulf, F., Berliner, P. *et al*. (2005) Psycho-social support to large numbers of traumatized people in post-conflict societies: An approach to community development in Guatemala. *Journal of Community & Applied Social Psychology* 15(2), 136–152.

Ash, A. (2020) First-grade agarwood can cost as much as $100,000 per kilogram. Why is it so expensive? *Business Insider,* Online. Available at: https://www. businessinsider.com/why-agarwood-is-so-expensive-oud-vietnam-2020-8 (accessed 6 February 2025).

Atal, C., Gupta, O. and Afaq, S. (1975) *Commiphora mukul*: Source of guggal in Indian systems of medicine. *Economic Botany* 29(3), 209–218.

Baral, N. and Heinen, J.T. (2005) The Maoist people's war and conservation in Nepal. *Politics and the Life Sciences* 24(1–2), 2–11.

Barstow, M. (2021) *Aniba rosodora*. The IUCN Red List of Threatened Species 2021: e.T33958A88301381. Available at: https://dx.doi.org/10.2305/IUCN.UK.2021-3.RLTS.T33958A88301381.en (accessed 2 February 2025).

Berhanu, Y., Vedeld, P., Angassa, A. and Aune, J.B. (2021) The contribution of frankincense to the agro-pastoral household economy and its potential for commercialization - a case from borana, Southern Ethiopia. *Journal of Arid Environments* 186, 104423.

Beyers, R.L., Hart, J.A., Sinclair, A.R., Grossmann, F., Klinkenberg, B. *et al.* (2011) Resource wars and conflict ivory: The impact of civil conflict on elephants in the Democratic Republic of Congo - the case of the Okapi Reserve. *PloS One* 6(11), e27129. DOI: 10.1371/journal.pone.0027129.

BIFAD (2019) *Board for International Food and Agricultural Development 180th Public Meeting. Meeting Minutes, Des Moines Downtown Marriott, Des Moines,Iowa, Tuesday*. USAID, Washington, DC.

Bode, M. (2006) Taking traditional knowledge to the market: The commoditization of indian medicine. *Anthropology & Medicine* 13(3), 225–236.

Boege, V. and Franks, D.M. (2011) Reopening and developing mines in post-conflict settings: The challenge of company-community relations. In: Lujala, P. and Rustad, S.A. (eds) *High-Value Natural Resources and Post-Conflict Peacebuilding*. Routledge, Abingdon, UK, pp. 87–120.

Bongers, F., Groenendijk, P., Bekele, T., Birhane, E., Damtew, A. *et al.* (2019) Frankincense in peril. *Nature Sustainability* 2, 602–610.

Boonman, C.C., Serra-Diaz, J.M., Hoeks, S., Guo, W.Y., Enquist, B.J. *et al.* (2024) More than 17,000 tree species are at risk from rapid global change. *Nature Communications* 15(1), 166.

Burgess, R., Miguel, E. and Stanton, C. (2015) War and deforestation in Sierra Leone. *Environmental Research Letters* 10(9), 095014.

Business & Human Rights Resource Centre (2020) Switzerland: Responsible Business Initiative Rejected at Ballot Box Despite Gaining 50.7% of Popular Vote. Available at: https://www.business-humanrights.org/en/latest-news/swiss-due-diligence-initiative-set-for-public-referendum-as-parliament-only-opts-for-reporting-centred-proposal/ (accessed 4 May 2025).

CESEFOR (n.d.) Guide to the Application of the Territorial Contract for Resin Tappers, an Innovative Institutional Support Formula for the Natural Resin Sector. Fundación Centro de Servicios y Promoción Forestal y de su Industria de Castilla y León (CESEFOR), Soria, Spain. Available at: https://oppla.eu/casestudy/20642 (accessed 9 February 2025).

Challender, D.W., Brockington, D., Hinsley, A., Hoffmann, M., Kolby, J.E. *et al.* (2022) Mischaracterizing wildlife trade and its impacts may mislead policy processes. *Conservation Letters* 15(1), e12832.

Clerici, N., Armenteras, D., Kareiva, P., Botero, R., Ramirez-Delgado, J.P. *et al.* (2020) Deforestation in Colombian protected areas increased during post-conflict periods. *Scientific Reports* 10, 4971.

Cooney, R., Challender, D.W., Broad, S., Roe, D. and Natusch, D.J. (2021) Think before you act: Improving the conservation outcomes of CITES listing decisions. *Frontiers in Ecology and Evolution* 9, 236. DOI: 10.3389/fevo.2021.631556.

Corporate Accountability Lab (2024) What Price for the King of Oils?: Exploitation and Abuse in doTERRA's Frankincense Supply Chain. Available at: https://corpaccountabilitylab.org/somaliland-report (accessed 5 February 2025).

Cunningham, A.B. (2001) *Applied Ethnobotany: People, Wild Plant Use and Conservation*. Earthscan, London.

Cunningham, A.B. and DeCarlo, A. (2022) Boswellia *Species in International Trade: Identification, Supply Chains, & Practical Management Considerations*. CITES Secretariat, Geneva, Switzerland.

Cunningham, A.B., Anoncho, V.F. and Sunderland, T. (2016) Power, policy and the *Prunus africana* bark trade, 1972–2015. *Journal of Ethnopharmacology* 178, 323–333.

Cunningham, A.B., Brinckmann, J.A., Kulloli, R.N. and Schippmann, U. (2018) Rising trade, declining stocks: The global gugul (*Commiphora wightii*) trade. *Journal of Ethnopharmacology* 223, 22–32. DOI: 10.1016/j.jep.2018.04.040.

Daiyoub, A., Gelabert, P., Saura-Mas, S. and Vega-Garcia, C. (2023) War and deforestation: Using remote sensing and machine learning to identify the war-induced deforestation in Syria 2010–2019. *Land* 12(8), 1509.

DeCarlo, A., Ali, S. and Ceroni, M. (2020) Ecological and economic sustainability of non-timber forest products in post-conflict recovery: A case study of the frankincense (*Boswellia* spp.) resin harvesting in somaliland (Somalia). *Sustainability* 12(9), 3578. DOI: 10.3390/su12093578.

Dejene, T., Lemenih, M. and Bongers, F. (2013) Manage or convert boswellia woodlands? Can frankincense production payoff? *Journal of Arid Environments* 89, 77–83.

dōTERRA (2019) Frankincense: A Co-Impact Sourcing Story. Available at: https://sourcetoyou.com/assets/downloads/frankincenseco-impact-storybrochureusv1singlepagewebversion-3-1.pdf (accessed 5 February 2025).

dsm-firmenich (n.d.) People, Planet, Progress and Ingredients. Available at: https://sustainability.dsm-firmenich.com/en/sustainability.html and https://www.dsm-firmenich.com/en/businesses/perfumery-beauty/ingredients.html (accessed 5 February 2025).

Ehrlich, P.R. and Raven, P.H. (1964) Butterflies and plants: A study in coevolution. *Evolution* 18(4), 586–608. DOI: 10.1111/j.1558-5646.1964.tb01674.x.

Elmqvist, B., Olsson, L., Mirghani Elamin, E. and Warren, A. (2005) A traditional agroforestry system under threat: An analysis of the gum arabic market and cultivation in the Sudan. *Agroforestry Systems* 64, 211–218.

Esseterre (n.d.) Sustainability, Responsibility, Competence. Esseterre, Dobrich, Bulgaria. Available at: https://esseterre.bg/en/activity/#ecology (accessed 5 February 2025).

FairWild Foundation (2023) FairWild Standard: Version 3.0. FairWild Foundation, Zurich, Switzerland. Available at: https://www.fairwild.org/news/launch-fw-standard-v3 (accessed 9 May 2025).

FAO (2016) Somalia Livelihood Profiles. East Golis Frankincense, Goats and Fishing (Zone S007), June 2016. Food Security and Nutrition Analysis Unit (FSNAU) and Famine Early Warning Systems Network (FEWSNET), pp. 91–102. Available at: https://fsnau.org/downloads/Somalia-Livelihood-Profiles-30-June-2016.pdf (accessed 9 May 2025).

Farah, M.H. (2008) Non-timber forest product (NTFP) extraction in arid environments: Land-use change, frankincense production and the sustainability of *Boswellia sacra* in Dhofar (Oman). PhD thesis, University of Arizona, Tucson, Arizona. Available at: https://api.semanticscholar.org/CorpusID:128809020 (accessed 5 February 2025).

Fox, M.W. (1999) India's sacred cow: Her plight and future. *Animal Issues* 3, 1–38.

Fuller, R.J. (2010) Beyond cliché – reclaiming the concept of sustainability. *Australian Journal of Environmental Education* 26, 7–18.

Galloway, F.B., Wynberg, R.P. and Nott, K. (2016) Commercialising a perfume plant, *Commiphora wildii*: Livelihood implications for indigenous Himba in north-west Namibia. *International Forestry Review* 18(4), 429–443.

Gaynor, K.M., Fiorella, K.J., Gregory, G.H., Kurz, D.J., Seto, K.L. *et al.* (2016) War and wildlife: Linking armed conflict to conservation. *Frontiers in Ecology and the Environment* 14(10), 533–542. DOI: 10.1002/fee.1433.

Gebrehiwot, K. (2003) Ecology and management of *Boswellia papyrifera* (Del.) Hochst. dry forests in Tigray, northern Ethiopia. Doctoral dissertation, University of Gottingen, Gottingen, Germany.

Gelaye, M.T. (2012) Restoration and sustainable management of frankincense forests in Ethiopia: A bio-economic analysis. PhD thesis, Katholieke Universiteit Leuven, Belgium.

Gelaye, M., Mathijs, E., Muys, B., Vranken, L., Deckers, J.A. *et al.* (2011) Valuing dry land forest ecosystem services: A case of rural households' willingness to pay and contribute labor for frankincense forest conservation in Ethiopia. In: *Workshop 6: Case Studies on the Practical Use of Ecosystem Services in Planning, Management and Capacity Building*. Ecosystem Services Partnership, Wageningen, Netherlands.

Gidey, T., Hagos, D., Juhar, H.M., Solomon, N., Negussie, A. *et al.* (2020) Population status of *Boswellia papyrifera* woodland and prioritizing its conservation interventions using multi-criteria decision model in northern Ethiopia. *Heliyon* 6(10), e05139.

Givaudan (n.d.) Transparency across Supply Chains with Sourcing4Good. Available at: https://www.givaudan.com/sustainability/communities/sourcing4good (accessed 5 February 2025).

Goshu, T. and Gebre, A. (2024) Privatisation of the commons: Land ownership, access, and exclusion among the Karrayu pastoralists of eastern Ethiopia. *Journal of Contemporary African Studies* 42(3), 371–387. DOI: 10.1080/02589001.2024.2346658.

Government of Assam (2021) The assam agarwood promotion policy 2020. Notification 18 december 2020. *The Assam Gazette, No. 25, Dipur, 11 January 2021*. Available at: https://industries.assam.gov.in/sites/default/files/swf_utility_folder/departments/industries_com_oid_4/portlet/level_2/gazette_notification_assam_agarwood_promotion_policy_2020.pdf (accessed 8 May 2025).

Habebo, T.T.W. (2011) Dryland resources, livelihoods and institutions: Diversity and dynamics in use and management of gum and resin trees in Ethiopia. PhD thesis, Wageningen University, Wageningen, Netherlands. DOI: 10.18174/188209.

Haq, S.M., Pieroni, A., Bussmann, R.W., Abd-ElGawad, A.M. and El-Ansary, H.O. (2023) Integrating traditional ecological knowledge into habitat restoration: Implications for meeting forest restoration challenges. *Journal of Ethnobiology and Ethnomedicine* 19(1), 33. DOI: 10.1186/s13002-023-00606-3.

Hassan, H. (2022) The price of purity. *VICE News* (Season 3, Episode 4) [Documentary].

Henig, D. (2012) Iron in the soil: Living with military waste in Bosnia-Herzegovina. *Anthropology Today* 28(1), 21–23.

Herms, D.A. and Mattson, W.J. (1992) The dilemma of plants: To grow or defend. *The Quarterly Review of Biology* 67, 283–335.

Homma, A. (1992) The dynamics of extraction in Amazonia: A historical perspective. *Advances in Economic Botany* 9, 23–31.

Homroy, S. and Rauf, A. (2024) Green washing in supply chains? Available at: https://ssrn.com/abstract=4446166 (accessed 5 February 2025).

Hoskins, M. (1990) *The Community Toolbox: The Idea, Methods and Tools for Participatory Assessment, Monitoring and Evaluation in Community Forestry*. FAO Community Forestry Field Manual 2, Food and Agriculture Organization (FAO), Rome. Available at: https://www.fao.org/4/x5307e/x5307e00.htm (accessed 5 February 2025).

Hudson, B., Hunter, D. and Peckham, S. (2019) Policy failure and the policy-implementation gap: Can policy support programs help? *Policy Design and Practice* 2, 1–14. Available at: https://doi.org/10.1080/25741292.2018.1540378

Inês, A., Diniz, A. and Moreira, A.C. (2023) A review of greenwashing and supply chain management: Challenges ahead. *Cleaner Environmental Systems* 11(8), 100136. DOI: 10.1016/j.indcrop.2023.117105.

Ingram, V. (2014) Win-wins in NTFP value chains: How governance impacts the sustainability of livelihoods based on Congo Basin forest products. PhD thesis, Amsterdam Institute for Social Science Research, Governance and Inclusive Development Group, University of Amsterdam, Amsterdam.

Jales, M. (2018) *Commodities at a glance: special issue on gum arabic*. UNCTAD, Geneva, Switzerland.

Janzen, D.H. (1975) Behavior of *Hymenaea courbaril* when its predispersal seed predator is absent. *Science* 189, 145–147.

Johnson, E. (2022) On the Hunt for the Mystery Herb. Royal Botanic Gardens, Kew, UK. Available at: https://www.kew.org/read-and-watch/silphium-mystery (accessed 5 February 2025).

Johnson, M.T. (2011) Evolutionary ecology of plant defences against herbivores. *Functional Ecology* 25(2), 305–311.

Johnson, S. and Bongers, F. (2024) Armed conflict as an underappreciated driver of conservation outcomes in frankincense (*Boswellia* spp.). *Trees, Forests and People* 15(2), 100490. DOI: 10.1016/j.tfp.2023.100490.

Johnson, S., DeCarlo, A., Satyal, P., Dosoky, N., Sorensen, A. *et al*. (2019) Organic certification is not enough: The case of the methoxydecane frankincense. *Plants* 8(4). Available at: https://doi.org/10.3390/plants8040088

Johnson, S., DeCarlo, A., Bongers, F. and Cunningham, A.B. (2023) Sustaining the global frankincense trade through CITES: Governance challenges and complexities. In: Segger, M.C.C., Wardell, D.A. and Harrington, A. (eds) *CITES as a Tool for Sustainable Development*. Cambridge University Press, Cambridge, UK, pp. 363–383.

Johnson, S., Davison, S.C., Ablard, K., Bongers, F., Cunningham, A.B. *et al*. (2024) Good intentions meet complex realities: CITES listing of diverse frankincense (*Boswellia species*) might do more harm than good. *Forest Policy and Economics* 163, 103207.

Kambewa, P.S., Mataya, B.F., Sichinga, W.K. and Johnson, T.R. (2007) *Charcoal - The Reality: A Study of Charcoal Consumption, Trade and Production in Malawi*. Technical Report to Forest Governance Learning Group, COMPASS II, USAID, Lilongwe, Malawi.

Kazora, J., Zhu, W., Kyaw, T.O., Sebaziga, J.N., Rusanganwa, F. *et al*. (2023) Enhancement of East African monsoon long rainfall (March to May) variability from weekly to annual scale by climatic extremes. *Atmospheric and Climate Sciences* 13(4), 491–506.

Kiehn, M. (2007) Silphion revisited. *Medicinal Plant Conservation* 13, 4–8.

Kiss, K. (2011) Rise and fall of the concept sustainability. *Journal of Environmental Sustainability* 1(1). DOI: 10.14448/jes.01.0001.

Knight, A.T., Cowling, R.M., Rouget, M., Balmford, A., Lombard, A.T. *et al*. (2008) Knowing but not doing: Selecting priority conservation areas and the research–implementation gap. *Conservation Biology* 22(3), 610–617. DOI: 10.1111/j.1523-1739.2008.00914.x.

Kusters, K., Ruiz Perez, M., Foresta, H., Dietz, T., Ros-Tonen, M. *et al.* (2008) Will agroforests vanish? The case of Damar agroforests in Indonesia. *Human Ecology* 36, 357–370.

La Montagna, D., Attorre, F., Hamdiah, S., Maděra, P., Malatesta, L. *et al.* (2023) Climate change effects on the potential distribution of the endemic *Commiphora* species (Burseraceae) on the island of Socotra. *Frontiers in Forests and Global Change* 6, 1183858.

Langenheim, J.H. (2003) *Plant Resins: Chemistry, Evolution, Ecology, and Ethnobotany* (No.620.1924 L275p). Timber Press, Portland, Oregon.

Langer, S., Tiede, D. and Lüthje, F. (2015) Long-term monitoring of the environmental impact of a refugee camp based on landsat time series: The example of deforestation and reforestation during the whole lifespan of the Camp Lukole, Tanzania. *GI_Forum - Journal for Geographic Information Science* 1.

Larson, A.M., Monterosso, I., Liswanti, N. and Tamara, A. (2023) What is forest tenure (in) security? Insights from participatory perspective analysis. *Forest Policy and Economics* 147, 102880. DOI: 10.1016/j.forpol.2022.102880.

Lawrence, A., Kinhal, G., Luintel, H., Molteno, S. and Gillett, S. (2008) *Participatory Science for Sustainable Wild Harvests: a Methods Handbook*. Environmental Change Institute (Oxford).

Leakey, R.R.B., Atangana, A.R., Kengni, E., Warhuhui, A.N., Usoro, C. *et al.* (2002) Domestication of dacryodes edulis in west and central Africa: Characteristics of genetic variation. *Forest, Trees and Livelihoods* 12, 57–71.

Lemenih, M., Arts, B., Wiersum, K. and Bongers, F. (2014) Modelling the future of *Boswellia papyrifera* population and its frankincense production. *Journal of Arid Environments* 105, 33–40.

Lemenih, M. and Kassa, H. (eds) (2011a) *Opportunities and Challenges for Sustainable Production and Marketing of Gums and Resins in Ethiopia*. CIFOR, Bogor, Indonesia.

Lemenih, M. and Kassa, H. (eds) (2011b) *Management Guide for Sustainable Production of Frankincense*. CIFOR, Bogor, Indonesia.

Leopold, A. (1949) *A Sand County Almanac: And Sketches Here and There*. Oxford University Press, Oxford, UK.

Liu, Y., Wang, Z., Zhao, F., Zeng, M., Li, F. *et al.* (2022) Efficient resin production using stimulant pastes in *Pinus elliottii* × *P. caribaea* families. *Scientific Reports* 12(1), 13129.

Lönngren, J. and Van Poeck, K. (2021) Wicked problems: A mapping review of the literature. *International Journal of Sustainable Development & World Ecology* 28(6), 481–502.

Mabberley, D.J. (2008) *Mabberley's Plant-Book: A Portable Dictionary of Plants, their Classification and Uses*. Cambridge University Press, Cambridge, UK.

Maděra, P., Vahalík, P., Hamdiah, S., Hušková, K., Sekava, J. *et al.* (2024) Distribution, ecology, and threats assessment of 11 endemic frankincense tree taxa (*Boswellia*) in the Socotra Archipelago (Yemen). *Plants, People, Planet* 6(6), 1552–1571.

Mane (n.d.) Purchasing: Sustainability. Available at: https://www.mane.com/sustainability/purchasing (accessed 5 February 2025).

McNeely, J.A. (2003) Conserving forest biodiversity in times of violent conflict. *Oryx* 37(2), 142–152.

Médard, J.F. (1992) Le 'Big man' en Afrique: Esquisse d'analyse du politicien entrepreneur. *L'Année Sociologique* 42, 167–192.

Michon, G., Foresta, H., Kusworo, A. and Levang, P. (2000) The damar agroforests of Krui: Justice for forest farmers. In: Zerner, C. (ed.) *People, Plants and Justice*. Columbia University Press, New York, pp. 159–203.

Miller, A.G. and Morris, M. (2004) *Ethnoflora of the Soqotra Archipelago*. Royal Botanic Garden Edinburgh, Edinburgh, UK.

Mithöfer, A. and Boland, W. (2012) Plant defense against herbivores: Chemical aspects. *Annual Review of Plant Biology* 63(1), 431–450.

Moore, G., Smith, G.F., Figueiredo, E., Demissew, S., Lewis, G. *et al.* (2010) Acacia, the 2011 Nomenclature Section in Melbourne, and beyond. *Taxon* 59(4), 1188–1195.

Mueller, B. (2020) Why public policies fail: Policymaking under complexity. *EconomiA* 21(2), 311–323.

Negash, E. and Birhane, E. (2023) War and siege inflicted damage on Tigray's ecosystem. *Ethiopia Insight*. Available at: https://www.ethiopia-insight.com/2023/02/22/war-and-siege-inflicted-damage-on-tigrays-ecosystem/ (accessed 6 February 2025).

Nexira (n.d.a) Beyond Sustainability. Available at: https://www.nexira.com/commitments/beyond-sustainability (accessed 14 May 2025).

Nexira (n.d.b) Nexira Launches the Third Phase of its Acacia Program to Promote Sustainability in Africa while Significantly Reducing Carbon Footprint. Available at: https://www.nexira.com/promoting-sustainability-reducing-carbon-footprint (accessed 14 May 2025).

Nour, L.A.N. (2002) Production and productivity of *Boswellia papyrifera* in Jebel Elgarrie area (Blue Nile State). MsC thesis, University of Khartoum. Available at: https://core.ac.uk/download/pdf/71672483.pdf (accessed 5 February 2025).

Office of Public Affairs (2017) Essential oils company sentenced for lacey act and endangered species act violations to pay $760,000 in fines, forfeiture, and community service, and to implement a comprehensive compliance plan. Office of Public Affairs, United States Department of Justice. Available at: https://www.justice.gov/opa/pr/essential-oils-company-sentenced-lacey-act-and-endangered-species-act-violations-pay-760000 (accessed 20 January 2025).

Ogbazghi, W. (2001) *The Distribution and Regeneration of Boswellia papyrifera (Del.) Hochst in Eritrea.* Tropical Resource Management Papers, No. 35. Wageningen University and Research Centre, Wageningen, Netherlands.

Ogbazghi, W., Rijkers, T., Wessel, M. and Bongers, F. (2006) Distribution of the frankincense tree *Boswellia papyrifera* in Eritrea: The role of environment and land use. *Journal of Biogeography* 33(3), 524–535.

Orr, D.W. (1991) *Ecological Literacy: Education and the Transition to a Postmodern World.* State University of New York Press, New York.

Ostrom, E. (1990) *Governing the Commons: The Evolution of Institutions for Collective Action.* Cambridge University Press, Cambridge, UK.

Oud Oil Trading (n.d.) Agarwood Chips. Available at: https://www.oudoiltrading.com/oud-oil/ (accessed 14 May 2025).

Perez, S., García-Robredo, F., Ayuga-Téllez, E. and Belda, C. (2013) Effects of the crisis in the resin sector on the demography of rural municipalities in Spain. *Forest Systems* 22(1). DOI: 10.5424/fs/2013221-02403.

Persoon, G.A. and van Beek, H.H. (2008) Growing 'the wood of the gods': Agarwood production in Southeast Asia. In: Snelder, D.J. and Lasco, R.D. (eds) *Smallholder Tree Growing for Rural Development and Environmental Services: Lessons from Asia.* Springer, Cham, Switzerland, pp. 245–262.

Peters, C. (1994) *Sustainable Harvest of Non-timber Plant Resources in Tropical Moist Forest: An Ecological Primer.* Biodiversity Support Program, Washington, DC.

Rajpoot, R., Adhikari, D., Verma, S., Saikia, P., Kumar, A. *et al.* (2020) Climate models predict a divergent future for the medicinal tree *Boswellia serrata* Roxb. in India. *Global Ecology and Conservation* 23(3), e01040. DOI: 10.1016/j.gecco.2020.e01040.

Rijkers, T., Ogbazghi, W., Wessel, M. and Bongers, F. (2006) The effect of tapping for frankincense on sexual reproduction in *Boswellia papyrifera*. *Journal of Applied Ecology* 43(6), 1188–1195.

Rivers, M., Newton, A.C., Oldfield, S., and Global Tree Assessment Contributors (2023) Scientists' warning to humanity on tree extinctions. *Plants, People, Planet* 5(4), 466–482.

Schrage, E.J. and Ewing, A.P. (2005) The cacao industry and child labour. *The Journal of Corporate Citizenship* 18, 99–112. Available at: http://www.jstor.org/stable/jcorpciti.18.99

Schultz, H. (2022) LT Say Third-Party Audit Verifies Sustainability of its Boswellia Sources. Nutra Ingredients USA. Available at: https://www.nutraingredients-usa.com/Article/2022/02/28/plt-says-third-party-audit-verifies-sustainability-of-its-boswellia-sources/ (accessed 9 February 2025).

Sedex (n.d.) Empowering Sustainable Supply Chains. Available at: https://www.sedex.com/ (accessed 5 February 2025).

Shackleton, S., Paumgarten, F., Kassa, H., Husselman, M. and Zida, M. (2011) Opportunities for enhancing poor women's socio-economic empowerment in the value chains of three African non-timber forest products (NTFPs). *International Forestry Review* 13(2), 136–151.

Soeharto, B., Budidarsono, S. and Noordwijk, M. (2016) Gaharu (eaglewood) domestication: Biotechnology, markets and agroforestry options. Working Paper no. 247. World Agroforestry Centre (ICRAF) Southeast Asia Regional Program, Bogor, Indonesia. DOI: 10.5716/WP16163.PDF.

Soehartono, T.R. (1999) Status and distribution of *Aquilaria* spp. in Indonesia, and the sustainability of gaharu trade. PhD thesis, Institute of Ecology and Resource Management, The University of Edinburgh, Edinburgh, UK.

Soju, S. and Meena, G.L. (2017) Dynamics of livestock population and output in rajasthan: a temporal analysis. *Journal of Animal Research* 7(2), 345–354.

Soliño, M., Yu, T., Alía, R., Auñón, F., Bravo-Oviedo, A. *et al.* (2018) Resin-tapped pine forests in Spain: Ecological diversity and economic valuation. *Science of the Total Environment* 625, 1146–1155. DOI: 10.1016/j.scitotenv.2018.01.027.

Sommerlatte, H. and van Wyk, B.E. (2022) Observations on the association between some Buprestid and Cerambycid beetles and black frankincense resin inducement. *Diversity* 14(1), 58.

Soumya, K.V., Shackleton, C.M. and Setty, S.R. (2019) Harvesting and local knowledge of a cultural non-timber forest product (NTFP): Gum-resin from *Boswellia serrata* Roxb. in three protected areas of the Western Ghats, India. *Forests* 10(10), 907. DOI: 10.3390/f10100907.

Stockdale, M. (2005) *Steps to Sustainable and Community-Based NTFP Management*. Non-Timber Forest Products-Exchange Programme for South and Southeast Asia, The Institute, Pokhara, Nepal.

Symrise (n.d.) Sourcing: Our Long-Term Objective Is to Source All of Our Raw Materials in a Way that Minimizes Any Negative Environmental and Human Rights Impact. Available at: https://www.symrise.com/sustainability/sourcing/#introduction (accessed 5 February 2025).

Thomas, M., Shrivastava, A. and Tomar, S. (2010) Guggul; production and conservation. *Jawaharalal Nehru Krishi Vishwa Vidyalaya Research Journal*.

Thompson, I.D., Lim, T. and Turjaman, M. (2022) *Expensive, Exploited and Endangered. A Review of the Agarwood-Producing Genera Aquilaria and Gyrinops: CITES Considerations, Trade Patterns, Conservation, and Management*. ITTO Technical Series No.51. International Tropical Timber Organization (ITTO), Yokohama, Japan.

Thulin, M. and Warfa, A.M. (1987) The frankincense trees (*Boswellia* spp., Burseraceae) of northern Somalia and southern Arabia. *Kew Bulletin* 2(3), 487–500.

Thulin, M., Decarlo, A. and Johnson, S.P. (2019) *Boswellia occulta* (Burseraceae), a new species of frankincense tree from Somalia (Somaliland). *Phytotaxa* 394(3), 219–224.

US Customs and Border Protection (2024) CBP Issues Withhold Release Order against Asli Maydi. Available at: https://www.cbp.gov/newsroom/national-media-release/cbp-issues-withhold-release-order-against-asli-maydi (accessed 1 November 2024).

Ved, D., Saha, D., Ravikumar, K. and Haridasan, K. (2015) *Commiphora wightii*. The IUCN Red List of Threatened Species. Available at: https://dx.doi.org/10.2305/IUCN.UK.2015-2.RLTS.T31231A50131117.en (accessed 1 December 2024).

Viale, M. and Vicol, M. (2023) Conserving traditional wisdom in a commodified landscape: Unpacking brand Ayurveda. *Journal of Ayurveda and Integrative Medicine* 14(1), 100667. DOI: 10.1016/j.jaim.2022.100667.

Viljoen, P.J. (1989) Habitat selection and preferred food plants of a desert-dwelling elephant population in the northern Namib Desert, South West Africa/Namibia. *African Journal of Ecology* 27(3), 227–240.

Wassie, E. (2011) The frankincense tree of ethiopia: Ecology, productivity and population dynamics. PhD thesis, Wageningen University, Wageningen, Netherlands.

Weeks, A. (2009) Evolution of the pili nut genus (*Canarium* L., burseraceae) and its cultivated species. *Genetic Resources and Crop Evolution* 56, 765–781.

Weeks, A. and Simpson, B.B. (2007) Molecular phylogenetic analysis of *Commiphora* (Burseraceae) yields insight on the evolution and historical biogeography of an 'impossible' genus. *Molecular Phylogenetics and Evolution* 42(1), 62–79. DOI: 10.1016/j.ympev.2006.06.015.

Wickens, G.E., Goodin, J.R. and Field, D.V. (1985) *Plants for Arid Lands*. George Allen and Unwin, London.

Williams, M., McKinsey, E. and Jenks, A. (2024) Acacia gum senegalia senegal (syn. *Acacia senegal, A. verek*) *Vachellia seyal* (syn. *A. Seyal*). *HerbalGram* 140.

Wollenberg, E. and Nawir, A. (2005) Turning straw into gold: Specialization among damar agroforest farmers in Pesisir, Sumatra. *Forests, Trees and Livelihoods* 15, 4317–4336.

Yusuf, B.A., Lelamo, L.L. and Abdi, M.M. (2023) Gum and resin production and marketing: Implications for pastoral livelihood in Adadle District, Somali Region. *Ethiopia International Journal of Forestry Research* 13. DOI: 10.1155/2023/8887494.

Zallen, J. (2019) *American Lucifers: The Dark History of Artificial Light, 1750-1865*. University of North Carolina Press, Chapel Hill, North Carolina.

9 Protection of Endangered Species

Ursula Moser*, Lisa Bradbury, Bruno Mainini and Mathias Lörtscher
International Species Protection in International Trade, Federal Food Safety and Veterinary Office (FSVO), Schwarzenburgstrasse 155, 3003 Bern, Switzerland

Abstract
CITES (the Convention on International Trade in Endangered Species of Wild Fauna and Flora) is a multilateral environmental agreement with the primary aim that no species of plant or animal goes extinct through unsustainable international trade. Since its inception in 1973 and its entry into force in 1975, the Convention has made great progress in this regard, and today more than 40,000 species of plants and animals benefit from its trade regulations. CITES has developed a large set of resolutions and processes that make sure that trade in those species is and remains sustainable. An ever-growing number of plant species that are used in the cosmetics and perfume industry are listed in the CITES appendices, thus directly influencing industry processes and putting a responsibility on the industry to make sure that the species used are harvested sustainably and that trade is legal and traceable. By doing so and supporting CITES and its processes, the industry also safeguards against trade threatening their survival.

The Need for CITES

CITES (the Convention on International Trade in Endangered Species of Wild Fauna and Flora) is an international, multilateral agreement between governments. Its aim is to ensure that international trade in specimens of wild animals and plants is sustainable, legal, traceable and does not threaten the survival of the species.

For many years CITES has been among the conservation agreements with the largest number of signatory countries, with now 184 Parties.[1]

Widespread information about the endangered status of many charismatic species, such as the tiger and elephants, might make the need for such a convention seem obvious. But at the time when the ideas for CITES were first formed, in the 1960s, international discussions about the regulation of wildlife trade for conservation purposes was something relatively new. In hindsight, the need for CITES is clear. Annually, the international wildlife trade is estimated to be worth billions of dollars and includes hundreds of millions of plant and animal specimens. The trade is diverse, ranging from live animals and plants to a vast array of wildlife products derived from them, including food products, exotic leather goods, wooden musical instruments, timber, tourist curios and medicines, to name a few. Levels of exploitation of some animal and plant species are high, and the trade in them, together with other factors, such as habitat loss, is capable of heavily depleting their populations and even bringing some species close to extinction. Many wildlife species in trade are not endangered, but the existence

*Corresponding author: Ursula.Moser@blv.admin.ch

DOI: 10.1079/9781800626690.0009

of an agreement to ensure the sustainability of the trade is important in order to safeguard these resources. There are many livelihoods that depend on the trade in wildlife, and it is a global responsibility to ensure that their resources stay around for future generations.

Because the trade in wild animals and plants often crosses borders between countries, the effort to regulate it requires international cooperation to safeguard certain species from over-exploitation. CITES was created in the spirit of such cooperation. Today, it accords varying degrees of protection to more than 40,000 species of animals and plants.[2]

The CITES Appendices I–III

Trade in over 40,900 species – including roughly 6,610 species of animals and 34,310 species of plants – is regulated by CITES to avoid over-exploitation through international trade.[3] They are listed in the three CITES appendices. The species are grouped in the appendices according to how threatened they are by international trade. The appendices include some whole groups, such as primates, cetaceans (whales, dolphins and porpoises), sea turtles, parrots, corals, cacti and orchids. But in some cases only a subspecies or geographically separate population of a species (for example the population of just one country) is listed. CITES has defined specific criteria for when a species should be added to the appendices. First and foremost, a given species must be in international trade, must have undergone strong reduction in its distribution or population size and it must be demonstrated that international trade is a main cause for this. In addition, a reason for listing may be if a species could be confused with an endangered species (look-alike reason). In these cases, the similar species are also included in the appendices.

Appendix I

Appendix I includes species threatened with extinction. Trade in specimens of these species is permitted only in exceptional circumstances. Almost 2,000 species are currently listed in appendix I, of which around 400 are flora species.

Appendix II

Appendix II includes species not necessarily threatened with extinction, but for which trade must be controlled in order to avoid utilization incompatible with their survival. The Conference of the Parties (CoP), which is the supreme decision-making body of the Convention and comprises all its Parties, has agreed in Resolution [Resolution Conf. 9.24 (Rev. CoP17)][4] on a set of biological and trade criteria to help determine whether a species should be included in appendix I or II. At each regular meeting of the CoP, Parties submit proposals based on those criteria to amend these two appendices. Those amendment proposals are discussed and decided upon, which is often through a vote. Currently around 39,000 species are listed in appendix II, of which around 33,800 are flora species.

Appendix III

This appendix contains species that are protected in at least one country, which has asked other CITES Parties for assistance in controlling the trade. Changes to appendix III follow a distinct procedure from changes to appendices I and II, as each Party is entitled to make unilateral amendments to it. Over 500 species are listed in appendix III, of which around 130 are flora species.

How CITES Works

In CITES the main decision-making body is the Conference of the Parties (CoP). It decides on the workplan for the time between its meetings, adopts or refuses changes to the species covered by the Convention, decides on Resolutions and the future direction of the Convention and elects members of the other bodies. It also decides over the budget of the Convention, in particular on the amount of the annual contributions of Parties and the utilization of this money. The

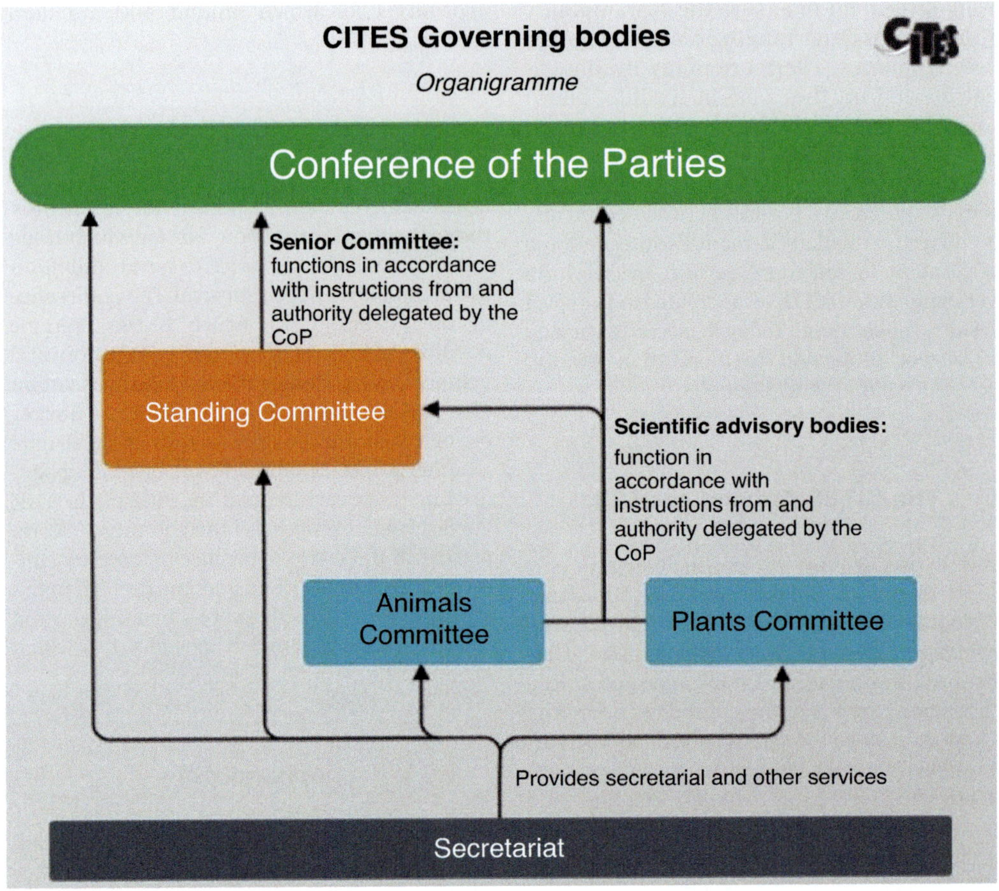

Fig. 9.1. The structure of CITES.[5] Author's own image.

additional bodies are the Standing Committee (SC), the political and strategic body in between the CoP and the scientific bodies, and the Animals and Plants Committees (AC/PC), which provide scientific advice to the other bodies of the Convention (see Fig. 9.1).

Conference of the Parties

Parties to the CITES Convention meet every 3 years at the CoP to review the implementation of the Convention and to decide on the next 3 years budget and workplan.[6] In addition, the Parties also:

- review progress for the conservation of the species listed in the appendices;
- review proposals for amendments to the lists of species in appendices I and II (and adopting them as appropriate);
- review documentation and reports from the Parties, the SC, the AC/PC and the Secretariat; and
- recommend measures to improve the effectiveness of the Convention.

At a CoP, not only do delegations of the CITES Parties participate, but so too do observers. These include representatives of states that are not parties to CITES, of UN organizations and of other international agreements. Observers from non-governmental organizations (NGOs) that deal with the conservation of biodiversity or trade may also participate at the discretion of the Parties. Observers have no voting rights but form a vital part of the discussions, often bringing important information and data to the table.

Standing Committee

The Standing Committee[7] provides guidance to the Secretariat on the implementation of the Convention and supervises the management of the Secretariat's budget. It also coordinates and supervises, as required, the work of other committees and working groups, carries out the tasks conferred to it by the CoP and drafts Decisions for consideration by the CoP. One very important task of the Standing Committee is the management of the CITES compliance procedures. CITES is one of the very few MEAs to have strong compliance regimes which allow it to exert pressures on the Parties to this Convention to abide by its rules.[8] This can end up in a suspension of trade for the Party concerned.

The members of the Standing Committee are Parties representing each of the six major geographical regions (Africa, Asia, Europe, North America, Central and South America and the Caribbean, and Oceania), with the number of representatives weighted according to the number of Parties within each region. The composition of the Standing Committee is reviewed at each ordinary session of the CoP.

In addition, the Standing Committee also includes a representative of:

- the depositary state (Switzerland);
- the party that hosted the last CoP; and
- the party that will host the next CoP.

Only regional representatives have the right to vote. The role of the depositary state is rather particular. In general, it does not have the right to vote, unless there is a tie. In such cases the depositary has to break the tie.

The members representing the regions elect the chair, vice-chair and the alternate of the vice-chair of the Standing Committee. All Parties that are not members of the Standing Committee have the right to send observers to its meetings. NGOs may participate in meetings of the Standing Committee as observers.

The Standing Committee normally meets only once a year, but also shortly before and after each meeting of the CoP. It may establish intersessional working groups as necessary to fulfil its functions.

Animals and Plants Committees

These expert committees[9] were established by the sixth meeting of the CoP in Ottawa, Canada (1987) to address gaps in biological and other knowledge about animal and plant species that are (or could be) subject to CITES trade controls. Their role is to provide technical scientific support for decision making. These two committees have a similar mandate, which includes, among other things, the following:

- providing scientific advice and guidance to the CoP, the other committees, working groups and the Secretariat;
- addressing nomenclature issues;
- conducting regular reviews of species to ensure that they are listed in the appropriate CITES appendices;
- advising on cases where certain species are not being traded sustainably and recommending remedial action (as part of a process known as a 'review of significant trade'); and
- drafting Resolutions on animal and plant issues for consideration by the CoP.

The Animals and Plants Committees meet twice between meetings of the CoP. They report to the CoP at its meetings and, when requested, advise the Standing Committee between those meetings.

The members of the Animals and Plants Committees come from the six major geographical regions (Africa, Asia, Europe, North America, Central and South America and the Caribbean, and Oceania), and there is one nomenclature specialist on each of the committees. They are elected at the meetings of the CoP, with the number of regional representatives being weighted according to the number of Parties in each. A chair and vice-chair are elected by the regional members of the committees. The zoological and botanical nomenclature specialists have no voting rights.

The Animal and Plant Committees may establish working groups to fulfil their tasks.

The CITES Secretariat

UNEP[10] as the leading global environmental authority, administers numerous Multilateral Environmental Agreements (MEAs) including CITES and provides the Secretariat in Geneva, Switzerland. The CITES Secretariat[11] plays a central role fundamental to the Convention, and its functions include, amongst others, playing a coordinating, advisory and supportive role in the implementation of the Convention, helping to implement the Convention and ensuring that its provisions are respected. The Secretariat organizes and manages all meetings of the various committees of CITES and the CoP as well. It makes recommendations for the implementation of the Convention, produces the annual report, issues appendices I–III when amendments are made by the CoP and adjusts Resolutions and Decisions.[12] In addition, the Secretariat provides information to assist identification of species listed in the appendices. The Secretariat distributes information to the Parties mostly through meeting documents and notifications.[13] Except in very few cases, all documents are made available in the three working languages of the Convention (English, French and Spanish) and are posted on the CITES website, which is administered by the Secretariat.

How CITES Is Implemented

Trade in CITES listed species

A specimen of a CITES-listed species may be imported into or exported (or re-exported) from a state party to the Convention only if the appropriate document has been obtained and presented for clearance at the port of entry or exit.[14] All countries must keep records of these imports and exports, and at the end of each year report the trade numbers to the CITES Secretariat. Based on this set of data, trade numbers can be analysed and where concern might arise that trade is increasing or taking on an unsustainable level, appropriate measures can be taken.

There is some variation of the requirements from one country to another and it is always necessary to check the national laws, which may be stricter.

Management Authority and Scientific Authority

Every party to CITES must appoint a Management Authority[15] and a Scientific Authority.[16] The Management Authority is responsible for implementing the Convention in its country. In particular, it is the body competent to grant import and export permits and re-export certificates on behalf of that party. The Management Authority is also responsible for communication with other Parties and the CITES Secretariat.

The Scientific Authority is responsible for providing technical and scientific advice to the Management Authority, in particular as to whether the export of a specimen will be detrimental to the survival in the wild of the species involved.

In particular, it monitors exports of specimens of appendix II-listed species and, whenever necessary, advises the Management Authority of suitable measures to be taken to limit such exports in order to maintain such species throughout their range at a level consistent with their role in the ecosystems and well above the level at which they would qualify for appendix I. This analysis is called the non-detriment finding[17] and it is required before an export or import permit may be granted for any specimen of an appendix I or appendix II species.

For appendix I species, the import of such specimens requires that import permits are only granted when the Scientific Authority of the state of import has advised that the import will be for purposes not detrimental to the survival of the species involved.

Specific examples

The following section explains, with reference to specific examples, why these species are listed in the CITES appendices and the impact on trade.

Example 1: Aniba rosaeodora *– Brazilian rosewood*

Aniba rosaeodora has been listed in the CITES appendices since 2010. The reason for its inclusion in CITES appendix II was the decline in wild populations. This species is a slow-growing tree and birds and insects eat the seeds. Furthermore, there was a strong international trade, putting additional pressure on the species.

Historically, the linalool-rich wood of *Aniba rosaeodora* was used extensively and the essential oil extracted from it was used as a fragrance in perfume production. The first records of its use date back to 1883 in French Guiana. The wood was harvested and transported to France, where the oil was extracted by distillation (Homma, 2003).

During World War I, transport to Europe was severely restricted and so the oil was distilled directly in French Guiana. However, the exploitation of the timber was not sustainable, so the natural stocks in the country were severely depleted and eventually wiped out. The same process occurred later in the Amazon, where the species was discovered in 1925. Brazil became the leading country for *Aniba rosaeodora* processing, replacing French Guiana on the world market after the species became extinct there.

Due to unsustainable harvesting, the government of the state of Amazonas issued a decree in 1932 that set the annual production quota of essential oil and required replanting of trees that were cut down.

Until the 1950s, the harvest increased and new areas were opened up for harvesting. With the development of synthetic linalool in the 1960s, demand declined as the synthetic version was cheaper. In the 1970s, new areas were opened up and the introduction of chainsaws in the region made harvesting more efficient and cost-effective. The worldwide oil crisis involving petroleum in the 1970s and the associated rise in petroleum prices had a negative impact on the production of synthetic linalool and led to a global demand for natural essential oil.

In the following years, research concentrated on finding a natural substitute for rosewood, and dozens of plants that produce linalool were found, such as bergamot, lemon, jasmine, basil and others. However, the use of these plants by the perfume industry was not practicable due to the undesirable fruit fragrance, and quite often these plants produced oils that cause irritation of the airways and skin. In that way, rosewood essential oil is unique (Homma, 2003).

In the 1980s, production suffered another market collapse due to a substitute product made from *Cinnamomum camphora*. The decline in production continued throughout the 1990s and 2000s. The reasons for this were decades of unsustainable harvesting, difficult access to the trees on the riverbanks, transport difficulties and the controls imposed by the Brazilian authorities.

New technologies like logging and timber transport equipment and machinery helped to open up new roads and made it possible to develop further areas. Since 2007, the Brazilian authorities have tightened controls on supply chains. The production of *Aniba rosaeodora* essential oil in Peru, Colombia and the Guianas has declined to negligible quantities due to the production of synthetic linalool.

An assessment conducted by the International Union for Conservation of Nature (IUCN) in 2021 (Barstow, 2021) indicated that research should be conducted into extracting essential oil from the leaves and other parts of the tree. Oil obtained in this way could be a suitable alternative to oil extraction from the wood, as only the leaves need to be collected, which can be managed more sustainably.

The population decline is not only the result of unsustainable timber harvesting but also, among other things, of land reclamation for roads and railways, housing and agriculture.

The listing of *Aniba rosaeodora* in appendix II was proposed by Brazil (CoP15 Prop. 29, 2010) at CoP15 in 2010. Its inclusion in appendix II with an amended annotation[18] was accepted (see Fig. 9.2). Annotations describe which parts and products are covered by the CITES provisions and which are excluded. Annotations should reflect the current commodities in international trade and their relative impact on the wild populations.

For the present case of *Aniba rosaeodora*, this means that logs, sawn wood, veneer sheets, plywood and extracts are covered by the provisions but finished products containing such extracts, including fragrances, are exempt.

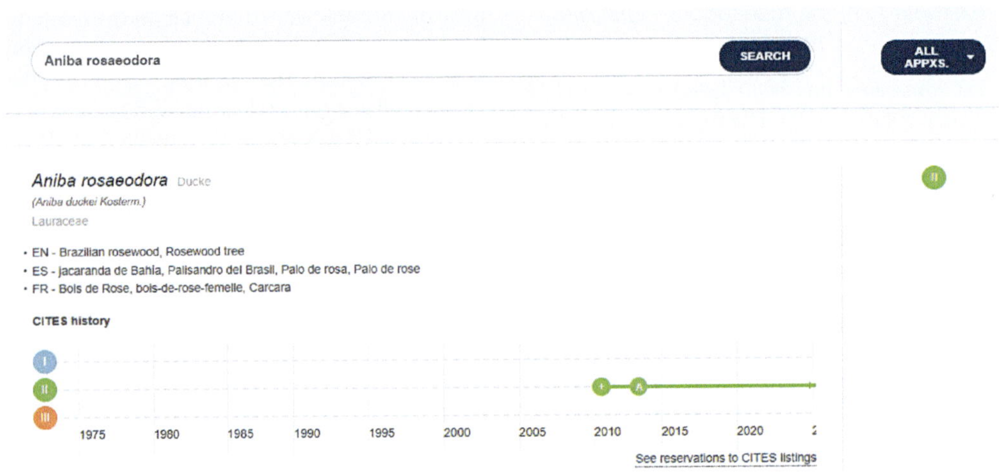

Fig. 9.2. Overview of listing history for *Aniba rosaeodora*.[19] Author's own image.

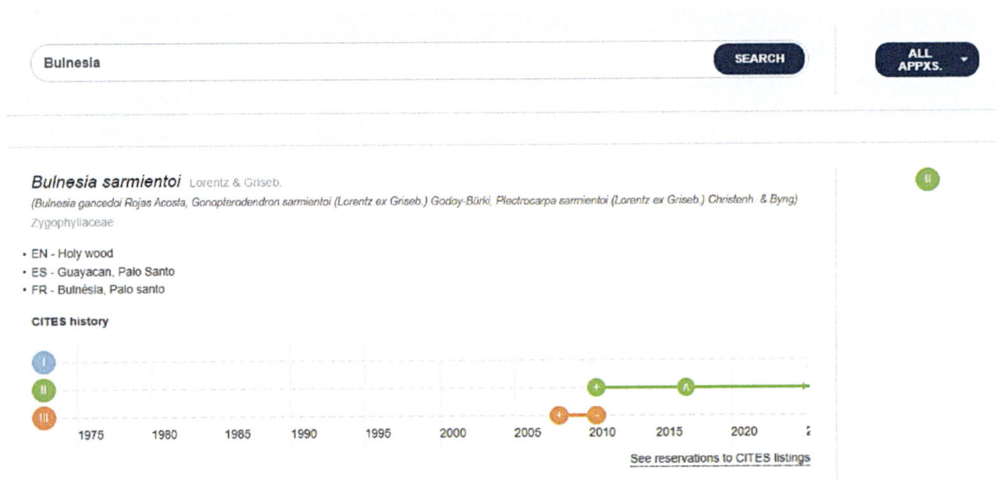

Fig. 9.3. Overview of listing history for *Bulnesia sarmientoi*.[21] Author's own image.

Example 2: Bulnesia sarmientoi *– holy wood or palo santo*

Bulnesia sarmientoi is native to the Gran Chaco region of South America, which extends across Argentina, Bolivia, Brazil and Paraguay. The species is widespread and currently has a large global population. However, a persistent boom in international trade in the wood and its essential oils is threatening the species. This, along with the deforestation of vast areas of Chaco Forest, has increased the threat to the species. Based on estimates of the loss of the Chaco Forest in Argentina, it is suspected that the global population is declining.[20]

In 2008, the species was included in appendix III at the request of Argentina, with an annotation #11 stating that logs, sawn wood, veneer sheets, plywood, powder and extracts should fall under the provisions of CITES (see Fig. 9.3).

In 2010, at CoP15 the species was transferred from appendix III to appendix II based on a proposal by Argentina (CoP15 Prop. 42, 2010), justified with the provisions of the Convention

(CITES, 1973) that appendix II is to include all specimens that, while not necessarily threatened with extinction, could become extinct if trade is not strictly regulated to avoid over-exploitation. Another criterion for listing species in appendix II is if it is known or can be predicted that regulation of trade in these species is necessary to avoid them being listed in appendix I in the near future.[22]

In 2016, at CoP17 the existing annotation was amended to harmonize with the existing annotation for *Aniba rosaeodora* because both species are often traded in the same form (i.e. extract) (CoP17 Prop. 62, 2016) which means that logs, sawn wood, veneer sheets, plywood and extracts are covered by the provisions but finished products containing such extracts, including fragrances, are exempt.

Example 3: Dipteryx *spp. including* Dipteryx odorata – *Cumaru, Tonka*

In 2022, at CoP19, Colombia, the European Union and Panama proposed to include the genus *Dipteryx* spp. in appendix II (CoP19 Prop.48, 2022) because several species, such as *Dipteryx alata, Dipteryx micrantha, Dipteryx odorata* and *Dipteryx oleifera*, are considered threatened. Due to look-alike reasons, it was proposed that the remaining species of the genus *Dipteryx* spp. should also be included in appendix II. Furthermore, a new annotation was proposed for this listing, designating 'logs, sawn wood, veneer sheets, plywood and transformed wood, and seeds'

The reason for this proposal was international demand for *Dipteryx* timber and its products. The genus is targeted for its valuable hardwood timber (often traded under the names cumaru or shihuahuaco), as well as its seeds, known as tonka beans, which are in demand internationally for use in the fragrance, tobacco and food industries. In several range states, *Dipteryx* spp. are also locally important for food, traditional medicine, charcoal, oil and as shade trees in cocoa agroforestry systems (CoP19 Prop.48, 2022).

Several *Dipteryx* spp. are also subject to intensive seed collection for the national and international markets (Kermath *et al.*, 2014; Bovell-Benjamin and Roberts, 2016; Government of Venezuela, 2020), and ongoing seed collection is considered to threaten survival of some *D. odorata* subpopulations in Brazil (Herrero-Jáuregui *et al.*, 2012).

The ecosystems in which all 14 species of *Dipteryx* occur have become increasingly threatened by habitat degradation and deforestation (Leisher *et al.*, 2013; Antongiovanni *et al.*, 2020). Habitat loss caused by conversion of forest, woodland and savanna for agriculture constitutes a serious threat to many *Dipteryx* spp. (Fleisswasser, 2014).

Seeds of *Dipteryx* spp. (in particular *D. odorata, D. oleifera, D. alata* and *D. punctata*) are used for food, medicine and oil locally (da Silva *et al.*, 2010; Vennetier *et al.*, 2012; Bovell-Benjamin and Roberts, 2016), and the timber is used for tools and charcoal (Putzel *et al.*, 2013; Kermath *et al.*, 2014).

Regarding the identification, according to the Thünen Institute of Wood Research in Germany, it is not possible to clearly distinguish the individual species within the genus *Dipteryx* using macroscopic and microscopic identification of wood anatomy (Koch, personal communication to Forest Trends, 2021). However, identification of *D. alata, D. ferrea, D. micrantha, D. odorata* and *D. punctata* using genetic markers is currently possible (Honorio Coronado *et al.*, 2020). Several species of *Dipteryx* are traded under the common names cumaru or shihuahuaco and are neither distinguishable nor identified to species level in trade (Aldana Gomero *et al.*, 2016).

At CoP19, the discussion of the proposal for listing the genus was very controversial and there was no consensus. Therefore, the chair called for a vote on the proposal with the suggested amendments to remove seeds from the annotation and to delay the entry into effect of the inclusion in appendix II by 24 months, to which the Parties agreed and the proposal was accepted (CoP19 Summary Records, 2022) (see Fig. 9.4). Covered under CITES regulations are logs, sawn wood, veneer sheets, plywood and transformed wood.

Example 4: Guaiacum *spp., including* Guaiacum officinale – *Brazil wood, guaiacum, lignum-vitae, pockwood*

Guaiacum officinale is native to most countries in the Caribbean including those from the Lesser

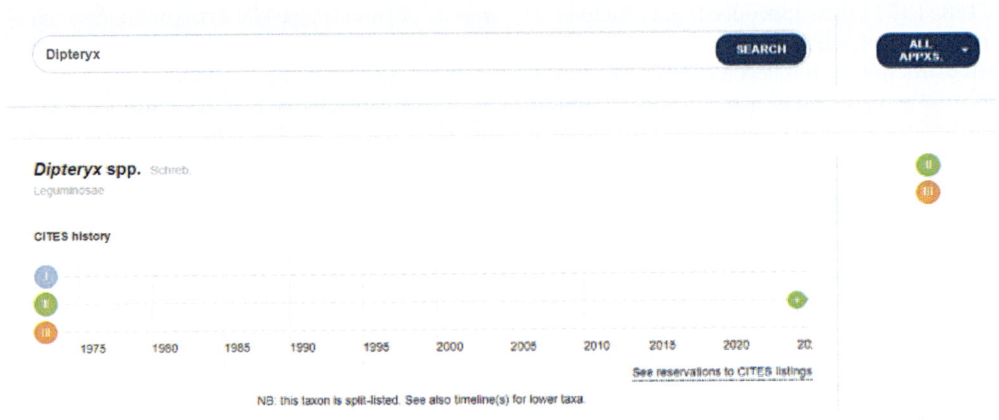

Fig. 9.4. Overview of listing history for *Dipteryx* spp.[23] Author's own image.

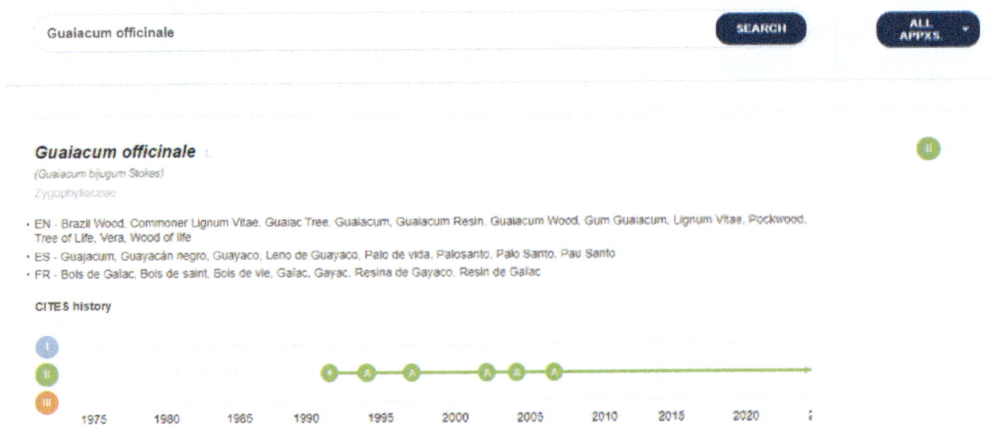

Fig. 9.5. Overview of listing history for *Guaiacum officinale*.[25] Author's own image.

and Greater Antilles Island groups and also to Colombia, Venezuela, Guyana, Suriname and Panama. The species has historically been over-exploited for timber and for medicinal properties. This has occurred over a period of 500 years and has caused *G. officinale* to become rare and close to extinction in most countries it is native to. It is likely the remaining population of the species is small and scattered, having declined by at least 50% over three generations. Therefore, the species is globally assessed as endangered.[24]

Around 1500, it was discovered that Native Americans were using guaiac wood to make a decoction to treat syphilis. This led to intensive international trade with Europe, and for more than two centuries there was a high demand for heartwood to produce the extract to treat syphilis. For trade as medicinal material it is named, for example, gum guaiac, guaiac resin, guaiacum, lignum vitae or lignum sanctum.

The wood is among the heaviest woods and its high density and the content of oily resin as well as its fine texture make it suitable for ship-building, mechanical engineering and turnery.

Guaiacum populations were massively reduced but over-exploitation continued nevertheless. The high demand drove up prices accordingly (CoP8 Prop. 97, 1992).

Fig. 9.6. Overview of listing history for *Guaiacum* spp.[26] Author's own image.

In 1992, at CoP8 the United States of America proposed to list *G. officinale* in appendix II, and all parts and derivatives of the plant were covered by the CITES regulations (see Fig. 9.5).

In 2002, at CoP12 it was decided to list the whole genus in appendix II (see Fig. 9.6). This decision was taken due to the unclear taxonomic classification and because trade in the 4–6 existing *Guaiacum* species was not sustainable. Taxonomic descriptions of *G. officinale* and *G. sanctum* were accepted but for *G. angustifolium*, *G coulteri*, *G. guatemalense* and *G. unijugum* the taxonomy was not yet accepted although all species of the genus *Guaiacum* were listed.

A new annotation was added which meant that CITES trade controls are required for all parts and derivatives of plants of the genus, with the exception of chemical derivatives and finished pharmaceutical products (CoP12 Prop. 54, 2002). This annotation was modified in 2007 at CoP14, and still all parts and derivatives (resin and oil as raw materials) require CITES certificates for trade but finished products are exempt from CITES regulations (CoP14 Prop. 27, 2007).

Mainly timber and pieces of heartwood are reported to be in trade. The products of *Guaiacum* advertised in trade, such as on the internet and in company brochures, include wood, bark, resin powder, fluid extract and tinctures. Germany has annually imported up to 40 t of heartwood resin and wood-chips as aromatic substances for use in the liqueur industry. Resin of the heartwood of *G. coulteri* was also imported to the United States to be used in diagnostic kits for detecting hidden gastrointestinal bleeding; however, verification of the nature and extent of international trade in specimens of *Guaiacum* for medicinal products has proved to be very difficult. It was still unclear where the processing of medicinal products takes place and which commodities constitute the international trade beyond timber (CoP14 Prop. 27, 2007).

Ongoing threats to the species are habitat loss due to burning or cutting down the natural vegetation, agriculture, introduction of non-native species and settlement construction. The main threat is deforestation and the wood is used as construction timber, for handicrafts or ornamental purposes. *Guaiacum* heartwood resin contains a number of substances which are or were used for medicinal purposes (CoP12 Prop. 54, 2002).

Notes

[1] CITES, 2024n.
[2] CITES, 2024n.
[3] CITES, 2024e.
[4] The Resolutions are generally intended to provide long-standing guidance over periods of many years. The Resolutions include the guidance provided by the CoP on how to interpret the provisions of the Convention but also include: the documents establishing the permanent committees; the budget and work programme of the Secretariat; rules for controlling the trade (such as issuing permits and marking specimens in trade); and the texts establishing long-term compliance processes. At each of its meetings, the CoP considers problems of implementation of the Convention and its effectiveness. The results of its deliberations are in the form of recommendations (Article XI, paragraph 3) that are recorded either in Resolutions or in Decisions of the CoP (CITES, 2024j).
[5] CITES, 2024m.
[6] CITES, 2024d.
[7] CITES, 2024k.
[8] CITES, 2024c.
[9] CITES, 2024a.
[10] https://wedocs.unep.org/bitstream/handle/20.500.11822/35352/UNEPOrg.pdf?sequence=4&isAllowed=y
[11] CITES, 2024l.
[12] An agreement between the Parties typically containing instructions to a specific committee, Parties or the Secretariat (https://cites.org/eng/dec/index.php). Decisions are typically intended to remain in effect for a short period only, usually until a particular task has been completed. Many Decisions require a report to be submitted at the meeting of the CoP following that at which they were adopted, and so would remain in effect from one meeting of the CoP to the next. The list of Decisions is therefore revised by the Secretariat after each such meeting, in accordance with Resolution Conf. 4.6 (Rev. CoP19), paragraph 4b), at which time it deletes the Decisions that are out of date.
[13] CITES, 2024h.
[14] CITES, 2024e.
[15] CITES, 2024f.
[16] CITES, 2024f.
[17] CITES, 2024g.
[18] When a species is included in one of the appendices, all parts and derivatives of the species are also included in the same appendix unless the species is annotated to indicate that only specific parts and derivatives are included. The symbol # followed by a number placed against the name of a species or higher taxon included in appendix II or III refers to a footnote that indicates the parts or derivatives of plants that are designated as 'specimens' subject to the provisions of the Convention in accordance with Article I, paragraph (b), subparagraph (iii) (https://cites.org/eng/app/interpret.php).
[19] CITES, 2024b.
[20] IUCN Red List, 2017a.
[21] CITES, 2024b.
[22] CITES, 2024i.
[23] CITES, 2024b.
[24] IUCN Red List, 2017b.
[25] CITES, 2024b.
[26] CITES, 2024b.

References

Aldana Gomero, D.R., Garcia Davila, C.R., Hidalgo Pizango, C.G., Flores Llampazo, G.R., del Castillo Torres, D. *et al.* (2016) Morphometric analysis of the species of *Dipteryx* in the Peruvian Amazon. *Folia Amazónica* 25(2), 101.

Antongiovanni, M., Venticinque, E.M., Matsumoto, M. and Fonseca, C.R. (2020) Chronic anthropogenic disturbance on Caatinga dry forest fragments. *Journal of Applied Ecology* 57(10), 2064–2074.

Barstow, M. (2021) *Aniba rosodora. The IUCN Red List of Threatened Species*. The IUCN Red List of Threatened Species 2021: e.T33958A88301381. Available at: https://dx.doi.org/10.2305/IUCN.UK.2021-3.RLTS.T33958A88301381.en

Bovell-Benjamin, A.C. and Roberts, J. (2016) Naturally occurring toxicants: Presence in selected commonly consumed fruits. In: Vishweshwaraiah, P., Olga Martín-Belloso, O., Keener, L., Astley, S., Braun, S. *et al*. (eds) *Regulating Safety of Traditional and Ethnic Foods*. Elsevier, Amsterdam Inc, pp. 247–282.

CITES (1973) Convention Text. Convention on International Trade in Endangered Species of Wild Fauna and Flora | CITES. Available at: https://cites.org/eng/disc/text.php

CITES (2024a) Animals and Plants Committees. Available at: https://cites.org/eng/disc/ac_pc.php (accessed 12 May 2025).

CITES (2024b) Checklist of CITES Species. Available at: https://checklist.cites.org/#/en/search/output_layout=alphabetical&level_of_listing=0&show_synonyms=1&show_author=1&show_english=1&show_spanish=1&show_french=1&scientific_name=Aniba+rosaeodora&page=1&per_page=20 (accessed 12 May 2025).

CITES (2024c) Compliance Procedures. Available at: https://cites.org/eng/prog/compliance (accessed 12 May 2025).

CITES (2024d) Conference of the Parties. Available at: https://cites.org/eng/disc/cop.php (accessed 12 May 2025).

CITES (2024e) How CITES Works. Available at: https://cites.org/eng/disc/how.php (accessed 12 May 2025).

CITES (2024f) National CITES Authorities. Available at: https://cites.org/eng/parties/country-profiles/national-authorities (accessed 12 May 2025).

CITES (2024g) Non-Detriment Findings. Available at: https://cites.org/eng/prog/ndf/index.php (accessed 12 May 2025).

CITES (2024h) Notifications to the Parties. Available at: https://cites.org/eng/notif/index.php (accessed 12 May 2025).

CITES (2024i) Resolutions. Available at: https://cites.org/sites/default/files/documents/COP/19/resolution/E-Res-09-24-R17.pdf (accessed 12 May 2025).

CITES (2024j) Resolutions of the Conference of the Parties in Effect after the 19th Meeting. Available at: https://cites.org/eng/res/index.php (accessed 12 May 2025).

CITES (2024k) Standing Committee. Available at: https://cites.org/eng/disc/sc.php (accessed 12 May 2025).

CITES (2024l) The CITES Secretariat. Available at: https://cites.org/eng/disc/sec/index.php (accessed 12 May 2025).

CITES (2024m) The Structure of CITES. Available at: https://cites.org/eng/disc/org.php (accessed 12 May 2025).

CITES, A. (2024n) What is CITES? Available at: https://cites.org/eng/disc/what.php (accessed 12 May 2025).

CoP12 Prop. 54 (2002) Inclusion of *Guaiacum* spp. in Appendix II. Available at: https://cites.org/sites/default/files/eng/cop/12/prop/E12-P54.pdf (accessed 12 May 2025).

CoP14 Prop. 27 (2007) Amendment of the Annotations for the Following Taxa to Read: *Guaiacum* spp. Available at: https://cites.org/sites/default/files/eng/cop/14/prop/E14-P27.pdf (accessed 12 May 2025).

CoP15 Prop. 29 (2010) Inclusion of *Aniba rosaeodora* Ducke in Appendix II, in Compliance with the Provisions of Article II, paragraph 2 (a), of the Text of the Convention, and Paragraph A of Annex 2 a. Available at: https://cites.org/sites/default/files/eng/cop/15/prop/E-15-Prop-29.pdf (accessed 12 May 2025).

CoP15 Prop. 42 (2010) Inclusion of *Bulnesia sarmientoi* in Appendix II, in Compliance with the Provisions of Article II, Paragraph 2 (a), of the Text of the Convention, and Resolution Conf. 9.24 (Rev. CoP14), Annex 2 a, Paragraph A. Available at: https://cites.org/sites/default/files/eng/cop/15/prop/E-15-Prop-42.pdf (accessed 12 May 2025).

CoP17 Prop. 62 (2016) Amendment of the Listing of Bulnesia sarmientoi in Appendix II. Amend Annotation #11. Available at: https://cites.org/sites/default/files/eng/cop/17/prop/060216/E-CoP17-Prop-62.pdf (accessed 12 May 2025).

CoP19 Prop.48 (2022) To Include the Species Dipteryx alata, Dipteryx micrantha, Dipteryx odorata and Dipteryx oleifera in Appendix II of CITES in Accordance with Article II, Paragraph (2)(a) of the Convention and Satisfying Criterion B of Annex 2a of Resolution Conf. 9.24 (Rev.CoP17). Available at: https://cites.org/sites/default/files/documents/E-CoP19-Prop-48.pd (accessed 12 May 2025).

CoP19 Summary Records (2022,17 November) CoP19 Com. I Rec. 5 (Rev. 1) – p. 1): 89. Proposals to amend Appendices I and II. Available at: https://cites.org/sites/default/files/documents/E-CoP19-Com-I-Rec-05-R1.pdf (accessed 12 May 2025).

CoP8 Prop. 97 (1992) Inclusion of *Guaiacum officinale* in Appendix II. Available at: https://cites.org/sites/default/files/eng/cop/08/prop/E08-Prop-97_Guaiacum.pdf (accessed 12 May 2025).

Da Silva, T.M., da Silva Jardim, F.C., da Serra Silva, M. and Shanley, P. (2010) O mercado de amendoas de *Dipteryx odorada* (cumaru) no estado do Para. *Floresta* 40(3), 603.

Fleisswasser, S. (2014) *Dipteryx oleifera* Benth. Plant Species Factsheet. Bundesamt fur Naturschutz. Bonn, Germany. Available at: https://www.bfn.de/fileadmin/BfN/artenschutz/Dokumente/Dipteryx_oleifera_Factsheet.pdf (accessed 13 October 2021).

Government of Venezuela (2020) La sarrapia, fruto histórico venezolano. Available at: https://www.mincyt.gob.ve/la sarrapia fruto historico venezolano (accessed 11 October 2021).

Herrero-Jáuregui, C., Sist, P. and Casado, M. (2012) Population structure of two low-density neotropical tree species under different management systems. *Forest Ecology and Management* 208, 31–39.

Homma, A.K.O. (2003) *História da agricultura na Amazônia: da era pré-colombiana ao terceiro milênio*. Embrapa Informação Tecnológica, Brasília.

Honorio Coronado, E.N., Blanc Jolivet, C., Mader, M., García Dávila, C.R. and Aldana Gomero, D. (2020) SNP markers as a successful molecular tool for assessing species identity and geographic origin of trees in the economically important South American legume genus *Dipteryx*. *Journal of Heredity* 1–11.

IUCN Red List (2017a) Assessment for *Bulnesia sarmientoi*. Available at: https://www.iucnredlist.org/species/32028/68085692 (accessed 12 May 2025).

IUCN Red List (2017b) Assessment for *Guaiacum officinale*. Available at: https://www.iucnredlist.org/species/33701/68085935 (accessed 12 May 2025).

Kermath, B., Bennett, B. and Pulsipher, L.M. (2014) *Food Plants in the Americas: A Survey of the Domesticated, Cultivated, and Wild Plants Used for Human Food in North, Central and South America and the Caribbean*. Available at: https://www.researchgate.net/publication/263888295_Food_Plants_in_the_Americas_A_Survey_of_the_Domesticated_Cultivated_and_Wild_Plants_Used_for_Human_Food_in_North_Central_and_South_America_and_the_Caribbean (accessed 12 May 2025).

Koch, G. (2021) Koch, G. (Thünen Institute of Wood Research, Hamburg, Germany) personal communication to Forest Trends, 21 April 2021.

Leisher, C., Touval, J., Hess, S.M., Boucher, T.M. and Reymondin, L. (2013) Land and forest degradation inside protected areas in Latin America. *Diversity* 5(4), 779–795.

Putzel, L., Padoch, C., Ricse, A., Harrison, Ó.S. and Herbohn, J. (2013) Putting back the trees: Smallholder silvicultural enrichment of post logged concession forest in Peruvian Amazonia. *Small Scale Forestry* 12, 421–436.

Resolution Conf. 9.24 (Rev. CoP17) (n.d.) Criteria for Amendment of Appendix I and II. Available at: https://cites.org/sites/default/files/documents/COP/19/resolution/E-Res-09-24-R17.pdf (accessed 12 May 2025).

Vennetier, C., Peltier, R. and Coimbra, J. (2012) Valorizar la almendra chiquitana, *Dipteryx alata* (Vogel), una estrategia para mitagar el impacto ambiental del desarrollo agropecuario en Bolivia? *Bois et Forêts Des Tropiques* 311, 35–48.

10 Future Global Market for Resins, Gums and Related Products

Hartwig Schulz[1]*, Victor S. Jayakumar[2,3] and Niranjan Prasad[4]

[1]Consulting and Project Management for Medicinal and Aromatic Plants, Waltraudstraße 4, Stahnsdorf, Germany; [2]St Thomas College of Engineering & Technology, Kozhuvalloor PO 689521 Chengannur, Kerala, India; [3]Deepthi, Amaravila PO, Thiruvananthapuram 695122, Kerala, India; [4]ICAR-National Institute of Secondary Agriculture, Namkum, Ranchi, India

Abstract

Today, farmers and their entire supply chains are under increasing pressure to ensure the availability of their natural raw materials and the products made from them at more affordable prices, while also complying with the often demanding regulations and certifications of many customers. However, as most of the resinous trees grow in remote areas, it usually takes some time for the farmers and the small-scale processing industry to adapt to the new challenges in terms of modern harvesting methods and extraction technologies, as well as the introduction of new quality standards and certification systems. This is not primarily about purchasing new machines, but rather about the individual training of farmers and processors. In this context, sustainability must be understood as a complex and dynamic balance between economic, ecological (renewable raw materials such as biomass and energy, waste reduction, biodegradability, biodiversity, conservation of natural resources) and social aspects (fair trade, profit sharing, etc.). In the future, research should generally focus more on genetic improvement and the selection of optimized genotypes that not only produce larger quantities of gums and resins, but also provide improved resistance to plant pests and drought stress. An increasingly important aspect in the gum- and resin-producing countries will be the development of sustainable jobs and of living conditions for the traditional rural population in order to stop migration to the bigger cities. The cultivation, harvesting and processing of plants have a significant impact on the well-being of rural populations. The world must therefore step up its efforts to provide jobs and income for rural communities, which are often located in very remote areas. Furthermore, regulators should be aware that the production of natural extracts from resins, gums and oleoresins is usually very complex and also subject to many imponderables, especially in today's world, with a view to the dramatic effects of climate change, economic globalization and rapidly increasing urbanization. On the other hand, the cost side must always be taken into account: if the regulations for the production of extracts from resins and gums are too strict, multinational companies in particular will no longer produce individual consumer goods because the costs associated with production will no longer be economically viable. These natural products would then simply be replaced by petrochemical and biotechnological products. Overregulation of natural extracts would thus destroy globally functioning supply chains and further consolidate the market-dominating position of a few large corporations.

*Corresponding author: hs.consulting.map@t-online.de

 Natural Gums and Resins: Botany and Sustainable Uses in Medicine, Nutrition, Perfumery and Cosmetics (eds Hartwig Schulz *et al.*)
DOI: 10.1079/9781800626690.0010

Introduction

Nowadays, gums and resins have a wide range of different applications, not only in the food and beverage sector, but also in the pharmaceutical and cosmetics industries (Tadesse *et al.*, 2007). As new applications are discovered and developed, demand for these natural products could continue to rise. Furthermore, the health and wellness trend is influencing various industries, including the food and beverage industry. In this context, gums and resins, which are believed to have health benefits, are expected to experience increased demand, especially if they can be labelled as natural and functional ingredients. Based on current trends, it can be assumed that the future market for resins and gums will be mainly influenced by the following aspects:

- Consumers are increasingly favouring natural and sustainable products.
- Gums and resins derived from trees and non-tree plants are often seen as natural and environmentally friendly, which will certainly boost the demand for these raw materials.
- Advances in extraction and processing technologies will improve the efficiency of extracting gums and resins from natural sources, making production more cost-effective and sustainable.

With the increasing focus on sustainability, companies will discover more environmentally friendly and socially responsible sourcing practices in this context and utilize these in their marketing strategies (Cronin *et al.*, 2011). This could include existing initiatives such as fair trade and agroforestry. However, it should be noted that the resin and gum industry may be facing one of its biggest challenges related to ongoing climate change, deforestation and other dramatic environmental factors that will generally impact the future availability and quality of raw materials. Furthermore, changes in regulations related to the production and trade of natural products could affect overall market dynamics. Therefore, compliance with standards and regulations will become a key factor for companies operating in this sector of industrial products in the near future. Furthermore, economic factors, including GDP (gross domestic product) per capita, inflation rates and global trade policies, are still expected to have a significant impact on the overall market conditions for natural products. It is fair to assume that the commodities market in the years to come will probably be influenced primarily by a combination of consumer trends, new technological innovations and regulatory changes (Quested *et al.*, 2010).

Players influencing the gum and resin industry

The production and trade of gums, resins, oleoresins, essential oils and solvent extracts involve a diverse set of companies. The industry is comprised of various players, including those focused on agriculture, harvested natural products, (hydro)distilled and extracted flavours and fragrances, and other natural products (Fig. 10.1).

In Table 10.1 a selection of those companies that are active in the above-mentioned market segments is presented.

The economic importance for various indigenous communities around the world

The indigenous forest tribes living in particular in India, Indonesia, Thailand and various South American countries will continue to rely on the harvesting of resins and gums as a viable source of income in the future. However, to ensure the sustainability of this source of income in the coming decades and to respect species conservation accordingly, cooperation between government agencies, volunteers and private companies is essential to provide the necessary support for the development of these communities.

In addition to some gum- and resin-producing plants that are only bought by the local population, there occur also species in India that are exported in large quantities to the rest of the world. The latter group includes in particular plants that produce gum karaya, gum ghatti, gum guggul and salai gum (Indian frankincense), whose raw materials will certainly be

Fig. 10.1. Khari Baoli Market in Old Delhi (India), the largest market for medicinal plants, including gums and resins, in South Asia. (By Anthony Cunningham.) Used with permission.

of particular commercial interest in the near future.

Species belonging to the *Acacia* family, such as *Acacia nilotica*, from which Indian gum arabic is produced, and *Acacia senegal*, from which the true gum arabic is isolated, will also contribute significantly to the preservation of indigenous communities in the future. In addition, species from the Pinaceae (colophony, amber), Leguminosae (copal) and Dipterocarpaceae (dammar) families will further strengthen the importance already existing today. In terms of future economic development prospects, the resins and gums listed below are likely to be the most important ones.

*Indian gum arabic (*Acacia nilotica*)*

Acacia nilotica gum (babul gum) is an exudate obtained from stems and branches of the *Acacia nilotica* tree. The *Acacia nilotica* tree, commonly known as the Indian gum arabic tree, has been recognized worldwide as a multipurpose tree. The gum is a water-soluble hydrocolloid which has been claimed to possess several beneficial physiological, nutritional and therapeutic properties due to the presence of various bioactive phytochemicals. It has been increasingly recognized as a potential commodity, having diverse pharmacological and industrial applications. The gum of *Acacia nilotica* is sometimes used as a substitute for gum arabic (obtained from *Acacia senegal*) although the quality is inferior. It has some properties similar to those of gum arabic. The gum of *Acacia nilotica* is odourless and tasteless but it can be used to improve the texture, appearance and stability of a final product. Ethnobotanical studies suggested the beneficial use of *Acacia nilotica* gum for the treatment of any irritation of the skin and smoothening of the inflamed membranes of the pharynx, the genito-urinary organs and the alimentary canal (Sonibare and Gbile, 2008). The gum is also applied as a suspending agent for insoluble drugs, for antiseptic preparations, tablet coating, jelly lubricants and to mask the unpleasant taste of syrups. Furthermore, it is used for the

Table 10.1. Selection of companies involved worldwide in the extraction, processing and trading of resins, gums and oleoresins. Author's own table.

Company	Market segment
Aromaaz International, India	essential oils, oleoresins and other natural products
Arian Trading Group, Iran	asafoetida gum and other natural raw materials
Bandish Enterprises, India	gums, resins and waxes (gum rosin, gum copal, shellac)
Biolandes, France	essential oils, oleoresins
Boswellia Co. Ltd., India	gum-resins, particularly frankincense
DRT-Firmenich, France	gums and resins, flavours and fragrances, essential oils
Ernesto Ventos S.A., Spain	essential oils, aroma chemicals, fragrance products
Frankincense Direct, UK	frankincense and myrrh resins
Gungxi Dinghong Resin Ltd., China	gum rosin and modifications of the gum
Mountain Rose Herbs, U.S.	organic herbs, spices, essential oils
Alban Muller International, Germany	natural botanical ingredients for cosmetic, pharmaceutical and food industries
Givaudan (formerly Naturex), Switzerland	natural ingredients for the food, health and cosmetic industries
Indukern, Spain	natural raw materials, essential oils, resins and gums
Nature is Unique, Germany	gum-resins, essential oils
Robertet, France	natural raw materials, essential oils, resins and gums
Roeper GmbH, Germany	gums, resins, oleoresins, plant extracts, natural colours
Synthite Industries Ltd., India	spice extracts, essential oils and natural food colours
Symrise AG, Germany	flavours and fragrances, spice extracts, essential oils and natural food colours
International Flavors and Fragrances, USA	flavours and fragrances, spice extracts, essential oils and natural food colours
Takasago, Japan	essential oils and fragrances
AS Resinas, Brazil	gum rosin, turpentine
ACM Resinas, Spain	gum rosin, turpentine
Martin Bauer, Germany	medicinal plants and herbs, gums and resins
Tradin Organic, Netherlands	gums and resins
Nexira/Starlight, France	gums and resins
Willy Benecke GmbH, Germany	gums and resins
Newport Industries Ltd., UK	gum rosin, rosin esters, phenolic resins
Roudenar Company, Iran	asafoetida gum, aroma plants such as thyme and cumin
William Bernstein Company, USA	gums and resins

preparation of facial masks, foam stabilizer, production of liquid soaps, lotions, protective creams, face powder, toothpaste, mouthwash and hair cream.

*Tamarind seed gum (*Tamarindus indica*)*

Tamarind seed gum is applied as a pharmaceutical excipient for oral drug delivery (Bansal *et al.*, 2014) and as a binder and emulsifier in pharmaceutical tablets. But tamarind seed gum is also gaining in importance in the food industry as a thickening, stabilizing and gelling agent. It is used in particular as an ingredient in processed foods to improve the texture and viscosity of the end products (e.g. in dairy products, ketchups, meat products, pet food, instant noodles, bread and biscuits).

Tamarind is available in numerous forms, such as concentrated pulp with seeds, canned paste, dried pods or powdered form. It is readily available in India and a few Asian markets.

The global market for tamarind extract was valued at US$408.06 million in 2022 and is expected to continue to grow positively at a compound annual growth rate (CAGR) of 6.5% until 2032 (Polaris Market Research & Consulting, 2024). Tamarind extract is mainly used as a flavour enhancer in various foods, but is also widely used for its antifungal, antibacterial, antioxidant and antiseptic properties. In addition, the extract is used as a skin-lightening and moisturizing agent in various cosmetic products. At present, the key driver in the tamarind extract market is the increased focus by governments in producer countries on the preparation and storage of tamarind extract. The largest companies operating in the tamarind extract market include ABC International (India), Baaeco V Food (Thailand), Jadli Foods (India), Magma Foods (India), MoonLite Corporation (India) and Revata Food Products (India).

*'Real' gum arabic (*Acacia senegal*)*

Gum arabic is the oldest known gum, having been used for diverse applications for more than 5000 years. The gum is produced from various species of the genus *Acacia*, mainly growing/planted in arid and semi-arid regions of Africa and Asia. Sudan is the leading producer and exporter of gum arabic in the world. The gum is a mixture of potassium, magnesium and calcium salts of polysaccharide acid and is easily soluble in water. As it possesses a unique biochemical structure and properties, it finds application as a stabilizer, emulsifier, bulking agent, thickener, carrier, glazing and firming agent, humectant and antioxidant. At present, people worldwide are feeling the need to use natural additives in products for human consumption, due to the sometimes adverse effects of some synthetic additives. Such a shift in the human mindset has created an additional demand for natural products like gum arabic produced in resource-constrained areas, where people are poor and need the creation of good options for improving their economic conditions. In addition, *Acacia senegal* trees contribute to improving soil health and mitigating climate change and desertification in arid and semi-arid regions. The increase in demand for gum arabic is motivating farmers in these regions to plant *Acacia senegal*, but in general, the search for additional applications in this important sector is urgently needed (Prasad *et al.*, 2022). Global gum arabic market demand is anticipated to be valued at US$897.3 million in 2022, with a forecast CAGR of 7.2% to be valued at US$1796.7 million by the end of the period 2022–2032. From 2016 to 2021 a CAGR of 5.9% was registered for the market (Future Market Insights, n.d.a).

Gum rosin

Gum rosin, also known as rosin or colophony, is collected from various pine species and some other conifers, and is obtained by heating fresh liquid resin to vaporize the volatile liquid terpene components. The market size was estimated at 655.45 kt in 2024, and is expected to reach 800.53 kt by 2029, growing at a CAGR of 4.08% during the forecast period 2024–2029.

The growing demand for bio-based adhesives and sealants and the burgeoning paper packaging industry are likely to be the main drivers of the gum rosin market's growth over the medium term.

The Asia-Pacific region is expected to account for the largest share of the market over the forecast period, owing to the huge consumption of countries such as China, India and Japan (Mordor Intelligence, n.d.).

Gum ghatti (Anogeissus latifolia)

Gum ghatti is a natural proteoglycan produced from axlewood or the ghatti tree (*Anogeissus latifolia*). Chemically, the gum occurs as a complex combination of several mono-sugars and uronic acids in the form of their calcium and magnesium salts. Gum ghatti comprises soluble and insoluble fractions, which form a thick colloidal solution in water. The gum is non-toxic, biocompatible with the human digestive system and biodegradable. The United States has granted GRAS (generally recognized as safe) status to gum ghatti since 1976, and it is also permitted as a food emulsifier in numerous Asian countries, including China, India, Japan, South Korea and Singapore. Gum ghatti is becoming increasingly important in the food industry as an emulsifier, stabilizer, binder, thickener, film-forming agent and raw material for the synthesis of hydrogels. It can be assumed that the current rethinking among consumers will generally lead to an increasing demand for natural gums in order to avoid the sometimes adverse effects of some synthetic substances. As ghatti trees are found in the forest or on non-agricultural lands, the collection and sale of the gum provide a sustainable livelihood to the economically backward and resource-constrained population inhabiting that area (Singh *et al.*, 2024). Products made from gum ghatti must meet regulatory standards set by international organizations such as the Food and Drug Administration (FDA) and the European Food Safety Authority (EFSA). In addition, for certain markets where compliance with Jewish and Islamic dietary laws is relevant, kosher and halal certifications are also required.

The market size for gum ghatti was valued at US$108.3 million in 2023 and is projected to reach US$148.7 million by 2030, growing at a CAGR of 8.7% during the forecast period 2024–2030 (Singh *et al.*, 2024; Verified Markets Reports, 2024).

Gum guggul (Commiphora wightii)

Chewing gum, whose main ingredient is guggul (*Commiphora wightii*), provides a suitable platform for the application of a wide range of different flavour blends, from the commonly known mint to various exotic fruit flavours. The chewing gum market is attracting investment from many companies due to its enormous development potential. Major market players have made significant investments, including Cadbury, Lotte, Perfetti Van Melle and Wrigley Company. In addition, chewing gum is used in a variety of areas, including health care and education. In 2022, for example, in collaboration with Givaudan, a global flavour and fragrance company, Sweet Victory developed a delicious children's chewing gum with plant-based ingredients that could be used to administer bitter-tasting pharmaceutical active ingredients. It is therefore to be expected that this trend will also have a positive impact on the future of the chewing gum industry worldwide. The global chewing gum market size was valued at US$25,778.53 million in 2022 and is expected to expand at a CAGR of 5.19%, reaching US$34,922.8 million by 2028 (Fig. 10.2).

Traditionally, gum guggul is used as incense, as a fixative in perfumes and in medicine. In India, *Commiphora mukul* is often used for its anti-inflammatory, antispasmodic, blood sugar-lowering, antiseptic, sedative, diaphoretic and expectorant properties (Bellamkonda *et al.*, 2011). Furthermore, gum guggul has been used to treat dyspepsia, endometritis, hypercholesterolaemia, hypertension, impotence, bronchitis, catarrh, gingivitis and hay fever.

Modern therapeutic applications of gum guggul include various nervous disorders, hemiplegia, leprosy, marasmus, muscle cramps, neuralgia, skin diseases, ischaemia, hypertension and urinary tract disorders. Research studies have shown that guggul is effective against various cardiovascular diseases. Since gum guggul is able to reduce the stickiness of blood platelets, pharmaceutical preparations containing the resin have proven to be an effective and cost-efficient alternative to conventional treatment of hyperlipoproteinaemia.

Gum Dammar (Shorea wiesneri)

Dammar designates a group of resins obtained from Indian or East Asian trees belonging to the families Dipterocarpaceae and Burseraceae and principally those of the genera *Shorea*, *Balanocarpus* or *Hopea*. The principal dammars of India are sal dammar, white dammar and black dammar. Trees of these families range from medium-sized to very large. They are widespread and are a very important source

Fig. 10.2. *Commiphora wightii* (gum guggul) for sale in Khari Baoli market, imported from Pakistan due to over-exploitation in India. (By Anthony Cunningham.)

of tropical hardwood throughout the Indian and South-east Asian regions. Most of the dammar gum-resins are produced by tapping trees, however, some are also collected in fossilized form from the ground. Dammars are solid resins, generally less hard and durable than the copals, and white to yellow in colour. Usually, the gum-resins are partly soluble in alcohol and acetone but completely soluble in benzene and turpentine. The raw material is used in Ayurvedic medicine as an astringent and cleansing agent and is administered for diarrhoea and dysentery.

Dammar gum-resin consists mainly of triterpenoid substances and its oxidation products such as dammarane, oleaonic acid, oleanane and dammarenolic acid. Due to its clouding properties, it is used in the food industry and in the incense stick industry because of its slight lemon scent. The major global manufacturers of gum dammar include Willy Benecke, Sennelier, Bandish Enterprises, Nexira Inc. and Starlight. In general, the global gum dammar market is benefitting from the rise in demand for natural products, especially in the food, cosmetics and pharmaceutical sectors. The market for dammar gum is currently experiencing significant growth, which is expected to continue in the future with only a few setbacks due to limited availability.

Copal resin (Agathis *sp.)*

Copal applies to a group of resins known for their hardness and relatively high melting point. The resin is produced from various *Agathis* species from different parts of the world (Africa, Asia and Central and South America). Mostly, copal is produced from *Agathis* tree species growing in South-east Asia: the Malay and Indonesian archipelagos in particular and, to a lesser extent, the Philippines. Today, the major use of copal is as a varnish for wood and paper. Earlier, much of the copal was collected in fossilized or semi-fossilized form. However, today, most copal is produced by tapping the trees. Further, nowadays copals are produced from trees raised on a plantation basis and there is little dependence on the wild forest for its production.

The global copal resin market is experiencing growth driven by its applications in various industries, such as those producing varnishes, coatings and adhesives, and by the rising demand for natural resins. The market is influenced by global economic trends, industrial development and the movement toward more sustainable, eco-friendly materials. In 2022, the global copal resin market was valued at around US$1.2 billion. It is expected that the market will grow at a CAGR of around 4.5–6% from 2023 to 2030. By 2030, the market is anticipated to reach between US$1.8 billion and US$2.0 billion, based on various growth factors such as industrial demand, sustainability trends and technological developments in resin processing (Industry ARC, n.d.). Similar to other natural resins, there is a growing emphasis on sustainable harvesting practices for copal resin. Ethical sourcing and certifications like 'Fair Trade' and 'Organic' are increasingly important, especially for manufacturers targeting markets in the EU and North America. The impact of climate change may also affect the supply of raw materials in copal-producing regions, influencing the availability and pricing of resin.

Technological innovations in resin processing are helping to improve the quality and versatility of the resin, allowing it to be used in more complex industrial applications. For instance, the development of modified natural resins to improve performance characteristics like water resistance and durability is expanding its use in high-performance coatings. Several manufacturers are also blending copal with other resins like dammar, shellac and rosin to enhance its properties for specific applications.

Amber resin

Amber is a fossil tree resin which is found in stable stage after loss of its volatile constituents and chemical change after burial in the ground. Amber is found throughout the world, but the largest and most significant deposits occur along the shores of the Baltic Sea in sands. Amber occurs in irregular nodules, rods or droplike shapes in all shades of yellow colour. Milky-white opaque varieties are called bone amber. The turbidity of some amber is caused by inclusions of many minute air bubbles. Due to its limited occurrence, amber is not used commercially, but presently modern research techniques aim to isolate and identify as many of the individual resin components as possible and ultimately to establish a genetic relationship between fossil resins and modern resin-producing trees. For example, infrared spectroscopy was used to prove that Mexican (Chiapas) amber is related to a modern legume tree from the Hymenaea family. Although it was previously believed that amber was completely amorphous, later X-ray diffraction studies have revealed crystalline components in some fossilized resins (Riquelme *et al.*, 2014). Furthermore, a recent study has significantly improved our understanding of the pathology of epilepsy and also shown that amber can be used to inhibit epileptic seizures (Zhu *et al.*, 2019) and to reverse damage to neurons in the cortex and hippocampus.

*Mastic gum (*Pistacea lentiscus*)*

Mastic gum, a resin extracted from the mastic tree (*Pistacia lentiscus*), has been used for generations for its various medicinal effects, which include improving digestive health, reducing inflammatory processes and freshening breath.

According to the forecast by CMI Consulting LLC (2024), the global mastic gum market was valued at US$350.8 million in 2022 and is expected to further increase to US$645.6 million by 2032, at a CAGR of 8.1% during the period 2023–2032.

The mastic gum market is being driven by rising consumer health consciousness, as it is viewed as a natural medicine with several health advantages. The resin has grown in popularity due to its supposed digestive health benefits, which include relieving stomach pain and improving dental health. Furthermore, its antibacterial and anti-inflammatory characteristics appeal to people looking for natural alternatives to traditional therapy. The current trend of customers favouring natural remedies such as mastic gum is likely to continue. This development will also be supported by a growing interest in traditional and holistic wellness techniques. The cosmetics industry is also increasingly focusing on products with natural ingredients, which is prompting manufacturers to use more mastic gum in their products.

*Moringa gum (*Moringa oleifera*)*

The polysaccharide exudates obtained from moringa gum are increasingly used in the food and pharmaceutical industries, as well as in other areas. The gum has traditionally been used to treat dental caries (Elgamily *et al.*, 2016) and was used as a rubefacient and astringent. The mixture of gum with sesame oil was quite often used to alleviate fevers, asthma, headaches, intestinal complaints and dysentery and to treat rheumatism and syphilis. It was also used as an abortifacient (Attah *et al.*, 2022). Murthy *et al.* (2011) reported the antifilarial activity of the natural product. The authors evaluated the antifilarial activity of the gum extract against the human lymphatic filarial parasite *Brugia malayi* using adult worms and microfilariae in two *in vitro* assays for viability and two animal models.

Badwaik *et al.* (2020) provide a comprehensive overview of the extraction, purification, structural elucidation and modification of moringa gum. In addition, they present detailed information on the various aspects of moringa gum and its applications in different industrial and biomedical fields.

In Senegal and India, moringa resin is used to treat fever, asthma and dysentery. The resin is sometimes also used as an astringent and skin tonic. On the Indonesian island of Java, moringa gum is used to treat intestinal complaints (Nussinovitch, 2009; Singhal *et al.*, 2012).

Sales of moringa extract have increased significantly in recent years due to rising health and wellness trends, as the plant extract appeals particularly to health-conscious consumers looking for natural and nutrient-rich supplements. As moringa has only recently been recognized as a 'superfood', the growing interest in moringa extract is expected to continue, particularly in the area of dietary supplements (weight control and appetite regulation).

*Gum karaya (*Sterculia urens*)*

The gum is predominantly used in various products and processes in the food industry (Setia *et al.*, 2010) but also in medicine. It has been found that gum karaya improves the quality of juice in the sugar industry and serves as an emulsifier and binder in meat processing. In the salad dressing industry, gum karaya is used as a stabilizer by increasing the water–oil emulsion. In a same way, the gum also acts as a stabilizer in sauces, spice bases, ketchups, sweet pickles and liqueurs by increasing the viscosity (Setia *et al.*, 2010). The resin is also used as a stabilizer in the production of various dairy products, frozen desserts, fruit ice and similar products with a low pH. Finally, gum karaya is also used as a binding agent in the production of low-calorie pasta products (such as noodles), bread and other baked goods.

In medicine, gum karaya is used primarily for its laxative effect, but also to treat liver cirrhosis. Furthermore, it has been found that gum karaya powder, when applied to open wounds, leads to an increase in normal granulation tissue and also to optimal epidermal growth. Finally, gum karaya is also used in dentine adhesives, medical adhesive tapes for the treatment of stomatitis and the production of pressure-sensitive masking tapes, medical jellies and pastes.

As recent research on the development of natural dosage forms for the oral administration of active ingredients with controlled release have shown, gum karaya, due to its extremely low hydration capacity, shows a very low tendency for drug release, with the erosion mechanism apparently playing a dominant role (Munday and Cox, 2000).

According to a study performed by Future Market Insights, the market for gum karaya is expected to grow by 5.9% per year and the karaya industry will reach a value of US$193.2 million by 2032. In the European Union, gum karaya is used in various culinary products and labelled with the number E416. In addition, the resin is used in medicine to treat diabetes, wounds and pressure ulcers and high plasma lipid levels. It is also used as an adhesive for dental implants and adhesive plasters. Currently, mergers and acquisitions of karaya manufacturers by end-user industries in the food and pharmaceutical sectors can be observed worldwide. This significant trend is expected to continue in the near future (Future Market Insights, n.d.b).

*Frankincense gum-resin (*Boswellia *sp.)*

Frankincense is known as the resinous exudate predominantly derived from seven of 24 known species of the genus *Boswellia*. The gum-resin is exuded by the tree in response to natural or intentional and selective wounding of the bark. The trees are widespread in arid and semi-arid ecosystems, ranging from West to East Africa, across the southern Arabian Peninsula to India. Tapping or scraping naturally exuded gum-resin from specific species has been a traditional practice of local communities for millennia. The sorted gum-resin is often used in religious and cultural practices. Further processed products, such as hydro-distilled essential oils and alcoholic extracts, are currently used in a wide range of pharmaceutical, aromatherapeutic and flavouring applications, with increasing use of the natural raw material.

Today, frankincense is frequently used in traditional medicine for the health benefits documented in various studies (Al-Yasiry and Kiczorowska, 2016). Finally, the gum-resin has a special significance due to its use in cultural and spiritual practices, as well as in the production of scented candles and other incense products. According to Coherent Market Insight (2023), in 2022 the global frankincense extracts market was valued at US$302.14 million, exhibiting a CAGR of 8.44% during the period 2023–2030.

The so-called 'Indian olibanum' or 'salai gum-resin' is obtained from the tree species *Boswellia serrata* growing mainly in India (Soumya *et al.*, 2019). It is a moderate to large branching tree, found in dry deciduous forests, being common in dry hills throughout India. Salai gum-resin is traditionally used as incense because of its unique fragrance. Furthermore, it is widely used in Ayurvedic formulations for treating asthma and arthritis. The gum-resin obtained from the tree has been shown to be as effective and, in many cases, better than drugs like phenylbutazone and other synthetic anti-inflammatory drugs. It is used in indigenous medicine for rheumatism, for nervous diseases, as a diaphoretic, as an astringent and as an ingredient in certain ointments (Basch *et al.*, 2004; Hussain *et al.*, 2022).

Frankincense is widely used today in incense sticks, perfumes, aromatherapy, cosmetics, traditional medicine and various other applications (Fig. 10.3).

International trade in frankincense gum is expected to exceed 6000–7000 t per year by 2023, which would be worth tens of millions of US dollars. After further processing into essential oil, resinoids and other extracts, the value added could even exceed US$ 1 billion (Cunningham and DeCarlo, 2022). The gum-resin thus serves as an important additional source of livelihood for hundreds of thousands of people in rural areas in the countries where *Boswellia* trees are found. However, as the aforementioned

Fig. 10.3. A. *Boswellia sacra* tree in Dhofar; B. frankincense trader in Oman showing 'green' graded medicinal quality *Boswellia sacra* resin collected from trees in the Hasik area of Dhofar. (By Anthony Cunningham.) Used with permission.

A B

Fig. 10.4. A. Adult *Boswellia carterii* tree; B. harvest of frankincense in Puntland-Somalia. Author's own image.

processing steps for frankincense do not take place in the resin-producing countries, they do not receive the resulting profits from the sale of these frankincense products (Cunningham and DeCarlo, 2022).

In the rural areas of Ethiopia, Sudan, Somalia and Oman, especially in the drylands, the collection (from *Boswellia papyrifera, B. neglecta, B. frereana, B. carterii* and *B. sacra*) and sale of frankincense gum-resins is one of the most important livelihood activities (Fig. 10.4).

As identified by Yussuf *et al.* (2023), the existing problems and constraints in the production and sale of gums and resins are primarily caused by recurrent droughts, lack of market access and market knowledge, and lack of institutional cooperatives, infrastructure and facilities. Therefore, the production and commercialization of frankincense needs to be significantly improved and strategies to adapt to climate change urgently need to be developed. Furthermore, local authorities and communities must work together in a better way to ensure the sustainability of the gum-producing tree species in the future.

Climate change and other socio-economic and cultural factors are changing the sustainable relationship between harvesters and natural tree regeneration. As it is one of the most ancient globally traded aromatic phytochemicals, the current global annual trade volume is putting high pressure on certain *Boswellia* populations and species in various areas.

The potential threats to any one frankincense harvesting area are multiple: climate change, changes in land ownership and use, urban migration, burning and wildfires, general deforestation, insect infestation, livestock and overgrazing, loss of expertise in sustainable harvesting practices and/or poverty pushing some to unsustainable harvesting practices. The sudden rise in demand and the volume of trade for *B. sacra* syn *carterii* from Somalia for essential oil production has also raised alarm and initiated a fact-finding study by the Convention on International Trade in Endangered Species (CITES), who lean towards a precautionary international regulatory approach (CITES 2016, 2020). Therefore, it will be extremely necessary to collect reliable data on the benefits and impact of and actual need for an international top-down approach, as opposed to investing in verifiable assessment data and support for the trees and communities (Groenendijk *et al.*, 2012; Lemenih *et al.*, 2014; Bongers *et al.*, 2019).

Intentional propagation or plantations have had limited success to date, so any further increase in trade will come from existing wild populations. A recent ground survey that found yet-to-be catalogued populations of *B. ogadensis* recommends downlisting it to endangered on the International Union for Conservation of Nature's (IUCN) Red List of Threatened Species. Like *B. occulta*, *B. globosa* in Somaliland, *B. pirrottae* in Ethiopia and *B. microphyllae* on the Ethiopian/Kenyan border, these unique, hardly harvested species exist in small, restricted ranges and may well benefit from active conservation.

The currently existing socio-political, economic and governance structures and security situations in some key range states make research, assessment and verification challenging. However, it must be recognized that the international trade provisions of CITES, which are aimed at protecting endangered *Boswellia* species, could potentially lead to a significant restriction or even a complete cessation of trade in several countries. In the future, a better balance must therefore be found between species protection and the local employment situation, and more voluntary incentives must be created overall for 'good harvest practices' by workers and traders, as well as for national forestry plans.

This skilful balance between ecology and economy certainly applies not only to frankincense, but also to other gums and gum-resins worldwide, and generally requires a great willingness for dialogue between the stakeholders involved.

It must be better understood in the future that the support and further development of proven methods in the management of *Boswellia* forests and in the harvesting and trade of frankincense not only secures the livelihoods of numerous communities and families, but also ensures the long-term survival of *Boswellia* trees, provided that the respective sustainability aspects are taken into account.

In order to be able to reliably distinguish the threatened *Boswellia* species from the more common ones in the future, suitable analytical methods are needed to ensure traceable and transparent supply chains prior to export. This data collection will also enable traders to work with communities and local and national governments to establish locally relevant species and site-specific sustainable management plans. Recently, Johnson *et al.* (2024) have also provided a comprehensive description of the complex situation and pointed out the need to develop new analytical methods to distinguish between gum-resin types from different regions. Certainly, this approach applies not only to *Boswellia* species, but also to other resin plant genera, where only certain species are threatened, not necessarily the entire genus. When searching for individual solutions, it is therefore important to keep a sense of perspective so that the arguments of both sides – species protection and trade – are taken into account in the producer countries, especially in the interest of the poor population whose livelihoods depend on frankincense collection. A general ban on the frankincense trade, as called for by CITES, has therefore been criticized as counterproductive (Johnson *et al.*, 2024).

In this context, consideration should be given to introducing internationally recognized quality standards such as International Organization for Standardization (ISO) certification for frankincense (and also for other resins and gums that have so far been insufficiently specified). The Technical Committee (TC54) of ISO, which is specifically responsible for the area of essential oils, resins, oleoresins and resinoids, was established many years ago with the aim of creating a reliable basis for international trade in natural products worldwide (ISO, n.d.). Currently, the only specifications in world trade are for the gum-resins produced from *Boswellia caterii* (olibanum) and *Boswellia serrata* (Indian frankincense or salai guggul). Furthermore, olibanum has not been listed in the newer pharmacopoeias for some time and there is only one monograph for Indian frankincense (*B. serrata*), which only specifies the titrimetric content of 'total boswellic acids'. Consequently, a clear distinction between *Boswellia serrata* and other *Boswellia* species is presently not required anywhere in the trade, and no ISO standard has yet been defined for *Boswellia* gum-resin. In order to better counter the threats to individual *Boswellia* species, new cultivation systems such as agroforestry must be developed, education and knowledge must be created at different levels and traditional knowledge and the safeguarding of sustainable livelihoods must be optimally combined.

Cunningham and DeCarlo (2022) point out that small-scale propagation experiments have already been carried out with various *Boswellia* species, but nevertheless, practically all frankincense traded today is derived from wild collections. In recent years, the increasingly poor conservation status and inadequate sustainability of the harvest of several *Boswellia* species have been criticized in connection with the results of several studies from Ethiopia, Sudan and Eritrea (Groenendijk *et al.*, 2012; Lemenih *et al.*, 2014; Bongers *et al.*, 2019).

*Asafoetida gum (*Ferula asafoetida*)*

The gum-resin obtained from *Ferula asafoetida* has long been used as a food flavouring and medicinal herb and even today it is a very important flavouring agent in many spice mixtures. The gum-resin is antispasmodic, carminative, expectorant, laxative, sedative and very useful in the treatment of respiratory disorders like asthma, bronchitis and whooping cough. Furthermore, asafoetida is sometimes used in the treatment of some nervous conditions.

In 2023, the global market for asafoetida gum was estimated at US$655 million and it is expected to reach US$1193 million by 2030, with a CAGR of 8.3% over the forecast period 2024–2030 (according to Valuates, 2024).

Ferula asafoetida is widespread from the Mediterranean region to Central Asia. In India, it is mainly cultivated in the Kashmir region and some parts of Punjab, but the majority of plants are imported into India from Afghanistan and Iran. Asafoetida gum-resin is also cultivated in North America, China and Japan. The main buyers of the plant-based raw material are North America, Latin America, Europe (Germany, France, Great Britain, Italy, Russia) and several Asian countries (China, Japan, South Korea, Taiwan, India).

There are two main varieties of asafoetida: Hing Kabuli Sufaid (so-called 'milky-white asafoetida') and Hing Lal (so-called 'red asafoetida'). Asafoetida has a hot and bitter taste and a strong, unpleasant, pungent odour due to the various sulfur compounds it contains. The white variety is water-soluble, while the dark variety is mainly oil-soluble. Since pure asafoetida is not preferred due to its strong taste, it is usually mixed with starch and other resins and mostly sold as 'composite asafoetida'. However, it is also available as a free-flowing powder or in tablet form. Asafoetida is used primarily as a spice in curries, sauces and pickles, but it is also used in the manufacture of chewing gum. Due to its antibiotic properties, it is also used in medicines.

The following manufacturing companies are mainly responsible for production and global distribution: Everest Masala (India), Laljee Godhoo (India), Yuvara Agro Impex (India), Arian Trading Group (Iran) and Roudenar Company (Iran).

Phytochemicals and nutraceuticals

A central problem that most developing countries have been facing for a long time is the often inadequate provision of basic medical care. The World Health Organization (WHO) has repeatedly pointed out that many countries, which together account for more than half of the world's population, are unable to afford adequate health care. This is mainly due to the fact that the poor population is unable to afford the healthcare services currently available. For these poor people, medicinal plants represent an alternative, affordable medicine that opens up enormous possibilities (Peters *et al.*, 2008).

For example, frankincense has been shown to provide a wide range of biological effects, making this herbal drug particularly well known for its anti-inflammatory, antimicrobial, antiasthmatic, antioxidant and antiplatelet effects (Almutairi *et al.*, 2022). The immunomodulatory effects have shown that frankincense improves the parameters of the immune response without overstimulating it and that it can reduce the risk of a cytokine storm. Frankincense-based formulations can therefore be used either as cost-effective and safe therapies or even as antiviral supplements to boost immunity and build tolerance of various viral infections (Gomaa *et al.*, 2021).

New challenges for research

It can be assumed that integrative therapy approaches that combine conventional medicine with alternative healing methods will become increasingly popular in the near future. Today, complementary and alternative medicine is recognized as an essential component of the healthcare system and is already being successfully used in developing countries, both within and outside the formal healthcare system (Mortada, 2024). In this context, some gums and resins will be of particular importance. Based on many years of experience with frankincense, perubalsam and other natural raw materials, the great potential of other plant exudates will certainly be further researched and utilized for the treatment of various diseases.

The production and use of genetically modified organisms for the production of natural resins and gums (NRGs) will open up the possibility of developing new product groups in the areas of phytopharmaceuticals, agrochemicals and food supplements. Furthermore, it is expected that numerous previously unknown, bioactive secondary metabolites occurring in resins and gums will be selected and identified by means of non-target metabolite profiling techniques (Tripathi *et al.*, 2024).

Sustainable development and utilization of gums and resins

In general, there is an increasing need to effectively improve rural livelihoods and sustainable land management in producing countries and, in particular, to counteract the increasing desertification worldwide through appropriate measures such as reforestation, improved harvesting methods (e.g. use of cyber-tracking) and the development of agroforestry systems. The growing global consumer interest in sustainably produced organic natural products is increasing the demand for these products, particularly in the cosmetics, phytopharmaceuticals and food sectors. In general, it can be assumed that these natural products, provided they fulfil certain ecological and quality requirements, will achieve significantly higher prices. Accordingly, the demand for gums and resins will increase significantly worldwide in the coming decades.

Some developing countries, particularly on the African continent, already have a significant share of the market for certain gums and gum-resins, such as frankincense and gum arabic. However, some NRG species will very soon be threatened with extinction due to advancing climate change, the spread of various plant diseases and largely uncontrolled harvesting of resin-producing trees and shrubs. To counteract this negative development, several regional and international institutions have recently been set up to help facilitate and support the procurement of funds for various national projects (e.g. Network for Natural Gums and Resins in Africa, NGARA) in order to effectively promote the provision of market information and networking. It should also be mentioned here that, in principle, sufficient labour is available in the gum and resin-producing countries to manage the available resources sustainably. However, with regard to sustainable production, improved training programmes will be required in the future in order to impart the necessary knowledge for sustainable practices, not only to farmers but also to those involved in the trade.

Global status of resins and gums

The global status of resins and gums is mainly determined by supply dynamics, market demand, environmental concerns and trade regulations. There is a growing preference for environmentally friendly and sustainable alternatives, which is driving demand for natural resins in certain markets. The market for natural gums is driven by demand in the food and beverage, pharmaceutical and cosmetics industries. Table 10.2 shows the countries involved in the production, processing and utilization of some of the most important NRGs.

Currently, scientists are evaluating the use of medicated chewing gum to control oral drug delivery. Thus, medical chewing gum could establish itself as a new and advanced method of drug administration. As research has continued to deepen, chewing gum as a drug delivery system has become more and more sophisticated, and higher demands have already been placed on the resins in the chewing gum base, which has simultaneously promoted the expansion of the market.

Table 10.2. Countries producing and processing resins and gums worldwide. (Giri *et al.*, 2008.) Used with permission.

Resins and gums	Major producing countries
Gum rosin	China, Indonesia, India, Portugal, Brazil, Russia.
Dammar	Indonesia, Thailand, India, Vietnam, Laos, Cambodia
Guar gum	India, Pakistan, Sudan, Australia, USA
Tamarind seed gum	India, Bangladesh, Burma, Sri Lanka, Kenya, Nigeria, Sudan, Tanzania, Uganda, Afghanistan, Brazil, Cambodia, China, Cuba, Egypt, Malaysia, Mexico, Thailand, Pakistan
Cassia gum	India, China
Gum arabic	Sudan, South Sudan, Cameroon, Chad, Mali, Nigeria, India
Gum ghatti	India, Sri Lanka
Acacia nilotica gum	India
Gum karaya	India, Sudan, Senegal
Moringa gum	India, Pakistan, Bangladesh, Afghanistan
Gum guggul	India, Pakistan, UAE
Salai gum	India, Pakistan
Asafoetida gum	Afghanistan, Iran

Socio-economic importance of production, processing and commercialization of NRGs

In principle, employment-orientated, inclusive agricultural growth is possible in disadvantaged regions of the world if the necessary processing of harvested products, added value, diversified product development and application and effective marketing along the value chain from production to consumption are guaranteed. Such an approach would help local farmers in particular to increase the value of their products and thus achieve sustainable growth in the production of NRGs. Since gums and resins cannot usually be marketed in freshly harvested form, the refinement of NRGs is of particular importance, not least to ensure a uniform product quality that meets the individually specified ranges (Kumar *et al.*, 2015).

Ecological and forestry impacts of breeding and sustainable cultivation of NRGs

Forests are not only an important source of livelihood, employment, income and raw materials for a number of industries, but are also of enormous importance for ecological balance and stability, preservation of biodiversity, food security and sustainable development in the respective regions. Vegetation outside contiguous forest areas also plays a crucial role in maintaining and regulating all individuals (animals and plants) of an ecosystem. Due to the growing interest in natural resins and gums, immense social, ecological and industrial development opportunities are increasingly being recognized worldwide. Consequently, human intervention in the form of plantations with economic value is essential for sustainable economic development in and outside forest areas as well as in deserts and areas with degraded soils. Interventions to date have largely focused on the exploitation of these resources and, if at all, on the planting of trees with lower economic value, such as fast-growing trees. The planting of trees with high economic value will therefore play a central role in sustainable development and environmental stabilization in the future (Fig. 10.5).

In order to support the African gum and resin producing countries in formulating and implementing a coordinated strategy for the sustainable development of their natural resources to improve livelihoods in rural areas

Fig. 10.5. Cultivation trials with: A. *Commiphora*; B. *Boswellia* species in a tree nursery in Kenya. (By Amin Chaanin, VitaPlant AG, Switzerland.) Used with permission.

and increase resilience to climate change and desertification, NGARA was established in 2000 at a regional workshop in Nairobi, Kenya. In this context, NGARA's mission is to bring together stakeholders in natural gums and resins, including farmers, gatherers, traders, governments, non-governmental organizations, exporters and importers, who generally share the desire to improve the production and quality of locally produced gums and resins for domestic, regional and global markets (NGARA, 2017).

Similarly, the Indian Council of Agricultural Research (ICAR) has developed a vision for the year 2050 that describes the expected developments in the agricultural sector, particularly with regard to sustainable production and the future processing of gums and resins in India (IINRG, 2013).

How can the large-scale cultivation of trees required for the commercial utilization of important NRGs be promoted and how can the basic principle of sustainability be respected accordingly?

The gross value of goods and services provided by the forestry sector in India is estimated to average about US$3000 million/year, or about 2.37% of GDP. Of the forestry revenues recorded, about 60% is accounted for by minor forest products such as gums and resins. The sector supports about 50 million people living in forests and sub-forest areas, and about 70% of employment in the sector is in minor forest products (about 1.6 million man-days). It is recognized that neither the forests nor the tribal communities and poor living people in them should be undermined if the environment is to be protected. The only approach seems to be to develop minor forest products such as natural resins and gum trees in these areas to support the poor and maintain the required forest cover or vegetation at the same time (Giri *et al.*, 2008)

Lemenih and Kassa (2011) assume that the production of gums and resins will continue to increase in the next few years, as national economic development initiatives increase the pressure on natural resources for intensive use. The focus is on the development and implementation of sustainable production systems, not only to conserve natural resources, but also to further increase their sustainable socio-economic and ecological significance. To achieve these two goals, it is essential to implement coordinated and integrated multi-dimensional management measures. The activities can be carried out jointly by various institutions and stakeholders at the federal, state and local levels, and with the involvement of the private sector.

There is the possibility of establishing gum tree plantations or planting gum trees on field margins to increase gum and resin production and thus provide an income opportunity for the local farmers. The expected income generated from different gum tree plantations is presented in Table 10.3.

Table 10.3. Scope of raising plantation of natural gum-producing trees and expected income. Used with permission.

Natural gum-producing tree	Spacing for plantation raising (m)	Number of trees per ha	Gum production per ha	Assumed price for gum (₹/kg)	Estimated income from gum production (₹/ha)
Acacia senegal	5×5	400	120	400	48,000
Acacia nilotica	5×5	400	120	300	36,000
Anogeissus latifolia	10×10	100	150	300	45,000
Sterculia urens	10×10	100	250	200	50,000
Butea monosperma	5×5	400	120	150	18,000
Buchanania lanzan	6×6	256	192	150	28,800

Promotion of R&D and technology projects in connection with the production and processing of NRGs worldwide

In view of the immense potential of natural resins in the near future for the production of non-toxic and biodegradable products, the ICAR recently decided to include these natural products in the regular research programmes of the Indian Institute of Natural Resins and Gums (IINRG) in Ranchi. The institute conducts research on various aspects of production, processing, product development and value addition of all vegetable resins, gums and gum-resins. The special focus of the research is on improving the quality and quantity of the natural products and on developing optimized storage conditions. Furthermore, the identification and development of methods to isolate potential phytochemicals from gums and resins for use in dietary supplements, functional foods and as antioxidants are priority research areas. It can be assumed that the research endeavours are suitable for further promoting sustainable development, particularly in disadvantaged areas of India.

The individual research activities focus on the tree species *Pinus roxburghii*, *Cyamopsis tetragonoloba*, *Sterculia urens*, *Acacia senegal*, *Commiphora mukul*, *Tamarindus indica* and *Pinus kesiya* as well as generally on the development of new agroforestry models for efficient and sustainable production of resins and gums.

In nature, only plants that are relatively old in evolutionary terms are usually able to produce natural resin, and the long-term harvesting of these resins is always destructive for the plants, which means that plant resins are generally comparatively rare (Ding *et al.*, 2018; Maděra *et al.*, 2020). Due to their great economic importance, resulting in particular from their use in the food, cosmetics and pharmaceutical industries, natural resin plants have been excessively exploited, with the result that many of these species are classified as endangered on the IUCN Red List of Threatened Species. Further research is therefore needed to determine the genetic background of these plants and the mechanisms underlying the formation of plant resins. The resulting scientific knowledge could possibly be used to develop certain induction technologies that could provide the technological basis for the sustainable production of natural resins, especially of endangered species. There is also a lack of knowledge about how resin-producing plants accumulate different compounds in their resin and what significance these might have, not only for the plant itself, but also for the respective ecosystems or for ecology in general.

These natural resins usually present application value in various industries and medicine, and their original plants also present unique traits that have not been explored extensively until now. Therefore, it will be necessary in the future to make further progress in this area, which will help to strengthen the theoretical and technological basis for the efficient use of natural resins and inspire more scientists to conduct research in this field. It is worth mentioning that only about 10% of species in the plant kingdom excrete a resin, and the vast majority of these species are found in the tropics (Ding *et al.*, 2024).

In recent years, with the rise of network sciences such as systems biology and network pharmacology, drug combination has also become a hot topic in modern drug research and is presently setting a new trend in the development of phytopharmaceutical products (Hopkins, 2008; Madani Tonekaboni *et al.*, 2018; Cao *et al.*, 2019).

In the near future, the following main issues will be at the centre of the economic utilization of NRGs, in line with the above considerations:

- identifying the optimal geographical/climatic conditions for a sustainable and most efficient cultivation of various resin-producing plants;
- researching new, sustainable techniques and methods in the field of postharvest and processing of natural resins;
- improving the individual processing of resins, identifying previously unknown resin ingredients, with different applications in industry;
- further research into the potential socio-economic development opportunities in forestry, industry, trade and environment;
- identification of new applications for NRGs as important raw materials of the future, particularly with regard to improving the livelihoods and quality of life of people in the producer countries; and
- positive impact on climate change: reduction of the carbon footprint by sustainable use of NRGs with positive contributions to the preservation of forests and their important function in carbon sequestration.

Strategy and framework

The research topics mentioned regarding the NRG sector could be brought into line with the following strategic framework in order to take NRGs to new heights worldwide. Optimal networking of all stakeholders (farmers, industry and non-profit organizations) could significantly facilitate the establishment of individual public–private partnerships (PPP) to realize the desired vision. Among other things, optimized resource and input management could establish environmentally friendly, sustainable production systems that would not only benefit the producing countries but also offer a significant advantage worldwide in the form of ecosystem services. Furthermore, it must be understood that a company's carbon footprint alone is by no means a guarantee of a sustainable contribution to biodiversity and biodegradability, fair trade, employment in rural areas and other indicators that could justify a balanced business model. Social criteria are also an essential part of the sustainability profile of any natural raw material and are therefore increasingly associated with an emotional value by many consumers in their individual purchasing decisions.

A recent study has shown that the production and marketing of gums and resins significantly improves the livelihoods of households and serves as an efficient strategy for risk prevention, adaptation and mitigation of climate change fluctuations such as unpredictable rainfall and droughts (Yussuf *et al.*, 2023). It was also found that market access, infrastructure, the availability of sufficient institutional production cooperatives and short- and long-term training are crucial to building reliable marketing chains.

Nevertheless, the use of forward-looking research areas such as genetic engineering, biotechnology and nanotechnology could also make an important contribution to revolutionizing production and consumption systems. In addition, a higher added value of the products (cascade utilization as described in bioeconomy) could also contribute to improving sustainability and quality (supporting quality management through the development of meaningful standards, testing procedures and high-quality testing facilities for NRGs). Furthermore, a further diversification of the use of NRGs in niche areas and the identification and development of particularly neglected or under-utilized gums and resins could bring further benefits.

Finally, it will certainly be beneficial to further develop synergistic interactions with all stakeholders worldwide in the development, deployment and adoption of individual technologies. In this context, the establishment of an international centre of excellence for natural resins and gums should also be an important goal.

In recent years, new scientific findings have provided important impetus for growth and technological development in agriculture and related areas, which have made it possible worldwide to significantly improve food security even in times of increasing natural disasters and armed conflicts. However, agricultural development today faces further challenges related to the conservation of natural resources, climate change, the use of agricultural land for non-agricultural purposes and international trade with agricultural products. In order to respond appropriately to these challenges and to find the right balance between the economy on the one hand and the sustainable use of natural resources and socio-economic aspects on the other, it will be particularly important in the future to formulate innovative research and development projects that take all these areas into account and place them in a broader context.

A study by Gidey *et al.* (2020) found that *Boswellia papyrifera* forests in northern Ethiopia are particularly threatened by overgrazing, overharvesting, pest infestations, agricultural expansion and poor management. However, the study showed that the application of a multi-criteria decision-making model can support the prioritization of intervention strategies for the sustainable conservation of *Boswellia papyrifera* forests by bringing different groups with competing interests into dialogue to maximize consensus and minimize conflicts in a timely manner. Furthermore, ongoing training of the individual interest groups has resulted in sustainable use and management of the forest, thus promoting the survival of the remaining dry *Boswellia* forests in northern Ethiopia and other locations.

As global interest in this sector grows, natural gums and resins will receive much more attention in the future, as they offer immense opportunities to advance social, ecological and industrial development in parallel, in line with the 'One Health' approach (connection between human, animal and environmental health), and thus ensure sustainable growth.

By sharing research results, resources and new ideas, it will be possible to identify key One Health needs more accurately in the near future and develop appropriate solutions for this field. As the global health landscape is constantly changing, it is more important than ever to better identify future One Health research needs, including the sustainable use of non-timber products. For example, sustainable management of resin trees could help not only to ensure a regular income for the resin tappers, but also to improve the health of the trees to be harvested and of the cattle grazing in the area.

Gachathi and Eriksen (2011) also emphasize that the collection of gums and resins can make an important contribution to greater social justice and ecological integrity and thus support sustainable management overall. If used sustainably, renewable resources such as natural resins and gums could not only secure household income but also contribute to the preservation of biodiversity and the functioning of ecosystems, thus increasing the overall productivity of various countries. However, this requires taking into account some political factors that currently restrict income, as well as improving some harvesting practices that currently still result in sometimes dramatic damage to resin trees. Conversely, sustainable use of forests would guarantee that goods and services can continue to be provided by the forest ecosystem within ecological, social and economic parameters. They could serve as raw materials for businesses based on them, thus creating jobs and significantly improving the socio economic status of poor population groups (FAO, 1995). Investments in forests based on non-timber products could be even more efficient and resilient, especially if certification is provided for such production systems. Keeping the option of collecting gums and resins as a livelihood activity is also in line with the global objective of mitigating ongoing climate change. Today, it is well known that forests that produce gums and resins have the potential to mitigate climate change through carbon sequestration or fixation, and furthermore provide many local environmental services, such as protection against erosion, shading and maintaining soil fertility.

References

Almutairi, M.B.F., Alrouji, M., Almuhanna, Y., Asad, M. and Joseph, B. (2022) *In-vitro* and *in-vivo* antibacterial effects of frankincense oil and its interaction with some antibiotics against multidrug-resistant pathogens. *Antibiotics* 11(11), 1591. DOI: 10.3390/antibiotics11111591.

Al-Yasiry, A.R. and Kiczorowska, B. (2016) Frankincense – therapeutic properties. *Postępy higieny i medycyny doświadczalnej (Advances in hygiene and experimental medicine) (Online)* 70, 380–391. DOI: 10.5604/17322693.1200553.

Attah, A.F., Akindele, O.O., Nnamani, P.O., Jonah, U.J., Sonibare, M.A. *et al.* (2022) *Moringa oleifera* seed at the interface of food and medicine: Effect of extracts on some reproductive parameters, hepatic and renal histology. *Frontiers in Pharmacology* 13, 816498. DOI: 10.3389/fphar.2022.816498.

Badwaik, H.R., Hoque, A., Kumari, L., Sakure, K., Baghel, M. *et al.* (2020) Moringa gum and its modified form as a potential green polymer used in biomedical field. *Carbohydrate Polymers* 249, 116893. DOI: 10.1016/j.carbpol.2020.116893.

Bansal, J., Kumar, N., Malviya, R. and Kumar Sharma, P. (2014) Extraction and evaluation of tamarind seed polysaccharide as pharmaceutical *in situ* gel forming system. *American-Eurasian Journal of Scientific Research* 9(1), 1–5. DOI: 10.5829/idosi.aejsr.2014.9.1.8259.

Basch, E., Boon, H., Heerema, T.D., Foppo, I., Hashmi, S. *et al.* (2004) Boswellia: An evidence-based systematic review by the Natural Standard Research Collaboration. *Journal of Herbal Pharmacotherapy* 4(3), 63–83. DOI: 10.1080/J157v04n03_06.

Bellamkonda, R., Rasineni, K., Singareddy, S.R., Kasetti, R.B., Pasurla, R. *et al.* (2011) Antihyperglycemic and antioxidant activities of alcoholic extract of *Commiphora mukul* gum resin in streptozotocin induced diabetic rats. *Pathophysiology* 18(4), 255–261. DOI: 10.1016/j.pathophys.2010.10.002.

Bongers, F., Groenendijk, P., Bekele, T., Birhane, E., Damtew, A. *et al.* (2019) Frankincense in peril. *Nature Sustainability* 2(7), 602. DOI: 10.1038/s41893-019-0322-2.

Cao, B., Wei, X.-C., Xu, X.-R., Zhang, H.-Z., Luo, C.-H. *et al.* (2019) Seeing the unseen of the combination of two natural resins, frankincense and myrrh: Changes in chemical constituents and pharmacological activities. *Molecules* 24, 3076. DOI: 10.3390/molecules24173076.

CITES (2016) Criteria for amendment of appendices I and II (Resolution Conf. 9.24 (Rev. Cop17). In: *Convention on International Trade in Endangered Species of Wild Fauna and Flora (CITES)*. Available at: https://cites.org/sites/default/files/document/E-Res-09-24-R17.pdf (accessed 5 May 2025).

CITES (2020) Boswellia trees (Boswellia spp.) (PC25 Inf. 3; p. 515). In: *Convention on International Trade in Endangered Species of Wild Fauna and Flora (CITES)*. Available at: https://cites.org/sites/default/files/eng/com/pc/25/Inf/E-PC25-Inf-03-Rev.pdf (accessed 5 May 2025).

CMI Consulting LLC (2024) Global Mastic Gum Market 2024–2033. Available at: https://www.custommarketinsights.com/report/mastic-gum-market/ (accessed 8 May 2025).

Coherent Market Insight (2023) Frankincense Extracts Market Analysis & Forecast: 2025–2032. Available at: https://www.coherentmarketinsights.com/market-insight/frankincense-extracts-market-1889 (accessed 8 May 2025).

Cronin, J.J., Smith, J.S., Gleim, M.R., Ramirez, E. and Martinez, J.D. (2011) Green marketing strategies: An examination of stakeholders and the opportunities they present. *Journal of the Academy of Marketing Science* 39, 158–174. DOI: 10.1007/s11747-010-0227-0.

Cunningham, A.B. and DeCarlo, A. (2022) Boswellia species in international trade: Identification, supply chains, & practical management considerations (Information Document CoP19 Inf. 10; p. 203). In: *Convention on International Trade in Endangered Species of Wild Fauna And Flora*. Available at: https://cites.org/eng/node/133556 (accessed 5 May 2025).

Ding, X., Mei, W., Huang, S., Wang, H., Zhu, J. *et al.* (2018) Genome survey sequencing for the characterization of genetic background of *Dracaena cambodiana* and its defense response during dragon's blood formation. *PloS One* 13, e0209258. DOI: 10.1371/journal.pone.0209258.

Ding, X., Jinhui, C. and Maděra, P. (2024) Plant natural resins: From formation mechanism to ecological significance. *Frontiers in Plant Science* 15, 1412805. DOI: 10.3389/fpls.2024.1412805.

Elgamily, H., Moussa, A., Elboraey, A., El-Sayed, H., Al-Moghazy, M. *et al.* (2016) Microbiological assessment of *Moringa oleifera* extracts and its incorporation in novel dental remedies against some oral pathogens. *Open Access Macedonian Journal of Medical Sciences* 4(4), 585–590. DOI: 10.3889/oamjms.2016.132.

FAO (1995) *Non-Wood Forest Products for Rural Income and Sustainable Forestry*. FAO, Rome. Available at: https://openknowledge.fao.org/server/api/core/bitstreams/4b242ee9-0229-4613-98d6-3cc6aadfdd25/content (accessed 5 May 2025).

Future Market Insights (n.d.a) Future Market Insights. Available at: https://www.futuremarketinsights.com/contacts (accessed 8 May 2025).

Future Market Insights (n.d.b) Karaya Gum Market Outlook 2025 to 2035. Available at: https://www.futuremarketinsights.com/reports/karaya-gum-market (accessed 8 May 2025).

Gachathi, F.N. and Eriksen, S. (2011) Gums and resins: The potential for supporting sustainable adaptation in Kenya's drylands. *Climate and Development* 3(1), 59–70. DOI: 10.3763/cdev.2010.0066.

Gidey, T., Berhe, D.H., Juhar, H.M., Solomon, N., Aklilu, N. *et al.* (2020) Population status of *Boswellia papyrifera* woodland and prioritizing its conservation interventions using multi-criteria decision model in northern Ethiopia. *Heliyon* 6(10), e05139. DOI: 10.1016/j.heliyon.2020.e05139.

Giri, S., Prasad, N., Pandey, S., Prasad, M. and Baboo, B. (2008) *Natural Resins and Gums of Commercial Importance - At a Glance*. ICAR- Indian Institute of Natural Resins and Gums, Ranchi, India.

Gomaa, A.A., Mohamed, H.S., Abd-ellatief, R.B., Mohamed, A. and Gomaa, M.A. (2021) Boswellic acids/*Boswellia serrata* extract as a potential COVID-19 therapeutic agent in the elderly. *Inflammopharmacology* 29, 1033–1048. DOI: 10.1007/s10787-021-00841-8.

Groenendijk, P., Eshete, A., Sterck, F.J., Zuidema, P.A. and Bongers, F. (2012) Limitations to sustainable frankincense production: Blocked regeneration, high adult mortality and declining populations. *Journal of Applied Ecology* 49(1), 164–173. DOI: 10.1111/j.1365-2664.2011.02078.x.

Hopkins, A.L. (2008) Network pharmacology: The next paradigm in drug discovery. *Nature Chemical Biology* 4(11), 682–690. DOI: 10.1038/nchembio.118.

Hussain, H., Rashan, L., Hassan, U., Abbas, M., Hakkim, F.L. *et al.* (2022) Frankincense diterpenes as a bio-source for drug discovery. *Expert Opinion on Drug Discovery* 17(5), 513–529. DOI: 10.1080/17460441.2022.2044782.

IINRG (2013) *IINRG Vision 2050*. Available at: http://14.139.215.35/~lpd/Text%20Vision%202050.pdf (accessed 8 May 2025).

Industry ARC (n.d.) Copal Market – Forecast (2025–2031). Available at: https://www.industryarc.com/Research/Global-Copal-Market-Research-511849 (accessed 8 May 2025).

ISO (n.d.) Aromatic Natural Raw Materials – Vocabulary. Available at: https://www.iso.org/obp/ui/en/#iso:std:iso:9235:ed-3:v1:en (accessed 8 May 2025).

Johnson, S., Canney Davison, S., Ablard, K., Bongers, F., Cunningham, A.B. *et al.* (2024) Good intentions meet complex realities: CITES listing of diverse frankincense (*Boswellia* species) might do more harm than good. *Forest Policy and Economics* 163, 103207. DOI: 10.1016/j.forpol.2024.103207.

Kumar, A., Jaiswal, A.K., Singh, A.K. and Yogi, R.K. (2015) *Advances in Lac Production, Processing, Product Development and Value Addition*. ICAR-IINRG, Ranchi, India.

Lemenih, M. and Kassa, H. (2011) *Opportunities and Challenges for Sustainable Production and Marketing of Gums and Resins in Ethiopia*. CIFOR, Bogor, Indonesia.

Lemenih, M., Arts, B., Wiersum, K.F. and Bongers, F. (2014) Modelling the future of *Boswellia papyrifera* population and its frankincense production. *Journal of Arid Environments* 105, 33–40. DOI: 10.1016/j.jaridenv.2014.02.006.

Madani Tonekaboni, S.A., Soltan Ghoraie, L., Manem, V.S.K. and Haibe-Kains, B. (2018) Predictive approaches for drug combination discovery in cancer. *Briefings in Bioinformatics* 19(2), 263–276. DOI: 10.1093/bib/bbw104.

Maděra, P., Forrest, A., Hanáček, P., Vahalík, P., Gebauer, R, *et al.* (2020) What we know and what we do not know about dragon trees? *Forests* 11, 236. DOI: 10.3390/f11020236.

Mordor Intelligence (n.d.) Gum Rosin Market Size & Share Analysis - Growth Trends & Forecasts (2025–2030). Available at: https://www.mordorintelligence.com/industry-reports/gum-rosin-market (accessed 8 May 2025).

Mortada, E.M. (2024) Evidence-based complementary and alternative medicine in current medical practice. *Cureus* 16(1), e52041. DOI: 10.7759/cureus.52041.

Munday, D.L. and Cox, P.J. (2000) Applications of gum karaya in drug delivery systems: A review on recent research. *International Journal of Pharmaceutics* 203, 179–192.

Murthy, P.K., Saxena, K., Lakshmi, V., Kushwaha, V., Verma, S.K. *et al.* (2011) Antifilarial activity of gum from *Moringa oleifera* Lam. on human lymphatic filaria *Brugia malayi*. *Chronicles of Young Scientists* 2(4), 201. DOI: 10.4103/2229-5186.93025.

NGARA (2017) *The Network for Natural Gums and Resins: Overview and Framework of Priorities, 2017–2030.* NGARA, Nairobi. Available at: https://afforum.org/forum/wp-content/uploads/2021/11/NGARA-FRAMEWORK-OF-PRIORITIES-ENGLISH-WEB-VERSION.pdf (accessed 8 May 2025).

Nussinovitch, A. (2009) *Plant Gum Exudates of the World: Sources, Distribution, Properties, and Applications.* CRC Press, Boca Raton, Florida.

Peters, D.H., Garg, A., Bloom, G., Walker, D.G., Brieger, W.R. *et al.* (2008) Poverty and access to health care in developing countries. *Annals of the New York Academy of Sciences 1136* 161–171. DOI: 10.1196/annals.1425.011.

Polaris Market Research & Consulting (2024) Enpowering Businesses with Data-Driven Market Insights for Smarter Decisions. Available at: https://www.polarismarketresearch.com (accessed 5 May 2025).

Prasad, N., Thombare, N., Sharma, S.C. and Kumar, S. (2022) Gum arabic – a versatile natural gum: A review on production, processing, properties and applications. *Industrial Crops and Products* 187, 115304. DOI: 10.1016/J.INDCROP.2022.115304.

Quested, T.E., Cook, P.E., Gorris, L.G.M. and Cole, M.B. (2010) Trends in technology, trade and consumption likely to impact on microbial food safety. *International Journal of Food Microbiology* 139(Suppl), S29–S42. . DOI: org/10.1016/j.ijfoodmicro.2010.01.043.

Riquelme, F., Hernández-Patricio, M., Martínez-Dávalos, A., Rodríguez-Villafuerte, M., Montejo-Cruz, M. *et al.* (2014) Two flat-backed Polydesmidan millipedes from the Miocene Chiapas-Amber Lagerstätte, Mexico. *PLOS One* 9(8), e105877. Available at: https://doi.org/10.1371/journal.pone.0105877

Setia, A., Goyal, S. and Goyal, N. (2010) Applications of gum karaya in drug delivery systems: A review on recent research. *Der Pharmacia Lettre* 2(5), 39–48.

Singh, R., Priya, H., Ranjan Kumar, S., Trivedi, D., Prasad, N. *et al.* (2024) Gum ghatti: A comprehensive review on production, processing, remarkable properties, and diverse applications. *ACS Omega* 9(9). DOI: 10.1021/acsomega.3c08198.

Singhal, A.K., Jarald, E.E., Showkat, A. and Daud, A. (2012) In vitro evaluation of *Moringa oleifera* gum for colon-specific drug delivery. *International Journal of Pharmaceutical Investigation*. DOI: 10.4103/2230-973X.96926.

Sonibare, M.A. and Gbile, Z.O. (2008) Ethnobotanical survey of anti-asthmatic in South Western Nigeria. *African Journal of Traditional Complementary and Alternative Medicine* 5, 340–345.

Soumya, K.V.M., Shackleton, C.R. and Setty, S. (2019) Harvesting and local knowledge of a cultural non-timber forest product (NTFP): Gum-resin from *Boswellia serrata* Roxb. in three protected areas of the Western Ghats, India. *Forests* 10(10), 907. DOI: 10.3390/f10100907.

Tadesse, W., Desalegn, G. and Alia, R. (2007) Natural gum and resin bearing species of ethiopia and their potential applications. *Forest Systems* 16(3), 211–221. DOI: 10.5424/srf/2007163-01010.

Tripathi, T., Gupta, S.C., Khan, Y.A., Bhunia, S.S., Gupta, A. *et al.* (2024) Metabolomics and anti-inflammatory activity of *Commiphora madagascariensis* jacq. leaves extract using in vitro and in vivo models. *Journal of Chromatography B* 1244, 124214. DOI: 10.1016/j.jchromb.2024.124214.

Valuates (2024) Asafoetida - Global Market Share and Ranking, Overall Sales and Demand Forecast 2024–2030. Available at: https://reports.valuates.com/market-reports/QYRE-Auto-22D8282/global-asafoetida (accessed 8 May 2025).

Verified Markets Reports (2024) Global Ghatti Gum Market, Report ID: 539440. Available at: https://www.verifiedmarketreports.com/product/ghatti-gum-market/ (accessed 8 May 2025).

Yussuf, B.A., Lelamo, L.L. and Abdi, M.M. (2023) Gum and resin production and marketing: Implications for pastoral livelihood in Adadle District, Somali Region, Ethiopia. *International Journal of Forestry Research, Article* ID, 8887494. DOI: 10.1155/2023/8887494.

Zhu, Z., Chen, C., Zhu, Y., Shang, E., Zhao, M. *et al.* (2019) Exploratory cortex metabolic profiling revealed the sedative effect of amber in pentylenetetrazole-induced epilepsy-like mice. *Molecules* 24(3), 460. DOI: 10.3390/molecules24030460.

Index

Note:- Page numbers in bold type refer to figures
Page numbers in italic type refer to tables.

Printed and bound by CPI Group (UK) Ltd, Croydon, CR0 4YY

24/08/2026

14947852-0002